Social odours in mammals

Social odours in mammals

Volume 1

Edited by

RICHARD E. BROWN
Department of Psychology, Dalhousie University,
Nova Scotia, Canada

and

DAVID W. MACDONALD
Department of Zoology, University of Oxford, England

CLARENDON PRESS · OXFORD

Oxford University Press, Walton Street, Oxford OX2 6DP
Oxford New York Toronto
Delhi Bombay Calcutta Madras Karachi
Petaling Jaya Singapore Hong Kong Tokyo
Nairobi Dar es Salaam Cape Town
Melbourne Auckland
and associated companies in
Beirut Berlin Ibadan Nicosia

Oxford is a trade mark of Oxford University Press

Published in the United States
by Oxford University Press, New York

© The several contributors listed on p. xi, 1985

First published 1985
Reprinted (with corrections) 1987

British Library Cataloguing in Publication Data

Social odours in mammals.
Vol. 1
1. Mammals—Behaviour   2. Smell
I. Brown, Richard E.   II. Macdonald, David W.
599.01'826   QL739.3
ISBN 0-19-857546-7

Library of Congress Cataloging in Publication Data

Main entry under title:

Social odours in mammals.

Bibliography: v. 1., p.
Includes index.
1. Mammals—Behavior.   2. Odors.   3. Smell.
4. Social behavior in animals.   I. Brown, Richard E.
II. Macdonald, David W. (David Whyte)
QL734.3.S6   1985   599'.059   83-17290
ISBN 0-19-857546-7 (v. 1)

Printed in Great Britain by
St Edmundsbury Press, Bury St Edmunds, Suffolk.

Preface

Generations of countryfolk, naturalists, and huntsmen have wondered at the acute sense of smell which opens a dimension to the lives of many other mammals, but from which we are barred. Some have doubtless cursed, rather than marvelled, at the sense which drives the once reliable dog besotted for a far-off bitch, or which thwarts the stalker at the whim of a breeze. It does not take a training in science to recognize our nose-blindness, or to hunger for a knowledge of what Kenneth Grahame, in *The wind in the willows*, called the mysterious fairy calls from out of the void, for which 'we have only the word smell, to include the whole range of delicate thrills which murmur in the nose of the animal night and day, summoning, warning, inciting, repelling'.

The realization that our own species is dismally ill-equipped to notice, far less to decipher, this profusion of mammalian odours, seems not to have deterred, but rather to have spurred biologists to excesses of curiosity. The result has been, over the past decade or so, a rapid expansion of information on these scents which function in mammalian communication, and which we shall call social odours. However, information on these social odours, their nature, sources, deployment, and functions, is widely scattered in the scientific literature and has become, in two senses, unusually fragmented. The first reason for this fragmentation is that the strength and the weakness of studies of olfactory communication is that they can be launched from diverse perspectives, involving the expertise of biologist or chemist, ecologist or histologist, field naturalist, endocrinologist or experimental psychologist. There is thus the opportunity for the fruitful combination of interdisciplinary forces in the common quest, or for their artifical division along the barriers of jargon. One practical consequence of these varied inputs is that it is not always obvious where to look in order to read about either recent advances or established background.

The second sense in which information on social odours is fragmented is a combined result of the tendency for biologists to study intensively only a few species, and the fact that members of thousands of mammalian species are each endowed with several sources of social odours. In consequence, a survey of current publications concerning social odours often gives an imbalanced impression of the prevalence of olfactory communication amongst mammals, disguising the fact that existing studies have tackled only the tip of the iceberg. Furthermore, the same focus on only a few species fails to direct the reader to the enormous variety of relevant snippets of information on a much wider array of mammals.

The growth of theoretically exciting ideas concerning social odours, the scattered publications and expertise, the feeling that a systematic review of literature on the whole Class could provide a realistic perspective and a useful reference work all prompted us to produce this two volume book. The structure of the book very much reflects our aims, as elaborated in our introductory chapter (pp. 1-18). Briefly, the two volumes are organized taxonomically, each chapter dealing with one or more orders. We hope it has an encyclopaedic quality which will be enduring useful, in that a vast store of literature is summarized in the text and, particular, in the summary tables. The possibilities for reference use are enhanced by comprehensive species and author indexes. At the same time, there is also an emphasis on reviewing modern ideas and even, in some chapters, otherwise unpublished data.

One of the most exciting fields of modern biology is the study of social behaviour; and the role of social odours in mammalian societies could hardly be more forcefully emphasized than in the chapters which follow. Indeed, we would argue that odours are a fundamental part of mammalian sociality and adaptation. While the secretions of the mammary glands may legitimately have pride of place, the scent-producing glands of the skin do not rank far behind in the list of qualities that are essentially mammalian.

We gratefully acknowledge all those who have helped in the production of these two volumes. In particular, at Dalhousie and Oxford, we were aided in proof-reading by Geoffrey Carr, Stefen Natynczuk, Jenny Ryan and Lilyan White, and in indexing by Gillian Kerby and Colin Pringle, together with Peter King and Dierdre Harvey of the Kellogg Health Sciences Library at Dalhousie. The composite drawings in Chapters 3, 6, 9, 10, 11, 12, 13, and 15 are by Priscilla Barrett.

Halifax and   R.E.B.
Oxford   D.W.M.
August, 1984

Contents

Plates

Plates fall between pp. 84 and 85 of the text.

Contributors

Richard E. Brown,
Department of Psychology,
Dalhousie University,
Halifax, Nova Scotia, Canada.

Peter Flood,
Department of Veterinary Anatomy,
Western College of Veterinary Medicine, Saskatoon,
Saskatchewan, Canada.

Zuleyma Tang Halpin,
Department of Biology,
University of Missouri,
St. Louis, USA.

Dietrich v. Holst,
Lehrstuhl für Tierphysiologie,
Universität Bayreuth,
W. Germany.

David W. Macdonald,
Department of Zoology,
University of Oxford, UK.

Eleanor M. Russell,
CSIRO Division of Wildlife Research,
Helena Valley,
Western Australia.

Uwe Schmidt,
Zoologisches Institut,
Universität Bonn,
W. Germany.

Introduction: The pheromone concept in mammalian chemical communication

DAVID W. MACDONALD and RICHARD E. BROWN

Apart from man, most other animals think through their noses. If we too were olfactory animals there would be no bird watchers, but in their place we would have mammal smelling societies.

Nisbett (1976, p. 30).

The importance of mammalian scent glands was first noted by perfumers who used three odours from mammalian scent glands: musk, civet, and castor. Musk is taken from the preputial glands of the musk deer (*Moschus moschatus*); civet from the anal glands of the civet (*Viverra civetta*), and castor from the castor gland of the beaver (*Castor canadensis*). A fourth mammalian source of perfume odour is ambergris, produced in the intestines of the sperm whale. According to Ellis (1962) the use of ambergris and musk is 'almost as old as civilization itself' (p. 40) while civet and castor were not used in Europe until the fifteenth and sixteenth centuries.

Although there are some early descriptions of mammalian scent glands, such as that of Pallas (1779), scientific study did not begin until the nineteenth and early twentieth centuries. Histological analysis and classification of mammalian skin glands were first conducted by Graff (1879) and Ranvier (1887). Later surveys of scent glands were published by Brinkmann (1912), Schiefferdecker (1922), and Eggeling (1931). From 1910 to 1945 R. I. Pocock described the scent glands of mammals kept in the London Zoological Gardens, and his prolific articles (listed in Hindle's (1948) bibliography) remain as the only descriptions of the scent glands of many species. The most complete modern reference for the skin glands of mammals is Schaffer's (1940) monograph, which is, unfortunately, difficult to obtain.

Some works on specific groups of mammals, such as Ewer's (1973) book on the Carnivora, contain chapters on scent glands and some reviews on the general aspects of skin glands have recently been published including those of Ortmann (1960) on the anal glands, Gabe (1967) on the research history and classification of skin glands, and Adams (1980) on the odour-producing organs of mammals; no attempt has been made to summarize all the information on mammalian scent glands since Schaffer's (1940) monograph. Today such a summary needs to be made in the context of evolutionary and ecological theory which did not exist in Schaffer's time. One aim of the authors of this book has been to compile such a summary.

Social functions of odours

A second purpose of this book is to examine the social functions of the odours produced by mammals. In *The descent of man* (1877) Darwin discussed two functions for scent glands and odours in communication: defence and reproduction. He states that:

With some animals, as with the notorious skunk of America, the overwhelming odour which they emit appears to serve exclusively as a defence. With shrew-mice (*Sorex*) both sexes possess abdominal scent-glands, and there can be little doubt, from the rejection of their bodies by birds and beasts of prey, that the odour is protective; nevertheless, the glands become enlarged in the males during the breeding season.

Darwin (1887, p. 528).

Darwin associated scent glands with reproduction because he noted that secretions occur more often in males than females, are more abundant in the breeding season, do not commence until adulthood and are inhibited by castration. He also noted species differences in scent glands among closely related animals.

It was clear to Darwin that the odours produced by the mammalian scent glands were important for sexual selection:

In most cases, when only the male emits a strong odour during the breeding season, it probably serves to excite or allure the female . . . . The odour emitted must be of considerable importance to the male, inasmuch as large and complex glands, furnished with muscles for everting the sack, and for closing or opening the orifice, have in some cases been developed. The development of these organs is intelligible through sexual selection, if the most odoriferous males are the most successful in winning the females, and in leaving offspring to inherit their gradually-perfected glands and odours.

Darwin (1887, p. 530).

After Darwin, the role of odorous secretions in courtship and sexual selection was emphasized as the primary function of the scent glands of mammals. Pycraft (1914), for example, gives the following anecdote

That these secretions play an important and perhaps variable part in the selection of mates seems demonstrated in the case of an incident related to me by my friend Mr. John Cooke, who some time ago was watching a flock of some three hundred sheep while it was being driven by the shepherd and his dogs into a field. As soon as they were securely shut in, and the shepherd had gone, three rams who were included in the flock at once began a three-cornered fight. One, presumably the youngest, was soon vanquished. The other two soon settled their differences, and the clashing of horns was at once followed by a very different performance. The master ram began to run in and out among the ewes, sniffing at each, and driving out those whose odour most pleased him. Having at last satisfied himself with a harem of about one hundred, the second ram was allowed to make a like choice, and behaved in a like manner, leaving the remainder to the ram which was first vanquished.

Pycraft (1914, pp. 69–70).

While many early zoology textbooks mentioned the scent glands of mammals and suggested possible functions for the odours of these glands, they were mainly speculative in nature. In *The Cambridge natural history*, for example, Beddard (1923) states the following:

It seems to be possible that the function of these various glands is at least two-fold. In the first place, they may serve, where predominant in one sex, to attract the sexes together. In the second place, the glands may be useful to enable a strayed animal of a gregarious species to regain the herd. It is perfectly conceivable too that in other cases the glands may be a protection, as they most undoubtedly are in the Skunk, from attacks. In connexion with the first, and more especially the second, of the possible uses of these glands, it is interesting to note that in purely terrestrial creatures, such as the Rhinoceros, the glands are situated on the feet, and would therefore taint the grass and herbage as the animal passed, and thus leave a track for the benefit of its mate. The same may be said of the rudimentary glands of Horses if they are really glands. The secretion of the 'crumen' of Antelopes is sometimes deposited deliberately by *Oreotragus* upon surrounding objects, a proceeding which would attain the same end. One may even perhaps detect 'mimicry' in the similar odours of certain animals. Prey may be lured to their destruction, or enemies frightened away. The defenceless Musk-deer may escape its foes by the suggestion of the musky odour of a crocodile. It is at any rate perfectly conceivable that the variety of odours among mammals may play a very important part in their life, and it is perhaps worthy of note that birds with highly-variegated plumage are provided only with the uropygial gland, while mammals with usually dull and similar coloration have a great variety of skin glands. Scent is no doubt a sense of higher importance in mammals than in birds. The subject is one which will bear further study.

Beddard (1923, pp. 13–14).

The role of olfaction in sexual selection in humans was examined by Havelock Ellis (1920) who concluded that smell plays only a small role in sexual selection in humans, except for some rare cases:

As the sole factor in sexual selection olfaction must be rare. It is said that Asiatic princes have sometimes caused a number of the ladies to race in the seraglio garden until they were heated; their garments have then been brought to the prince, who has selected one of them solely by the odor. There was here a sexual selection mainly by odor. Any exclusive efficacy of the olfactory sense is rare, not so much because the impressions of this sense are inoperative, but because agreeable personal odors are not sufficiently powerful, and the olfactory organ is too obtuse, to enable smell to take precedence of sight.

Ellis (1920, p. 66).

Ellis's major contribution to the study of olfactory communication in man lay not in his discussion of sexual selection, but in his identification of the sources of human odours and their information content. Ellis also noted differences in human odours due to race, sex, and age and made the important observation that a person's odour could be altered by his emotional state and by hormonal changes at puberty.

It is a significant fact, both as regards the ancestral sexual connections of the body odors and their actual sexual associations today, that, as Hipprocates long ago noted, it is not until puberty that they assume their adult characteristics. The infant, the adult, the aged person, each has his own kind of smell, and, as Monin remarks, it might be possible, within certain limits, to discover the age of a person by his odor. In both sexes puberty, adolescence, early manhood and womanhood are marked by a gradual development of the adult odor of skin and excreta, in general harmony with the secondary sexual developments of hair and pigment. Venturi, indeed, has, not without reason, described the odor of the body as a secondary sexual character. It may be added that, as is the case with the pigment in various parts of the body in women, some of these odors tend to become exaggerated in sympathy with sexual and other emotional states.

Ellis (1920, p. 63).

While there are many examples of changes in odour associated with sexual arousal in mammals, there are fewer examples of odours associated with other emotions but Bedichek (1960, p. 139), for example, states that:

There is a lot of evidence that the odour generated by fear in the human body stimulates the dog to attack. Cowboys, generally, believe that the odour of fear exhaled by the human body excites and often renders a horse unmanageable.

That physiological changes due to emotional, hormonal, and genetic factors could lead to changes in bodily odour was clearly pointed out by Bethe (1932) in his article on 'neglected hormones'. Bethe discussed the findings that there are species, family, and individual differences in odours, examined the changes in the odours of female mammals when they come into heat and discussed the role of odours in mother–infant attachment.

Scent marking

Another function of this book is to examine the social function of scent marking in mammals. The early ethologists recognized the importance of odour marks in an animal's environment: 'territory marking' using urine, faeces and the secretions from skin glands was described for pine martens by Goethe (1938), for bears and dogs by Bilz (1940) and for other mammals by Hediger (1944, 1949). According to Hediger (1950) scent marks make an animal's territory familiar as well as warding off intruders. Hediger (1955) summarizes the importance of these scent marks as follows:

A particularly important type of locality in the animal's territory should not be overlooked, the so-called demarcation places, found with deer, and many other mammals. These exist usually on prominent twigs or branches, or tree stumps, or stones, to which the owner of the territory applies its own property marks, so to speak, in the form of a self-produced scent. We must remember that most mammals are macrosmatic, i.e., they have a literally superhuman sense of smell, by means of which they recognize faint traces of scent, which are quite beyond our powers of detection, as conspicuous signals.

Whilst human beings usually demarcate their buildings and homes optically by means of signboards and street numbers, macrosmatic animals naturally use scents. These are produced in parts of the body varying with the species concerned. In deer and among antelopes, the gland above the eye, the so-called antorbital gland, produces a strong-smelling, oily substance, a small quantity of which is rubbed off on to branches and the like. In this way the whole living space is virtually impregnated with the individual scent of the owner. Any other member of its own species is thus warned off by these scent signals, as soon as it enters an occupied territory.

Hediger (1955, p. 19).

The value of scent in warning off intruders was referred to very explicitly by Hediger (1955, p. 23):

Scents, whether of dung, urine, or glandular secretion, are detachable from the body. They can literally be separated to act as place-reservations. The particle of dung, or the trace of secretion on the marking place, becomes as Bilz shows, the *pars pro toto* (part for the whole), and continues to be efficacious even in the absence of its author.

Bilz, discussing the demaraction behaviour of bears and dogs (1940, p. 285), states: 'Excrement and the image conveyed by it, frighten and even terrify the intruder'. It was this concept of territorial marking and the implied aggressiveness of the scent marks which attracted the interest of the classical ethologists. Tinbergen (1953), for example, made only one reference to olfactory communication in mammals and that is in a chapter on fighting:

Not all threat is visual. Many mammals deposit 'scent signals' at places where they meet or expect rivals. Dogs urinate to that purpose; Hyaenas, Martens, Chamois, various Antelopes and many other species have special glands, the secreta of which are deposited on the ground, on bushes, tree stumps, rocks, etc. The Brown Bear rubs its back against a tree, urinating while it does so.

Tinbergen (1953, pp. 58–9).

Lorenz (1954) viewed scent marking in the same way as Tinbergen, stressing the relationship between scent marking and aggressive motivation:

The leglifting of a dog has a very definitive meaning which is, paradoxically, exactly the same as that of a nightingale's song: it means the marking of the territory, warning off all intruders by telling them as clearly as their senses can perceive it that they are trespassing on the ground owned by somebody else. Nearly all mammals mark their territory by means of scent, as being one of their strongest sense faculties.

Lorenz (1954, p. 94).

This theme is expanded by Lorenz (1963) in *On aggression*:

Among mammals, which 'think through their noses', it is not surprising that marking of the territory by scent plays a big role. Many methods have been tried; various scent glands have been evolved, and the most remarkable ceremonies developed round the depositing of urine and faeces; of these the leglifting of the domestic dog is the most familiar. The objection has been raised by some students of mammals that such scent marks cannot have anything to do with territorial ownership because they are found not only in socially living mammals which do not defend single territories, but also in animals that wander far and wide; but this opinion is only partly correct. First, it has been proved that dogs and other pack-living animals recognize each other by the scent of the marks, and it would at once be apparent to the members of a pack if a non-member presumed to lift its leg in their hunting-grounds. Secondly, Leyhausen and Wolf have demonstrated the very interesting possibility that the distribution of animals of a certain species over the available biotope can be effected not only by a space plan but also by a time plan. They found that, in domestic cats living free in open country, several individuals could make use of the same hunting-ground without ever coming into conflict, by using it according to a definite timetable, in the same way as our Seewiesen housewives use our communial washhouse. An additional safeguard against undesirable encounters is the scent marks which these animals—the cats, not the housewives—deposit at regular intervals wherever they go. These act like railway signals whose aim is to prevent collision between two trains. A cat finding another cat's signal on its hunting-path assesses its age, and if it is very fresh it hestitates, or chooses another path; if it is a few hours old it proceeds calmly on its way.

Lorenz (1963, p. 27).

Eibl-Eibesfeldt (1970) has emphasized both the threatening function of scent marks and their function in reassuring the territory owner:

The scent marks are chemical property signs. They aid the territory owner, first of all, as signs of recognition. They help in orientation and make the area familiar. A badger that becomes agitated or frightened in a strange environment can be calmed by letting it sniff an object that it had marked previously. A male hamster that enters the territory of a female during the mating season will mark this strange territory before it actually begins to court. It is probable that this also has a repelling function for others. Strange scent marks have an aggression-releasing effect in hamsters, which show threat behavior when sniffing strange scent marks.

Eibl-Eibesfeldt (1970, p. 311).

While most ethologists emphasized the role of odours in territory marking, aggressive and sexual encounters, Schloeth (1956) described the importance of olfactory investigation between mammals encountering strangers for the first time. Schloeth defined three basic patterns of olfactory investigation of conspecifics: naso-anal, naso-genital, and naso-nasal (see Fig. 1), and pointed out that animals had specific 'contact points' at which olfactory investigation most often occurred. Schloeth considered the investigation of scent marks to be an 'indirect' olfactory investigation as opposed to the more direct naso-anal, naso-genital, or naso-nasal postures.

Recent years have seen mounting acknowledgement, and somewhat increased understanding, of the complexity of mammalian societies. Nevertheless, proof of some enduring interpretations of the functions of social odours has remained sparse (e.g. the *pars pro toto* model, cf. Gosling (1982)). Olfactory communication turns out to be as complex and flexible as are the societies within which it operates: Kruuk (1972), for example, found that spotted hyaenas, *Crocuta*

crocuta, from two adjacent populations deployed their faeces and associated glandular odours in quite different patterns. As intra-specific variations in social organization come increasingly to light, so corresponding variation in the pattern of scent marking between populations and among individuals is emphasized (Macdonald 1980).

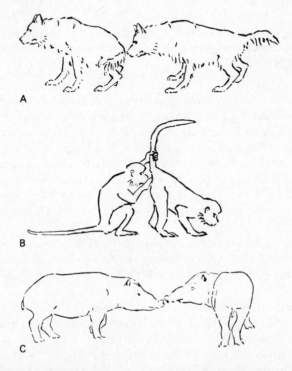

Fig. 0.1. Typical encounters involving olfactory investigation. A. Naso-anal investigation in the brown hyaena (*Hyaena brunnea*). B. Naso-genital investigation in the Philippine macaque (*Macaca irus*). C. Naso-nasal investigation in the Brazilian tapir (*Tapirus terrestris*). (From Schloeth 1956.)

Odour classification

As the knowledge of olfactory communication in mammals has increased, there have been a number of attempts to develop classification systems for the social odours. The first such system was developed by Bethe (1932) who distinguished between *endohormones* which are secreted into the individual's body (and termed hormones in 1905 by Starling) and *ectohormones* which are excreted from the exterior of the body as follows:

Zu unterscheider wären dann wieder zwei Hauptklassen: Die Endohormone, die im produzierenden Organismus selbst zur Wirksamkeit gelangen, und die

Ektohormone, die nach aussen abgegeben auf andere Lebewesen einen für den Produzenten oder die Art nützlichen Einfluss ausuben.[1]

Bethe (1932, p. 178).

Bethe then divided ectohormones into two subgroups, depending on whether their action was intraspecific (*homoiohormones*) or interspecific (*alloiohormones*):

Wir wollen sie schon hier unterteilen in die Homoiohormone und in die Alloiohormone, jenachdem sich ihre Wirkung auf Individuen der gleichen Art oder einer anderen Art ersteckt.[2]

Karlson and Lüscher (1959) coined the term *pheromone* for those substances called *homoiohormones* by Bethe, and Brown (1968) coined the term *allomone* for the *alloiohormones*. Later, Brown, Eisner, and Whittaker (1970) discriminated between *allomones*, defined as chemical signals which, when received by an animal of another species, evoke a response favourable to the emitter of the odour; and *kairomones*, defined as a 'transspecific chemical messenger, the adaptive benefit of which falls on the recipient rather than on the emitter' (Brown *et al.* 1970, p. 21). Since no benefit accrues to the transmitter, kairomones cannot be thought of as an agent of communication (*sensu* Dawkins and Krebs 1978).

Because of its completeness, Kirschenblatt's (1962) terminology is worth noting (see Table 0.1), but its complexity and the fact that many chemicals may not fall exclusively into one category have kept this system from being used more widely. The terms pheromone and allomone are currently the most widely used (see Eisenberg and Kleiman 1972 and Wilson 1975, p. 231). Otte (1974) has examined the relationships between hormones, pheromones, and allomones and how such signalling systems might have evolved (see Fig. 0.2).

Pheromones may act through the central nervous system to stimulate rapid behavioural changes (releaser, signalling, or trigger effects) or through the neuro-endocrine system to stimulate physiological changes resulting in a delayed behavioural change (primer or pump effects, see Fig. 0.3) (Adler 1974; Bronson 1968; Whitten and Bronson 1970; Parkes and Bruce 1961).

Wilson and Bossert (1963) originally applied the terms primer and releaser to the *effects* of pheromones and stated that 'it is quite possible for the same pheromone to be both a primer and releaser' (p. 675). The terms primer and releaser later became applied to the pheromones themselves, rather than their effects, leading to attempts to enumerate the number of different primer and releaser (signalling) pheromones (Bronson 1968, 1971).

---

[1] We differentiate the two principal classes: the *endohormones*, which are effective in the producing organism itself, and the *ectohormones*, which are given off externally to other organisms to exert an advantageous influence for the producer or its species.

[2] We intend at this time to subdivide these into the *homoiohormones* and the *alloiohormones* according to whether their action affects individuals of the same species or those of other species.

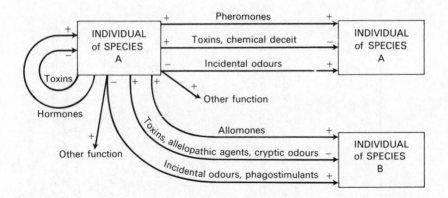

Fig. 0.2. Summary of the roles of chemical compounds in mediating interactions within individuals and between individuals. The terms *hormone, pheromone,* and *allomone* are restricted to cases where a net benefit accrues to the emitter and receiver as a result of chemical transmission. The symbols + and − indicate whether the fitness of emitters or receivers increases or decreases as a result of odour transmission. (From Otte 1974.)

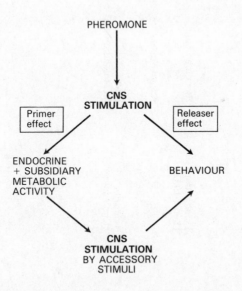

Fig. 0.3. The distinction between releaser effects and primer effects is that in the latter case the behavioural change is stimulated indirectly through neuro-endocrine changes rather than by direct stimulation of the central nervous system. (From Wilson and Bossert 1963.)

Table 0.1. A chronology of terminology for chemical secretions

Hormone (Starling 1905)	Chemical message secreted by one organ in the body and acting on another organ within the body of the producing organism		
Endohormone (Bethe 1932)			
Ectohormone (Bethe 1932)	Homoiohormone		A chemical secreted to the exterior of the body of one organism which acts on another organism of the same species
	Alloiohormone		A chemical secreted to the exterior of the body of one organism which acts on an organism of another species
Telergone (Kirschenblatt 1962)	Homotelergone (homoiohormones)	Gonophyone	Acting on the development of the sex glands and the formation or modification of the sex characteristics of other organisms
		Gamophyone	Stimulates the maturation of the sex glands and the onset of the reproductive process in individuals of the other sex
		Epagone	Attracts other individuals of the same species, most frequently of the opposite sex
		Odmichnione	Deposited on the ground and other surrounding objects and serves for marking the locality and facilitating orientation
		Thorybone	Excite alarm reactions in other individuals of the same group

Heterotelergone (alloiohormones)	
Amynone	Protective or repulsive substance protecting animals against the attacks of their enemies
Prohaptone	Poisonous substance paralysing or killing the prey
Lichneumone	Substance which attracts symbiotic species
Xenoblaptone	Produced by parasites and has specific influences on the structure and physiology of their hosts
Pheromone (Karlson and Lüscher 1959)	A general term for homoiohormones and all subclasses of homotelergones
Ectohormone or telergone	
Releaser (Wilson and Bossert 1963)	Pheromones which directly stimulate the central nervous system resulting in a rapid behavioural response
Primer (Wilson and Bossert 1963)	Pheromones which stimulate neuro-endocrine changes and indirectly result in behavioural changes
Allomone (Brown 1968)	A general term for alloiohormones and those heterotelergones which evoke responses from animals of other species which benefit the animal emitting the odour
Kairomone (Brown, Eisner, and Whittaker 1970)	Those heterotelergones which benefit the receiver rather than the emitter

Because of the criticisms of the use of the term 'pheromone' in describing mammalian chemical signals (Beauchamp, Doty, Moulton, and Mugford 1976) and the fact that responses to these odours are not stereotyped, but depend on developmental and experiential variables, Brown (1979) suggested that the term 'social odours' would be more appropriate than 'pheromones' and classified these odours according to their information content as shown in Table 0.2. Whether a behavioural or physiological response is elicited depends on the nature of the odour and the ability of the receiver to extract the information about the sender from this odour. In this classification system the signalling and priming effects are determined by the interaction between the chemical signal, the receiver and the social relationship between the sender and receiver.

Table 0.2 A classification of mammalian social odours

Identifier odours: those produced by the body's normal metabolic processes that are stable for long periods of time
  Individual: odours that are unique to each individual animal
  Colony: odours that are characteristic to all members of a colony, nest, den, deme, or living group; in some cases family odours
  Species typical: odours that are characteristic of all members of a species or subspecies
  Age specific: odours that distinguish animals of different age classes; such as preweaning (infant), postweaning (juvenile), and adult
  Sex specific: odours that distinguish males from females

Emotive odours: those produced or released only in special circumstances
  Rut: odours distinguishing animals in breeding condition from those that are not
  Social status: odours that distinguish subordinate from dominant animals, usually males
  Stress: odours released in response to fear- or alarm-inducing stimuli; also odours released in frustration situations
  Maternal: odours distinguishing lactating from non-lactating females, and possibly pregnant from non-pregnant females

---

(From Brown 1979.)

Recent advances in the study of mammalian olfactory communication

The first attempt to produce a taxonomic review of the role of scent glands as 'odours of expression' in mammals was that of Fiedler (1964). Since this review, a wide range of books and articles have explored the importance of skin glands and scent marking in mammalian communication. These include a chapter on scent marking in Ewer's (1968) book on mammalian ethology; Tembrock's (1968) chapter on communication in land mammals; Altmann's (1969) book on urination and defecation patterns in mammals; and, most recently, Eisenberg (1981) has examined the taxonomy of olfactory communication in mammals.

The physiological effects of mammalian odours (the primer effects) were discovered by Whitten in 1956, and reviewed by Whitten (1966). Gleason and Reyneirse's (1969) review of vertebrate pheromones was the first review of olfactory communication in a psychology journal and served the important function of integrating psychological, ethological, and physiological research and pointing out the importance of olfactory signals in behavioural studies.

The First International Symposium on Olfaction and Taste (ISOT) was held in 1962 and the publication of the first three of these symposia (Zotterman 1963; Hayashi 1967; Pfaffmann 1969) provided an important stimulus to research on olfactory communication. In 1970 the first book devoted entirely to communication by chemical signals was published (Johnson, Moulton, and Turk 1970).

Some of the recent reviews include those of Mykytowycz (1970); Cheal and Sprott (1971); Ralls (1971); Eisenberg and Kleiman (1972); Johnson (1973); Whitten and Champlin (1973); Christiansen and Doving (1974); Cheal (1975) Thiessen and Rice (1976); and Brown (1979). Books focusing on olfactory communication have been written by Stoddart (1976, 1980b) and Shorey (1976) and edited by Birch (1974), Doty (1976), Sebeok (1977), Ritter (1979) and Stoddart (1980a). A continuing series of international meetings on chemical signals in vertebrates has been held three times and the results edited by Müller-Schwarze and Mozell (1977), Müller-Schwarze and Silverstein (1980) and Müller-Schwarze and Silverstein (1983).

The European Chemoreception Research Organisation (ECRO) (founded in 1970) fosters research on olfactory communication in mammals through specialized symposia and the Association for Chemoreception Sciences (AChemS) founded in 1980 has also promoted interest in olfactory communication in mammals. A special abstracting service (*Chemoreception Abstracts*) started to catalogue research on chemical communication in 1972 and a journal devoted to chemical communication (the *Journal of Chemical Ecology*) was introduced in 1975.

Years have passed since the first speculations were penned as to the functions of social odours within mammalian societies. A few remarkable studies have proved the information content of some social odours and a handful of others have demonstrated their functions within societies. A major and continuing stumbling-block has been our limited understanding of mammalian societies and hence of how social odours may function within them. Although clearly fruitful in many respects, the focus of attention on a very few species has limited our perspective on olfactory communication in mammals and constrained the search for its functional roles. Each of the authors contributing to this book seeks to set knowledge of social odours in a given taxon against the perspective of the tremendous diversity of mammalian radiations and behavioural ecology. In this sense the book is a synthesis, but at the same time each chapter develops different themes. To a large extent these themes were thrust upon us by the distinct flavour of the studies directed at each family, e.g. members of some families are frequently the subjects of experimental work whereas others rarely are. Consequently the chapters are structured to capitalize on the available material. For this reason some families are covered in a single review chapter whereas others (e.g. primitive eutherians and primates) are dealt with in two sections, one of which concerns the general material while the other focuses on a case study which serves as the most fully understood representative of its taxon. Similarly, where an author's presentation was made more complete by previously unpublished data, these are included.

Each chapter merges two aspects, summaries and catalogues of information on the anatomical, chemical, and behavioural adaptations to olfactory communication are interwoven with a discussion of the possible functions of the odours. Doubtless while the constituents of some hypotheses will survive, others will prove more volatile but we hope that the encyclopaedic element of the book will prove an enduringly useful reference work.

The literature on social odours lies widely scattered in the vaults of several disciplines, and spans more than one era of biological thinking. A profusion of loose ends juts out from this literature to overwhelm the reader; by considering all the mammalian families within these two volumes we hope that some of these loose ends will be knotted together, while the remainder will be considerably untangled.

> They haven't got no noses,
> Those fallen sons of Eve;
> Even the smell of roses
> Is not what they supposes;
> But more than mind discloses
> And more than men believe.
>
> And Quoodle here discloses
> All things that Quoodle can,
> They haven't got no noses,
> They haven't got no noses,
> And goodness only knowses
> The noselessness of Man.
>
> G. K. Chesterton
> (from 'Song of Quoodle').

References

Adams, M. G. (1980). Odour-producing organs of mammals. *Symp. zool. Soc. Lond.* **45**, 57–86.

Adler, N. T. (1974). The behavioral control of reproductive physiology. In *Reproductive behavior* (ed. W. Montagna and W. A. Sadler), pp. 259–86. Plenum Press, New York.

Altmann, D. (1969). *Harnen und Koten bei Säugetiere.* Ziemsen Verlag, Wittenberg Lutherstadt.

Beauchamp, G. K., Doty, R. L., Moulton, D. G., and Mugford, R. A. (1976). The pheromone concept in mammalian chemical communication: A critique. In *Mammalian olfaction, reproductive processes, and behavior* (ed. R. L. Doty), pp. 143–60. Academic Press, New York.

Beddard, F. E. (1923). *Mammalia. The Cambridge natural history, Vol. 5.* Macmillan and Company, London.

Bedichek, R. (1960). *The sense of smell.* Michael Joseph, London.

Bethe, A. (1932). Vernachlässigte Hormone. *Die Naturwissenschaften* **11**, 177–81.

Bilz, R. (1940). Pars pro toto. *Ein Beitrag zur Pathologie menschlicher Affekte und Organfunktionen.* Thieme, Leipzig.

Birch, M. E. (ed.) (1974). *Pheromones.* North Holland, Amsterdam.

Brinkmann, A. (1912). Die Hautdrüsen der Säugetiere. *Entwicklungsgesch* **20**, 1173–231.

Bronson, F. H. (1968). Pheromonal influences on mammalian reproduction. In *Perspectives in reproduction and sexual behavior* (ed. M. Diamond), pp. 341–61. Indiana University Press, Bloomington.

— (1971). Rodent pheromones. *Biol. Reprod.* **4**, 344–57.

Brown, R. E. (1979). Mammalian social odors: A critical review. In *Advances in the study of behavior* (ed. J. S. Rosenblatt, R. A. Hinde, C. Beer, and M.-C. Busnel), Vol. 10, pp. 103–62. Academic Press, New York.

Brown, W. L., Jr. (1968). An hypothesis concerning the function of the metapleural glands in ants. *Am. Nat.* **102**, 188–91.

— Eisner, T., and Whittaker, R. H. (1970). Allomones and Kairomones: Transspecific chemical messengers. *Bioscience* **20**, 21–2.

Cheal, M. L. (1975). Social olfaction: A review of the ontogeny of olfactory influences on vertebrate behavior. *Behav. Biol.* **15**, 1–25.

— and Sprott, R. L. (1971). Social olfaction: A review of the role of olfaction in a variety of animal behaviors. *Psychol. Rep.* **29**, 195–243.

Christiansen, E. and Doving, K. B. (1974). Communication in small rodents. *Fauna* **26**, 17–30.

Darwin, C. (1887). *The descent of man and selection in relation to sex.* (2nd ed, revised). John Murray, London.

Dawkins, R., and Krebs, J. (1978). Animal signals: information or manipulation? In *Behavioural ecology, an evolutionary approach.* (ed. J. R. Krebs and N. B. Davies) p. 282–309. Blackwell Scientific Publications, Oxford.

Doty, R. L. (ed.) (1976). *Mammalian olfaction, reproductive processes, and behavior.* Academic Press, New York.

Eggeling, H. v. (1931). Hautdrüsen. *Handbuch verg. Anat. Wirbeltiere* **1**, 663–92.

Eibl-Eibesfeldt, I. (1970). *Ethology: the biology of behavior.* Holt, Rinehard, and Winston, New York.

Eisenberg, J. F. (1981). *The mammalian radiations: an analysis of trends in evolution, adaptation and behavior.* University of Chicago Press.

— and Kleiman, D. G. (1972). Olfactory communication in mammals. *A. Rev. Ecol. Systemat.* **3**, 1–32.

Ellis, A. (1962). *The essence of beauty: A history of scent.* Collier Books, New York.

Ellis, H. H. (1920). *Studies in the psychology of sex. Vol. 4. Sexual selection in man.* Davis, Philadelphia, Pennsylvania.

Ewer, R. F. (1968). *Ethology of mammals.* Elek Science, London.

— (1973). *The carnivores.* Cornell University Press, Ithaca, New York.

Fiedler, W. (1964). Die Haut der Säugetiere als Ausdrucksorgan. *Studium General* **17**, 362–90.

Gabe, M. (1967). Le tegument et ses annexes. In P.-P. Grasse (ed.), *Traite de Zoologie: Anatomie, Systematique, Biologie,* **16**, 1–223.

Gleason, K. K. and Reyneirse, J. H. (1969). The behavioral significance of pheromones in vertebrates. *Psychol. Bull.* **71**, 58–73.

Goethe, F. (1938). Beobachtungen uber das Absetzen von Witterungsmarken beim Baummarder. *Deutsche Jäger* **13**, 211–13.

Gosling, L. M. (1982). A reassessment of the function of scent marking in territories. *Z. Tierpsychol.* **60**, 89–118.

Graff, J. G. (1879). Vergleichend anatomische Untersuchungen uber den Bau der Hautdrüsen der Haussäugetiere. Thesis, Lepzig (cited by Gabe 1967 and Schaffer 1940).

Hayashi, T. (ed.) (1967). *Olfaction and taste* II. Pergamon Press, Oxford.

Hediger, H. (1944). Die Bedeutung von Miktion und Defakation bei Wildtieren. *Z. Psychol.* **3**, 170–82.

— (1949). Saugetier-Territorien und ihre Markierung. *Bijd. Dierk.* **27**, 172–84.

— (1950). *Wild animals in captivity.* Butterworth, London.

— (1955). *The psychology and behaviour of animals in zoos and circuses.* Butterworths Scientific Publications, London.

Hindle, E. (1948). Reginald Innes Pocock. *Obituary Notices of Fellows of the Royal Society* **6**, 189–211.

Johnson, J. W., Jr., Moulton, D. G., and Turk, A. (ed.) (1970). *Advances in chemoreception,* Vol. I. Appleton-Century-Crofts, New York.

Johnson, R. P. (1973). Scent marking in mammals. *Animal Behav.* **21**, 521–35.

Karlson, P. and Lüscher, M. (1959). 'Pheromones': A new term for a class of biologically active substances. *Nature, Lond.* **183**, 55–6.

Kirschenblatt, J. (1962). Terminology of some biologically active substances and validity of the term 'pheromones'. *Nature, Lond.* **195**, 916–17.

Kruuk, H. (1972). *The spotted hyaena: a study of predation and social behavior.* University of Chicago Press.

Lorenz, K. (1954). *Man meets dog.* Methuen, London.

— (1963). *On aggression.* Methuen, London.

Macdonald, D. W. (1980). Patterns of scent marking with urine and faeces amongst carnivore communities. In *Olfaction in mammals* (ed. D. M. Stoddart). *Symp. zool. Soc. Lond.* **45**, 107–39.

Müller-Schwarze, D. and Mozell, M. M. (1977). *Chemical signals in vertebrates.* Plenum Press, New York.

— and Silverstein, M. M. (1980). *Chemical signals: vertebrates and aquatic invertebrates.* Plenum Press, New York.

— and — (1983). *Chemical signals III.* Plenum Press, New York.

Mykytowycz, R. (1970). The role of skin glands in mammalian communication. In *Advances in chemoreception, Vol. I, Communication by chemical signals* (ed. J. W. Johnston Jr. *et al.*), pp. 327-60. Appleton-Century-Crofts, New York.

Nisbett, A. (1976). *Konrad Lorenz.* Dent, London.

Ortmann, R. (1960). *Die Analregion der Säugetiere. Handb. Zool.* 8:26, 3(7), 1-68.

Otte, D. (1974). Effects and functions in the evolution of signalling systems. *A. Rev. Ecol. Systemat.* 5, 385-417.

Pallas, P. S. (1779). *Spicilegia Zoologica*, fas. xiii, p. 23. (Referred to in Darwin, 1887).

Parkes, A. S. and Bruce, H. M. (1961). Olfactory stimuli in mammalian reproduction. *Science, N.Y.,* 134, 1049-54.

Pfaffmann, C. (ed.) (1969). *Olfaction and taste III.* Rockefeller University Press, New York.

Pycraft, W. P. (1914). *The courtship of animals* (2nd edn). Hutchinson, London.

Ralls, K. (1971). Mammalian scent marking. *Science, N.Y.* 171, 443-9.

Ranvier, L. (1887). Le mecanisme de la secretion. *J. Micrographie* 11, 1-146.

Ritter, F. J. (1979). *Chemical ecology: Odour communication in animals.* Elsevier/North Holland Biomedical Press, Amsterdam.

Schaffer, J. (1940). *Die Hautdrüsenorgane der Säugetiere.* Urban & Schwarzenberg, Berlin.

Schiefferdecker, P. (1922). Die Hautdrüsen des Menschen und der Säugetiere. *Zoologica* 27, 1-154.

Schloeth, R. (1956). Zur Psychologie der Begegnung zwischen Tieren. *Behaviour* 10, 1-80.

Sebeok, T. A. (ed.) (1977). *How animals communicate.* Indiana University Press, Bloomington.

Shorey, H. H. (1976). *Animal communication by pheromones.* Academic Press, New York.

Starling, E. H. (1905). The chemical correlation of the functions of the body. Lecture 1. The chemical control of the functions of the body. *Lancet* 2 339-41.

Stoddart, D. M. (1976). *Mammalian odours and pheromones.* Arnold, London.

— (1980*a*) (ed.). Olfaction in mammals. *Symp. zool. Soc. Lond.* 45, 1-363.

— (1980*b*). *The ecology of vertebrate olfaction*, Chapman and Hall, London.

Tembrock, G. (1968). Land mammals. In *Animal communication.* (ed. T. A. Sebeok), pp. 338-404. Indiana University Press, Bloomington.

Thiessen, D. D. and Rice, M. (1976). Mammalian scent gland marking and social behavior. *Psychol. Bull.* 83, 505-39.

Tinbergen, N. (1953). *Social behaviour in animals.* Methuen, London.

Whitten, W. K. (1956). Modifications of the oestrous cycle of the mouse by external stimuli associated with the male. *J. Endocrinol.* 13, 399-404.

— (1966). Pheromones and mammalian reproduction. In *Advances in reproductive physiology*, I (ed. A. McLaren), pp. 155-78. Academic Press, New York.

— and Bronson, F. H. (1970). The role of pheromones in mammalian reproduction. In *Advances in chemoreception* Vol. I (ed. J. W. Johnson, Jr., D. G. Moulton and A. Turk), pp. 309-26. Appleton-Century-Crofts, New York.

— and Champlin, A. K. (1973). The role of olfaction in mammalian reproduction. In *Handbook of physiology*, Sect. 7, Vol. II. Pt 1 (ed. R. O. Greep), pp. 109-23. American Physiological Society, Washington, D.C.

Wilson, E. O. (1975). *Sociobiology*. Belknap Press of Harvard University Press, Cambridge, Massachusetts.
— and Bossert, W. H. (1963). Chemical communication among animals. *Rec. Prog. Horm. Res*. **19**, 673–710.
Zotterman, Y. (ed.) (1963). *Olfaction and taste* I. Pergamon Press, Oxford.

1 Sources of significant smells: the skin and other organs

PETER FLOOD

This chapter will be concerned with the sources of the volatile compounds that are inevitably released into the environment by mammals in the normal course of their existence. These sources are diverse and have been investigated to varying extents; they are summarized graphically in Fig. 1.1 and include the skin and its associated glands, the oral and nasal cavities with their glands, the lungs, the urinary and genital tracts, and the anal region. Other sources probably include secretions associated with the eye and perhaps even those of the external ear. Urine, faeces, and the secretions of skin are of special behavioural importance in that they can be deposited in the environment and form a semi-permanent record. The skin and its glands have received much attention and will be described first. The other potential odour sources will be dealt with in less detail, mainly because much less is known about them.

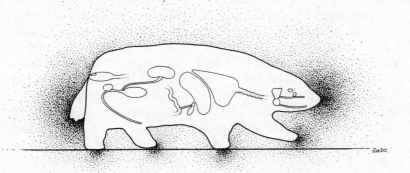

Fig. 1.1. A summary diagram of an imaginary animal illustrating a variety of possible odour sources. The stippling indicates a potential distribution of odorants in the immediate environment. The following are shown: salivary glands and glands associated with the eye; the lungs and trachea; the liver, gall-bladder, bile duct, and portion of small intestine; the kidney, ureter, bladder, urethra, and male accessory gland; the rectum; and an anal sac. The female genital system could be readily substituted for the male.

The skin

The skin is generally regarded as the largest single organ in the body and in the newborn puppy represents about a quarter of its body weight though this value is reduced to about 16 per cent in the adult beagle (Warner and McFarland 1970). The skin is primarily a protective structure and resistance to frictional, osmotic, chemical, thermal, and photic environmental damage is provided by the outer, avascular epidermis, which is formed from the ectoderm of the embryo. The dermis, which is mesodermal, lies immediately beneath the epidermis and is intimately attached to it. It consists of a dense vascular feltwork of collagen and elastin fibres and provides protection from major mechanical insults. When the dermis has been suitably preserved we make use of its remarkable physical properties as leather.

Returning to the epidermis: this consists of a continuous cohesive layer of cells—an epithelium—overlying the dermis (Fig. 1.2). These cells proliferate rapidly and all but a few that continue to divide are forced towards the surface. As they move they are transformed into thin tough microscopic scales or squames, rich in the structural protein keratin. This process is known as keratinization and because of it the epidermis has a multilayered appearance and the epidermal epithelium becomes known as a stratified squamous epithelium.

This epithelium is the basic universal component of the epidermis and it invariably shows deep and complicated invaginations that penetrate the dermis and even the loose connective tissue beneath. The epidermal cells lining each invagination are modified to form either a hair or one of two different kinds of gland. The growth of hair will receive little attention in this chapter but some of the microanatomy of a hair and its follicle (sheath) is shown in Fig. 1.2. Hair, no doubt, greatly influences the odour information transmitted by an animal by controlling the pattern of air circulation to the skin surface, by increasing the area available for the evaporation of informative compounds and by providing a substrate for the growth of odour-producing micro-organisms. Hair also acts as a brush with which odorants are applied to objects in the environment and sometimes as a palate on which volatile compounds normally produced by the skin are mixed with those from urine, saliva, faeces, and the like. Specially roughened hairs to which secretions adhere strongly have been found over the tarsal scent gland of the black-tailed deer (Müller-Schwarze, Volkman, and Zemanek 1977). These osmetrichia, as they have been called, possess divergent cuticular scales that sometimes bear groups of tiny knobs and project from the main shaft of the hair. In all these roles, hair is essentially a passive substrate, but the small involuntary muscles associated with the hair follicles (*musculi arrectores pilorum*) may actively express secretion from the follicle and the surrounding glands as well as performing their main task of pulling the hair-shaft closer to the perpendicular in relation to the skin surface (Fig. 1.2). Some direct evidence for this comes from the interdigital glands of deer where the hair is relatively immobile but the arrector muscles are particularly well developed and partially envelop the sebaceous glands (Quay and Müller-Schwarze 1970). Erection of the hair is normally a response to cold air-flow over the skin surface and hence reducing, presumably, the release of odorants. However, in some areas that are particularly

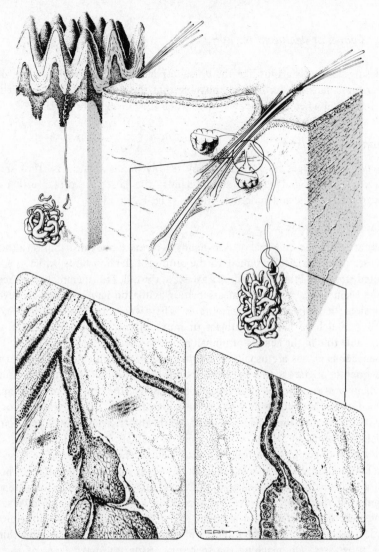

Fig. 1.2. *Top*—A composite illustration showing the major features of the skin. The block to the left illustrates an area of foot pad with a thickened, papillate epidermis and an eccrine gland deep in the dermis. The glandular coil shows both the thick secretory part and a narrow convoluted section of the duct. The duct has a characteristic spiral form where it penetrates the epidermis. The block to the right shows an area of hairy skin with a compound follicle in section that, for the sake of simplicity, shows only two secondary hairs. Two sebaceous glands are present with ducts opening into the cylindrical cavity containing the hair-shafts known as the pilosebaceous canal. The single apocrine sweat gland also opens into it. The arrector muscles of the hair extend from the follicle towards the skin surface and are usually related to the sebaceous glands. ($\times$ 25.)

Left inset—Detail of the openings of sebaceous and apocrine ducts into the pilosebaceous canal. Some secretory ascini of the sebaceous glands are also shown.

Right inset—The junction between the secretory part of the apocrine gland and its duct.

well supplied with glands like the dorsal aspect of the tail in canids, the dense under-fur is lacking and it seems possible that the stiff guard-hairs permit freer air circulation when they are erect than when they lie close to the skin.

Cutaneous glands

The two basic types of skin gland may readily be distinguished—flask-shaped sebaceous glands and tubular sweat glands: the latter are often further subdivided into eccrine and apocrine varieties (Fig. 1.2 and Plate 1.1).

The sebaceous glands

These are the most universal of skin-glands in terms of body distribution although they are morphologically similar to the uropygial gland of birds, which is a large isolated structure lying on the dorsal aspect of the tail. True mammalian sebaceous glands seem to have evolved in association with the hair coat and to have no equivalent in poikilothermic vertebrates. They were first observed by Eichorn in 1826 and have been the subject of much attention since because of their intractable role in the genesis of human acne.

Sebaceous glands are usually found in association with a hair follicle but may subsequently migrate so that the gland and the hair grow independently. Sebaceous glands may also be found in areas lacking hair like the lips and in such improbable sites as the buccal mucosa, larynx, oesophagous, and the cervix of the uterus. Classically, the sebaceous gland lies in the triangle formed between the follicle, the skin, and the arrector muscle of the hair (Fig. 1.2) and has a short duct which opens into the follicle a little below its cutaneous orifice (Fig. 1.2, inset). Commonly several sebaceous glands open into a single follicle. Though each gland is basically flask-shaped, the 'flask' may be more or less lobulated and irregular so that glands isolated by dissection look like tiny cauliflowers. The lobules are often referred to as ascini. The stratified squamous epithelium of the skin is continued into the follicle and thence into the duct, becoming thinner as it progresses. On reaching the secretory tissue, the basal layer of the epithelium becomes continuous with a proliferative cell layer lining each gland (Plate 1.2). New cells generated by this layer force the older cells centrally; these become particularly rich in lipids and disintegrate completely as they approach the lumen. In consequence the lumen is often rather ill-defined and filled with tissue debris, the consistency of which is hard to judge microscopically (Plate 1.2). The life of a sebaceous cell from formation to disintegration is about eight days in the rat (Bertalanffy 1957). This seemingly extravagant process, in which whole cells are converted into a secretory product, is known as holocrine secretion and the sebaceous glands are by far the best example of the process in mammals. The secretory product, sebum, doubtless varies greatly in detailed composition but is always oily and rich in lipids, and often has antimicrobial properties (see for example, Pillsbury and Rebell 1952).

Reported sebum secretion rates are always slow and vary from about 0.005 g/m^2 per h for the back of the human hand (Johnson 1952) to 0.1 g/m^2 per h for the bovine flank (Smith and Jenkinson 1975) and 0.6 g/m^2 per h for the human forehead (Cunliffe and Shuster 1969). None the less, Smith and Jenkinson (1975) suggested that the rate of secretion they observed was too high to be accounted for solely in terms of holocrine secretion and that some additional mechanism was therefore involved. Unfortunately, they did not provide direct morphological evidence of this and it is possible that their results could be interpreted in other ways.

The lipid content of sebaceous cells (Plate 1.3) is evident ultrastructurally as abundant membrane-bound vesicles (Plate 1.4) that appear first around the nucleus. Other notable features of differentiating sebaceous cells are abundant smooth endoplasmic reticulum reminiscent of that seen in steroidogenic tissues, and well-developed Golgi membrane systems. The fully developed cells are packed with sebum vesicles separated by narrow strands of cytoplasm and their nuclei and cell membranes are often misshapen and fragmented (Plate 1.5).

The sebaceous acini are well vascularized at their periphery but they have no contractile elements and the secretion presumably reaches the skin surface as a result of general body movement, rubbing the skin and grooming, the capillary action of the keratinized skin surface and the contraction of the arrector muscles of the hair. These muscles are under direct nervous control by autonomic adrenergic fibres (Uno 1977) and their contraction produces the familiar response known as 'goose flesh'. Evidence for their role in the expression of sebum has already been mentioned.

It is well established (see Montagna and Parakkal 1974) that sebaceous glands depend for their development and activity on androgenic steroids from the testis, ovary or adrenal cortex; they show little activity prior to puberty, then grow rapidly to reach maximal activity during adulthood and cannot be further stimulated by exogenous androgens. More recently, α-melanocyte-stimulating hormone from the intermediate lobe of the pituitary has been shown to stimulate sebaceous glands strongly and full secretory activity can be restored to hypophysectomized rats by administration of this hormone and testosterone. Neither of these compounds is completely effective alone. Thyroxine and the corticosteroids also stimulate sebaceous secretion to a minor degree (Thody and Shuster 1975).

In general it has proved impossible to demonstrate nerves supplying sebaceous glands though networks of cholinesterase-containing nerves have been seen around the large and specialized sebaceous glands of the eyelid (Montagna and Ford 1969) and around the sebaceous glands of cattle (Jenkinson, Sen Gupta, and Blackburn 1966). It seems likely that these nerves are sympathetic and cholinergic but their significance is not yet known.

What function sebaceous glands may have apart from their contribution to general mammalian odour is by no means clear though it is reasonable to speculate that they have a role in maintaining the quality of the pelage, the suppleness of the skin, and the status of the cutaneous microflora. Though sebum output is greatest in the adult male, it may be that even in the young sebum excretion is sufficient to be of some significance. It is perhaps appropriate to end this section with Montagna and Parakkal's (1974) intuitive statement that, 'Aside from axillary odour, the aroma that distinguishes different clean human bodies is sebum' and the recollection that the smell of the sebum-permeated wool of the ram stimulates ovulation in the ewe (Knight and Lynch 1980).

Tubular glands

The tubular skin-glands—first recognized by Malphigi (1628–94)—are also referred to as sudoriferous or sweat glands; their secretion is always watery and they can be divided into two main types according to the histology of the secretory cells, the position of the gland opening and the neural mechanisms influencing their activity. When the distinction between the two types of glands was first recognized at the beginning of the century it was supposed, on the basis of light microscopy, that they differed radically in their mode of secretion: in one kind, fragments of cytoplasm seemed to be shed from the apices of the cells into the lumen and these were referred to as apocrine glands; in the other, secretion apparently occurred without cytoplasmic loss and these became known as eccrine glands. Subsequent ultrastructural studies commonly failed to support the distinction between these two methods of secretion and in both types fluid production apparently occurred without loss of cell cytoplasm. Despite this, the distinction between the two types of gland proved to be a useful one and the original terminology persists even if it is a little misleading. The kind of secretion originally thought to occur in the eccrine glands is also known as merocrine or ectocrine secretion and when these terms are applied to sweat glands, they are synonymous with 'eccrine'. I have decided to adopt the term 'eccrine' here for no better reason than its almost universal acceptance in the current dermatological literature. The mammary gland is the only epidermal derivative showing unambiguous ultrastructural features indicative of apocrine secretion as originally conceived; in it large lipid droplets are budded off from the cell surface enveloped in a cell membrane and a thin rim of cytoplasm that may occasionally include mitochondria and other organelles. The existence or otherwise of true apocrine secretion in other cutaneous structures is still a matter of some controversy (see below).

Typical eccrine glands are confined to the foot-pads in most species, where they have been particularly well investigated in carnivores but they are almost universal in the skin of man and the anthropoid apes where they are responsible for typical thermoregulatory sweating. In man, apocrine glands are mainly confined to the pubic, perineal, and axillary regions but in most mammals they are widely distributed over the hairy skin and it is these that produce the characteristic foamy, thermoregulatory sweat of the horse.

All sweat glands consist of a duct and a secretory coil (Plate 1.11). The duct traverses the epidermis and the outer dermis to reach the secretory portion lying in the inner dermis or the loose connective tissue beneath it, the hypodermis. The ducts of eccrine glands open directly on to the skin surface, traversing the epidermis by a specialized helical canal; the ducts of apocrine glands, on the other hand, normally open into the pilosebaceous canal though they can be independent of it as occurs occasionally in man, often in apes and monkeys, and always in lemurs (Montagna and Parakkal 1974). The duct is usually lined by a double layer of roughly cuboidal cells with scattered myoepithelial cells (see below) separating them from the basement membrane. The junction between the duct and the secretory coil is abrupt, the coil being two or three times thicker and very irregular (Fig. 1.2, inset). It may divide into separate channels and fuse again or have bridges between adjacent loops; it is occasionally branched (Fig. 1.2).

Histology of apocrine glands

In apocrine glands the secretory coil is lined by a relatively orderly single layer of cells that vary in shape from columnar to low cuboidal, and even to the simple squamous form according to the degree of dilation of the lumen (Plate 1.6). The nuclei are usually near the middle of each cell forming a regular pattern and the apical cytoplasm is often rich in granules that are cleary visible with the light microscope and stain with the periodic acid–Schiff reaction, a method that usually colours the hydroxyl groups of sugars. The lumen of apocrine glands is commonly distended with secretion which is coloured by conventional histological stains. Often the apical parts of the cells show balloon-like projections or 'blebs' extending into the lumen (Plates 1.6, 1.7, and 1.9). These are usually thought to be artefacts produced by fixation and to have no counterpart in the living cell, none the less it seems that some apocrine cells have a particular predisposition to form them and it is probably this characteristic that gave them their name. Some investigators contend that the blebs are real in some cases (see, for example, Kneeland 1966) and that they represent genuine ultrastructural evidence of apocrine secretion. Their argument is based on the observation that the blebs are present in frozen tissue and tissue fixed by a variety of methods but absent from essentially similar but possibly less active cells. The problem of apocrine secretion in the bronchial epithelium has been exhaustively investigated recently by Etherton, Purchase, and Corrin (1979) who conclude that it does indeed occur.

Aside from the confused issue of the apical blebs and standard cytological features like mitochondria, Golgi apparatus, and rough and smooth endoplasmic reticulum, the main features of apocrine sweat cells concern the cell surface and cytoplasmic vesicles. The cells are firmly attached to one another around their luminal borders by junctional complexes while desmosomes also inter-connect the cells and attach them to the myoepithelium (Plate 1.9). All the cell surfaces possess a variable number of irregular microvilli; those on the free border project into the lumen, those of adjoining cells tesselate with one another, and those on the basal border project into the potential space between the secretory cells and the myoepithelium. The cytoplasm of the cells always contain membrane-bound secretory vesicles but their precise nature varies from species to species. In cattle they contain a flocculent material that is somewhat denser than the luminal contents (Jenkinson, Montgomery, and Elder 1979) while in the human axilla the granules are dense, varied in form and contain fat droplets (Bell 1974). The vesicle contents are discharged into the lumen by exocytosis. Jenkinson *et al.* (1979) noted that in the fatigued apocrine glands of heat-stressed sheep and goats, the supply of vesicles was not necessarily exhausted but the intercellular spaces and the cavities of the endoplasmic reticulum were much dilated and that some cells, possibly old ones, had degenerated completely and been shed into the lumen. These authors make the important point that only a portion of the secreted fluid can be accounted for by exocytosis of the vesicle contents and that additional fluid transport occurs at an unknown site by an unknown mechanism. The ultrastructural evidence therefore suggests that fatigue may alter the consistency, composition, and odour of apocrine sweat.

Histology of eccrine glands

Eccrine, unlike apocrine glands, have nuclei that are situated irregularly at all levels in the secretory lining of the coil (Plates 1.12 and 1.13). This is partly because eccrine glands are rarely distended with fluid so the cells appear huddled together and partly because they are usually composed of two distinctive cell types. All the cells are small and of a misshapen cuboidal form but some stain darkly with basic dyes while others stain little. The pale-staining 'light' or 'clear' cells have an extensive attachment to the peripherally located myoepithelial cells and their nuclei are basal in position. The 'dark' cells only have rather tenuous attachments to the myoepithelium but occupy much of the luminal surface of the duct; their nuclei are located distally. They have short microvilli on their luminal surfaces and their cytoplasm is packed with rough endoplasmic reticulum and membrane-bound granules: these are typical mucus-secreting cells. The clear cells lack any obvious secretory granules, they have abundant smooth, as opposed to rough, endoplasmic reticulum and are often rich in glycogen. Their rather restricted luminal surfaces have a few stumpy microvilli but where two clear cells contact each other the cell surfaces are thrown into an elaborate system of interlocking folds and microvilli. The most significant ultrastructural features of these cells are the intercellular canaliculi that extend from the lumen almost as far as the basal border. In essence, they are narrow extensions of the lumen and they are separated from the ordinary intercellular spaces by junctional complexes. These cells evidently produce the great bulk of the watery component of sweat and they are morphologically very like cells having a similar function in lacrimal and salivary glands. The cytoplasmic glycogen reserves are depleted by prolonged sweating (see Ellis 1967; Montagna and Parakkal 1974).

Myoepithelial cells

The myoepithelial cells (Plates 1.8, 1.10, 1.13, and 1.14), to which I have already frequently referred, form a dense meshwork of contractile elements that incompletely separate the secretory epithelium from the basement membrane of the tubule in both eccrine and apocrine glands. The thickest region of each cell contains the nucleus from which it tapers into delicate processes that normally are parallel to the long axis of the tube. The processes contain a few mitochondria but are otherwise almost completely occupied by contractile myofibrils.

The basement membrane, as seen in suitable stained sections with the light microscope, is a thin acellular layer completely surrounding each gland. It is a mucopolysaccharide structure reinforced on the outside by a delicate reticulum of elastic fibres that mainly encircle the tubule (Montagna and Giacometti 1969) crossing the outer surfaces of the myoepithelial cells more or less at right angles. These two structures combine to ensheath the gland in a resilient meshwork that protects it from overdistension by the pressure of its own secretion.

It will be seen from the preceding morphological considerations that secretion may arrive on the skin surface following a variety of events. The myoepithelial cells may contract expelling the accumulated sweat in a sudden pulse without actual secretion. Secretion may occur in the absence of myoepithelial contraction when the rate of sweat delivery to the skin surface would be expected to rise and fall gradually. Obviously both events may also occur simultaneously. The pattern of sweat secretion may have profound effects on the odorous penumbra surrounding the animal and some of the possible control mechanisms involved will be alluded to next.

Neural and humoral control of tubular glands

The long-term regulation of tubular skin glands is not extensively documented but two main points should be mentioned. First, human eccrine glands increase their potential sweat output by a factor of about three during two to three weeks' exposure to high ambient temperatures (see Precht, Christophersen, Hensel, and Larcher 1973). This process is usually referred to as acclimatization and its control mechanisms are only poorly understood. Second, the apocrine glands of the human axillary organ only reach full function with the development of coarse axillary hair at puberty. Once functional, however, their activity seems to be independent of gonadal hormones.

The short-term neuroendocrine control of sweat glands is intricate and difficult to summarize but I have attempted to draw together some of the salient features below and, in doing so, I have drawn heavily on reviews by Robertshaw (1977) and Montagna and Parakkal (1974) together with a few more recent sources.

Eccrine sweat glands

Eccrine sweat glands are surrounded by a dense meshwork of nerves that are rich in cholinesterase indicating that their transmitter substance is acetylcholine. This is unusual though not unique, as the nerves belong to the sympathetic division of the autonomic system which normally acts via noradrenalin. Uno (1977) found that the eccrine glands of man and the macaque were also supplied by adrenergic fibres using a histofluorescence method and was able to confirm this in ultra-structural studies. However, the electron micrographs also showed the cholinergic terminals to be the most numerous.

Denervation of an area of skin abolishes sweating and local application of acetylcholine to eccrine glands stimulates them. Administration of adrenergic substances either locally or systemically also causes eccrine sweating, especially on the frictional surfaces of the hands and feet, with associated improvement in grip. The effect of locally administered adrenalin can be blocked by α-adrenergic blocking agents and the systemic effect by β-blockers, indicating that separate pathways are involved. To what extent these effects are mediated through the secretory or myoepithelial components is not clear, but it is often suggested that the sudden emotionally induced sweating of the palms and soles is a myoepithelial event. However, Sato (1973) has shown that isolated eccrine glands from monkey palms will secrete equally well in response to acetylcholine and adrenalin even though their myoepithelium will only contract after treatment with acetylcholine and not with adrenalin (Sato, Nishiyama, and Koboyashi 1979). Apparently the essential role of the myoepithelium is the prevention of the stagnation of sweat rather than its active expulsion.

Control of sweating is further complicated by the fact that, as in so many organs, the primary secretory product is altered before final expulsion by reabsorption of critical components in the duct. The reabsorption of electrolytes and hence the production of hypotonic sweat is under the control of aldosterone, an adrenocortical steroid most commonly thought of in connection with renal function.

Apocrine sweat glands

Cholinesterase-reactive nerves are much rarer around human axillary apocrine glands than they are around the eccrine glands, but they are abundant around the thermoregulatory apocrine glands of the horse. Uno (1977) was able to find adrenergic fibres sporadically in the connective tissue around human axillary apocrine glands and a few adrenergic terminals were seen in electron micrographs of the periglandular space. Cholinergic terminals were occasionally seen around the gland ducts. Direct innervation of the general-body apocrine glands of the domestic ruminants and the dog and cat have yet to be demonstrated, but cutaneous denervation greatly curtails or abolishes thermoregulatory sweating in these species. In response to this apparent contradiction, Robertshaw (1977) formed the novel hypothesis that neurotransmitter substances are perhaps released into the circulation close to the glands and thus transported to them. This confused situation is exacerbated by the fact that innervation is not uniform throughout the skin: in the dog, for example, perianal apocrine glands are surrounded by a profusion of cholinesterase-reactive fibres but those elsewhere have none (Machida, Giacometti, and Perkins 1966).

Intradermal administration of adrenalin is far more effective than acetylcholine in inducing apocrine sweating in the general body skin of the horse but it is impossible to generalize from these findings to other species. Human axillary apocrine glands respond to both cholinergic and adrenergic stimuli and the isolated glands secrete when treated with either agent (Sato and Sato 1979); on the other hand the myoepithelium of human apocrine glands contracts only in response to adrenalin, the effect being inhibited by α-adrenergic blocking agents.

Although it is clear that innervation has a primary role in the regulation of apocrine secretion, sweating can be induced in denervated skin by stresses such as exercise (donkeys) or asphyxiation (dogs) and this can be shown to be associated with increased blood adrenalin concentrations and to be dependent on the integrity of the innervation of the adrenal medulla. Thus it appears that circulating adrenalin augments neural effects.

Although the preceding functional account may be far from clear, the intention is that it should be sufficient to form a background against which biologists can generate ideas in the field.

Other organs

Because the skin, and particularly its glandular elements, have been dealt with in some detail it should not be assumed that these necessarily contribute the most significant odour to the environment. Specialized skin glands are readily recognized, relatively easy to study, often associated with behaviour patterns that attract attention and are especially well adapted for the deposition of long-lasting marks. They have therefore received much attention possibly at the expense of other structures; no doubt even the non-glandular skin surface makes its contribution. Some other potential odour sources are mentioned below and indicated in Fig. 1.1. The examples given are indeed examples and I have not attempted to give an exhaustive account of each kind of phenomenon.

The respiratory system

None who have been associated with horses can doubt the olfactory significance of the intent nostril-to-nostril snuffing that occurs during courtship and when unfamiliar individuals meet (illustrated by Vavra 1979). The message value seems obscure but it is clearly of the most intense interest to the animals concerned. In our own species the smell of breath is probably more consciously informative than any other body odour: it clearly betrays excessive bacterial proliferation in the mouth, pharynx, or sinuses; the use of tobacco, alcohol, and aromatic herbs and spices; and the raised blood ketone levels often attendant on starvation. This is to say nothing of the frequent, though perhaps uncritical, references to the sweet breath of lovers—'Violets dim, but sweeter than . . . Cytherea's breath' (*The Winter's Tale*).

Though the alveolar membrane clearly represents the shortest route by which volatile metabolites can pass from the blood to the environment it seems to have been little studied. There are probably technical reasons for this and an understandable concentration on long-term, rather than very short-term responses. There is anecdotal evidence suggesting that sufficient steroids are exhaled by pregnant and menstruating research workers to interfere with sensitive radio-immunoassay systems (see discussion following Johansson 1970); such amounts would almost certainly be detectable by more sensitive noses than our own. Veterinarians dealing with dairy cattle regularly diagnose bovine ketosis by the characteristic smell of exhaled acetone and estimations of exhaled ethanol are permissable evidence in the courts of many countries.

The odour of breath is potentially augmented from sources in the upper respiratory tract though specific examples are hard to find. The nasal mucosa clearly has a potential role because of its large area, secretory lining, and extreme vascularity; air entering the nose at −100 °C is almost fully saturated and at body temperature when it reaches the trachea (King and Brown 1976). The dog and many other species possess a large gland that lies lateral to the nasal cavity and discharges into it (Evans and Christensen 1979; Adams, Deyoung, and Griffith 1981) and the nostrils of the Equidae have a skin-lined diverticulum (Getty 1975) that functions primarily to permit it to flare but may also contribute volatiles. The nasolacrimal ducts also open at the entrance to the nasal cavity and influence breath but their role will be discussed later.

The digestive system

The production of androstenol and androstenone by the mandibular salivary glands (Patterson 1968) of the boar and its role in courtship and mating (Melrose, Reed, and Patterson 1971) has become central to any discussion of mammalian chemical ecology. Steroid metabolism has also been recognized in the mandibular glands of other species (e.g. Baldi and Charreau 1972) and androgen-dependent sexual dimorphism of salivary glands has been recognized for many years (Lacassagne 1940) and much investigated (see Dorey and Bhoola 1972*a,b*). There is also occasional, striking behavioural evidence of communication by saliva like that seen in the hedgehog which deposits blobs of strong-smelling foamy spittle on its spines when young and during courtship (Brockie 1976). Thus saliva may be considered a prime source of socially important odour that may be used directly or applied to the coat while self-grooming or grooming others.

Regurgitation of gastric contents is a rare event in non-ruminants and so does not lend itself to scent marking; furthermore the material is valuable. Even in ruminants the cud does not seem to be used as a marking material though it is sufficiently expendable to be used as a weapon among the Camelidae. Possibly the odour of rumen contents is so dependent upon food-stuffs and rumen microorganisms common to all members of the group that it has little message value.

The power and message value of faeces are well recognized though some reports are possibly exaggerated: the 'bonnacon' (bison), it is said, 'Emits a fart with the contents of his large intestine which covers three acres, any tree that it reaches catches fire and its pursuers are driven off with the noxious excrement'; the faeces of the 'cocodryllus' (crocodile) on the other hand 'Provide an oint-ment with which old and wrinkled whores anoint their figures and are made beautiful, until the flowing sweat of their efforts washes it away' (Anon. c. 1150).

The information carried by faeces is probably derived in large part from stored metabolic information derived from the liver via the bile. While microbial fermentation in the gut probably destroys some significant compounds, it may liberate or enhance others. Steroids are secreted in bile and though no doubt resorbed in part and altered by bacteria, they certainly appear in the faeces of the sheep (Terqui, Rombauts, and Fèvre 1968). Oddly, the faeces of female lactating rats contain a compound that attracts their sucklings. It is produced, apparently in the caecum, in response to prolactin-induced changes in bile and can be induced in non-lactating rats by injecting the bile of attractive mothers into their caeca (Moltz and Leidahl 1977). The liver is probably not the only organ supplying unique volatiles to the faeces but because of its almost uni-versal metabolic importance it would be expected to be dominant. Other information may be applied to faeces by the anal gland complexes that are so widespread in mammals and by the admixture of urine.

The urogenital system

A discussion of the details of renal metabolism is quite beyond the scope of this account though very relevant to the informational value of urine. Suffice it to say that all the volatile compounds present in blood will appear in the glomerular filtrate at the beginning of the renal tubule. Water and other valuable materials are then resorbed during its passage down the tubule to become urine. Thus a very wide variety of metabolic events will contribute to the vola-tile constituents of urine and, unlike those of bile, they will normally be pre-sented in the environment without further bacterial alterations. Again steroid hormones provide a ready example: their metabolites are greatly concentrated in urine, they are sufficiently volatile to be odorous and they can be expected to convey information about maturity, reproductive state, and degree of stress.

In the male the deferent ducts (from the testes) and the accessory sex glands open into the pelvic part of the urethra and may add to the urine. In sexually inactive males, sperm are gradually expressed into the pelvic urethra and can be

found in the urine; it seems likely that some of the secretion of the male accessory glands enters the urine in the same way and may contribute to its odour.
The male accessory glands, which normally comprise the prostate, the ampullary
glands, the vesicular glands, and bulbo-urethral glands, are highly androgen
dependent and normally contribute about 90 per cent of the ejaculate by
volume. They have been reviewed by Brandes (1974) and Eckstein and Zuckerman (1956). Initial evidence of a pheromonal role for them was provided by Haug
(1971) and expanded by Jones and Nowell (1973*a,b*). The coagulating glands
(part of the prostate) of the mouse apparently produce an aversive and aggression-
inhibiting pheromone but some interaction with urine seems to be necessary for
it to be effective.

In many species a little urine remains in the preputial cavity following urination where it is attacked by micro-organisms and becomes a potent source of
odorous volatiles and some species have preputial hair tufts that might be
expected to dissipate the smell. Often the prepuce seems to be elaborated to
facilitate this process and the pig possesses a distinctive dorsal diverticulum of
the prepuce which reaches the size of an orange in the mature boar (Getty
1975). Its lining is non-glandular and it normally contains an exquisite mixture
of fermenting urine, desquamated cells, and possibly a little semen. This material
provides yet another incentive to copulation in the sow (Signoret 1970). Glands
that discharge their contents close to or within the preputial cavity are very
widespread (Schaffer 1940; Brown and Williams 1972). They reach their maximum development as the castoreum glands of the beaver (Schaffer 1940; Svendsen 1978) and though castoreum was used to bait traps for centuries, the first
experimental evidence of the pheromonal activity of preputial glands was provided by Bronson and Caroom (1971) using the mouse.

There is abundant behavioural evidence of the importance of vaginal and
vestibular smells in the recognition of oestrus. Convention dictates that the
vagina extends from the cervix to the urethral opening and that the vestibule
runs caudally from there to the external opening termed the vulva. A true
vestibule is absent in many species including rodents. In most species the peri-
toneum, the ovarian follicles, and the uterus probably make only minor and
intermittent contributions to the vaginal fluids but the lining of the cervix is
highly secretory and forms, in the cow for example, both the dense gluey
material that plugs the cervix in pregnancy and the bright elastic mucus that
often hangs from the vulva at oestrus. The vagina is lined by a non-glandular
stratified squamous epithelium that is capable of producing fluid directly by a
process of transudation during sexual arousal (see Wagner and Levin 1980)
and undergoes cyclical cornification in many species. In the vestibule the vaginal
fluids are augmented by urine and the secretions of the female homologues of
the male accessory sex organs: the major and minor vestibular glands. These are
briefly described for the domestic animals in the standard texts including Nickel,
Schummer, Seiferle, and Sack (1979) who in the original 1959 German edition

refer to them as a potential source of 'Brunstduftstoffe' or heat-scent sub-stance. The vestibular glands are sensitive to oestrogens in cattle (Kroes, Ruiten-berg, and Berkvens 1970) and are reported to show morphological changes during the oestrous cycle by some authors (Reutner and Morgan 1948) but not others (Friess 1972). They no doubt behave similarly in other species. Clearly, the behaviourally important volatiles of the female genital tract have a wide and interesting variety of possible sources but most are likely to be controlled by ovarian hormones.

Sense organs

The remaining natural orifices are those of the eyes and ears, the nose having been dealt with in the context of respiration. The ears seem to be of little significance though they can be extremely malodorous in disease and close attention to the eyes is generally resented. However, the fluid covering the eye and occupying the conjunctival sac is conveyed to an area just within the openings of the nostrils by the nasolachrymal ducts: here it can be easily smelt. The lachrymal gland, the superficial and deep (Harderian) glands of the third eyelid (Nomina Anatomica Veterinaria 1972), the goblet cells of the conjunc-tival membrane and the tarsal glands all contribute to the tear film. Of these, the lachrymal gland is usually the largest, usually dorsal to the eye and usually discharges its contents into the dorsal part of the conjunctival sac by several minute ducts. Small accessary lachrymal glands are often found at other sites in the orbit.

The deep and superficial glands of the third eyelid are, as the name implies, more or less associated with the third eyelid and open via small ducts on its inner surface. They vary greatly in development from species to species and where both are present, as in the pig, they differ histologically. The lacrimal glands and the glands of the third eyelid are responsive to androgens and hypo-physeal hormones and were suggested as possible pheromone sources because of their ability to concentrate exogenous androstenone (see Ebling, Ebling, Randall, and Skinner 1975). Such pheromones have now been demonstrated in the deep (Harderian) gland of the third eyelid in the hamster (Payne 1977) and the Mongolian gerbil (Thiessen, Clancy, and Goodwin 1976), where they provoke increased aggression and attraction respectively.

The tarsal glands are a series of elongated sebaceous units that lie in the free border of the eyelids and discharge where the margins of the eyelids contact the cornea. They provide the outer lipid layer of the corneal fluid covering. Goblet cells are most numerous in the deeper parts of the conjunctival sacs and supply mucus. These last do not seem to have been implicated in pheromone production.

In conclusion it seems that, in mammals, any informative odour-source that can be used will be used and that once in use it will be subjected to selective pressure and potentially enhanced. No odour-source is too obscure to be unworthy of consideration; in the present state of knowledge we can probably detect the equivalent of the dazzling smile or the shaken fist of the human visual repertoire: what of the momentary furrowing of the brow or the wrinkling of the corner of an eye?

References

Adams, D. R., Deyoung, D. W., and Griffith, R. (1981). The lateral nasal gland of dog: its structure and secretory content. *J. Anat.* **132**, 29–37.

Anon. (c. 1150). *The Bestiary* (transl. and ed. T. H. White). Putnam, New York.

Baldi, A. and Charreau, E. H. (1972). 17β-Hydroxysteroid dehydrogenase activity in rat submaxillary glands: Its relation to sex and age. *Endocrinology* **90**, 1643–6.

Bell, M. (1974). The ultrastructure of human axillary apocrine glands after epinephrine injection. *J. invest. Derm.* **63**, 147–59.

Bertalanffy, D. F. (1957). Mitotic activity and renewal rate of sebaceous gland cells in the rat. *Anat. Rec.* **129**, 231–9.

Brandes, D. (1974). *Male accessory sex organs.* Academic Press, New York.

Brockie, R. (1976). Self anointing by wild hedgehogs *Erinaceous europaeus* in New Zealand. *Anim. Behav.* **24**, 68–71.

Bronson, F. H. and Caroom, D. (1971). Preputial gland of the male mouse: attractant function. *J. Reprod. Fert.* **25**, 279–82.

Brown, J. C. and Williams, J. D. (1972). The rodent preputial gland. *Mammal. Rev.* **2**, 105–47.

Cunliffe, W. J. and Shuster, S. (1969). The rate of sebum excretion in man. *Br. J. Derm.* **81**, 697–704.

Dorey, G. and Bhoola, K. D. (1972*a*). Ultrastructure of acinar cell granules in mammalian submaxillary glands. I. *Z. Zellforsch.* **126**, 320–34.

—— —— (1972*b*). Ultrastructure of duct cell granules in mammalian submaxillary glands. II. *Z. Zellforsch.* **126**, 335–47.

Ebling, F. J., Ebling, E., Randall, V., and Skinner, J. (1975). The effects of hypophysectomy and of bovine growth hormone on the responses to testosterone of prostate, preputial, Harderian and lachrymal glands and brown adipose tissue in the rat. *J. Endocr.* **66**, 401–6.

Eckstein, P. and Zuckerman, S. (1956). Morphology of the reproductive tract. In *Marshall's physiology of reproduction*, Vol. 1, Part 1, pp. 43–147. Longman, London.

Ellis, R. A. (1967). Eccrine, sebaceous and apocrine glands. In *Ultrastructure of normal and abnormal skin* (ed. A. S. Zelickson). Lea and Febiger, Philadelphia.

Etherton, J. E., Purchase, I. F. H., and Corrin, B. (1979). Apocrine secretion in the terminal bronchiole of mouse lung. *J. Anat.* **129**, 305–22.

Evans, H. E. and Christensen, G. C. (1979). *Anatomy of the dog*, 2nd edn. Saunders, Philadelphia.

Friess, A. E. (1972). Histochemical investigation on the glandulae vestibulares majores (Bartholini) of cattle. *Acta histochem.* **44**, 62–70.

Getty, R. (1975). *The anatomy of the domestic animals*, 5th edn. Saunders, Philadelphia.

Haug, M. (1971). Rôle probable des vésicules séminales et des glandes coagulantes dans la production d'une phéromone inhibitrice du comportement agressif chez la souris. *C.r. hebd. Séanc. Acad. Sci., Paris D* **273**, 1509–10.

Jenkinson, D. McE. Montgomery, I., and Elder, H. Y. (1979). The ultrastructure of the sweat glands of the ox, sheep and goat during sweating and recovery. *J. Anat.* **129**, 117–40.

— Sen Gupta, B. P., and Blackburn, P. S. (1966). The distribution of nerves, monoamine oxidase and cholinesterase in the skin of cattle. *J. Anat.* **100**, 593–613.

Johansson, D. B. (1970). A simplified procedure for the assay of progesterone. *Acta endocr.* Suppl. **147**, 188–203.

Johnson, S. G. (1952). Quantitative determination of skin lipid secretion in the dorsal region of the hand. *Acta derm.-vener., Stockh.* **32**, 168–73.

Jones, R. B. and Nowell, N. W. (1973a). The coagulating glands as a source of aversive and aggression inhibiting pheromone in the male albino mouse. *Physiol. Behav.* **11**, 455–62.

— — (1973b). Effects of preputial and coagulating gland secretions upon aggressive behaviour in male mice: a confirmation. *J. Endocr.* **59**, 203–4.

King, A. S. and Brown, C. M. (1976). *A guide to the physiological and clinical anatomy of the head.* Published by the authors, Department of Veterinary Anatomy, University of Liverpool.

Kneeland, J. E. (1966). Fine structure of the sweat glands of the antebrachial organ of *Lemur catta. Z. Zellforsch.* **73**, 521–33.

Knight, T. W. and Lynch, P. R. (1980). Source of ram pheromones that stimulate ovulation in the ewe. *Anim. reprod. Sci.* **3**, 133–6.

Kroes, R., Ruitenberg, L. J., and Berkvens, J. M. (1970). Histological changes in the genital tract of the female calf after the administration of diethyl-stilbestrol and hexestrol. *Zentbl. Vet. Med. Reihe A* **17**, 440–50.

Lacassagne, A. (1940). Mesure de l'action des hormones sexuelles sur la glande sous-maxilliaries de la souris. *C.r. Soc. Biol. Paris* **133**, 227–9.

Machida, H., Giacometti, L., and Perkins, E. (1966). Histochemical and phramaco-logical properties of the sweat glands of the dog. *Am. J. vet. Res.* **27**, 566–73.

Melrose, R. D., Reed, H. C. B., and Patterson, R. L. S. (1971). Androgen steroids associated with boar odour as an aid to the detection of oestrus in pig arti-ficial insemination. *Br. vet. J.* **127**, 497–502.

Moltz, H. and Leidahl, L. C. (1977). Bile, prolactin and the maternal pheromone. *Science, NY* **196**, 81–3.

Montagna, W. and Ford, D. M. (1969). Histology and cytochemistry of human skin. XXXIII. *Archs Derm.* **100**, 328–35.

— and Giacometti, L. (1969). Histology and cytochemistry of human skin. XXXII. The external ear. *Archs Derm.* **99**, 757–67.

— and Parakkal, P. F. (1974). *The structure and function of skin,* 3rd edn. Academic Press, New York.

Müller-Schwarze, D., Volkman, N. J., and Zemanek, R. F. (1977). Osmetrichia: specialized scent hair in black-tailed deer. *J. ultrastruct. Res.* **59**, 223–30.

Nickel, R., Schummer, A., Seiferle, E., and Sack, W. O. (1979). *The viscera of the domestic animals,* 2nd English edn. Paul Parey/Springer, Berlin.

Patterson, R. L. S. (1968). Identification of 3α-hydroxy-5α-androst-16-ene as the musk odour component of boar submaxillary salivary gland and its relationship to the sex odour taint in pork meat. *J. Sci. Fd Agric.* **19**, 434–8.

Payne, A. P. (1977). Pheromonal effects of Harderian gland homogenates on aggressive behaviour in the hamster. *J. Endocr.* **73**, 191–2.

Pillsbury, D. M. and Rebell, G. (1952). The bacterial flora of the skin. *J. invest. Derm.* **18**, 173–86.

Precht, H., Christophersen, J., Hensel, H., and Larcher, W. (1973). *Temperature and life.* Springer, Berlin.

Quay, W. B. and Müller-Schwarze, D. (1970). Functional histology of integumentary glandular regions in black-tailed deer (*Odocoileus hemionus columbianus*). *J. Mammal.* **51**, 675–94.

Reutner, T. F. and Morgan, B. B. (1948). A study of the bovine vestibular gland. *Anat. Rec.* **101**, 193–211.

Robertshaw, D. (1977). Neuroendocrine control of sweat glands. *J. invest. Derm.* **69**, 121–9.

Sato, K. (1973). Sweat induction from an isolated eccrine sweat gland. *Am. J. Physiol.* **225**, 1147–52.

— Nishiyama, A., and Koboyashi, M. (1979). Mechanical properties and functions of the myoepithelium in the eccrine sweat gland. *Am. J. Physiol.* **237**, 177–84.

— and Sato, F. (1979). Pharmacology and function of the isolated human apocrine gland *in vitro. Clin. Res.* **27**, 535A.

Schaffer, J. (1940). *Die Hautdrüsenorgane der Säugetiere.* Urban & Schwarzenberg, Berlin.

Signoret, J. P. (1970). Reproductive behaviour of pigs. *J. Reprod. Fert.* Suppl. **11**, 105–117.

Smith, M. E. and Jenkinson, D. McE. (1975). The mode of secretion of sebaceous glands in cattle. *Br. vet. J.* **131**, 610–18.

Svendsen, G. E. (1978). Castor and anal glands of the beaver. *J. Mammal.* **59**, 618–20.

Terqui, M., Rombauts, P., and Fèvre, J. (1968). Répartition urinaire et fécale de l'excrétion des oestrogènes chez la truie et la brebiz. *Annls Biol. anim. Biochim. Biophys.* **8**, 33–348.

Thiessen, D. D., Clancy, A., and Goodwin, M. (1976). Harderian gland pheromone in the Mongolian gerbil *Meriones unguiculatus. J. chem. Ecol.* **2**, 231–8.

Thody, A. J. and Shuster, S. (1975). Control of sebaceous gland function in the rat by α-melanocyte stimulating hormone. *J. Endocr.* **64**, 503–10.

Uno, H. (1977). Sympathetic innervation of the sweat glands and piloarrector muscles of macaques and human beings. *J. invest. Derm.* **69**, 112–20.

Vavra, R. (1979). *Such is the real nature of horses.* William Morrow, New York.

Wagner, G. and Levin, R. J. (1980). Electrolytes in vaginal fluid during the menstrual cycle of coitally active and inactive women. *J. Reprod. Fert.* **60**, 17–27.

Warner, R. L. and McFarland, L. Z. (1970). Integument. In *The beagle as an experimental dog* (ed. A. C. Anderson). Iowa State University Press, Ames.

2 The prototherians: order Monotremata

ELEANOR M. RUSSELL

Introduction

Although the relationship of the monotremes to the rest of the mammals continues to be a matter for debate (Griffiths 1978), it is not disputed that the mammalian line which gave rise to the monotremes has been distinct from the metatherian–eutherian line for a very long time, probably since the late Triassic (Clemens 1977). The extreme specialization of the three surviving genera (the echidnas *Tachyglossus* and *Zaglossus* and the platypus *Ornithorhynchus*), their obvious reptilian affinities and the singular lack of fossil monotremes have made them objects of considerable interest to zoologists ever since they were first discovered. The enormous early literature on the anatomy and embryology of monotremes and more recent work on physiology are reviewed by Griffiths (1968, 1978) and Augee (1978), but they have little to say about olfactory communication. Although the visual, auditory and tactile senses have recently been investigated in some detail (Gates, Saunders, Bock, Aitken, and Elliott 1974; Bohringer 1976; Bohringer and Rowe 1977; Gates 1978), olfactory capabilities are still a matter for speculation. A study of instrumental learning in *Tachyglossus* found typical mammalian learning curves, and performances on spatial discrimination learning tasks comparable with rats—as Buchmann and Rhodes (1978) suggest, the animated pincushion image is misleading.

This chapter attempts to investigate the importance of olfaction in the social behaviour of monotremes by collecting together the admittedly sparse information on skin glands, olfactory receptors, and behaviour. Most of this information concerns *Tachyglossus* and *Ornithorhynchus* since virtually nothing is known about *Zaglossus*.

Occurrence of scent glands

Schaffer (1940) in his wide-ranging monograph on the skin glands of mammals summarized earlier, mostly nineteenth-century, observations of skin glands in monotremes, including the important work of von Eggeling (1900, 1901) and added some observations of his own; there is little more recent work. Griffiths (1978) in his book on the biology of monotremes ignores skin glands other than sebaceous and sweat glands, which are found in all three genera of monotremes. The echidna *Tachyglossus* has very few, but *Zaglossus* from New Guinea appears to sweat (Dawson, Fanning, and Bergin 1978). The platypus, *Ornithorhynchus anatinus*, has widely distributed functional sweat glands (Augee 1976) and Schaffer quotes a report by Klaatsch (1888) of well-developed 'sweat glands' in the soles of hind- and fore-feet.

Sebaceous glands occur, as is usual in mammals, in association with hairs or spines, including the hairs in the mammary area, and Griffiths, McIntosh, and Coles (1969) have shown that there is a concentration in the mammary area of *Tachyglossus*. Von Eggeling (1900) found very well-developed coiled glands with wide tubes in the cartilaginous ear opening of *Tachyglossus*, and Schaffer says that Burne (1909) confirmed this, finding there few sebaceous glands, but many large tubular glands, the nature of whose secretion is not known. Augee (personal communication) found abundant sebaceous glands associated with hair follicles and even more abundant ceruminous glands opening both directly to the surface and into hair follicles.

MacKenzie and Owen (1919) describe a scent gland (the 'cervical sex gland') in *Ornithorhynchus* with 'a markedly penetrating odour somewhat resembling that of an onion'. The gland lies beneath the skin dorsolaterally on the neck, just anterior to the fore-limb (Fig. 2.1). In males it is easily felt beneath the

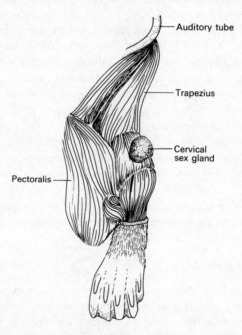

Fig. 2.1. Diagram of dissection of ventral surface of *Ornithorhynchus* to show position of cervical scent gland. (From MacKenzie and Owen (1919).)

skin, and may weigh up to 8 g; although it occurs in females, it is much smaller (up to 0.5 g). They could find no sign of a duct, but Burrell (1927) illustrates a platypus with a stain on the fur in the mid-line of the sternal region which he attributes to the scent gland, and which he says increases in size in the male during the breeding season. In section, the whole gland is composed of vesicles lined with tall columnar cells, and filled with a viscous fluid. MacKenzie and Owen (1919) described no comparable gland in *Tachyglossus*. In the axillary region, Augee (personal communication) found only a few sebaceous glands and occasional coiled glands, possibly sweat glands.

In view of the possible cloacal marking behaviour observed by Dobroruka (1960) and Strahan and Thomas (1975) (see p. 42), glands in the cloacal region are of most interest. Schaffer and Hamperl (1926) review the early literature, and produce a composite picture for *Tachyglossus* (Fig. 2.2) based on their own observations and the work of von Eggeling (1900, 1901) and Keibel (1902, 1904). They show tubular proctodaeal glands and complex sebaceous glands near the cloacal opening, some associated with hairs and apocrine glands, and some free but there is no indication of the nature of the glandular secretions.

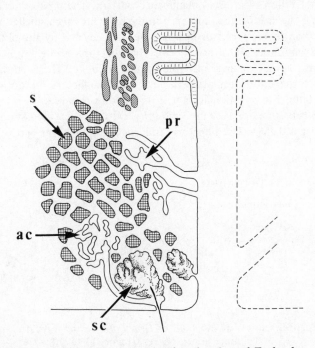

Fig. 2.2. Diagram of longitudinal section of anal region of *Tachyglossus* showing circumanal glands. (After Ortmann (1960).)
ac, circumanal apocrine glands; *pr*, proctodaeal glands; *s*, sphincter muscle; *sc*, circumanal sebaceous glands.

Disselhorst (1904) also found nothing other than large sebaceous glands in the region of the cloaca. N. T. Allen (personal communication) has confirmed these observations from fresh material, and found no sign of any glands comparable with the paracloacal or anal glands of marsupials or eutherians. Schaffer and Hamperl (1926) and Ortmann (1960) consider the proctodaeal glands homo-

logous with proctodaeal glands in marsupials and eutherians (see Fig. 2.2,
p. 39). Schaffer (1940) reviewed much earlier work on anal glands in *Ornitho-*
rhynchus; Home (1802) originally identified Cowper's glands as anal glands but
Schaffer and other subsequent workers failed to find any glands opening into the
lower rectum or cloaca. Several darkly pigmented apparent gland openings at
the area of transition from rectum to cloaca proved to be crypts of lymph
nodules (Schaffer 1940). As in *Tachyglossus*, there are circumanal sebaceous and
apocrine glands.

Scent reception

The organization of the olfactory organs and olfactory pathways in the brain of
monotremes follows the usual mammalian pattern. The area of olfactory epithelium and the size of olfactory areas in the brain are much smaller in *Ornitho-*
rhynchus than in *Tachyglossus* in which the area covered by the olfactory epithelium is 'enormous' (Griffiths 1978). The olfactory epithelium is carried on
the turbinal bodies arising from the ethmoidal bone. These ethmoturbinals
originate at the cribriform plate through which olfactory fibres pass to the
olfactory bulb. Ethmoturbinals vary in number from one or two in primates,
five in the cats, three or four in *Ornithorhynchus* (Fig. 2.3) to twelve in *Tachy-*
glossus (Negus 1958). The olfactory bulb makes up a larger part of the brain

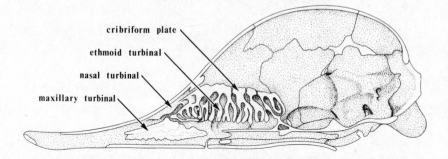

Fig. 2.3. Sagittal section of the skull of *Ornithorhynchus* indicating position of
turbinate bones. (After Gregory (1947).)

volume in *Tachyglossus* (4 per cent) than in *Ornithorhynchus* (1 per cent)
(Pirlot and Nelson 1978), and the area of the forebrain which is most involved
in processing olfactory information (paleocortex and amygdaloid region) is also
much larger (20 per cent compared with 6 per cent), although Pirlot and Nelson
point out that this area can be quite large even where only moderate use of
olfaction is made. In *Ornithorhynchus* an enormous area of the cortex is connected with inputs from the tactile organs of the muzzle (Bohringer and Rowe
1977) which is consistent with its reliance on tactile sensory information when

submerged and feeding, since the eyes, ears, and olfactory organs are all cut off from the outside world.

The vomeronasal organ in monotremes is more developed than in marsupials and eutherians (Broom 1897). It is not yet known whether monotremes have the two separate olfactory systems which in marsupials and eutherians result from the separate connection of the vomeronasal organ via the vomeronasal nerve to the accessory olfactory bulb and projection to the amygdaloid nucleus and thence to the medial hypothalamus (Scalia and Winans 1975). Nor is it known whether the vomeronasal organ in monotremes has a specific function in sexual behaviour as it appears to have in marsupials and eutherians (Estes 1972; Scalia and Winans 1976). Its importance in chemoreception in reptiles, in tracking prey, sex recognition, and courtship (Burghardt 1970), suggests that it may have some or all of these functions in monotremes.

Behaviour

The problems of maintenance in captivity, and of observing a nocturnal aquatic animal are so great that little is known of the details of behaviour of *Ornithorhynchus*. It appears to be a more or less sedentary, solitary animal, with several animals having overlapping home ranges in areas of suitable habitat (Grant and Carrick 1978). There is certainly overlap of feeding areas in lakes or rivers, but more than half the day is spent in an extensive burrow in the bank. Animals are generally found alone in these resting burrows, according to Burrell (1927) who noted that the burrows of males had a definite foxy odour. Burrow usage by marked animals has not been followed, so it is not known whether individuals have exclusive use of particular burrows. The male does not appear to be involved in any aspect of nesting, burrow construction, or incubation of eggs. Strahan and Thomas (1975) described an element of the courtship behaviour of a pair of platypus in captivity as marking. The male swam to the bottom of the tank, settled over a stone or other object and everted the cloaca, from which was produced a yellow mucilaginous liquid which settled on the object; the male then swam away. The female did not at any time show interest in the marks.

Tachyglossus is larger, more readily kept in captivity and observed in the wild, and some aspects of its behaviour have been described. The degree of development of the olfactory system in *Tachyglossus* may in part be related to the diet of ants and termites, which may be detected by smell, but there is also evidence suggesting that olfaction is important in bringing animals together for breeding, and perhaps in individual recognition.

The home range of animals followed by radio-tracking in the field (Augee, Ealey, and Price 1975) was approximately 800 m in diameter and changed little in the course of two years. Up to 13 conspecifics were sighted in the home-range of a radio-tracked individual over a period of 10 days, and there was no evidence of defence or exclusive use of even a small part of the home-range, nor was there any evidence of a regular den-site. However, after the young begins to grow spines and is no longer carried in the female's pouch it is left in a burrow to which the female returns about once a day to suckle. Echidnas are solitary for most of the year, although there are a few records of small groups of up to six echidnas being found together in the wild, normally several males and one female (Augee *et al.* 1975; Griffiths 1978; Johnson 1978). All such aggregations have occurred during what Griffiths (1978) considers is the breeding season, and it has been assumed that olfaction plays a part in bringing males and female together, since echidnas are generally silent apart from sniffing noises (Brattstrom 1973). Johnson (1978) describes four echidnas moving in a line nose to tail, with the cloaca of the leader being sniffed by the animal following. Dobroruka (1960) twice observed mating in captivity, but does not mention any cloacal sniffing as such, although he described the male attempting to overturn the female by pushing his snout underneath her. He described, and illustrated clearly, cloacal marking, in which the cloaca was everted and wiped on the substrate, which was left shiny and smelling of echidna. He suggested that these animals marked their territory but this has not been supported by field observations; Dobroruka did not suggest that the cloacal glands of the female might be used in attracting males for copulation, a statement which Augee, Bergin, and Morris (1978) ascribed to him. Semon (1894, p. 7) noted that 'especially in the breeding season, both sexes release a pronounced smell, which probably serves in mutual detection and in sexual arousal. It is this smell too, which bestows a characteristic flavour to the flesh of an animal roasted in its skin.' Elsewhere (*ibid.*, p. 8) he mentions a chest gland, but says no more about it.

Brattstrom (1973) did not observe any marking behaviour in large numbers of echidnas kept in restricted (2.2 × 2.2 m) crowded (33 animals) conditions. In ·such conditions he found that a dominance hierarchy became established which was also found in captivity by Augee *et al.* (1978). Although such a hierarchy may have little relevance to a solitary animal in the wild, its establishment in captivity must be based on individual recognition. Brattstrom considered that vision was important in social relationships, but he also said that much of the contact behaviour consisted of one animal sniffing another. An echidna may be sniffed at by another echidna anywhere on the body, but Brattstrom described a social challenge in which two approaching echidnas stop and sniff each other in the axillary region by which he suggests each determines the dominance status of the other. No particular glandular areas have been found in the axillary region (Augee, personal communication); MacKenzie and Owen (1919) who found a scent gland in this area in *Ornithorhynchus*, also investigated *Tachyglossus* and found no similar gland. Brattstrom described 'ear sniffing', a mutual smelling into each others ears, in the course of an encounter. He suggested that the auditory component of the sniff might be important, but the presence of large coiled tubular glands opening into the ear passage suggests (p. 38) that the sniffing may have an olfactory function.

Conclusion

The anatomy of the monotremes, especially of *Tachyglossus*, suggests that olfaction is an important sense. There are obviously sources of odour from the body of individuals, and some evidence that these are investigated in social encounters. What is needed is a study of the behaviour of animals kept in something closer to a natural environment than was used in earlier studies, together with a systematic study of skin glands. It would be naive to try to infer anything about the olfactory senses of Mesozoic mammals from the highly specialized modern monotremes, but it will nevertheless be interesting to make comparisons with other modern mammals when more information becomes available.

References

Augee, M. L. (1976). Heat tolerance of monotremes. *J. therm. Biol.* **1**, 181–4.

— (ed.) (1978). Monotreme biology. *Aust. Zool.* **20**, 1–257.

— Bergin, T. J., and Morris, C. (1978). Observations on behaviour of echidnas at Taronga Zoo. *Aust. Zool.* **20**, 121–9.

— Ealey, E. H. M., and Price, I. P. (1975). Movements of echidnas, *Tachyglossus aculeatus*, determined by marking-recapture and radio-tracking. *Aust. wildl. Res.* **2**, 93–101.

Bohringer, R. C. (1976). Bill receptors in the platypus, *Ornithorhynchus anatinus*. *J. Anat.* **121**, 417.

— and Rowe, M. J. (1977). The organization of the sensory and motor areas of cerebral cortex in the platypus (*Ornithorhynchus anatinus*). *J. comp. Neurol.* **174**, 1–14.

Brattstrom, B. H. (1973). Social and maintenance behaviour of the echidna *Tachyglossus aculeatus*. *J. Mammal.* **54**, 50–70.

Broom, R. (1897). A contribution to the comparative anatomy of the mammalian organ of Jacobson. *Trans. R. Soc. Edinb.* **39**, 231–55.

Buchmann, O. L. K. and Rhodes, J. (1978). Instrumental learning in the echidna *Tachyglossus aculeatus setosus*. *Aust. Zool.* **20**, 131–45.

Burghardt, G. (1970). Chemical perception in reptiles. In *Advances in chemoreception*, Vol. 1 (ed. J. W. Johnston, D. G. Moulton, and A. Turk). Appleton-Century-Crofts, New York.

Burne, B. H. (1909). A gland upon the ear conch of *Dasyurus maugei*. *J. Anat. Physiol.* **42**, 312–13.

Burrell, H. (1927). *The platypus*. Angus and Robertson, Sydney.

Clemens, W. A. (1977). Phylogeny of the marsupials. In *The biology of marsupials* (ed. B. Stonehouse and D. P. Gilmore). Macmillan, London.

Dawson, T. J., Fanning, D., and Bergin, T. (1978). Metabolism and temperature regulation in the New Guinea monotreme *Zaglossus bruijni*. *Aust. Zool.* **20**, 99–103.

Disselhorst, R. (1904). Die männlichen Geschlechtsorgane der Monotremen und einige Marsupialen. *Denkschr. med.-naturw. Ges. Jena* **6** (2), 121–50.

Dobroruka, L. J. (1960). Einige Beobachtungen an Ameisenigeln, *Echidna aculeata* Shaw (1792). *Z. Tierpsychol.* **17**, 178–81.

Eggeling, H. von (1900). Über die Hautdrüsen der Monotremen. *Anat. Anz.* **14**, 29–42.

— (1901). Über die Stellung der Milchdrüsen zu den übrigen Hautdrüsen. 2. Mitt. Die Entwicklung der Mammardrüsen, Entwicklung und Bau der übrigen Hautdrusen der Monotremen. *Denkschr. med.-naturw. Ges. Jena* 7, 173–204.

Estes, R. D. (1972). The role of the vomeronasal organ in mammalian reproduction. *Mammalia* 36, 315–41

Gates, G. R. (1978). Vision in the monotreme echidna (*Tachyglossus aculeatus*). *Aust. Zool.* 20, 147–69.

— Saunders, J. C., Bock, G. R., Aitken, L. M., and Elliott, M. A. (1974). Peripheral auditory function in the platypus, *Ornithorhynchus anatinus*. *J. acoust. Soc. Am.* 56, 152–6.

Grant, T. R. and Carrick, F. N. (1978). Some aspects of the ecology of the platypus *Ornithorhynchus anatinus*, in the Upper Shoalhaven River, New South Wales. *Aust. Zool.* 20, 181–99.

Gregory, W. K. (1947). The monotremes and the palimpsest theory. *Bull. Am. Mus. nat. Hist.* 88, 7–52.

Griffiths, M. (1968). *Echidnas.* Pergamon Press, New York.

— (1978). *The biology of monotremes.* Academic Press, New York.

— McIntosh, D. L., and Coles. R. E. A. (1969). The mammary gland of the echidna *Tachyglossus aculeatus* with observations on the incubation of the egg and on the newly hatched young. *J. Zool., Lond.* 158, 371–86.

Home, E. (1802). A description of the anatomy of the *Ornithorhynchus paradoxus*. *Phil. Trans. R. Soc. Lond.* 1802, 67–84.

Johnson, K. (1978). Behaviour in a group of wild echidnas. *Victorian Nat.* 95, 241–2.

Keibel, F. (1902). Zur Anatomie des Urogenitalskanals der *Echidna aculeata* var. *typica*. *Anat. Anz.* 22, 301–5.

— (1904) Zur Entwicklungsgeschichte des Urogenitalapparates von *Echidna aculeata* var. *typica*. *Denkschr. med-naturw. Ges. Jena* 6 (2), 153.

Klaatsch, H. (1888). Zuer Morphologie der Tastballen. *Morph. Jb.* 14, 407.

MacKenzie, W. C. and Owen, W. J. (1919). *The glandular system in monotremes and marsupials.* Jenkins Buxton, Melbourne.

Negus, V. (1958). *The comparative anatomy and physiology of the nose and paranasal sinuses.* Livingstone, Edinburgh.

Ortmann, R. (1960). Die Analregion der Säugetiere. *Handb. Zool.* 8:26, 3(7), 1–68.

Pirlot, P. and Nelson, J. E. (1978). Volumetric analyses of monotreme brains. *Aust. Zool.* 20, 171–9.

Scalia, F. and Winans, S. S. (1975). The differential projections of the olfactory bulb and accessory olfactory bulb in mammals. *J. comp. Neurol.* 161, 31–57.

— — (1976). New perspectives on the morphology of the olfactory system: olfactory and vomeronasal pathways in mammals. In *Mammalian olfaction, reproductive process and behavior* (ed. R. L. Doty). Academic Press, New York.

Schaffer, J. (1940). *Die Hautdrüsenorgane der Säugetiere.* Urban und Schwarzenberg, Berlin.

— and Hamperl, H. (1926). Über Anal-und Circumanal-drüsen. 3. Mitteilung: Marsupialier. *Z. wiss. Zool.* 127, 527–89.

Semon, R. (1894). Beobachtungen über die Lebenweise und Fortpflanzung der Monotremen nebst Notizen über ihre Körpertemperatur. *Denkschr. med.-naturwiss. Ges. Jena* 5, 1–15.

Strahan, R. and Thomas, D. E. (1975). Courtship of the platypus *Ornithorhynchus anatinus*. *Aust. Zool.* 18, 165–78.

3 The metatherians: order Marsupialia

ELEANOR M. RUSSELL

Introduction

Our knowledge of chemical communication in marsupials is fragmentary. Published experimental evidence of the importance of chemical signals in social behaviour is limited to Schultze-Westrum's well-known study of *Petaurus breviceps* (1965). Hediger's (1958) discussion of the behaviour of marsupials mentions marking behaviour only in *Trichosurus vulpecula*, and refers to the strong smell of *Phalanger, Petaurus*, and *Hypsiprymnodon*. This chapter will attempt to collect together the considerable body of scattered information on the occurrence of skin glands and observations of behaviour which suggests that these glands have some significance in communication. Collectively this information indicates that in the marsupials olfactory communication has a potential significance which we have not hitherto realized. Studies which investigate the nature and significance of glandular secretions in relation to the social organization of a species are rare—even Schultze-Westrum's work cannot be related to a detailed knowledge of the pattern of social behaviour of *Petaurus* in the field. Eisenberg and Golani's (1977) review of communication in Metatheria discussed olfactory communication briefly, but discussed only a few species and did not consider odour production nor the presence of scent glands.

This chapter will consider in turn known or potential scent-producing glands, scent reception, and behavioural evidence of olfactory communication and its significance in social behaviour. Most of the latter evidence is merely observation of what is assumed to be scent marking, but I shall attempt to place this behaviour in its social context where possible.

The classification of marsupials used follows that proposed by Kirsch (1977*a*) and the species names used are those of Kirsch and Calaby's complete list of living marsupial species (1977). In cases of doubt or unfamiliar names, readers are referred to Kirsch's (1977*b*) notes on nomenclature.

Occurrence of scent glands

Odour-producing glands are widespread in marsupials and Table 3.1 summarizes their occurrence. A glandular region is included in the table if (i) it is known to produce an odour detectable by man; (ii) it is known to be sniffed by conspecifics; (iii) it occurs in a region clearly involved in marking behaviour, e.g. various glands in the cloacal region; (iv) in related species, glands in the same area are included under (i), (ii), or (iii) above; (v) it has no other known function.

Table 3.1. Occurrence of scent glands in marsupials

	Labial	Ear	Sternal	Pouch	Paracloacal Oil Cell	Paracloacal Unknown	Proctodaeal	Circumanal	Eccrine Interdigital	Other	References
Family Didelphidae											
Caluromys philander											1
Didelphis aurita					1pr						2
Didelphis virginiana		+	+			+					3,4
Marmosa cinerea		+	+								5
Marmosa elegans		+	+								5
Marmosa mexicana			+								5
Marmosa murina											5
Marmosa robinsoni			δ>♀								5
Monodelphis dimidiata		+	+								6
Lutreolina crassicaudata				+		3pr		+			3
Philander opposum				+		+		+			1,3
Family Dasyuridae											
Antechinus apicalis						+					7
Antechinus bilarni						+		+			8
Antechinus stuartii			+			?		+			8,9
Dasyuroides byrnei	?					+					10
Dasyurus hallucatus								+			8
Dasyurus viverrinus		+									1
Phascogale tapoatafa			+			+					11
Sarcophilus harrisii		+	−					+			8,12
Sminthopsis crassicaudata	+			1pr	1pr	2pr		+	+		13
Sminthopsis leucopus											8
Family Myrmecobiidae											
Myrmecobius fasciatus			δ,♀			1pr					14,15,16

Family Peramelidae

Species	1	2	3	4	5	6	7	Notes	References
Chaeropus ecaudatus		—							30
Echymipera rufescens	+	+							30
Isoodon macrourus		δ,♀							8,30
Isoodon obesulus	+	δ,♀		3pr		+	+		8,13,30
Perameles gunnii		δ,♀							30
Perameles nasuta		δ,♀	1pr						17,27,30

Family Thylacomyidae

Species	1	2	3	4	5	6	7	Notes	References
Macrotis lagotis		♀							30

Family Burramyidae

Species	1	2	3	4	5	6	7	Notes	References
Cercatetus concinnus	+	+	1pr	1pr			+		13
Acrobates pygmaeus		—		2pr		+			16

Family Petauridae

Species	1	2	3	4	5	6	7	Notes	References
Petaurus breviceps	+	δ		2pr		+	+	a(δ)	17,18
Pseudocheirus dahli		δ							8
Pseudocheirus peregrinus	+	—		2pr		+			17,19
Petauroides volans									20

Family Phalangeridae

Species	1	2	3	4	5	6	7	Notes	References
Phalanger maculatus				3pr					1
Phalanger orientalis				+					7
Trichosurus arnhemensis			1pr	1pr					8
Trichosurus caninus		+	1pr	1pr		+			21
Trichosurus vulpecula	+	+	1pr	1pr		+	+	b	13,17,22,27

Family Vombatidae

Species	1	2	3	4	5	6	7	Notes	References
Vombatus ursinus				1pr		—	+		4

Family Phascolarctidae

Species	1	2	3	4	5	6	7	Notes	References
Phascolarctos cinereus	+					+		c	1,7,17

Table 3.1 (*cont.*)

	Labial	Ear	Sternal	Pouch	Paracloacal Oil	Paracloacal Cell	Paracloacal Unknown	Proctodaeal	Circumanal	Eccrine Interdigital	Other	References
Family Macropodidae												
Aepyprymnus rufescens							2pr					8,23
Bettongia gaimardi							+					31
Bettongia penicillata							2pr					24
Dendrolagus bennettianus			+				+				b	1,25
Dorcopsis luctuosa											b	1
Dorcopsis veterum											b	1
Macropus agilis							+					8
Macropus antilopinus							+					8
Macropus eugenii							+					3
Macropus giganteus			♂>♀	+			1pr			+		1,26,31
Macropus robustus			♂<♀	+			+			+		26
Macropus rufogriseus							1pr					3,31
Macropus rufus			♂>♀	+			+			+		26
Onychogalea unguifera							+					8
Petrogale brachyotis							+					8
Petrogale penicillata							+					27
Petrogale burbidgei							+					8
Potorous tridactylus							+		+			1
Setonix brachyurus					1pr							8
Thylogale billardieri							+	—	+			4
Thylogale stigmatica							1pr					31
Thylogale thetis							1pr					31
Wallabia bicolor			♂				1pr			+		26,31
Family Notoryctidae												
Notoryctes typhlops											d	28
Family Tarsipedidae												
Tarsipes rostratus							2pr					8,29

+: Present, sex unspecified. −: Absent. ♂: In male only. ♂,♀: in ♂ and ♀. ♂>♀: larger in ♂ than in ♀. 1,2 pr: one or two pairs of glands. Other glands: (a) frontal; (b) chin; (c) cheek; (d) flank.

References:

1. Schaffer (1940).
2. Munhoz and Merzel (1967).
3. Schaffer and Hamperl (1926).
4. Ortmann (1960).
5. Barnes (1977).
6. Beddard (1888).
7. van den Broek (1904).
8. N. Allen (personal communication).
9. Braithwaite (1974).
10. Aslin (1974).
11. Cuttle (1978).

12. Flynn (1910).
13. Green (1963).
14. Beddard (1887).
15. Ford (1934).
16. Hill (1900).
17. Bolliger and Hardy (1945).
18. Schultze-Westrum (1965).
19. Thomson and Owen (1964).
20. Tyndale-Biscoe (1973).
21. Thomson and Pears (1962).

22. Winter (1977).
23. van den Broek (1910).
24. Christensen (1980).
25. Waite (1894).
26. Mykytowycz and Nay (1964).
27. Bolliger and Whitten (1948).
28. Sweet (1907).
29. Rotenberg (1928).
30. Stoddart (1980).
31. K. A. Johnson (1977).

Table 3.1 does not include ordinary holocrine sebaceous glands normally associated with hairs or apocrine sudoriferous glands, both of which are widespread over the body and contribute to individual odour. Nor are eccrine sweat glands included; they occur in the ventral region of the pads of the manus and pes where they are the only sweat glands present (Green 1963). Salivary glands, which are present in all mammals, are not included, although as Schultze-Westrum, (1965) has shown for a marsupial, saliva can convey olfactory information, as it also can in some eutherian mammals.

Probable scent glands have been recorded so far in 13 out of 16 families, and in 63 species in 40 out of 73 genera. The remaining species do not necessarily lack glands, but have not been investigated.

Frontal gland

Frontal glands occur only in the marsupial gliders of the genus *Petaurus*. Their histology in *Petaurus breviceps papuanus* was described by Schultze-Westrum (1965) and similar glands occur in *P. norfolcensis* and *P. australis*. The gland occurs only in males, and is situated in the midline of the head, just anterior to a line joining the anterior margins of the ear openings. The dark median dorsal strip of the pelage divides around the gland which is visible in a fully grown male as a clear area of 16 × 18 mm. The gland appears to be made up of holocrine and apocrine elements in approximately equal numbers with large peripheral sebaceous glands. The secretion of the gland is clearly visible, can be wiped off on filter paper, and has a pungent musky smell. There is a yellow-brown oily component, and a light fluid component.

Glands of the ear region

Extensive glandular areas have been found lying close to the external ear, and seem to be particularly obvious in dasyurids and peramelids.

Schaffer (1940) quoted a report by Burne (1909) of a sebaceous glandular mass c. 20 × 10 mm covering the ear muscle of the native cat *Dasyurus viverrinus*, opening into the auditory passage. He could find no similar gland in the cuscus *Phalanger maculatus* or 'a kangaroo'. Wood-Jones (1949) found a similar well-developed 'conchal' gland in another dasyurid, *Dasycercus cristicauda*. It is a yellowish-white sebaceous mass, and lies closely applied to the posterior and lower aspect of the concha at its junction with the external auditory meatus, and opens into the cavity of the external ear (Fig. 3.1). The small carnivore *Sminthopsis crassicaudata* has massive accumulations of sebaceous glands deep in the dermis of the external auditory canal, as well as sebaceous glands associated with hairs (Green 1963).

In her comprehensive investigation of cutaneous glands of marsupials Green (1963) found glandular concentrations in the external auditory meatus of all four species (from different families) which she investigated (see Table 3.1). In the brush-tail possum *Trichosurus vulpecula* the skin covering the cartilaginous processes of the external auditory meatus carries hairs, with each of which are associated large tubular apocrine glands, and a large accumulation of sebaceous gland cells in a saccular formation. The same region in the pigmy possum *Cercartetus concinnus* is hairless, but carries large typical sebaceous and apocrine glands opening on to the skin surface. In the bandicoot *Isoodon obesulus*, she found large compound sebaceous glands opening on to the skin surface, often several glands sharing a common duct. It is not quite clear whether these are the subauricular glands described by Stoddart (1980) which occur in bandicoots of the genera *Isoodon, Perameles*, and *Echymipera* but are absent in *Chaeropus*; the glandular area lies slightly below and behind the ear (Fig. 3.2) and consists

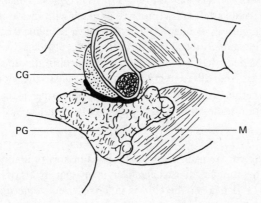

Fig. 3.1. Position of ear (conchal) gland of the small carnivorous dasyurid marsupial *Dasycercus cristicauda*. Right side of the face, with the concha removed from the external auditory meatus. (After Wood-Jones (1949).) CG = conchal gland. PG = parotid gland overlying the masseter muscle M.

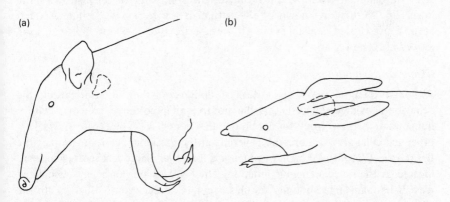

Fig. 3.2 Position of auricular glands of bandicoots (a) sub-auricular gland, adult male *Isoodon obesulus*; (b) interauricular gland, female *Macrotis lagotis*. (From Stoddart (1980).)

of enlarged sebaceous acini overlying sudoriferous tubules, both opening into the follicle of the overlying guard hair. Activity of the complex, especially of the sebaceous element, increases markedly in the breeding season, particularly in males, when the pungent watery secretion is sufficiently copious to impart a damp look to the fur of the neck. In the rabbit-eared bandicoot *Macrotis* the glandular complex lies between the ears and is absent in males; the sebaceous acini are less hypertrophied than in *Isoodon* or *Perameles* and the sudoriferous glands are also reduced.

Salivary glands

The morphology of salivary glands in marsupials is typically mammalian (Forbes and Tribe 1969) and variation is correlated as much with diet as with taxonomic position (Young and van Lennep 1978). The saliva contains a complex mixture of organic and inorganic substances, including glycoproteins. Schultze-Westrum (1965) demonstrated that *Petaurus* could distinguish the smell of saliva of individuals, at least from different groups, but the origin of the odorous substances is not clear—whether from food, from licking some other gland area, from the saliva itself, or from some other gland in the mouth area.

Glands of the chin

Glands described on the neck–chin or throat–chin area probably belong to the series of glands described as sternal glands in many marsupials (p. 000). The only definite report of glands on the chin is in *Trichosurus vulpecula*. Although no glandular region is obvious externally, Winter (1977) reported observations of W. J. Freeland that the lower margin of the chin, particularly of adult males, has well-developed sebaceous glands, even larger than those described for the sternal gland by Bolliger and Hardy (1945). These form an almost uninterrupted layer of glandular tissue in the region of the dark patch of fur just below the lower lip. Nothing is known of the nature of the secretion. Although Green (1963) investigated the upper lip of *T. vulpecula*, she does not appear to have looked at the lower lip.

Glands of the mouth region

Although there is not a great deal of definite anatomical information, the behavioural evidence suggests that the mouth is an important source of olfactory information. Apart from food odours, the odour of expired air and saliva, other glandular areas are present, generally at the mucocutaneous junction of the lips (Green 1963). In *Cercartetus concinnus* many very large sebaceous glands are located at the mucocutaneous junction. These glands are not always associated with hairs, and large apocrine tubules occur in conjunction with them. In the upper lip region of a female *Isoodon obesulus*, 8-10 large sebaceous glands not related to hairs open by short ducts at the mucocutaneous junction. The upper lip of *Sminthopsis crassicaudata* shows only apocrine tubules and no large sebaceous glands. The small hairs of the upper lip of *T. vulpecula* are completely encircled with a sebaceous appendage; larger hairs and vibrissae have apocrine tubules and smaller sebaceous glands. Inside the mouth, Biggins (1979) found that the inner surfaces of the posterior aspects of the upper and lower lips are lined with rows of small filamentous projections which expand beyond the angle of the mouth. These are glandular in nature, although the type of gland was not specified. These labial glands produce a very complex mixture of lipids, which is different from that produced by sternal and paracloacal glands. A similar glandular area is found in the small carnivore *Phascogale tapoatafa* (Cuttle 1978). The walls of the nasal vestibule contain either muco-serous glands (*T. vulpecula*) or apocrine tubules (*Isoodon, Cercartetus*). Martin (1836) described a cheek (subzygomatic) gland in the koala *Phascolarctos*, a pea-sized gland on the masseter. The detailed study of the anatomy of the head and neck by Sonntag (1921) makes no mention of it, and there seems to be no further information about it.

Sternal glands

There are many reports of glandular patches or pigmented areas in the sternal region of marsupials (Table 3.1), and of associated marking behaviour (Tables 3.3-3.8) but only a few have been investigated histologically. In some species, there is an almost hairless secretory area (*Myrmecobius*: Ford 1934; *Petaurus*: Schultze-Westrum 1965), in others there is no obvious bare patch (kangaroos, *Macropus* spp: Mykytowycz and Nay 1964; *Trichosurus*: Bolliger and Hardy 1945). In most species with sternal glands the hairs of this area are stained yellow-brown or red especially in the mature male. Schultze-Westrum (1965) found that in the sternal area of male *Petaurus*, apocrine glands were few and that the sebaceous glands produced a yellow-brown secretion that smelt like the secretion of the frontal gland. In *T. vulpecula*, the patch of brownish-yellow coloured hairs is larger in males than in females. The sebaceous and apocrine sudoriferous glands of the sternal area are very much larger and more active than those on the general body surface. The region produces a moist, frequently copious secretion with a distinct but not disagreeable odour (Bolliger and Hardy 1945; Green 1963). The murine opposum *Marmosa robinsoni* appears similar to *T. vulpecula* in that the gland is larger in the male than the female, and enlarged apocrine and sebaceous glands are present (Barnes 1977). The early report by Beddard (1887) on the sternal gland of the numbat *Myrmecobius* suggested that it was larger and more complex than that found in any other marsupial, but Ford (1934) has shown that the gland-bearing skin patch of sebaceous and apocrine glands overlies the submaxillary gland, which Beddard identified as a sternal gland. In the kangaroos *Macropus rufus*, *M. fuliginosus*, and *M. robustus* highly convoluted, enlarged apocrine glands occur in the sternal and axillary region, to a greater extent in males than in females, but sebaceous glands are not enlarged (Mykytowycz and Nay 1964). Males have a distinct aromatic odour.

There appears to be seasonal variation in production of pigment and other secretions from the sternal gland in *Macropus, Trichosurus,* and probably many other species. The pigmentation of the sternal patch may have no olfactory significance. Nicholls and Rienits (1971) identified the colour as due to hair pigment. In the species they examined (*M. rufus, T. vulpecula,* and *Dendrolagus goodfellowi*), hair from all parts of the body had pigment adhering to the outside of the hair-shaft as well as internal pigment. Hair from the sternal area carried large flakes of the same pigment, sometimes as a solid mass matting the hairs together. The pigments, which were tryptophan derivatives, were secreted from the sweat glands.

Biggins (1979) investigated the chemical nature of the lipids present in the secretion of the sternal gland of *T. vulpecula.* He found a complex mixture of the major classes of lipids, and gas chromatography showed the presence of vast numbers of compounds. Sternal gland secretion was different from labial and paracloacal gland secretions in the same individual, and there were differences in the sternal gland secretions from different animals. The most significant point on which the secretion of the sternal gland differed from those of the labial and paracloacal glands was the presence of characteristically higher concentrations of more volatile components in sternal gland secretion.

Not all species possess sternal glands. Green (1963) failed to find any particular enlargement or specialization of glands of the sternal region in *Sminthopsis crassicaudata, Isoodon obesulus,* or *Cercartetus concinnus.* Thomson and Owen (1964) found no sternal gland in the ring-tail possum *Pseudocheirus peregrinus,* which is in the same family as *Petaurus.*

Glands of the pouch

Where histological studies have been done (see Table 3.1) the pouch of the female marsupial is found to contain both apocrine and sebaceous glands. Mykytowycz and Nay (1964) found a preponderance of enlarged apocrine glands in the pouch of *Macropus rufus, M. fuliginosus,* and *M. robustus.* On the other hand, the lining of the pouch of *Petaurus* is thickly packed with sebaceous glands (Schultze-Westrum 1965). Pigmentation of the pouch is similar to that of the sternal region and the degree of pigmentation changes with the reproductive state of the female (Sharman and Calaby 1964; Barnes 1977), but this may not be significant for olfactory behaviour.

Glands of the anal region

The naming and the descriptive morphology of the glands occurring in the anal region of marsupials is a matter of considerable confusion, partly because much early work has not been reported in English, but also because some of the early and much quoted descriptions were based on ancient material inadequately preserved. These glands have been discussed in some detail by Schaffer and Hamperl (1926), Schaffer (1940), Bolliger and Whitten (1948), and Green

(1963). Ortmann (1960) gives a comprehensive comparative account of the glands in the anal region of mammals and recognizes several different types: rectal, proctodaeal, circumanal (apocrine tubular and sebaceous), paracloacal (Fig. 3.3). The rectal glands are typical intestinal glands and will not be discussed further.

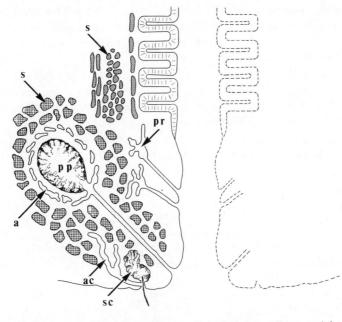

Fig. 3.3. Diagram of longitudinal section of anal region of marsupial, showing proctodaeal, circumanal, and one pair of paraproctal glands. (From Ortmann (1960).)
pr, protodeal glands; s, sphincter muscle; pp, paracloacal gland; ap, apocrine tubular glands surround paracloacal gland; ac, circumanal apocrine glands; sc, circumanal sebaceous glands.

Proctodaeal glands are tubulo-aveolar, often branched glands, which open into the anal canal. Their ducts do not have the myoepithelial cells typical of apocrine sudoriferous glands. Nothing is known of their function. Where detailed histological studies have been done, they have been found in the following marsupials: *Petaurus* (Schultze-Westrum 1965); *Didelphis virginiana* and *Trichosurus vulpecula* but not *Thylogale billardieri* or *Vombatus ursinus* (Ortmann 1960).

Circumanal glands. Typical apocrine and sebaceous glands occur at the point where the anal canal opens to the outside, surrounding the anus itself. The glands usually lie within the anal sphincter muscle, and their contents may be squeezed out with the faeces, or they may add to the combined odour of the cloacal region. They are particularly well developed in dasyurids (N. T. Allen, personal communication).

Paracloacal glands.   These glands have also been called rectal glands, anal glands, paraproctal glands, and anal pouches (Analbeutel), but this chapter follows the terminology of the most recent papers in English (Bolliger and Whitten 1948; Green 1963).

The reviews of Schaffer and Hamperl (1926) Schaffer (1940) and Ortmann (1960) speculate at length on the homology of marsupial paracloacal glands with other glands of the anal region in mammals. The present chapter is concerned chiefly with their structure and function.

From one to four pairs of cloacal glands occur in most marsupials (Table 3.1). Although in several species there is no report of their occurrence, there is no definite report of their absence in any species, and they are the most widespread and specialized of the integumentary glands.

The paracloacal glands are small, cyst-like glands lying close to the lateral cloacal walls, and opening into the cloaca by slender ducts (Fig. 3.4). In a male

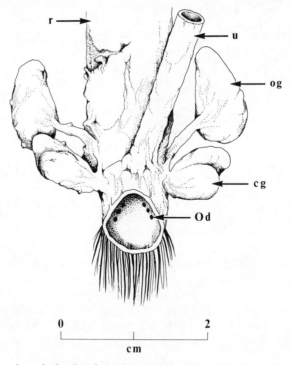

Fig. 3.4. Paracloacal glands of the brush-tail possum *Trichosurus vulpecula*. The egg-shaped oil glands and the frequently bilobed cell glands open into the cloaca. (From Bolliger and Whitten (1948).)

cg, cell gland; od, openings of ducts into cloaca; og, oil gland; r, rectum; u, ureter.

Trichosurus vulpecula, which may weigh 2–3 kg, the largest (ventral) pair of glands are about 1 cm long, yellow and egg shaped; the dorsally located pair of glands are slightly smaller, generally bi- or trilobed. Histologically, the basic structure of the glands is similar in all species so far described. The thin outer wall shows a superficial arrangement of circularly disposed bundles of striated muscle, with intervening apocrine tubules. The internal glandular substance is separated from the outer muscle-apocrine gland layer by a thin capsule of connective tissue, from which numerous septa carrying blood vessels project into the lumen of the cyst-like gland. Several layers of secretory cells are arranged along the projecting septa (Plate 3.1).

Schaffer and Hamperl (1926), Schaffer (1940), and Ortmann (1960) and the earlier work they reviewed discuss all paracloacal glands as being of the same type producing a secretion containing cells. Ortmann (1960) did not refer to the work of Bolliger and Whitten (1948) in which they described in detail two types of paracloacal gland in *Trichosurus vulpecula*, nor did he make the distinction in the material from *T. vulpecula* which he himself examined. Bolliger and Whitten distinguished between oil-secreting glands and cell-secreting glands. The oil-secreting glands (the ventral pair in *T. vulpecula*) produce a cream-coloured secretion of oily droplets emulsified in an aqueous phase with an odour variously described as pungent, musky, resembling rotten onions or garlic. The oily secretion is expressed from the gland, released in the urine, deposited on the faeces, or released when the animal is handled, as well as being released in specific marking behaviour. The cell-secreting glands produce a secretion which consists of cells suspended in a small volume of an aqueous medium, and this has no odour. This secretion is washed away in the urine in which intact cells may be found, or deposited on the faeces.

The more detailed histological work of Green (1963) confirmed Bolliger and Whitten's observations on *T. vulpecula* and extended to species of three other families. The oil-secreting gland has a thin connective tissue capsule with polygonal secretary cells which degenerate, losing their nucleus and resulting in the appearance of fatty globules which are the final product when the cells disintegrate. The cell-secreting gland has a thicker, more muscular wall with less lumen than the oil-secreting gland. Its capsule of connective tissue is denser and the lumen is more densely packed with septa carrying secretory cells. The round secretory cells degenerate, and are shed (as oval cells) into the lumen of the gland. Fatty globules are present in the exfoliated cells. The apocrine tubules surrounding the glands are open both directly into the cloaca and into the ducts of the paracloacal gland. Presumably they produce the aqueous phase of the secretion for both types of gland. The apparent difference in the type of secretion produced by the two types of gland which show structural similarities is puzzling. Cells in both types of gland contain a fatty secretion, but in the cell-gland they are shed intact, while in the oil-gland a fatty secretion is produced.

As yet there is not enough information about a large number of species to be able to see any clear pattern in the occurrence of oil- and cell-secreting glands. Both oil-secreting and cell-secreting glands are found in *Cercartetus concinnus* and *Sminthopsis crassicaudata* (Green 1963) and also in the mountain possum *Trichosurus caninus* (Thompson and Pears 1962). Among the bandicoots, *Perameles nasuta* (one pair of glands) produced only a cellular secretion (Bolliger and Whitten 1948) while in *Isoodon obesulus*, the three pairs of glands conformed more closely to the structural pattern of the sebaceous glands of the skin, albeit with the surrounding layer of muscle and apocrine tubules, and produced only an oily secretion (Green 1963). In the family Petauridae, *Pseudocheirus peregrinus* has two pairs of glands which produce an oily secretion with no cellular elements (Bolliger and Whitten 1948). They could palpate two pairs of glands in *Petaurus breviceps papuanus* and found only oily secretion, and no cell bodies in the urine. Schultze-Westrum (1965) does not say how many paracloacal glands he found in *P. b. papuanus*, but he describes two different odour substances, a white, oily, strongly musk-smelling secretion in the paracloacal glands and a pungent secretion mixed with the urine, about the origin of which he does not comment. Only one pair of glands is found in almost all kangaroos that have been investigated, but the nature of the secretion is not clear. The figures and descriptions of Schaffer and Hamperl (1926) and Ortmann (1960) suggest that they were cell-producing glands in the species they looked at. A strong odour emanating from the paracloacal glands is reported in a few species of macropodid (*Dendrolagus bennettianus*: Waite 1894; *Macropus rufogriseus*: Barone and Blavignac 1965). Three pairs of paracloacal glands were described by Van den Broek (1910) in the rufous rat kangaroo *Aepyprymnus* (= *Hypsiprymnus*) of the subfamily Potoroinae. He remarked on their difference from other macropodids, and his drawing shows the ducts of one pair of glands opening at the tip of a cloacal papilla similar to that described by Christensen (1980) in *Bettongia penicillata*, another rat-kangaroo, which has two pairs of paracloacal glands. It is possible that the third pair found by Van den Broek in *Aepyprymnus* were in fact male accessory glands.

In the opossum *Didelphis aurita* from Brazil, Munhoz and Merzel (1967) describe one pair of paracloacal glands which produce a scent secretion. The glands appear to be of the cell-secreting type, with the muscular layer investing the gland only on its outer half, and the typical layer of apocrine tubules lying between the layers of connective tissue and muscle. The secretion stored in the central cavity is composed of desquamating epithelial and mast cells and amorphous material. Histochemically, it appears that the amorphous material derives from the mast cells and is polysaccharide in nature. Proteins and lipids (unsaturated glycerides) derive from the epithelial cells, and the apocrine glands produce a carbohydrate–protein complex secretion.

Biggins (1979) investigated the chemical nature of the lipids present in the oil-secreting and cell-secreting glands of *T. vulpecula*. He found a vast number of compounds of the major types of lipids. The secretions of the two glands were not markedly different qualitatively in the lipids present, but they differed with respect to the relative concentrations of the various components. Neither secretion had high concentrations of the more volatile components which were present in sternal gland secretion. Determination of the mass spectrum of some of the more abundant compounds present suggested that the major components of the oil-gland secretion are low molecular weight fatty acids and alcohols. Autrum, Fillies, and Wagner (1970) investigated the secretion of the paracloacal glands of *Petaurus breviceps papuanus*, which does not appear to produce a cellular secretion (Bolliger and Whitten 1948; Schultze-Westrum 1965). They identified substances by thin-layer chromatography and by infrared and nuclear magnetic resonance spectroscopy. Substances present in the secretion included: (i) n-fatty acids, from C_6 to C_{22}, saturated and unsaturated; (ii) methylesters of saturated and unsaturated C_6-C_{18} fatty acids; (iii) triglycerides (about 89 per cent of total secretion); (iv) cholesterol and cholesterol esters; (v) lanosterol; (vi) squalene (about 75 per cent of the unsaponifiable matter of the secretion).

Whether it is the complexity of such a mixture which is itself important, or whether a small number of individual components are particularly important has not been investigated. It is not unlikely that some substances act as fixatives, prolonging the odour of more volatile, behaviourally significant compounds. Squalene, for example, is used as a commercial fixative.

The paracloacal glands are an important source of odour in marsupials, but information about their distribution, structure, and function is inadequate. It is worth pointing out that as a result of the morphology of marsupials, with urinary and genital ducts and rectum opening into a common cloaca, both urine and faeces may carry the odours of paracloacal glands and genital ducts.

Other glands

The marsupial mole, *Notoryctes typhlops*, has a paler area on the hinder part of the back, where the fur has the appearance of being matted (Wood-Jones 1923). Sweet (1907) found that in this area, the epidermis is thickened with a concentration of sebaceous glands. Little is known of the biology of *Notoryctes* and one can only speculate that in a blind burrowing mammal, olfaction may be of great significance.

The eccrine glands, which occur interdigitally and in the non-hairy ventral skin of the paws of marsupials have been described as typically mammalian by Green (1963) and Mykytowycz and Nay (1964). Fortney (1973) found that in the eccrine sweat glands of the opossum *Didelphis virginiana* one type of secretory cell forms proteinaceous non-mucoid secretory granules which are released into the gland lumen. Such cells are not yet known in other mammals and the significance of this secretion is not known.

The perception of chemical signals

To understand the part that olfaction plays in behaviour, it is important to take into account the capacity of the animal to collect and process olfactory information. The general reviews of this area (Allison 1953; Negus 1958; Moulton and Beidler 1967; Steinbrecht 1969; Graziadei 1971, 1976; Parsons 1971) give little specific information about the olfactory system of marsupials.

The structure of typical olfactory cells in the olfactory mucosa of marsupials does not appear to differ significantly from that of other mammals, and there is evidence that receptor and supporting cells may be replaced (Kratzing 1978), as in eutherians (Graziadei 1976). The area of olfactory epithelium is extensive (Fig. 3.5) in those species which have been studied (Negus 1958; *Isoodon macrourus*: Kratzing 1978; *Trichosurus vulpecula*: Biggins 1979). In *Isoodon* the ethmoturbinate, covered by predominantly olfactory epithelium, lies partly within the space of the maxillary sinus and a limited amount of olfactory epithelium occurs dorsally in the sinus itself (Kratzing 1978). Similarly in *T. vulpecula*, the turbinate bones extend into expanded nasal sinuses. Such expansion of turbinates covered with olfactory epithelium into nasal sinuses themselves bearing sensory epithelium is characteristic of macrosmatic animals (Negus 1958; Parsons 1971). In *Isoodon*, Kratzing located a septal olfactory organ, consisting of separate areas of olfactory epithelium (Fig. 3.5) on the lateral aspects of the free ventral flange of the septum, and projecting into the stream of respiratory air. Kratzing suggested that it may serve to monitor the inhaled air-stream in quiet respiration and possibly, because of its bilateral nature, to appreciate direction of odour source. Similar structures are found in rodents and *Didelphis* (Masera 1943).

The olfactory bulbs, which receive the input from the olfactory receptors are exceptionally prominent in all marsupials, in some cases making up almost half the forebrain, and are essentially similar to those of other mammals in structure and in their central connection to the cerebral cortex (J. I. Johnson 1977). There is some evidence of regional specialization within the bulb (Phillips and Michels 1964).

The vomeronasal or Jacobson's organ is well developed in marsupials (Röse 1893; Broom 1896). This second area of sensory epithelium is enclosed in a tube on each side of the base of the nasal septum, in the anterior part of the nasal cavity (Fig. 3.5). Anteriorly, this blind tube opens into the nasopalatine duct or directly into the nasal cavity, as in the rodents, lagomorphs and the rat-kangaroo *Aepyprymnus*. The small area of olfactory epithelium in the vomeronasal organ led to its early dismissal as a 'rudimentary accessory olfactory organ' (Röse 1893). McCotter (1912) showed that in *Didelphis* and several eutherian mammals, processes from the vomeronasal receptors form a distinct nerve which terminates in the accessory olfactory bulb, which although located in the dorsal surface of the olfactory bulb remains separate from it. In *Didelphis*, the rat,

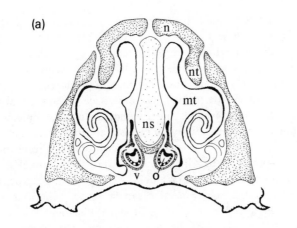

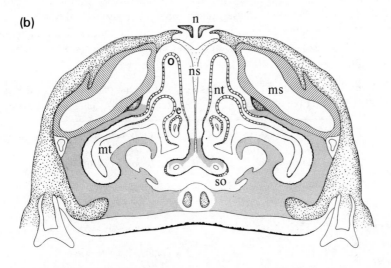

Fig. 3.5 (a) and (b). Diagrammatic cross-section of the snout of a bandicoot showing position of vomeronasal organ, distribution of olfactory epithelium, and position of turbinate bones: (a) at the level of the third incisor; (b) at the level of the first molar. (From Kratzing (1978).)
e, ethmoturbinate; m, maxilla; mo, maxillary sinus; mt, maxilloturbinate; n, nasal bone; ns, nasal septum; nt, nasoturbinate; so, septal olfactory organ; v, vomeronasal organ.

and the rabbit, the accessory olfactory bulb does not project to the principal parts of the olfactory cortex, but to the medial amygdaloid nucleus and the posteromedial sector of the cortical amygdaloid nucleus (Scalia and Winans 1975) and thence most significantly to the medial hypothalamus. Scalia and Winans (1976) consider that the vomeronasal pathway provides a significant portion of the olfactory input to the medial hypothalamic areas, which control sexual behaviour in a variety of species. Estes (1972) suggested that in macrosmatic animals, the vomeronasal organ functions as a specialized receptor for excreted sex hormones which indicate reproductive status. There is circumstantial evidence (Estes 1972; Eisenberg and Kleiman 1972) that the facial grimace called 'flehmen' performed by some male mammals when presented with female urine is involved in the functioning of the vomeronasal organ, although flehmen is not a necessary prerequisite for the function of the vomeronasal organ. Compounds of low volatility which would not reach the olfactory epithelium are transported in a liquid medium into the vomeronasal organ, and may stimulate the receptor cells (Wysocki, Wellington, and Beauchamp 1980). The importance of the vomeronasal organ in sexual behaviour was demonstrated by Powers and Winans (1975). Broom (1896) and others (Table 3.2) found vomeronasal organs in all species of marsupials they sampled and Broom considered that the vomeronasal organ was relatively well developed, to a greater extent in small species. In three genera of the family Petauridae which he examined the organ is well developed, but in the smaller *Petaurus*, the sensory area is proportionally larger.

Table 3.2. Occurrence of vomeronasal organ in Marsupials. (Data from Röse (1893), Broom (1896), Sweet (1907), Kratzing (1978), Gaughwin (1979*a*).)

Family **Didelphidae**	Family **Vombatidae**
Didelphis virginiana	*Vombatus ursinus*
Marmosa murina	*Lasiorhinus latifrons*
Family **Dasyuridae**	Family **Petauridae**
Dasyurus viverrinus	*Petaurus breviceps*
Dasyurus maculatus	*Pseudocheirus peregrinus*
Phascogale tapoatafa	*Petauroides volans*
Family **Peramelidae**	Family **Phalangeridae**
Perameles nasuta	*Trichosurus vulpecula*
Isoodon macrourus	
	Family **Macropodidae**
Family **Notoryctidae**	*Aepyprymnus rufescens*
Notoryctes typhlops	*Macropus* sp.
	Macropus giganteus
Family **Phascolarctidae**	*Wallabia bicolor*
Phascolarctos cinereus	*Petrogale penicillata*

Whitten and Bronson (1970) used the term 'primer pheromone' for a substance acting on the endocrine system, probably through the central nervous system, to produce a response which may take some time to develop. There is evidence that the vomeronasal organ is a receptor for the primer pheromone in male mouse urine (Kaneko, Debski, Wilson, and Whitten 1980).

Olfaction in social behaviour

Apart from a few experimental studies with captive animals, most information about behaviour involving olfactory signals comes from descriptive studies of the behaviour or ecology of various species in the field or in captivity. Studies of social behaviour and social organization are difficult at the best of times, especially if they are to have any relevance to the field situation. They are even more difficult if the animals are small and nocturnal, even arboreal. Consequently, although social behaviour and social organization is well known in a few species, about some species we have only very limited information and about most others we know almost nothing. Sometimes we have a detailed description of sniffing in social interactions and marking behaviour, sometimes we have only a brief mention. In this section I have summarized this information, family by family, and related it to the occurrence of scent glands.

In Tables 3.3–3.8 methods of distribution of chemical traces are summarized for each family of marsupials for which there is information. The tables also record regions of the body which are sniffed in social encounters, whether marked objects in the environment are sniffed, and by whom. Most of the categories of distribution are self-explanatory, and differences in the behaviour involved are explained in the discussion of each family. In addition to the various specific marking behaviours there are of course many other ways in which odour is released into the air—from urine, faeces, expired air, and the body in general. Much information is presumably gained by general exploratory sniffing, as well as by sniffing at specific areas of the body of another animal. One type of behaviour of uncertain significance is included in the tables. Face washing involved in grooming of the head and fore-limbs is described in a wide range of marsupials. This behaviour involves repeated rubbing of the fore-paws along the side of the head and snout, from behind the ears, along the side of the mouth to the top of the snout. Both paws move together and periods of such rubbing are interspersed with periods of licking of the paws (Plate 3.2). Eisenberg and Golani (1977) suggest only the spreading of saliva as a possible self-marking function of this behaviour, but the variety of glandular regions in the head makes it potentially a more significant form of odour distribution.

Family Didelphidae

Behaviour involved in odour distribution and response to odour are shown in Table 3.3. As far as is known from trapping studies the didelphids are basically solitary animals, not territorial in the sense of defending a particular area, but extremely aggressive to any other individual who approaches the animal wherever it may be. Females are more sedentary than males, and during the reproductive season can be found nesting in a particular den-site or moving from den to den within a reasonably well-defined home range (Hunsaker and Shupe 1977). There is not much information about the social behaviour of any species.

Table 3.3. Family Didelphidae—distribution of odour and response to odour

(a) Distribution of odour: modes of chemical deposition

	Urine dribble	Cloacal drag	Chin rub	Sternal rub	Chew twigs	Saliva drool	Cheek, mouth rub	Face wash	Reference
Didelphis virginiana ♂			+*		lick	+	+	+	1,2,3
♀						+	+	+	
Marmosa robinsoni ♂	+	+	+	+	+		+	+	1,2
♀	+	+	+		+		+	+	
Philander								+	2
Caluromys								+	2,4

*On back of female.

(b) Response to odour—areas of the body sniffed during encounters, and marked objects sniffed

	Mouth, N/N	Cheek, mouth angle	Ear	Neck	Cloaca	Pouch	Urine	Sternum	Marked twigs	Reference
Didelphis virginiana ♂	+				+				+	1,2
♀	+				+					1,2
♂	+				+					
♀	+				+					

References

1. Eisenberg and Golani (1977).   2. Hunsaker and Shupe (1977).   3. McManus (1970).   4. Altmann (1968).

In tables 3.3–3.5 (+) means observed, (−) means recorded as not observed; no sign means not observed to date, and no record of absence. N/N: nasonasal sniffing.

Table 3.4. Family Dasyuridae. Distribution of odour and response to odour

(a) Distribution of odour—modes of chemical deposition.

	Urine dribble	Cloacal mark	Chin rub	Sternal rub	Chew twigs	Cheek/ mouth rub	Face wash	Sand bathe	Deposit faeces	Reference
Antechinomys laniger		+	+				+	+		1
Antechinus flavipes		+		♂>♀	+		+	+		6
Antechinus stuartii	?	♂	+	+		+	+	+		2,3
Dasyuroides byrnei	?	÷	+	+	+		+	+	+	1,4,5
Dasyurcus cristicauda	?	+	+		+	+	+	+	+	1,5,6
Dasyurus viverrinus	+	+	+	+			+	+		1
Ningaui sp.			+	+			+	+	+	11
Planigale gilesi		+	+	+	+	+	+		+	11
Planigale maculata		+			−	−	+	−	+	12
Planigale tenuirostris		+			+	+	+		+	11
Phascogale tapoatafa		+			+	+	+			7
*Sarcophilus harrisii**	+	+	+†				+	+		1,8,9
Sminthopsis crassicaudata	+	+			+	+	+	+	+	10
Sminthopsis macroura	+	+				+				12
Sminthopsis murina	+						+		+	11

Table 3.4 (cont.)

(b) Response to odour—areas of the body sniffed during encounters, and marked objects sniffed

	Mouth nose	Cheek mouth angle	Ear	Neck	Cloaca	Pouch	Sternum	Urine	Marked twigs	Other	Reference
*Antechinomys laniger**	+		+					+		body	1
Antechinus stuartii	+								+		2,3
*Dasycercus cristicauda**			+	+	+						1
Dasyuroides byrnei	+	+	+	+	+						1,4
Dasyurus geoffroii	+	+	+	+	+					body	13
*Dasyurus viverrinus**	+					+					1
Dasyurus maculatus	+	+			+						11
Planigale gilesi		+			+						11
Planigale ingrami		+			+						14
Planigale maculata		+			+				+	body	12
Planigale tenuirostris	+	+			+				+		11
Phascogale tapoatafa		+	+					+	+	eye	7
*Sarcophilus harrissii**	+	+	+					+			1,9
Sminthopsis crassicaudata	+	+		+	+					shoulder	1,10
Sminthopsis macroura		+									11
Sminthopsis murina	+	+			+						11

* For ♂–♀ encounters only.
† On back of ♀.

References

1. Eisenberg and Golani (1977).   4. Aslin (1974).   7. Cuttle (1978).   10. Ewer (1968a).   13. Archer (1974).
2. Braithwaite (1974).   5. Sorenson (1970).   8. Buchmann and Guiler (1977).   11. Croft (1982).   14. McKay (1974).
3. Rigby (1972).   6. Ewer (1968b).   9. Eisenberg, Collins, and Wemmer (1975).   12. Van Dyck (1979).

Table 3.5. Distribution of odour and responses to odour: families Peramelidae, Thylacomyidae, Vombatidae, Phascolarctidae, Macropodidae, Tarsipedidae

(a) Distribution of odour—modes of chemical deposition

	Urine dribble	Cloacal drag	Chin rub	Sternal rub	Chew twigs	Cheek/ mouth rub	Face wash	Saliva drool	Other	Reference
Family Thylacomyidae										
Macrotis lagotis	+	+					+		●	1
Family Vombatidae										
Vombatus ursinus									Rubs rump	2
Family Phascolarctidae										
Phascolarctos cinereus				+						3
Family Macropodidae										
Aepyprymnus rufescens	+*	+		+			+			20
Dendrolagus dorianus	+*		+†	+*†			+			4
Macropus giganteus				+*			+		Grass pulling	5,6
Macropus rufus				+*			+	+	Grass pulling	7
Macropus parryi				+*			+	+	Grass pulling	9
Macropus agilis				+*			+	+	Grass pulling	1
Macropus rufogriseus				+†			+			10
Petrogale puella										8
Thylogale thetis							+		Grass pulling	21
Family Tarsipedidae										
Tarsipes rostratus	+	+					+			1

* Specialized.
† ♂ marks ♀.

Table 3.5 (*cont.*)

(b) Response to odour-areas of the body sniffed during encounters and marked objects sniffed

	Mouth N/N	Cheek mouth	Ear	Neck	Cloaca	Pouch	Sternum	Urine	Other	Reference
Family Peramelidae										
Isoodon obsesulus					+"					11
Perameles nasuta					+					11
Family Thylacomyidae										
Macrotis lagotis	+				+*	+		+		1
Family Vombatidae										
Lasiorhinus latifrons					+			+	piles of dung	12,13
Family Macropodidae										
Aepyprymnus rufescens	+	+			+	+				20
Bettongia lesueur	+				+*	+				14
Dendrolagus dorianus	+	+	+		+	+	+	+*	flank	4
Macropus giganteus	+			+	+	+		+	head and shoulders	6,15
Macropus parryi	+				+	+		+	scrotum by ♂	9
Macropus robustus	+				+	+	+	+		1,19
Macropus rufus	+				+	+		+	scrotum by ♂	1,7,16
Macropus rufogriseus	+				+	+		+	scrotum by ♂	10
Petrogale puella	+				+	+		+		8
Potorous tridactylus										15
Sentonix brachyurus	+				+*		+†			18
Thylogale billardieri	+				+*	+				17
Family Tarsipedidae										
Tarsipes rostratus	+				+	+	+			1

* Also licks.
† ♀ sniffs ♂.

References

1. Author observation.
2. McIlroy (1973).
3. Smith (1980b).
4. Ganslosser (1979).
5. Eisenberg and Golani (1977).
6. Grant (1974).
7. Sharman and Calaby (1964).
8. W. Davies (personal communication).
9. Kaufmann (1974a).
10. La Follette (1971).
11. Heinsohn (1966).
12. Gaughwin (1979a).
13. Wells (1978).
14. Stodart (1966b).
15. Veselovsky (1969).
16. Russell (1970).
17. Morton and Burton (1973).
18. Packer (1969).
19. Croft (1981b).
20. Johnson (1980).
21. K. A. Johnson (1977).

Table 3.6. Family Petauridae. Olfactory signals in *Petaurus breviceps papuanus*. (From Schultze-Westrum (1965).)

Source of odour	Object marked	Mode of deposition	Implied context and function
Frontal gland (♂ only)	Conspecifics	Rubbed on sternum (especially of ♀)	Community odour establishment*
Sternal gland (♂ only)	Conspecifics; substrate	Rubbed on partner or substrate	Community odour establishment* Territory marking Response to strange odour (rare)
Pouch glands (♀ only)	Self and pouch young	Passive transfer	Sexual status* Individual recognition of mother by young ? Location of pouch by new born young
Foot (eccrine glands)	Substrate	Wiping and scratching while oral marking	? Territory marking Response to strange ♂ odour
Flank (no specific gland)	Substrate	Rubbing against	Response to strange ♂ odour
Saliva (and perhaps other glands in mouth)	Self-body	Self-grooming	Territory marking in response to strange ♂ odour
	Substrate	Chewing objects with copious saliva flow	
Urogenital area and proctodaeal glands	? Substrate	Deposition with faeces	?
Paracloacal glands	Substrate; self	(i) in urine, especially at specific sites— nest hollow (ii) release of white oily secretion	Community odour* and territorial marking ? submissive, fear
Faeces	Substrate	No discrete loci of deposition	? community odour

* And basis of individual recognition.

Hunsaker and Shupe (1977) quote unpublished observations by Boggs of the murine opossum *Marmosa robinsoni* in which dominant males in a captive group marked the habitat with the viscous oily secretion of the sternal gland. Eisenberg and Golani (1977) include cloacal dragging, urine dribbling, and distribution of saliva by *Marmosa* in a tabulation of marking behaviour, but there is no indication of the context of this behaviour. According to Hunsaker and Shupe (1977) there is no obvious reaction in *Marmosa* when another male or female smells a marked object.

Table 3.7. Family Phalangeridae. Summary of contexts in which chinning and chesting behaviour were performed by *Trichosurus vulpecula* (from all-night observations). (From data of J. W. Winter, by permission.) (In vicinity = in same tree or <20 m away on ground)

Marker	Context	Percent of observations
♂	No other possum in vicinity	
(*N*=159)	Travelling on the ground	44
	In tree	4
	Exploring den-tree other than own	6
		Total 54
	Own den tree—dusk	3
	—dawn	14
		Total 17
	Oestrous ♀ in vicinity	
	Following ♀ on ground	5
	In tree with ♀	14
	In tree with ♀ and ♂	3
		Total 22
	Other	8
♀	No other possum in vicinity	22
(*N*=22)	Den-tree (own and other)	36
	Adult ♀ in area	28
	Joey following	5

Licking and rubbing of objects in the environment is very obvious in the common opossum *Didelphis virginiana* in captivity. It was first described by Reynolds (1952), and is mentioned by other authors (McManus 1970; Hunsaker and Shupe 1977; Eisenberg and Golani 1977). The animal alternately licked an object, or a corner of a den, and rubbed the sides of his head on it. This continued for five minutes or more and the hair on the side of the head was matted with saliva, which seemed to be produced copiously. This behaviour was seen most in males, especially during the breeding season, but was also seen at other times of the year, and in females. Other males responded to marked objects by licking and rubbing them, and with what Reynolds (1952) called the 'fighting dance'. This behaviour also occurred in aggressive interactions between males (McManus 1970). The male assumed a posture with fore-legs rigid, head up, and the hindquarters depressed, bringing the inguinal region close to the ground—the tail was extended straight out from the body and kept close to the ground, and in this manner the animal shuffled along slowly. It would be interesting to know whether this behaviour involved cloacal-marking.

Since fixed territories do not seem to characterize didelphid social organization, we must look for the function of marking behaviour elsewhere, perhaps

Table 3.8. Family Phalangeridae. Summary of olfactory communication in *Trichosurus vulpecula*. (Based on unpublished data of J. W. Winter and J. G. Biggins)

Source of odour	Mode of deposition	Object marked	Performer	Possible information carried	Response
Mouth: labial and chin glands; saliva	(a) Chinning; rubs chin, mouth angle, inner labial area (b) Chewing and licking	Trees, ground, especially branches	♂>♀	Identity, social status, location, sex, time of marking, age	(a) Over-marking by chinning, chesting or chewing, (b) Approach or withdrawal depending on relative status
Sternal gland	Chesting	Ground, trees, especially base	♂>♀	As above	As above (a) and (b)
Paracloacal (cell) gland	Urine dribbling	Substrate: tree, ground	♂♀	? as above (long-term signal)	No apparent response (Winter)
	Passive (urine and faeces)	Substrate	♂♀	? as above (long-term signal)	Avoidance or use of area depending on relative status
Paracloacal (oil) gland	Cloacal dragging	Substrate	♂♀	? as above	Avoidance or use of area depending on relative status
	Passive (urine and faeces)	Substrate	♂♀		Avoidance or use of area depending on relative status
Cloacal area (other than paracloacal glands)	Copious emission by loser in fight	Self	♂♀	Fear, acting as submissive signal	Inhibition of aggression
	Passive emission	Substrate; self	♀	Sex, sexual status	♂ approach if oestrous
Pouch	Passive emission	Self Pouch Young	♀	Sex, sexual status Location of pouch Identity of mother and young	♂ approach if oestrous Reinforce mother–young bond

as advertisement of presence, marking of particular features of home-ranges or dens, or saturation of the home-range with familiar odours. Olfactory investigation is an important component of exploratory behaviour (Altmann 1968; McManus 1970; Hunsaker and Shupe 1977). The exchange of olfactory information in interactions between individuals seems to be restricted to sniffing of the head and cloacal regions (McManus 1970; Eisenberg and Golani 1977), and individual recognition presumably occurs, since dominance hierarchies were formed in captivity in *M. robinsoni* (Boggs, in Hunsaker and Shupe 1977).

Family Dasyuridae

Marking and sniffing behaviour in the family Dasyuridae are summarized in Table 3.4. Croft's (1982) review of communication in the Dasyuridae provides information on several previously little-known species. I have included data from Eisenberg and Golani's (1977) tables on marking and sniffing although for most of their data there is no description of the behaviour and it is hard to equate the areas where skin glands are present with the areas of the body that they name (particularly muzzle, mouth, and cheek). The angle of the mouth is an important area (Ewer 1968a; Cuttle 1978; personal observation) which may be considered as muzzle, cheek, or mouth. Chin and sternum may also be confused.

The cloaca and the mouth appear to be most important sources of odours. The most obvious form of marking in the dasyurids is cloacal marking, generally by dragging the cloacal opening over the substrate. The deposition of a whitish material on the walls of cages was observed by Ewer (1968a) and Aslin (1974). In some species (*Antechinus stuartii, Sarcophilus harrisii*) a circular or figure-of-eight pattern is described (Braithwaite 1974; Buchmann and Guiler 1977). Chewing or rubbing the mouth against objects frequently deposits saliva plus the secretions of the various glandular regions in and around the mouth. Face-washing is also important in spreading mouth and ear secretions over the head, and probably occurs in all species. In some species, chiefly from arid areas, sand-bathing occurs, and in species such as *Dasyuroides byrnei* with a sternal glandular area, this may have the functions of self and substrate marking as well as a grooming function. Not surprisingly, the main body areas to which social sniffing is directed are the mouth and the cloaca. Urine appears to be an important source of odour in *Sarcophilus* (Eisenberg, Collins, and Wemmer 1975).

For only a few species is there some information about the context of marking behaviour. *Sminthopsis crassicaudata* is a small (c. 20 g) carnivore of open grassland and desert environments. From a long-term mark–recapture study, Morton (1978) suggested that animals appeared to inhabit large unstable, 'drifting' home ranges that overlapped those of other individuals both during and after breeding. During breeding (*S. crassicaudata* is polyoestrous), individuals inhabited nests by themselves, except for transient pairing of males and oestrous females. In the non-breeding period a high proportion shared nests in temporary groups of 2–8 individuals which appeared to be random aggregations. He thought that the only animals which might have been territorial were females with advanced young, and they probably defended only a small area around the nest site until the young dispersed at weaning. Males did not appear to defend territories at any time, but monopolized oestrous females. Ewer (1968*a*) who studied *S. crassicaudata* in captivity suggested that the mouth smell is the main factor involved in individual recognition. Two strangers approach cautiously to sniff each other's mouths carefully, from tip of snout to the angle of the jaw. A group of individuals kept together paid very little attention to each other, until a stranger was introduced, whereupon every encounter was preceded by tense mouth-sniffing until familiarity was re-established. Animals appeared to be unable to recognize each other by sight. Marking by chewing and mouth-rubbing occurred infrequently and there was no obvious response of other animals to a chewed object. Cloacal-marking was elicited by a strange object or by a familiar object given a strange smell, such as a clean food dish. The cloaca was briefly applied to the substrate with no obvious dragging movements. Enclosures were routinely marked, by the male only, if a male and female were kept together, or by a lone male or female. Ewer suggested that the main function of such marking was the reassurance of a familiar smell which identified a familiar area.

In another small (c. 30 g) carnivore, the forest-dwelling *Antechinus stuartii*, females are monoestrous and almost all males die after the mating period in September–October. After weaning, females and their young inhabit overlapping home-ranges, and members of a family may share a nest until the juveniles begin to disperse and the sexes become solitary in May–June. Males are territorial from June until mating, and their territories overlap the home-ranges of several females. Less is known about the details of behaviour in *Antechinus* than in *Sminthopsis* but it does appear that more object marking by sternal-rubbing, chewing and mouth-rubbing occurs. Captive animals marked objects in their environment by rubbing the chin or dragging the body across them rapidly three or four times (Rigby 1972), and they marked strange objects in the same way. Ewer (1968*b*) described similar behaviour in *Antechinus flavipes*, which has the same life-history pattern, and said that sternal marking of branches was often combined with chewing; the sternal gland is larger in the male than in the female. Braithwaite (1974) staged encounters between wild-caught male *A. stuartii* throughout the year. In this situation he observed cloacal-marking performed only by highly aggressive dominant animals in late September (the usual time of mating) interspersed between bouts of chasing and wrestling. Cloacal-marking seems to be much more vigorous than in *Sminthopsis*—the animal propelled itself forward by a vigorous and synchronous sculling action of the laterally-held limbs while pressing the cloaca to the substrate, moving in a circular or figure-of-eight path. Marking lasted for up to 90 seconds. He observed similar behaviour in the field on one occasion, and suggested that cloacal-marking thus performed marks the site of a victory and may function as a threat for some time. In *Antechinus swainsonii*, similar cloacal-marking occurs in agonistic encounters; sawdust from the cage of a strange animal elicited greatly increased exploration, but no marking (G. J. Hocking, personal communication).

Dasyuroides byrnei is a rat-sized (c. 130 g) carnivore of the arid inland of Australia, frequently found living in burrow complexes in stony desert with a sparse cover of low shrubs. There is no information from the field on social behaviour or social organization in this species, but Aslin (1974) has studied them in captivity and described marking behaviour and its context clearly. Animals housed together in captivity did not injure each other seriously, but showed little contact-promoting behaviour which would be likely to lead to the formation of groups in the field. Aslin suggested that mouth-scent was important in individual recognition—the mouth of another animal was sniffed at meeting. Nasocloacal contacts were common in encounters especially if a male was involved. Environmental objects were marked by deposition of faeces and urine, and by cloacal dragging. Following defecation or micturition the animal briefly touched the cloacal area to the substrate. Males deposit faeces and urine throughout their cage, and on new objects introduced. Females use faeces and urine to mark near the home burrow. The sternal gland is best developed in females with large detached unweaned young. Object-marking with the sternal gland occurred, but Aslin suggested that substrate-marking during sandbathing was the most important means of marking with this gland. The animal pushed itself along in loose sand by vigorous leg movements, with its chin and ventral surface in contact with the ground. Sandbathing occurred as a routine activity in the home-cage, and in response to objects marked by conspecifics. Animals placed in strange surroundings sandbathed vigorously, and sandbathing occurred during encounters between strangers. The sandbathing loci used by each of the animals in an encounter frequently overlapped. There was significantly more marking (of all types) in same-sex rather than in opposite-sex encounters. The most active marking was by females with large young, which were also most aggressive towards human interference. Marking of the burrow in this way may prevent attempts by conspecifics to use the burrow and may help the young remain near and return to the home burrow during their first excursions alone, which may occur before the eyes open. As is general in dasyurids, the young become detached from the teat while still blind and naked, and since the pouch is generally not large enough to contain them they are left in the nest while the mother forages.

Phascogale tapoatafa is a carnivore similar in size to *Dasyuroides* (c. 150 g), living in similar forest habitat to *Antechinus* spp. It is more arboreal than *Antechinus* and females construct quite elaborate arboreal nests. Like *Antechinus*, *P. tapoatafa* is monoestrous, and the young of the single litter remain in the maternal nest until the following breeding season, when females disperse to their own mutually exclusive range during mating, gestation, and lactation. Males do not appear to be territorial, but they become aggressive towards other males during the breeding season, and Cuttle (1978) suggested dispersion is based on mutual avoidance during activity periods, even though males may share a common range. Routine marking by captive animals of objects in their enclosure was performed by both males and females, which mouthed objects and dragged the perineal region over them or along the ground. Novel objects were also marked in the same way. Although the occurrence of paracloacal glands in *Phascogale* has not been recorded, Cuttle mentioned the presence of a glandular area inside the lips near the mouth angle.

Starting at some time between February and April the sternal gland in the male begins to produce a substance which stains the fur dark yellow until after mating in June–July. During the breeding season, males used the sternal gland in both routine and novel object marking. The chest was lowered onto the object and the fore-paws used to drag the body forwards while the chest was pressed against the object. Captive males marked vigorously with the sternal gland over previous marks when allowed access to an unfamiliar enclosure in which another male had been marking, or after a strange male had marked in their familiar enclosure. Females did not mark sites marked by familiar or unfamiliar males. Cuttle suggested sternal marking may alert males to the presence of other males and help them to avoid each other. Olfactory investigation of the head and cloaca appeared to be used in individual identification. At a meeting of familiar animals, mouth angle, eye, and ear were sniffed, and in unfamiliar animals and during the breeding season the cloaca was also sniffed. Marking during agonistic encounters as described for *Dasyuroides* did not occur in *P. tapoatafa*.

The smaller Eastern native cat *Dasyurus viverrinus* lives much more amicably with conspecifics in captivity. Cloacal dragging occurs frequently after defecation and micturition, but appears to be unrelated to agonistic behaviour, dominance relations or the urine or faeces of a stranger (V. N. Moss, personal communication). The context of marking with secretions from the mouth area (Eisenberg and Golani 1977; Table 3.4(a)) is not known.

The largest living marsupial carnivore, the Tasmanian devil *Sarcophilus harrisii* appears to be more of a scavenger than an active predator on large mammals (Guiler 1970) with a diet of carrion, supplemented by a wide variety of small items such as insects. In captivity, they are very aggressive towards their own species and fight intensely over food. Guiler (1970) found considerable overlap of home-ranges, and Buchmann and Guiler (1977) suggested that they feed communally on large items of carrion, and that within these feeding groups, dominance ('truce') relationships are established. There is some evidence from the field that such groups are at least partly closed to outsiders, and that strangers are attacked. *Sarcophilus* has a very strong smell (Fleay 1952; Buchmann and Guiler 1977), of which the origin or function is not known, although N. T. Allen (personal communication) has found paracloacal glands are present. In their discussion of the communication system of *Sarcophilus*, Eisenberg *et al.*

(1975) made only brief reference to cloacal dragging, and sniffing of a female's urine by a male. Eisenberg and Golani (1977) also referred to sniffing and licking regions of the head during male–female encounters. Buchmann and Guiler (1977) described cloacal dragging as being closely associated with agonistic behaviour. It is not the brief touching of the perineal region to the substrate as described in *Sminthopsis*, but a more prolonged, stereotyped action, in which the perineal region is dragged jerkily, with the animal more or less sitting, through a circular and semicircular track, sometimes with a zig-zag component (Fig. 3.6(a)). Dragging may be performed during threat displays, by the victor after a fight or by the loser.

This summary of observations of marking and sniffing behaviour in dasyurids suggests that the complex of odours in the head region is important for individual recognition. Marking by cloacal dragging and sternal rubbing does not fit any clear designation as territory marking, as has sometimes been assumed. Along with chewing and mouth-rubbing, they serve to mark objects in the environment and the substrate itself, and besides demonstrating an individual's presence in an area, they may serve to identify a familiar area. Odours from the cloacal region are also important in recognition of sexual status. It is interesting that no obvious flehmen behaviour has been described, although the vomero-nasal organ is well developed, especially in the smaller species.

Family Peramelidae

Although we have good information on the occurrence of skin glands in the Peramelidae (Bolliger and Hardy 1945; Green 1963; Stoddart 1980), there is practically no information to indicate how these glands function in social behaviour (Table 3.5). Bandicoots are largely insectivorous, with a body weight of about 1000 g, and field-studies have shown that they are solitary living, socially intolerant animals. Heinsohn (1966) found that the home-ranges of *Perameles gunnii* overlapped extensively, with males having much larger home ranges (c. 26 ha) than females (c. 3 ha). Home-ranges of *Isoodon macrourus* (1-5 ha) overlap, but the core areas, the areas of most activity, are usually discrete. Male home-ranges are larger and usually overlap the range of several females (Gordon 1974). Young develop very rapidly and are weaned and disperse almost as soon as they leave the pouch; there are no permanent nest sites. Clearly there must be some form of communication which mediates spacing in such large home ranges. Bandicoots are not very vocal animals (Stodart 1966a), and it is probably reasonable to assume that the communication is olfactory, since the anatomical equipment for it exists (Table 3.1).

Observations of *Perameles nasuta* (Stodart 1966a; Heinsohn 1966) in captivity emphasize the agonistic nature of encounters between males. In a small enclosure, two males cannot survive together. Interactions were rare between females and even between males and females, except close to oestrus when the male followed the female for a few days. Heinsohn (1966) described the male sniffing

Fig. 3.6 (a). Cloacal dragging in the large dasyurid marsupial, the Tasmanian devil *Sarcophilus harrissii*. Cloacal dragging occurs in most dasyurids, but is most prolonged and stereotyped in *Sarcophilus*. With the animal more or less sitting, the perineal region is dragged along the ground. (Drawing by Othmar Buchmann.) (b) Sexual following of the bandicoot *Perameles nasuta*. The male holds the tail of the female, and his nose is close to her rump. His lips are parted and curled at the corners of the mouth, suggestive of 'flehmen'. (From Stodart (1966*a*).) (c) Adult male koala scent-marking. The base of a tree is usually marked; the male hugs the tree and rubs his chest, with its sternal gland, up and down several times. (From Smith (1980*b*).) (d) Marking by male brush-tail possum, *Trichosurus vulpecula*. Chinning, rubbing the glandular area of the chin and inside mouth angle, generally used on smaller branches in trees. This captive male is marking a branch. Note that the chest is held well clear of the branch and the lower lip pulled back by the rubbing action, exposing the lower teeth. (e) Chesting, rubbing the sternal glandular area on the substrate, in the wild generally the base of a tree. This captive male is chesting and chinning at the same time. He is pulling his chest hard against the tree with his forepaws. As in (d) his bottom lip is pulled back and the side of the mouth pressed against the substrate. (From photographs by J. W. Winter.)

round the cloacal region of the female, but Stodart's (1966*a*) description of following and mating says that the male made no attempt to smell the pouch or urogenital opening. However, she described (and illustrated) the male following the female closely with 'lips parted and curled at the corners of the mouth' (Fig. 3.6(b)) which has all the appearance of flehmen. Although Ewer (1968*b*) considered that flehmen did not occur in marsupials, it has now been described in wombats (Gaughwin 1979*a*) and several kangaroos (Coulson and Croft 1981).

Family Thylacomyidae

Little is known about the ecology and behaviour of the rabbit-eared bandicoot *Macrotis lagotis*, which is now rare. This animal lives in arid and semi-arid areas and avoids high temperatures by using deep complex burrows during the day. Groups of burrow systems occur in areas of suitable habitat, and in such an area a number of *Macrotis* may live. Each burrow system houses an individual or a male and female with one or two juveniles (Watts 1969). In captivity, *Macrotis* appears to be far less aggressive than *Perameles* and *Isoodon*. Groups of animals can be kept together in relatively small areas, and females will give birth to, and rear, young (Hulbert 1972). Observations of *Macrotis* in captivity (author, unpublished observations) have shown that there is considerable social interaction between individuals. One male and female formed a long-lasting pair, and always used the same nest-hollow, which was shared with the offspring of the female, but not with the third adult (male) in the cage. In 23 hours of observation, 6 per cent of the time was spent in social interaction involving sniffing (of the nose, body, and cloaca) of another individual (Table 3.5). Much of the sniffing was directed by the two males to the female who was in oestrus during the period of observation (Plate 3.3), but all other possible interactions occurred. Most other types of interaction were preceded by nose–nose sniffing. Cloacal dragging was performed only by the dominant male, chiefly during two weeks in the middle of which the female was in oestrus. When the male left the nest-box, he marked the substrate outside by dragging his cloaca along the ground for several seconds, sometimes in a circle, sometimes in a straight line. Thus although *Macrotis* is much more tolerant of other individuals than are other bandicoots, as would be expected in a species where a number of animals must live together in small widely separated colonies in patches of suitable environment, the burrows of an individual or a pair appear to be used exclusively and are scent-marked by the male.

Family Vombatidae

Wombats are large grazing herbivores with adult body weight of about 25–30 kg. The common wombat *Vombatus ursinus* is restricted to the humid, temperate south-east of Australia. It lives chiefly in forest areas, although it often grazes at night in open pastoral areas. The hairy-nosed wombat *Lasiorhinus latifrons* lives in arid or semi-arid areas. Its distribution appears to depend on the occurrence of suitable soil conditions for burrow construction, and burrows are grouped together in warrens. Despite these large ecological differences in the two species, there are many similarities in social behaviour and organization. Adult *Vombatus* are solitary; young wombats remain with their mothers for some months after weaning but disperse before they reach sexual maturity at 2-3 years. Home-ranges (4-23 ha) usually overlap, but actual feeding areas appear to be exclusive. Burrows tend to be clumped in suitable areas, and more than one wombat may use each burrow, generally at different times; simultaneous use may occur, with evidence of agonistic behaviour. Encounters in feeding areas are generally agonistic—intruders are rapidly chased away, and McIlroy (1973) made no mention of any preliminary sniffing encounters. However, he suggested that olfactory signals were largely responsible for the maintenance of separate feeding areas in such large home ranges, with vocal threats and attack operating at close quarters. As evidence of this he described two types of apparent scent-marking (Table 3.5). He saw *Vombatus* rubbing its back and rump on logs, low branches and the base of trees, particularly along tracks and near frequently used feeding areas. These features were subsequently found to carry the characteristic strong smell of wombat, encountered at frequently used burrows or in handling a trapped animal. Wombat dung was often deposited on prominent objects such as stones, logs and rocks, along tracks, or on bare earth with scratch marks nearby.

In *Lasiorhinus*, there is similarly no evidence of individual burrow ownership. Despite their appearance of gregariousness, Gaughwin found that avoidance at a distance was the main type of social interaction. Wombats grazed adjacent to their warren and clearly avoided each other when grazing (Wells 1978; Gaughwin (1979b). According to Wells (1978) objects such as logs and stumps in the grazing areas were surrounded by heaps of faeces. These areas were regularly revisited by wombats which spent considerable time examining the object and piles of faeces. They rubbed the area with their hindquarters, defecated, sometimes urinated, and scratched the ground before leaving. These defecation sites were more frequent close to large warrens where trails converged. Fresh marks received considerable attention, and animals sometimes moved rapidly away from a fresh mark. Wünschmann (1966) observed rubbing of the rump against objects in captive *Lasiorhinus* but he was not sure if it had any marking significance. Encounters between his male and female were generally extremely agonistic, and he did not describe any form of sniffing as a preliminary to attack. Males sniff the cloaca and urine of oestrous females and Gaughwin (1979a) described flehmen in *Lasiorhinus*, a lateral upward retraction of the upper lip, in between bouts of intense sniffing.

Thus in both species of wombat, individuals share a scarce resource, burrows, but maintain a pattern of solitary dispersion in which visual signals, at least, play little part, and olfaction appears to be very important.

Family Phascolarctidae

Despite the fascination which the koala, *Phascolarctos cinereus*, seems to hold for the public, zoos, and Australian Wildlife Authorities, its social behaviour in the wild is not very well known, although there has been some observation of the animals under zoo conditions (Robinson 1978; Smith 1980*a,b*). Eberhard's (1978) study of their ecology gives some idea of social organization. Adult koalas are resident in a home-range which consists of a group of trees. Although home-ranges overlap spatially, one koala seldom uses a tree that is commonly used by another koala. Observations in captivity (Smith 1980*a*) suggest that most vocalizations and agonistic encounters occurred when an animal attempted to enter an already occupied tree. Maturing juvenile koalas disperse, unless there are enough suitable, unused trees for them to establish themselves close to their mother's range. Eberhard (1978) suggested that an average koala spends 85 per cent of its time in 14-15 trees in its home-range of 1-2.5 ha and in one of these trees it spends about 35 per cent of its time. Animals move from tree to tree a number of times during the night (Robbins and Russell 1978). They climb down to the ground and walk to another tree, and I have seen them sniff at the bases of trees passed and before climbing. Smith (1980*b*) described adult male koalas actively marking the base of trees by rubbing them with their sternal gland (Fig. 3.6(c)) but it is not known how marking is distributed among preferred and non-preferred trees in the wild—Smith mentioned that marking occurred when males were moved to an unfamiliar environment, and he also mentioned mouthing and chewing of strange objects (Table 3.5). In any case, the same pattern of dispersion appears to apply to females which do not have a sternal gland; the restricted use of a few trees for resting and feeding would mean that faeces and urine could be concentrated round these trees, effectively marking them passively.

Family Tarsipedidae

Recent studies of the biology of the honey possum, *Tarsipes rostratus*, a tiny (10-15 g) arboreal, highly specialized, nectarivore (Wooller, Renfree, Russell, Dunning, Green, and Duncan 1981) have given us some knowledge of its behaviour in the field and in captivity. Females with young in the pouch are much more sedentary than females with no young or males, and there appears to be little overlap in the range of such breeding females, although other females and males overlap in range. In captivity, females (especially those with young) are very aggressive, especially towards males. Despite this, all animals often huddled together during periods of inactivity, possibly for warmth. It is not unlikely that females with young maintain exclusive use of a tree in flower which would provide food for the period when they are considerably encumbered by pouch young—a female of 10 g may carry three young each weighing 2 g by the time it first leaves the pouch. Olfaction is clearly very important to *Tarsipes* both for feeding and in social interactions (Table 3.5). In captivity, most encounters between familiar animals were nasonasal sniffing, while the cloaca of a stranger was also sniffed. Sniffing towards another animal was very obvious, from 2–3 cm distance. Cloacal dragging performed by both males and females was seen infrequently in many hours of observation, but no pattern of occurrence was discerned. There was certainly no obvious active marking of new objects by any method.

Family Macropodidae

The kangaroos are a large and diverse family, with adult males ranging in size from 1 kg to over 60 kg. The behaviour of the larger species is best known, but there has been no detailed investigation of olfactory communication.

There have been few long-term field-studies of known individuals, and so generalizations about the social organization of the Macropodidae (Kaufmann 1974b; Russell 1974) are somewhat speculative. Many of the smaller species appear to be solitary (except for females with unweaned juveniles) such as *Macropus parma* (Maynes 1977) and *Bettongia penicillata*, which Christensen (1980) found to have a nest area which was exclusive to an individual. Kaufmann (1974a) found *M. parryi* lived in permanent discrete mobs of up to 50 individuals of mixed age and sex, which were generally found in small subgroups, living in an area which overlapped little with the range of neighbouring mobs. He also found this pattern in *M. giganteus*, (Kaufmann 1975) and it may well apply to other medium and large kangaroos that live in stable open forest and open woodland habitat. *M. rufus* of the arid and semi-arid open plains, lives in smaller unstable groups which are usually nomadic (Croft 1981a). *M. robustus erubescens* of rocky, hill country in arid regions is more sedentary, group size is smaller, and groups are unstable except for adult females with associated younger females (Croft 1981b). In most of these larger gregarious species there appear to be no permanent pair-bonds, mating is promiscuous and performed by the larger dominant males; not a great deal of time is spent in social interaction. However, there are some exceptions to this pattern of promiscuous mating. Rock wallabies of the genus *Petrogale* generally are found in colonies, typically in mountain country with rough rocky areas. In *Petrogale puella* (North Queensland) W. Davies has found (personal communication), that males maintain territories in the rock area which is their day-time refuge; within these territories one or more females maintain a long-term bond with the male. *Bettongia lesueur* also appears to form harems (Stodart 1966b). Kitchener (1973) found long-lasting associations between male and female *Setonix brachyurus*.

The occurrence of sniffing and potential marking behaviour is shown in Table 3.5. A blank space in this table indicates lack of information, not absence of behaviour. Odours of the mouth and nose area are clearly important in social interactions. Most studies describe a nasonasal touching, sniffing or approach,

either as an interaction on its own or as a preliminary to other forms of interaction (Fig. 3.7(a)). Johnson (1980) describes how for a few days after a strange
female was introduced into an established group of the rufous rat kangaroo
Aepyprymnus rufescens, 'the members of a once stable group of animals no
longer recognized one another, so that at each meeting of any of the animals,
identity had to be established by nose and mouth sniffing'. In the course of

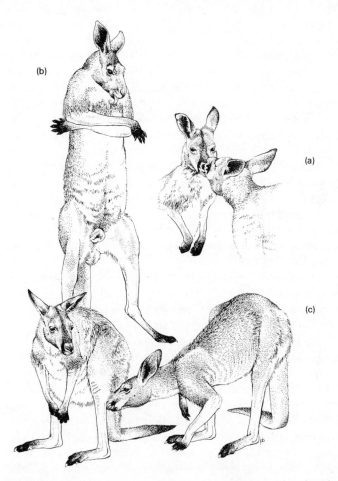

Fig. 3.7 (*a*). Nose-to-nose sniffing between two kangaroos; by far the most
frequent social interaction in marsupials, and a preliminary to many other
interactions. (b) High-standing threat of large adult male kangaroo, *Macropus
rufus*. The head is moved from side to side, licking and biting the fur of the
sternal glandular region, and spreading saliva on the chest. (Based on Sharman
and Calaby (1964).) (c) Sexual checking of a female by a male kangaroo. Adult
males smell the pouch and cloaca of any female they meet, and remain to follow
any female approaching oestrus.

grooming, the fore-paws are licked and rubbed along the head from behind the ears to the tip of the muzzle; I do not consider this significantly different, functionally, from face-washing in other marsupials, although Eisenberg and Golani (1977) listed it as modified. Secretions from the mouth area are therefore spread widely over the head. Allogrooming may serve as a means of passive transfer of odour from one individual to another, especially if it involves licking or biting the fur of another animal. Allogrooming is not frequently seen in the Macropodidae, except between females and their young, and occasionally between adult females (La Follette 1971; Grant 1974; Russell and Giles 1974) and male and female (Stodart 1966b). It is thus in general not a significant mode of odour transfer. However Davies (personal communication) found in *Petrogale puella*, in which males marked the females in their harem by sternal rubbing, that there was also a significant amount of time spent in allogrooming, and he suggests that the transfer of odour from the mouth to the groomed animal adds to the marking of the females. Copious saliva production occurs in hot weather, or when animals become overheated during exercise, or during periods of stress, for example in a red kangaroo introduced into a strange group in captivity. This saliva is rubbed on the fore-limbs (and occasionally pouch and hind-limbs), and may add to the general distribution of odour on the body, but does not appear to give rise to any particular olfactory investigation. Copious salivation and rubbing of saliva on the fore-limbs also occurs in the context of agonistic male–male encounters, along with rubbing and licking of the sternal gland area.

In the larger Macropodidae, the sternal gland of the male is one of the more obvious skin glands producing a secretion which stains the fur of the chest and has a pungent odour (Table 3.5). Behaviour in which these secretions are further spread on the animal's body is part of spectacular stylized displays, but whether it retains any olfactory significance is not known. This behaviour is best described in threat displays of *Macropus rufus* (Sharman and Calaby 1964; Veselovsky 1969; Croft 1981a). In a high-standing threat posture the male stretches up to his fullest extent, resting on the tip of the tail and toes. The fore-limbs are generally extended in front of the body, sometimes crossed, and the head is moved from side to side, licking or biting the fur of the sternal region and shoulders (Fig. 3.7(b)). Sharman and Calaby (1964) gave the impression that biting and licking occur in all such displays, but I have not observed such licking and biting of the sternal region as a frequent component of high-standing threats in *M. rufus* in the wild. Croft (1981a) also observed it rarely, and suggested that it was more likely a conflict activity than an important part of the threat display. Grant (1974) found that the chest and arms were 'sometimes' licked in similar threat displays of *M. giganteus*. A more typical lower intensity threat is a high-standing posture with head thrown well back, one fore-arm stretched out and the other scratching or touching the sternal region. Croft (1981a) observed this in *M. robustus* and *M. fuliginosus*, and Grant (1974) in *M. giganteus*. These

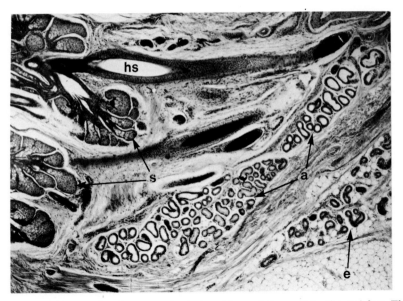

Plate 1.1. The junction between the hairy skin and the pad in the red fox. The skin surface is to the left of the illustration. Hair shafts (hs) and sebaceous (s), apocrine (a) and eccrine (e) glands are visible. Paraffin section. Haematoxylin and eosin. (X 44.)

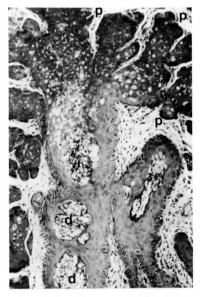

Plate 1.2. Sebaceous gland from the inguinal region of a male rabbit. Cell proliferation occurs around the periphery (p) of the ascini. Differentiation, then degeneration occurs as the cells move towards the ducts (d). Paraffin section. Haematoxylin and eosin. (X 102.)

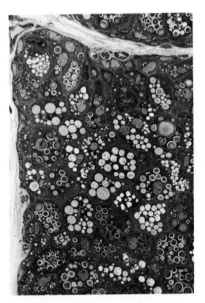

Plate 1.3. As Plate 1.2. Various stages in the differentiation of the cells and the development of their lipid droplets can be seen. Epoxy resin section (1 μm). Toluidine Blue. (X 428.)

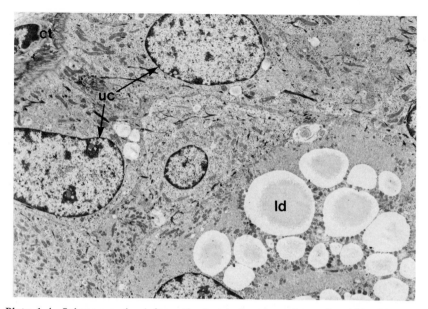

Plate 1.4. Sebaceous gland from the inguinal region of a male rabbit. Electron micrograph showing part of the periphery of an ascinus. Some periglandular connective tissue (ct) is visible in the upper left. Part of a differentiating cell is seen in the lower left; it contains large lipid droplets (ld) surrounded by mitochondria. The remainder of the field is occupied by undifferentiated cells (uc). (× 3600.)

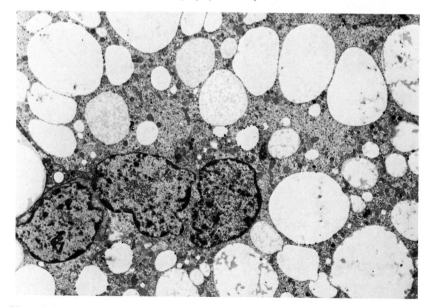

Plate 1.5. As above. Electron micrograph showing degenerating cells. The lipid droplets are beginning to coalesce, the organelles are degenerating, and no cell walls can be seen. (× 3492.)

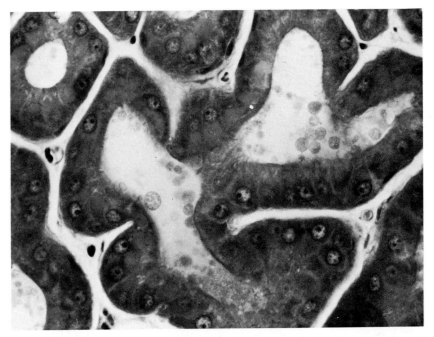

Plate 1.6. Apocrine glands from the inguinal region of a male rabbit. Several tubules seen in section. Fixed in Helly's fluid. Haematoxylin and eosin. (× 540.)

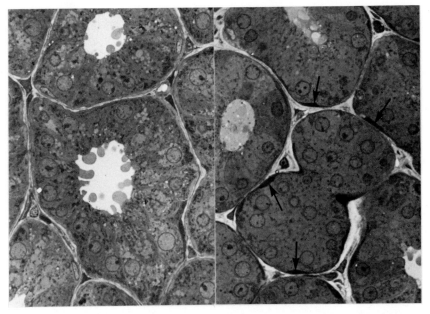

Plate 1.7. As above. Note the orderly arrangement of the nuclei, the secretory granules and the apical 'blebs'. Toluidine Blue. (× 540.)

Plate 1.8. As above. This section shows the myoepithelial cells (arrows) particularly clearly. Epoxy resin section. Toluidine Blue. (× 540.)

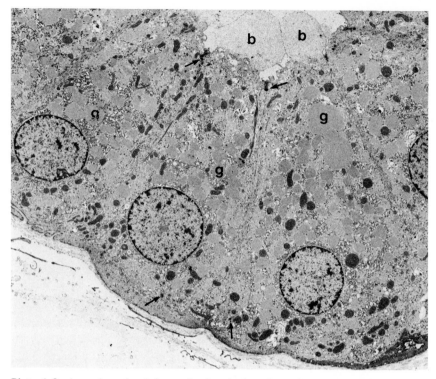

Plate 1.9. Apocrine gland from the inguinal region of a male rabbit. Electron micrograph showing three secretory cells. Note secretory granules (g) and apical blebs (b). Junctional complexes connect the cells at their apical border and desmosomes connect the secretory cells to the myoepithelial cells (arrows) (× 3000.)

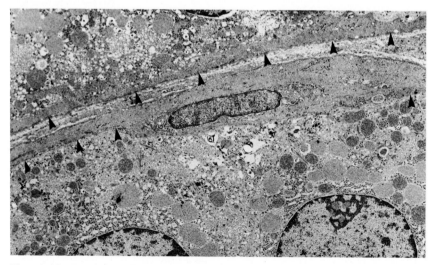

Plate 1.10. As above. The centre of the illustration is occupied by the thickened central part of a myoepithelial cell. The arrows show contractile processes of this cell and another in an adjacent tubule. (× 5175.)

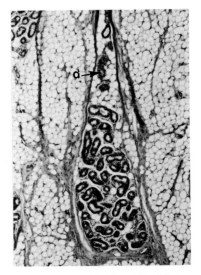

Plate 1.11. An eccrine gland from the foot-pad of a red fox. The main secretory coil is central and the duct (d) can be seen passing towards the skin surface. The gland is surrounded by fatty tissue. Paraffin section. Haematoxylin and eosin. (× 42.)

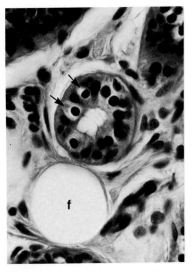

Plate 1.12. As Plate 1.11. Transverse section of a single eccrine tubule showing light and dark cells. In this case the mucus-secreting dark cells (arrows) are lightly stained and the intervening light cells are dark. A large fat cell (f) is also prominent. Paraffin section. Haematoxylin and eosin. (× 410.)

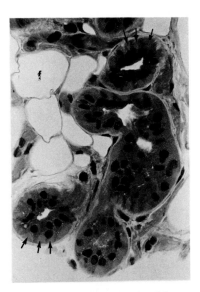

Plate 1.13. An eccrine gland from the foot-pad of a dog. Light and dark cells are not readily distinguishable in this preparation but the distribution of the myo-epithelial cells is particularly clear in cross-section (arrows). Fat cells (f) are present. Methacrylate section. Haematoxylin and eosin. (× 410.)

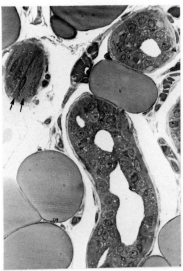

Plate 1.14. As Plate 1.13. Myoepithelial cells are seen in longitudinal section (arrows) in a tangentially sectioned tubule in the upper left. The lipid material has been retained in the fat cells in this preparation. Epoxy resin section. Toluidine Blue. (× 410.)

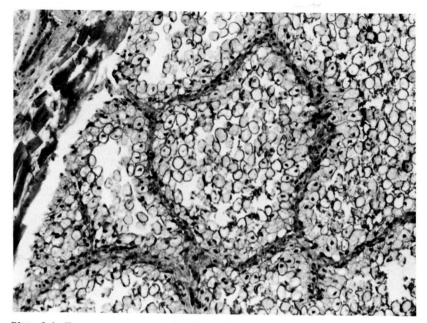

Plate 3.1. Transverse section of cell gland of male brushtail possum *Trichosurus vulpecula*. The muscular wall of the gland is visible at the left. The lumen of the gland is divided by septa carrying secretory cells (nucleated). Cells which have been shed into the lumen of the gland have no nucleus. (Slide and photograph presented by N. T. Allen.)

Plate 3.2. Face-washing in the honey-possum *Tarsipes rostratus*. In most marsupials the relatively unspecialized fore-limb is much used in grooming, rubbed from behind the ear to the tip of the snout, distributing the secretions of the many glandular regions of the head.

Plate 3.3. Cloacal sniffing of oestrous female by male rabbit–eared bandicoot. *Macrotis lagotis*.

Plate 3.4. Grass-pulling display of a male agile wallaby *Macropus agilis*. Similar displays are seen in several of the larger kangaroos. Grass or bushes are picked up and held near the glandular area of the chest. The origin of such display may be in sternal marking.

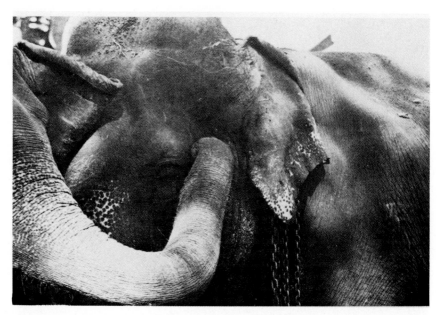

Plate 7.1. A female Asiatic elephant contacting a male's temporal gland with her trunk. (From Eisenberg *et al.* (1971) Plate X.)

Plate 7.2. Two African elephant cows, showing the temporal gland secretion. One cow is touching the other 'to reassure her in a moment of stress'. (From Douglas-Hamilton and Douglas Hamilton (1975).)

Plate 7.3 (a) A male tree hyrax, *Dendrohyrax arboreus* calling exictedly, under circumstances where the white hairs of the dorsal gland (b) are displayed. (Photo courtesy of Dr. H. Hoeck.)

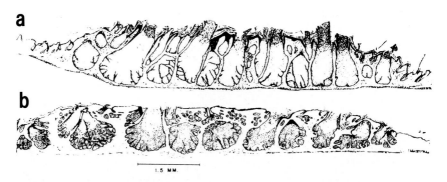

Plate 9.1. Comparative histological structure of the dorsal holocrine skin gland of: (a) an adult male *Dipodomys merriami* from Arizona (parasagittal section) and (b) an adult male *Dipodomys spectabilis* from Arizona (parasagittal section). (From Quay (1954*a*) © Alan R. Liss & Co.)

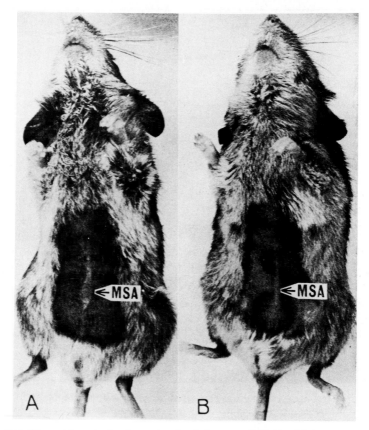

Plate 9.2. The mid-ventral sebaceous glandular area (MSA) of (a) a mature male and (b) a mature female *Peromyscus maniculatus bairdii*. The fur has been shaved to expose the glands. (From Doty and Kart (1972) © American Society of Mammalogists.)

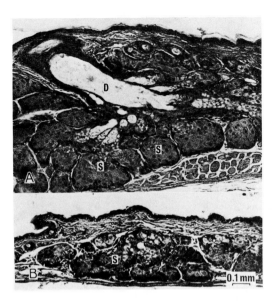

Plate 9.3. Transverse sections of the mid-ventral sebaceous area of (a) a 45-day-old male and (b) a female *Peromyscus maniculatus bairdii* stained with haematoxylin and eosin. S = sebaceous gland complex; D = duct. (From Doty and Kart (1972) © American Society of Mammalogists.)

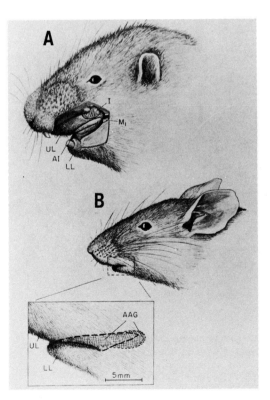

Plate 9.4. Anatomical relations of glands in the oral angle of (a) *Aplodontia rufa* and (b) *Neotoma lepida*. The distal parts of the incisors (AI) and part of the cheek of the *Aplodontia* have been removed in order to show the extent of the upper and lower lips (UL and LL) and the location of the 'insel' (I), a bristle-bearing pad containing enlarged sebaceous glands, and first lower molariform tooth (M_1). The enlarged insert from *Neotoma* shows the inner boundary of the lower lip and angulus oris (dashed line) and the extent of the area of occurrence of apocrine sudoriferous glands (AAG). (From Quay (1965*b*) © American Society of Mammalogists.)

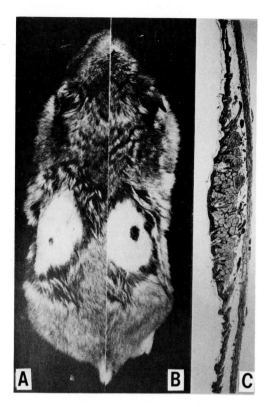

Plate 9.5. The pigmented flank organ (costovertebral scent gland) of (a) a fully mature female and (b) a fully mature male Syrian hamster. (c) a longitudinal section through the centre of a flank organ from a fully mature male hamster showing typical hypertrophied pilosebaceous complexes. (From Algard, Dodge, and Kirkman (1964) © Alan R. Liss & Co.)

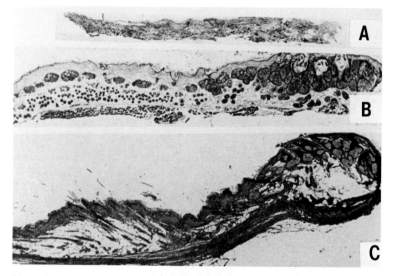

Plate 9.6. A comparison of the flank glands of: (a) a female hamster ovariectomized at weaning and not treated with androgen; (b) a female hamster ovariectomized at weaning and subsequently injected with androgen; and (c) a normal adult male. The right side of each section passes through the pigmented costovertebral spot. The left side shows the adjacent integument. All photographs were taken at a magnification of × 10. (From Hamilton and Montagna (1950) © Alan R Liss & Co.)

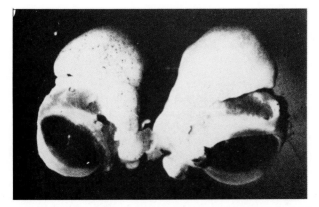

Plate 9.7. Eyes from an adult female (left) and an adult male (right) hamster showing the Harderian glands. (Magnification × 4). (From Christensen and Dam (1952) © Karolinska Institutet.)

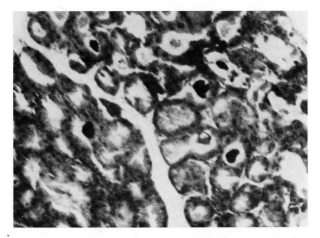

Plate 9.8. A paraffin section through the Harderian gland of a female hamster showing pigment granules. (Haematoxylin & eosin stain; magnification × 102.) (From Christensen and Dam (1952) © Karolinska Institutet.)

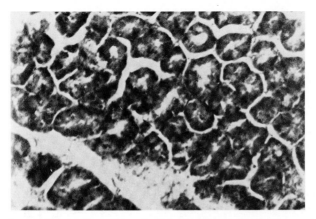

Plate 9.9. A paraffin section through the Harderian gland of a male hamster which has no pigment granules. (Haematoxylin & eosin stain; magnification × 70.) (From Christensen and Dam (1952) © Karolinska Institutet.)

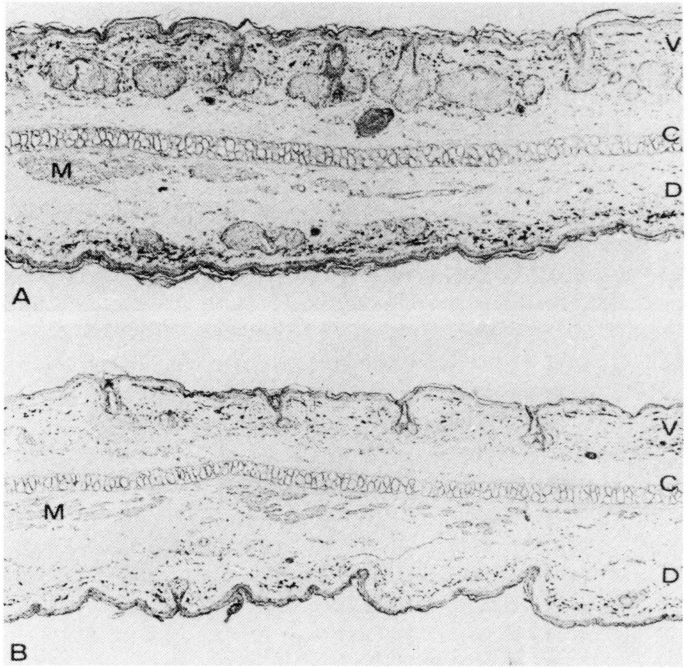

Plate 9.10. The sebaceous glands in the earlobe of (A) a male and (B) a female Syrian hamster. (H & E, × 34.) The males have larger glands on the ventral side than females. V = ventral; D = dorsal; C = cartilage; M = muscle. (From Plewig and Luderschmidt (1977) © Williams and Wilkins Co.)

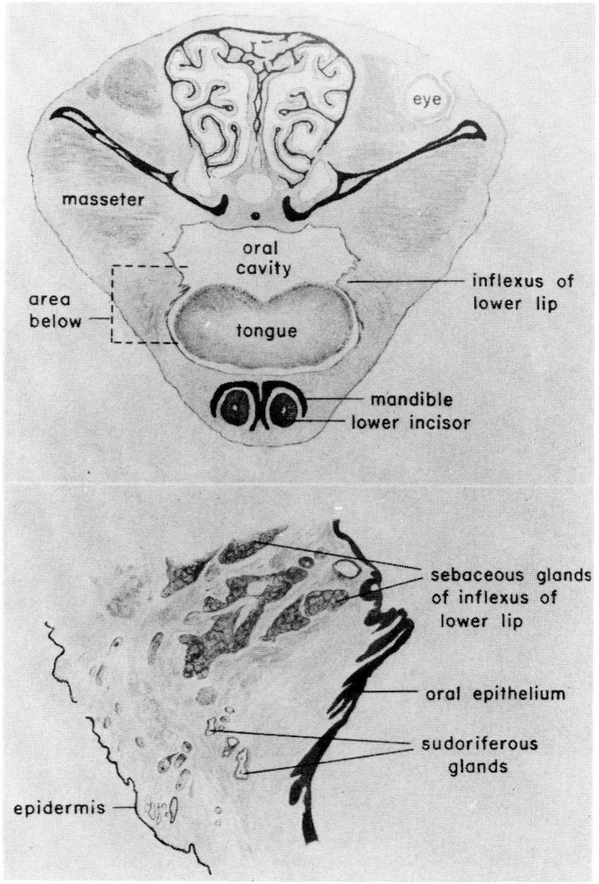

Plate 9.11. The location of the sebaceous and sudoriferous glands of the angulus oris of *Clethrionomys occidentalis* as seen in a cross-section of the head. (From Quay (1962) © American Society of Mammalogists.)

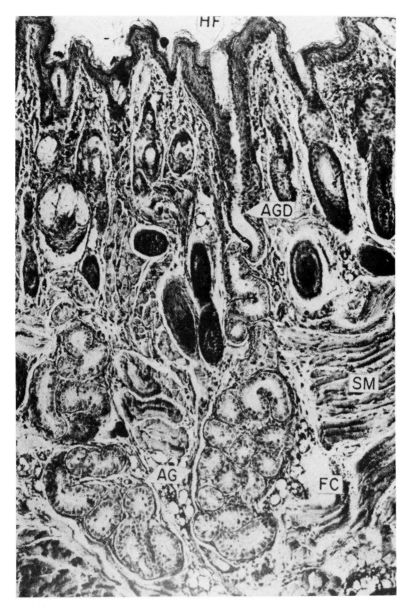

Plate 9.12. A longitudinal section of the two apocrine sudoriferous glands (AG) at the angulus oris of *Microtus californicus* showing the confluence of the duct (AGD) with a hair follicle (HF). FC = fat cells; SM = skeletal muscle fibres. Stained with periodic acid-Schiff and haematoxylin (From Quay (1962) © American Society of Mammalogists.)

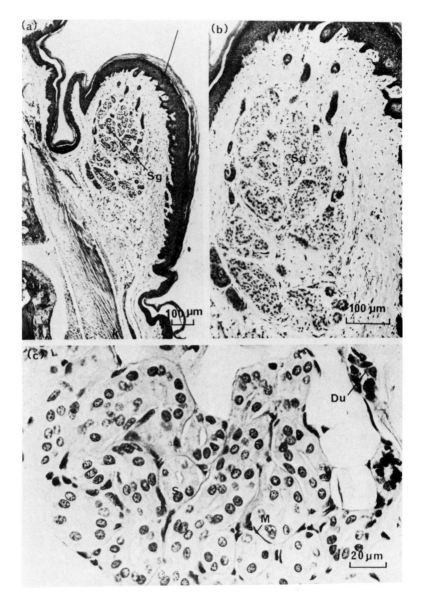

Plate 9.13. (a) (b) Longitudinal section through a foot pad of the bank vole (*Clethrionomys glareolus*) showing the keratinized epithelium indented by dermal papillae. The secretory portions of the sweat glands (Sg) are in the hypodermis surrounded by adipose connective tissue and connected to the surface by long ducts. The arrow indicates the passage of a duct through the stratum corneum. (c) A higher power view of the secretory portion of the eccrine glands showing myoepithelial cells (M), secretory cells (s) and ducts (D). (From Griffiths and Kendal (1980*a*) © The Zoological Society of London.)

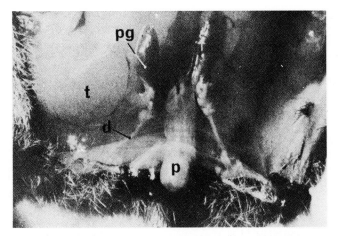

Plate 9.14. The preputial glands of an adult male bank vole, *Clethrionomys glareolus* showing whitish sebum in the ducts. d = duct of preputial gland; p = penis; pg = preputial gland; t = testicle. (From Christiansen, Wiger, and Eilertsen (1978) © The Scandinavian Society Oikos.)

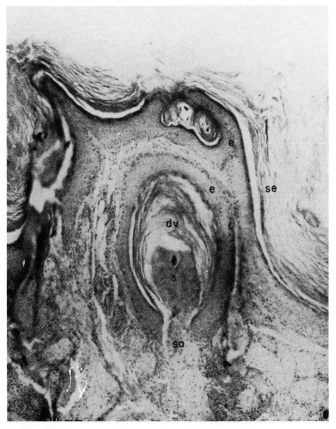

Plate 9.15. The flank organ of an adult male *Arvicola terrestris*. e = epidermis; se = sloughed epidermis; dv = dermal vesicle; s = secretion; sa = sebaceous acini. (From Stoddart (1972) © The Zoological Society of London.)

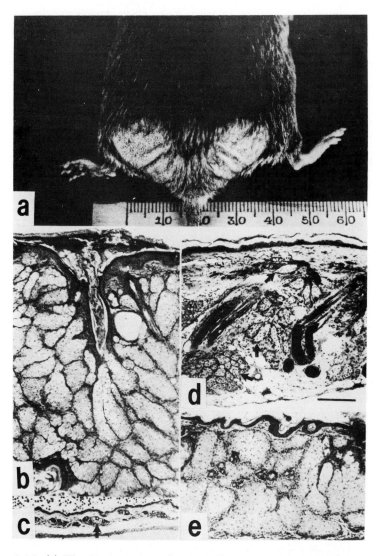

Plate 9.16. (a) The hindquarters of a sexually mature, intact, laboratory bred vole, *Microtus agrestis*, showing the bare sebaceous patches with folded skin on which the openings of large sebaceous ducts can be seen.

Figures b to e are haematoxylin and eosin stained sections of sebaceous patches from voles in different physiological states and are at the same magnification. The scale line in (d) is 200 μm.

(b) Biopsy from the animal in (a) showing sebaceous alveoli, one of which is opening into the large duct running to the skin surface which is filled with sebaceous secretion.

(c) One sebaceous alveolus (arrow) from a field-caught *M. agrestis* during the non-breeding season.

(d) The sebaceous alveoli (arrow) are fewer and smaller eight weeks after castration (same animal as a and b).

(e) Large alveoli of a field-caught *M. agrestis* during the breeding season.
(From Clarke and Frearson (1972) © The Biochemical Society.)

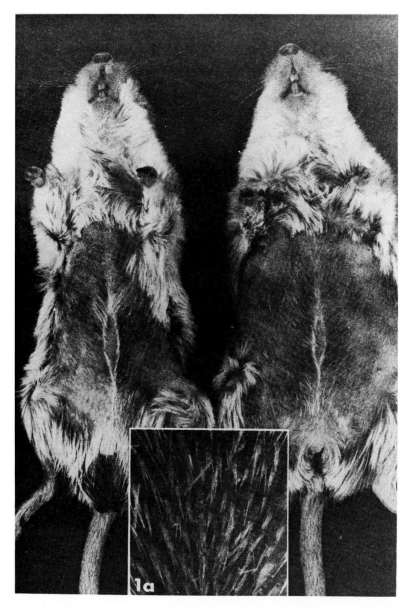

Plate 9.17. The ventral sebaceous gland of a mature male (left) and female (right) Mongolian gerbil (*Meriones unguiculatus*). The fur has been snipped to show the glands. The inset (magnification × 8) shows modified hairs emerging from the gland ducts (From Glenn and Gray (1965) © Williams and Wilkins Co.)

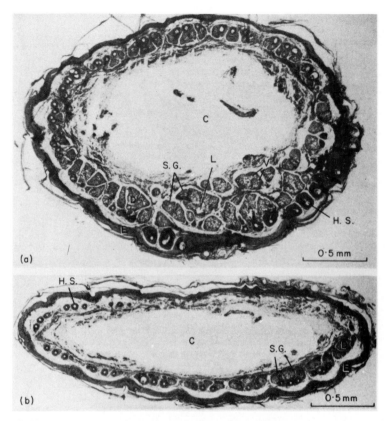

(a)

S.G.

L

C

H. S.

0·5 mm

(b)

H.S.

C

S.G.

0·5 mm

Plate 9.18. Transverse sections through the tail of (a) an adult male and (b) an adult female *Apodemus sylvaticus*, 10 mm from the base. C = position of the caudal vertebrae; D = dermis; E = epidermis; HS = hair shaft; L = lumen; SG = sebaceous glands; V = vacuolated cells. (From Flowerdew (1971) © The Zoological Society of London.)

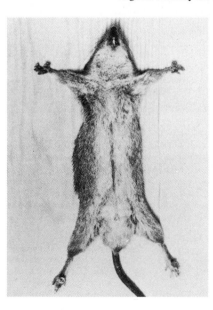

Plate 9.19. An adult male *Rattus exulans* from Hawaii showing the location and extent of the midventral glandular area. (From Quay and Tomich (1963) © The American Society of Mammalogists

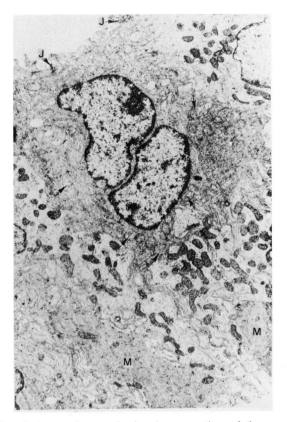

Plate 9.20. An electron micrograph showing a portion of the secretory coil of one plantar sweat gland from a rat (*Rattus norvegicus*). The columnar secretory cells contain either electron-lucent (arrows) or electron-opaque mitochondria. M = myoepithelial cells. J = junctional complexes. (Karnovsky's glutaraldehyde-paraformaldehyde; X 6968). (From Munger and Brusilow (1971) © Alan R Liss, Inc.)

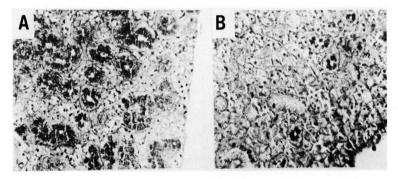

Plate 9.21. Cross-section of the submaxillary glands of (a) an adult male and (b) an adult female mouse. The male gland has a prevalence of terminal tubules in relation to acini, intense acidophilic granulation and large tubule diameter. The female gland has a prevalence of acinar tissue, small tubule size, and acidophilic granulation in peripheral tubules. (Zenker-Mallory, X 64.) (From Harvey (1952) © The University of Chicago Press.)

Plate 11.1. The male capybara has a large gland, the morrillo, on its snout. (Photograph: D. W. Macdonald.)

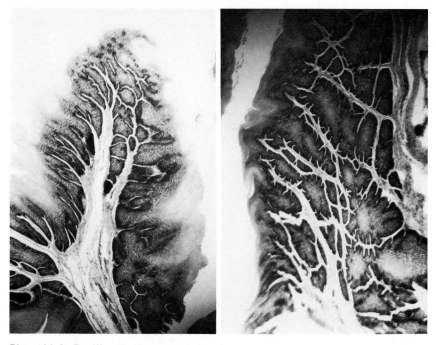

Plate 11.2. Papillae in the anal pouch of the male mara are penetrated by a central canal (*left*) from which a multitude of small channels ramify towards surface pores (*right*). (Photograph: D. W. Macdonald.)

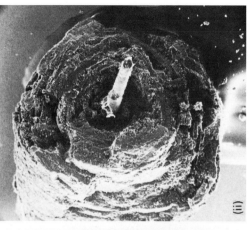

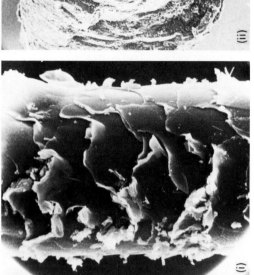

Plate 11.4. (i) SEM of bristle from anal sac of male capybara (cleaned of secretions). (ii) Broken, proximal end of bristle and its coating from anal pocket of a male capybara. Where the material has fractured the layered structure is clearly visible (X 18). Each major annulus is composed of thinner layers. (Photographs: courtesy of K. Kranz.)

Plate 11.3. (i) Male capybara marking an overhanging bough with its morrillo following a mild agonistic encounter in which it displaced the young male to its left. (ii) Large male capybara rubs with his morrillo on the neck and chin of a female. (Photograph: D. W. Macdonald.)

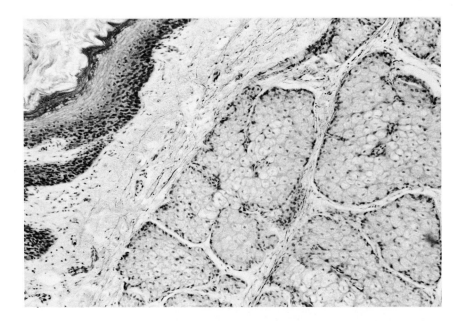

Plate 11.5. Densely packed sebaceous acini underlying the anal pocket of an adult female capybara. (Photograph: D. W. Macdonald.)

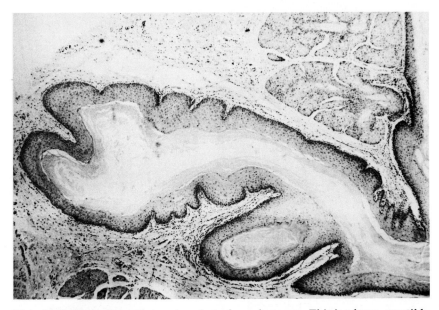

Plate 11.6. Tissue lining the anal pocket of a male copyu. This is a large, evertible structure which lies just below the anus. (Photograph: D. W. Macdonald.)

two displays appear to occur only in the large kangaroos (*M. rufus, M. robustus, M. giganteus, M. fuliginosus*), and were not described by Kaufmann (1974*a*) for *M. parryi* or for any other species of macropodid so far studied. A third threat display which appears to originate in marking behaviour even if it does not now serve that function was called 'grass-pulling' by Kaufmann (1974*a*) who saw it frequently in *M. parryi*, mostly in agonistic situations. The displaying animal vigorously pulled or dug up pieces of turf, while standing crouched. Sometimes small bushes or pieces of bark were so handled, and in extreme cases, the grass was lifted up to the chest, with the head thrown back. I have seen similar behaviour in *M. agilis* (Plate 3.4) and *M. rufus*, not always in the context of threat, but sometimes performed by subordinate males in company with a dominant male and an oestrous female. Croft (1981*a*) described what he terms 'object manipulation' in *M. rufus* and Germann-Meyer (1974) described a male *M. giganteus* clasping a log to its chest and neck. In *M. r. erubenscens*, Croft (1981*b*) described a striking 'bush display' usually performed by a large adult male, which hugged a low shrub to his chest and wiped his neck back and forth across it. He might then pull the bush through under his body and rise to full height on hind-limbs and tail, with one or both fore-arms held above his head. The display occurred most in groups of males with an oestrous female, and 18 of 46 performances were by the consort male. The display was also performed by a male approaching a male already with a female—the displaying male usually supplanted the other male. Response to the display by other animals was no more than brief orientation.

Other examples of behaviour may be more easily interpreted as marking with the sternal gland, including marking of the female by the male. Grant (1974) described as a typical component of the sequence of sexual interaction in *M. giganteus* the male standing in front of the female and rubbing his chest on her head, or grasping her head and forcing it against his chest. Ganslosser (1979) described definite object marking with the sternal gland in the tree kangaroo, *Dendrolagus dorianus*, very different from the various grass-pulling behaviours. The behaviour is seen only in males, which stand over an object on the ground or lie along a thick branch and rub the chest, throat, and chin with strong movements backwards and forwards on the object for up to five minutes (Fig. 3.8). He also described chin-rubbing in the sequence of sexual behaviour in which the male from time to time approached the female obliquely from behind and laid his chin on her back. The clearest example of sternal-gland marking of female by male is in *Petrogale puella* (W. Davies, personal communication) in which males marked the females of their harem with sternal-gland secretion. They approached the female frontally, clasped her around the neck and forced the top of her head and neck into contact with the sternal gland, while moving their body from side to side, up to 20 times in 30 seconds. It seems likely that the various grass-pulling behaviours seen in other species of macropod derive from object-marking or female-marking behaviour.

Fig. 3.8. Sternal and chin marking by the tree kangaroo *Dendrolagus dorianus*. (From photographs by Ulrich Schürer.)

Although many of the Macropodidae are known to possess paracloacal glands, and odour dispersal from this source presumably occurs, active cloacal marking does not appear to be widespread. It has certainly not been observed in any of the larger species which have been studied closely (*Dendrolagus dorianus, Macropus giganteus, M. parryi, M. robustus, M. rufus*). So far the only definite report of active cloacal marking is in the rufous rat kangaroo, *Aepyprymnus rufescens* of the subfamily *Potoroinae*, in which the paracloacal glands appear to differ from those of the Macropodinae (p. 56). Johnson (1980) frequently observed marking of substrate and of introduced objects in an enclosure. In marking, the animal arched its tail, squatted on its hind-feet with the cloacal region on the ground, and then slowly moved off in a shuffling hop. The function of marking is not clear. Males were so intolerant of other males that two could not be kept in the same enclosure. *Aepyprymnus* is not gregarious in the wild, although one male and several females lived together quite amicably in captivity. In this group, most marking was performed by females, which does not suggest that marking has a territorial function. Another rat-kangaroo, *Potorous tridactylus*, maintains large exclusive individual feeding areas which become passively marked with urine and faeces. In laboratory experiments, O. Buchmann (personal communication) has shown that the smell of urine and faeces from a stranger is investigated much more than an animal's own smell. Animals which were subordinate in encounters avoided areas marked with the smell of a dominant animal, with indications of fear. Dominant animals entered areas marked with the smell of a subordinate with evidence of agonistic behaviour. It is not known whether any deliberate marking occurs.

Urine marking is rare; twice in the wild I have seen very large male *M. rufus* who were threatening other large males, standing on tip of tail and toes with fore-arms outstretched and head thrown back, directing a stream of urine at their opponent. Large captive males also do this as part of threat displays towards humans, and W. E. Poole (personal communication) reports that male *M. fuliginosus* spray urine on the ventral surface of the body during aggressive and precopulatory displays.

Olfactory stimuli are clearly important in the recognition of the sexual status of females. For most of the species in Table 3.5, sniffing both of the cloaca and the pouch is reported (Fig. 3.7(c)). Many authors speak of males making a daily check of all females in their vicinity—sniffing the urogenital opening and the pouch opening of all those they encountered and smelling the ground where females had urinated, or placing their nose in the stream of urine (Sharman and Calaby 1964; Grant 1974; Kaufmann 1974a; Ganslosser 1979). Most authors do not remark on the presence or absence of flehmen. Kaufmann (1974a) recorded that he did not see it; Ewer (1968b) quoted Sharman as saying that he had not seen it in *M. rufus*. However, it seems that flehmen does occur, but because it is of short duration and not at all obvious, it has not generally been noticed. Ganslosser (1979) described nasocloacal checking and urine licking in *Dendrolagus dorianus*, in which a male with open mouth and lips drawn back over the teeth stuck his tongue out and licked the stream of urine. This is similar to behaviour described as flehmen in several species of *Macropus* (*agilis, antilopinus, fuliginosus, giganteus, robustus, rufogriseus*, and *rufus*) by Coulson and Croft (1981). Females appeared to urinate in response to cloacal sniffing by the male, who then inserted his nose into the urine stream. As he withdrew his head, he frequently licked his lips and shook his head slightly. Sometimes the lips were retracted, exposing the incisors. In most macropodids with no permanent pair-bond, there is a consort period during which a dominant male stays with a female approaching oestrus. Under such circumstances early detection of oestrus by the male is important. A subordinate male may have a chance to mate if more females are in oestrus together than one male can defend for himself, provided he can locate one quickly.

Increase in levels of testosterone and luteinizing hormone in male *M. eugenii* in the presence of females at the start of the breeding season were demonstrated by Catling and Sutherland (1980). The concentrations rose at least two weeks before mating occurred, and Catling and Sutherland suggested that the most likely stimulus is a chemical stimulus from the female, coinciding with the start of the first oestrous cycle of the breeding season, acting as a 'primer pheromone' in the sense of Whitten and Bronson (1970).

The possible significance of odour in reproductive isolation was suggested by Kirsch and Poole (1972). The closely related eastern and western grey kangaroos (*M. giganteus* and *M. fuliginosus*) overlap in a small part of their range. In captivity, only hybrids from *giganteus* female—*fuliginosus* male crosses are found; the reverse matings do not occur. *Giganteus* males tend to ignore *fuligi-nosus* females in oestrus, while *fuliginosus* males show interest in *giganteus* females even through an intervening fence. Kirsch and Poole suggest that the distinctive strong smell of the cloacal region of the female, peculiar to each species, may indicate an olfactory basis for this unidirectional hybridization. In addition, *M. fuliginosus* males have a very characteristic smell, strongest in the breeding season (W. E. Poole, personal communication), giving rise to their local name of 'Mallee stinkers'. Coulson and Croft (1981) describe the occurrence of flehmen in female *M. fuliginosus* when investigated by males of that species.

Family Petauridae

Little is known of the biology of most of the species of this family which in-cludes the gliding phalangers and the ringtail possums. One species, *Petaurus breviceps*, has been the subject of detailed investigation of olfactory communica-tion in captivity (Schultze-Westrum 1965), but first I shall briefly discuss field studies of two other species of the family, the ringtail possum *Pseudocheirus peregrinus* and the greater glider *Petauroides volans*.

Pseudocheirus (c. 800 g) is herbivorous, nocturnal, arboreal, and gregarious, living in forest or dense scrub habitat. The study by Thomson and Owen (1964) gave some idea of social organization, but the types of social interactions were not described. The behaviour of *Pseudocheirus* was centred upon communal nests or dreys. One or more nests might be shared by an adult male, one or two adult females with dependent young, and the immature offspring of the previous year. Group members were not always found together in the same nest. Nests of one group might be interspersed with nests of another group. Any stranger to the group which attempted to enter an occupied nest was repulsed without fighting. However, it is not very clear how stable these groups were; it is signifi-cant that less than 2 per cent of groups contained more than one adult male. The largest member of the family, *Petauroides*, is a forest-dwelling arboreal folivore (c. 1200 g). It lives solitarily for most of the year, and the pattern of dispersion suggests active repulsion between individuals, especially those of the same sex in overlapping home ranges (Tyndale-Biscoe and Smith 1969). It has been suggested that the basis of the patterns of dispersion of these two species is olfactory. Both species have paracloacal glands, but *Pseudocheirus* is known to have no sternal gland. Thomson and Owen (1964) assumed that the sternal gland in *Trichosurus* was used to mark a defended territory, and that since *Pseudocheirus* had no such gland, either it used the secretion of the paracloacal gland for the same purpose or was not territorial. Tyndale-Biscoe (1973) sug-gested that since *Petauroides* is generally silent, the basis of spacing must be olfactory, since the paracloacal glands are well developed, especially in males. These somewhat sweeping assumptions deserve further investigation in view of the known uncertainties of the function of paracloacal glands in other species (see pp. 56–9).

The complexity which may be found in olfactory communication is shown by the work of Schultze-Westrum (1965) on *Petaurus breviceps.* Recent work by Suckling (1980) on the ecology of *P. breviceps* allows Schultze-Westrum's observations to be related to social organization in the field. This small (140 g) glider is nocturnal, arboreal, and omnivorous, eating some nectar and pollen but mainly insects and the carbohydrate-rich sap of *Eucalyptus* spp. and gum of *Acacia* spp. Some of the population lived solitarily, but most lived socially in groups of up to seven adults (male and female) with dependent and immature young, which shared a communal nest, generally a tree hollow which was changed from time to time. Juveniles were generally expelled from the group at about the time of sexual maturity. Many of them continued to live solitarily, while others came together as the nucleus of a new group. Adult males which disappeared from a group were generally replaced by solitary males from outside the group. Group members shared a common home-range, but did not necessarily move together when feeding. Animals from neighbouring groups had overlapping home-ranges, but certain trees used by the group as sources of sap were defended (Suckling 1980). In captivity there was a clear dominance order among adult males and one or two dominant males performed most of the marking of group members and the territory, territory patrolling and defence, and mating. All group members, not just their mothers, retrieved young displaced from the nest-box (Schultze-Westrum 1965).

In Table 3.6, I have summarized the sources of odours for *P. breviceps*, the ways in which they are distributed, and the context and possible function of the odour deposition. Although Schultze-Westrum described the various glands and methods of marking, he says very little about the response to odour in normal behaviour, despite his many experiments which recorded marking and sniffing in response to odour presented on filter paper. Thus the context and function of the odour as listed in Table 3.6 is probably somewhat speculative. Although the proctodaeal gland was listed as a source of odour by Schultze-Westrum, he gives no further information on the secretion it produces or its function. Nor did he pay much attention to the different secretions of the paracloacal glands which he observed. It is not clear whether both are involved in his tests of the smell of the urogenital area. He does not specify whether one or more pairs of paracloacal glands are present, and it is not clear whether the odour in the urine and in the white oily secretion of the cloacal region come from the same or different glands. He suggested that the white, oily musky secretion has no intraspecific function, but is a defence against predators, since he found that it was not obviously used for marking, but that young and old animals produced big drops of secretion when in danger, when fighting, and when handled. This is in contrast with observations on *Trichosurus vulpecula* and *T. caninus* where it has been suggested that the secretion is not so much a defence against predators as a submissive signal (Winter 1977; Biggins 1979). There is no experimental evidence that the secretion is a defence against predators in *P. breviceps*, and indeed, Schultze-Westrum found that mice smeared with secretion were eaten as usual by the gliders. The function of paracloacal gland secretions will be discussed more fully in discussion of work on *T. vulpecula* (p. 92).

Schultze-Westrum used a technique which involved adaptation to one odour before presentation of another odour to demonstrate that animals could distinguish between frontal, sternal, and cloacal odours of the same individual, between different males on the basis of any one of these odours, and between the pouch odour of different females. Much the same type of marking behaviour occurred in response to odour taken from any region of a stranger, mainly chewing and licking, with foot-marking and flank-rubbing. Schultze-Westrum argued that there is no functional difference between the three important male odours (frontal, sternal, cloacal) and that the important message is the identity of the sender. One may then ask why there is such a complex system of sending the one message? It seems equally probable that different messages are conveyed which were not detected by the experimental technique used, but it is certainly clear that much of the pattern of social behaviour within the group depends on individual recognition which is largely olfactory. Much less is known of *Petaurus australis* and *P. norfolcensis.* Russell (1979) observed the same group of six individuals of *P. australis*, including at least three males, on several occasions in the wild. He described as a common interaction, an initiating animal thrusting its muzzle beneath the recipient's tail, and rubbing the forehead vigorously against the cloaca. Frontal and sternal glands occur in all species of *Petaurus.*

Family Phalangeridae

Two of the three genera of phalangers are very poorly known. The cuscuses, *Phalanger*, are found chiefly in New Guinea, with two Australian species and are very restricted in distribution. They have been much kept in zoos, but apart from the records of sniffing and marking in Eisenberg and Golani's (1977) tables little is known about them, especially in the wild. Even less is known about *Wyulda*, the scaly-tailed possum of rocky habitats in north western Australia. The third genus *Trichosurus* (commonly called possums in Australia) is well known, and *Trichosurus vulpecula* is the only marsupial for which there is good information on social organization and behaviour, including olfactory communication, from field as well as laboratory studies. The trapping studies of Dunnet (1956, 1964) and How (1978) are illuminated by the observation of known animals in the field by Winter (1977), and by Biggins's (1979) investigation of olfactory communication in the laboratory. Much of this work is as yet unpublished but with the permission of the authors, the following account is based on the work of J. Winter and J. G. Biggins.

The brush-tailed possum *T. vulpecula* is a nocturnal, arboreal folivore which weighs 2000–3000 g and lives in a variety of habitats, in particular open eucalypt forest. Studies suggest that the patterns of dispersion found at the lower densities (< 1 animal per ha) of naturally occurring populations in Australia may differ from the patterns found at the high population densities (> 10 per ha) reached by populations in New Zealand, where the species was introduced and has thrived to reach pest status (Dunnett 1956, 1964; How 1972; Crawley 1973; Winter 1977).

An important resource for possums is the den—generally a hollow in a *Eucalyptus* tree, in which they rest during the day and for parts of the night. Most individuals used from one to three dens regularly and 90 per cent of observations were of a single adult in a den. The remaining 10 per cent were mostly females whose young were slow to disperse or females approaching oestrus, with a consort male. Male and female home-ranges were superimposed and independent of each other. Males had larger home-ranges (mean: 3.7 ha) than females (1.7 ha), travelled further each night, and changed dens more frequently. Many females used the same den every night for months. A clear pattern of dominance rank was correlated with age and size in males (and probably in females). The older, dominant, males (> 4 years old) were resident on established home ranges which changed very little, and included a core-area which overlapped little with the core-areas of neighbouring established males. Young males dispersed at 9–16 months and established home-ranges that might completely overlap those of older males, but that tended not to overlap each other. These home-ranges could change and their persistence seemed to depend on the availability of a den not in use by an established male or female. A female's home-range also became more exclusive in relation to other established females as she grew older, and she appeared to be established by the end of her third year. Juvenile females establish home-ranges in, or adjacent to, the home-range of their mother. Thus the exclusive areas of resident, adult males and females of high status were completely overlapped by lower status individuals. One male's range probably overlapped the range of more than one female. A female might be accompanied, for 30–40 days before oestrus, by one, sometimes two consort males, during which time the male might share the female's den. Younger females might not undergo a consort period; in which case mating was often aggressive and attracted several males. Male and female possums, therefore, show a marked tendency to become territorial, in the sense of an exclusive area (Pitelka 1959). Dens, which occur in the core area of the home range are the only areas which are obviously defended from other individuals of the same sex, although a tree with more than one den may house a male and a female.

Winter has suggested that olfactory communication was largely responsible for the maintenance of this pattern of dispersion. The most frequent forms of marking he saw—in 1160 hours of spotlight observations over four years— were chinning and chesting (333 observations). *Chinning* is the depositing of secretions from the mouth area, including saliva and secretions of labial glands and is equivalent to the *labial sliding* of Biggins (1979) and chin and cheek sliding of Wemmer and Collins (1978). The anterior part of the mandible is rubbed on the substrate, with a forward and slightly rotating movement of the head, pushing back the lower lip. The whole side of the head may be in contact with the substrate, with the lip everted (Fig. 3.6(d)). Chinning may vary from a single light wipe to a repeated vigorous rubbing. Biggins also distinguished licking, gnawing, and rubbing the lips, usually of objects or areas previously odour marked, but suggested that they might serve for chemoreception, as well as odour deposition. *Chesting* brings the sternal gland into contact with the substrate by a forward rubbing movement, with the chin stretched forward and clear of the substrate, and the legs bent to bring the chest low to the substrate (Fig. 3.6(e)). To chest the base of a tree, the possum hugged the tree. *Chinning* and *chesting* may occur together, which generally indicates a high level of arousal. Active marking with the white oily secretion of the paracloacal oil gland was never seen in the field and urine-dribbling, presumably distributing the secretion of the paracloacal glands, was seen only 17 times in 1160 hours. However, passive distribution of the secretions of the paracloacal glands occurs whenever the animal urinates or defecates, according to Bolliger and Whitten (1948) and Biggins (1979).

Most marking by the secretions of the mouth and sternal regions was performed by males, in the contexts summarized in Table 3.7. Although more than 50 per cent of marking by males was done when no other possum was nearby, almost all chinning and chesting was performed in areas (trees) with a previous history of use by possums. Chinning was most used in trees, sometimes with chesting, and occurred throughout the night. Chesting was most used to mark the base of trees, and was most frequent at dusk, when the male moved on the ground from its den-tree to a feeding-tree, and at dawn, when it marked its own den-tree on return. Although marking continued all year it was highest during periods of breeding activity. Marking of all kinds (including passive marking by paracloacal glands) was scattered throughout the home-range. There was certainly no boundary marking, but the den-trees used by resident males and females were foci for this activity. Little overt positive or negative response to a scent mark was seen except that males tended to mark where a male had marked before. Scent-marks did not appear to act as a deterrent to other males. Obvious sniffing of an area was mostly performed by males searching for, following or in company with an oestrous female. Males were able to track an out-of-sight oestrous female, and locate which tree she was in by smelling its base, even though she had not actually marked it. Even a consort male could not at first approach a female close enough to sniff at her cloaca, and most checking of females was done by sniffing of branches where a female had sat.

In laboratory experiments, Biggins found that adult male *T. vulpecula* marked a novel environment, chiefly by chinning and chesting. They responded to the odour of conspecifics with olfactory investigation and marking, and could distinguish the odour of familiar and unfamiliar conspecifics; the latter elicited more marking. Differences in the sniffing and marking elicited by different secretions were apparent—sternal and labial gland secretions elicited more marking than paracloacal gland secretions, which were sniffed, but not marked. In encounters between two animals a dominant–subordinate relationship was very rapidly established, with the dominant animal performing most marking, by mouth and sternal secretions, while the subordinate animal marked little, with urine dribbling and cloacal dragging. When a possum was vigorously attacked, it immediately adopted a submissive posture, falling onto its side with all four limbs outstretched and the cloaca covered with the white oil-gland secretion; this posture appeared to appease the attacker. Castration led to a decline in the structural integrity and secretory activity of paracloacal and sternal glands, and castrated males did not scent mark, although their olfactory investigation of conspecific odours was not affected. No such changes occurred in castrates given replacement testosterone. The sternal glands were most rapidly affected and Biggins suggested that different testosterone levels in aggressive dominant males and submissive, subordinate males may produce differences in sternal gland secretions which enable recognition of status. The difference in the type of marking performed by dominant and subordinate males may also enhance the recognition of status.

The level of aggression in encounters was higher with unfamiliar than with familiar opponents, and experiments which gave an unfamiliar male the smell of a familiar male and vice versa demonstrated that the smell of the opponent was important in distinguishing it as familiar or unfamiliar, although visual recognition of familiar animals also occurred. Anosmic animals were not less aggressive to familiar opponents, but anosmic unfamiliar animals were less aggressive to each other than they were when their olfactory systems were intact. Biggins suggested that the presence of unfamiliar odours was important in determining the response to another animal. The success of an animal in encounters was positively correlated with his degree of familiarity with the place in which the encounter took place. A resident animal in a cage in which it had been for long enough for the cage to become thoroughly saturated with its own odours won encounters, even if in previous encounters in a novel environment it had been the loser. However, these experiments did not separate the effect of familiar odour alone from familiarity with the whole environment (including odour).

In Table 3.8, I have summarized information for *T. vulpecula* on the source of odours and their possible significance in communication. The most puzzling source of odours are the paracloacal glands. Passive deposition of secretion with urine and faeces seems the most common mode of distribution, since Winter saw urine dribbling and cloacal dragging so rarely in the field, although it is possible that the occurrence of less vigorous urine dribbling could pass unnoticed. In captivity, urine dribbling, cloacal dragging and paracloacal gland evacuation undoubtedly occur (Wemmer and Collins 1978; Biggins 1979). It seems clear from Bolliger and Whitten (1948) and Biggins (1979) that secretions of oil and cell glands are deposited together in these forms of passive marking. The persistence of the cells in urine of *T. vulpecula* for several weeks (Bolliger and Whitten 1948) led Kean (1967) to suggest that they acted as a slow-release source of odour. This suggestion is supported by the demonstration by Biggins (1979) and Allen (personal communication) that the secretions of the two types of paracloacal gland are very similar. Thus it may be that the combined secretion has a short-term, immediately effective, component from the oil-gland, and a long-term component from the cell-gland, which is gradually released as the cells break down. The use of the same secretion of the oil-gland, apparently as a sign of fear or submission is interesting as an example of dual function of a single secretion. An olfactory rather than a visual signal may be advantageous in a nocturnal animal, effectively swamping any other odours of the producer. The view of Thomson and Pears (1962) and Kean (1967) that the anal-gland secretion serves as the main territorial marker must be modified in the light of this more recent work. It is clearly an important component of the olfactory environment.

Winter suggested that the social function of marking behaviour is to advertise the presence of the marker to rivals for a den or an oestrous female. Scent is distributed, not round the border of the home-range, but in those areas used most frequently, which thus have the densest and freshest scent. Thus a possum travelling through an area has some measure of the probability of meeting another possum, and assuming individual recognition of scents, he can identify its dominance status relative to himself. Dominance is established in actual encounters. Where individuals which may meet are of equal status, mutual avoidance will lead to the development of exclusive areas. When there is a difference in status as between a resident and a young male, avoiding action by the subordinate will allow the two to coexist, with the subordinate using areas and dens not currently in use by the established individual. On the basis of his experimental data, Biggins saw the importance of marking as saturating the home range with familiar odour. In this familiar home ground, the animal is confident and most likely to be dominant in conspecific encounters. Foreign marks and odours which an animal perceives as well as its own are also important. Biggins's experiments suggest that the mutual avoidance which Winter described may be based on animals staying in areas with familiar odours and avoiding areas of strange odours. The exclusive home ranges of resident adult females were maintained without the use of sternal and labial secretions—presumably paracloacal gland secretion was enough.

As far as is known, the scent glands of the congeneric *Trichosurus caninus* are similar to those of *T. vulpecula* (Thomson and Pears 1962), but their importance in social behaviour is not known. *T. caninus* also lives in the eucalypt forests of south-eastern Australia, in tall open and closed forest, where it probably excludes *T. vulpecula*. There is evidence (How 1978) of very close overlap of the home-ranges of the male and female of a pair, which suggests that long-lasting pairs may form and occupy exlusive areas. Juveniles do not disperse until they are two years old, more than a year later than in *T. vulpecula*. A comparison of olfactory communication in the two species would be very interesting.

The grey cuscus *Phalanger gymnotis* which was studied in captivity by Wemmer and Collins (1978) showed no sign of marking by chinning or chesting. The most obvious form of marking was the production of copious white creamy secretion from the paracloacal gland. At the same time the hips were swayed rhythmically from side to side, with the result that a sinuous fine line of secretion was left on the substrate. Cloacal dragging occurred both in this context and after urination. The significance of these behaviours in the social behaviour of *Phalanger* spp. in the wild is unknown.

Olfaction in the mother–young relationship

The importance of olfaction in the relationship between mother and young has not been investigated. What guides the newborn marsupial as it crawls to the pouch from the urogenital opening is not clear. Some authors (Frith and Calaby 1969; Tyndale-Biscoe 1973) have assumed that since structures related to olfaction are the only obviously developed sense organs in the neonate the cues to navigation must be olfactory. The nostrils are open, olfactory epithelium is present, and the olfactory lobes of the brain are present, although there is no cortical development. Other sense receptors are, however, present, such as the free nerve endings in the snout, sensitive to tactile and thermal stimuli (Langworthy 1928; Hill and Hill 1955). There have been many suggestions that the apocrine glands of the pouch serve to guide the newborn young to the pouch. Although Cannon, Bakker, Bradshaw, and McDonald (1976) claim that negative geotropism, perhaps mediated by muscle stretch receptors, is the only directing mechanism, and that young only climb vertically upwards, there are reports of young moving other than vertically (Lyne, Pilton, and Sharman 1959; Sharman and Calaby 1964) and in some species the young's crawl to the pouch is horizontal (Poole 1975). Certainly, olfaction is not ruled out and may be involved in final orientation to the pouch opening and teat-finding, even if it does not direct the earlier journey to the pouch. It has been suggested that the glands of the pouch serve to impart an odour to the young which aids in individual recognition. This is certainly possible, but there is little evidence of individual recognition between mothers and young, in macropodids at least, before the young leaves the pouch. Transfers of young between females have been made successfully, even of young which were beginning to leave the pouch (Reynolds 1952; Merchant and Sharman 1966), and there are reports of young kangaroos getting into the pouch of the wrong mother and being accepted (Maynes 1973; Merchant 1976). I have seen female *M. eugenii* with two large young in the pouch. As I have suggested elsewhere (Russell 1973), the fact that the young is carried in the pouch eliminates the need for early development of specificity of recognition. After the young leaves the pouch, all the other avenues of individual recognition are open to it as well as pouch odour, although it has obviously had an excellent opportunity to learn its mother's odour.

Conclusion

The morphology of the olfactory system in marsupials and eutherians is so similar, even down to the fine details of the olfactory receptors and connections between olfactory bulb and cortex, that it is reasonable to assume that these morphological similarities derived from a common ancestor. Thus it would seem that there was considerable development of the olfactory sense in early mammals.

The marsupials may be considered macrosmatic (Negus 1958), with extensive areas of olfactory epithelium and a well-developed vomeronasal organ. The importance of smell in the behaviour of marsupials is further indicated by two types of evidence discussed in this chapter. First, the diversity of potential scent glands and their distribution throughout almost all marsupial families. Second, the occurrence of behaviour involving these glands, such as olfactory investigation and various marking behaviours. It is unfortunate that it is only possible to describe the context of such behaviour and only for a very few species can we suggest what may be its function in social behaviour. The same stimulus situation may be responded to by more than one kind of marking, and one kind of marking may involve secretions from more than one source. This is especially so in marsupials where both urine and faeces may carry their own smell, together with secretions from genital ducts and one or more pairs of paracloacal glands. We tend to think of scent marking in terms of presence or absence of smell, but in fact the possibilities for variation in the signal are great. In the two species for which most information is available, *Petaurus breviceps* and *Trichosurus vulpecula*, it is clear that messages of several types are conveyed by olfactory signals. Many marsupials live and forage alone or in small groups, patterns of dispersion in which chemical communication allows exchange of information at a distance or *in absentia*. We are still a long way from understanding the subtleties of this complex communication in the same way as we can understand visual displays. Even where we can suggest a function for marking, we can say nothing about motivation. In so many cases, the report of the occurrence of a scent gland or observation of marking states or assumes that the function is in territorial defence. What behavioural evidence there is suggests that this is not so, and that as Eisenberg and Kleiman (1972) suggest, we should think of scent as a means of exchanging information, orienting the movements of individuals, and integrating social and reproductive behaviour. The majority of marsupials are nocturnal, and auditory and olfactory communication may be expected to be more significant than visual communication. Eisenberg *et al.*

(1975) found that marsupial vocalizations exhibit a complexity equivalent to the calls of many eutherian mammals. I would suggest that olfaction is also a highly significant channel of communication.

Acknowledgements

I am most grateful to all those people who gave me access to unpublished material and provided illustrations; in particular: N. T. Allen, J. G. Biggins, O. L. Buchmann, D. B. Croft, P. Cuttle, U. Schürer, G. Suckling, and J. W. Winter. E. R. Hesterman, I. C. R. Rowley and W. F. Whitten criticized the manuscript. The Division of Wildlife Research, CSIRO, provided facilities during the preparation of this chapter; I am particularly grateful to P. de Rebeira who drew Figs. 3.3-3.5 and 3.10, and to the librarian Barbara Staples for her help in locating much of the material for this chapter.

References

Allison, A. C. (1953). The morphology of the olfactory system in vertebrates. *Biol. Rev.* **28**, 195-244.

Altmann, D. (1968). Bemerkungen über ein Wollopossum, *Caluromys laniger* (Desmarest). *Zool. Gart. (NF)* **35**, 22-9.

Archer, M. (1974). Some aspects of reproductive behaviour and the male erectile organs of *Dasyurus geoffroii*, and *D. hallucatus* (Dasyuridae: Marsupialia). *Mem. Qd Mus.* **17**, 63-7.

Aslin, H. (1974). The behaviour of *Dasyuroides byrnei* (Marsupialia) in captivity. *Z. Tierpsychol.* **35**, 187-208.

Autrum, H., Fillies, K., and Wagner, H. (1970). Squalen im Sekret der Analbeutel von *Petaurus breviceps papuanus* Th. (Phalangeridae, Marsupialia). *Biol. Zbl.* **89**, 681-94.

Barnes, R. D. (1977). The special anatomy of *Marmosa robinsoni*. In *The biology of marsupials* (ed. D. Hunsaker). Academic Press, New York.

Barone, R. and Blavignac, B. (1965). Observations sur les glandes de Cowper et les glandes anales de *Protemnodon rufogriseus*. *Mammalia* **29**, 613-14.

Beddard, F. E. (1887). Note on a point in the structure of *Myrmecobius*. *Proc. zool. Soc. Lond.* **1887**, 527-31.

— (1888). Note on the sternal gland of *Didelphys dimidiata*. *Proc. zool. Soc. Lond.* **1888**, 353-5.

Biggins, J. G. (1979). Olfactory communication in the brush-tailed possum, *Trichosurus vulpecula* Kerr, 1792 (Marsupialia: Phalangeridae). Unpublished Ph.D. thesis, Monash University, Victoria, Australia.

Bolliger, A. and Hardy, M. H. (1945). The sternal integument of *Trichosurus vulpecula*. *Proc. R. Soc. NSW* **78**, 122-33.

— and Whitten, W. K. (1948). The paracloacal (anal) glands of *Trichosurus vulpecula*. *Proc. R. Soc. NSW* **82**, 36-43.

Braithwaite, R. W. (1974). Behavioural changes associated with the population cycle of *Antechinus stuartii*. (Marsupialia). *Aust. J. Zool.* **22**, 45-62.

Broom, R. (1896). The comparative anatomy of the organ of Jacobson in marsupials. *Proc. Linn. Soc. NSW* **21**, 591-623.

Buchmann, D. K. L. and Guiler, E. R. (1977). Behaviour and ecology of the Tasmanian devil *Sarcophilus harrisii*. In *The biology of marsupials* (ed. B. Stonehouse and D. Gilmore). Macmillan, London.

Burne, B. H. (1909). A gland upon the ear conch of *Dasyurus maugei*. *J. Anat. Physiol.* **43**, 312–13.

Cannon, J. R., Bakker, H. R., Bradshaw, S. D., and McDonald, I. R. (1976). Gravity as the sole navigational aid to the newborn quokka. *Nature, Lond.* **259**, 42.

Catling, P. C. and Sutherland, R. L. (1980). Effect of gonadectomy, season and the presence of female tammar wallabies (*Macropus eugenii*) on concentrations of testosterone, luteinizing hormone and follicle-stimulating hormone in the plasma of male tammar wallabies. *J. Endocr.* **86**, 25–33.

Christensen, P. E. S. (1980). The biology of *Bettongia penicillata* Gray, 1837, and *Macropus eugenii* Desmarest, 1804, in relation to fire. *Bulletin* 91. Forests Department of Western Australia, Perth.

Coulson, G. M. C. and Croft, D. B. (1981). Flehmen in kangaroos. *Aust. Mammal.* **4**, 139–40.

Crawley, M. C. (1973). A live-trapping study of Australian brush-tailed possum, *Trichosurus vulpecula* (Kerr), in the Orongorongo Valley, Wellington, New Zealand. *Aust. J. Zool.* **21**, 75–90.

Croft, D. R. (1982). Communication in the Pasyuridae (Marsupialia): a review. In *Carnivorous marsupials* Vol. I (ed. M. Archer). Royal Zoological Society of NSW, Sydney.

— (1981*a*). Behaviour of red kangaroos, *Macropus rufus* (Desmarest, 1822), in north western New South Wales. *Aust. Mammal.* **4**, 5–58.

— (1981*b*). Social behaviour of the euro, *Marcropus robustus* (Gould), in the Australian arid zone. *Aust. wildl. Res.* **8**, 13–49.

Cuttle, P. (1978). The behaviour in captivity of the dasyurid marsupial, *Phascogale tapoatafa* (Meyer). Unpublished M.Sc. thesis, Monash University, Victoria, Australia.

Davies, W. (1979). Social organization of the rock wallaby (*Petrogale puella*) in North Queensland. *Bull. Aust. mamm. Soc.* **6** (1), 18.

Dunnet, G. M. (1956). A live-trapping study of the brush-tailed possum *Trichosurus vulpecula*. *CSIRO wildl. Res.* **1**, 1–18.

— (1964). A field study of the local populations of the brush-tailed possum *Trichosurus vulpecula* in eastern Australia. *Proc. zool. Soc. Lond.* **142**, 665–95.

Eberhard, I. H. (1978). Ecology of the koala *Phascolarctos cinereus* (Goldfuss) Marsupialia: Phascolarctidae in Australia. In *The ecology of arboreal folivores* (ed. G. G. Montgomery). *Proc. Nat. Zool. Park Symp.* No. 1. Smithsonian Institute Press, Washington, DC.

Eisenberg, J. F., Collins, L. R., and Wemmer, C. (1975). Communication in the Tasmanian devil (*Sarcophilus harrisii*) and a survey of auditory communication in the Marsupialia. *Z. Tierpsychol.* **37**, 379–99.

— and Golani, I. (1977). Communication in Metatheria. In *How animals communicate* (ed. T. Sebeok). Indiana University Press, Bloomington.

— and Kleiman, D. G. (1972). Olfactory communication in mammals. *A. Rev. Ecol. Syst.* **3**, 1–32.

Estes, R. D. (1972). The role of the vomeronasal organ in mammalian reproduction. *Mammalia* **36**, 315–41.

Ewer, R. F. (1968*a*). A preliminary survey of the behaviour in captivity of the dasyurid marsupial *Sminthopsis crassicaudata* (Gould). *Z. Tierpsychol.* 25, 319–65.

— (1968*b*). *Ethology of mammals.* Logos Press, London.

Fleay, D. (1952). The Tasmanian or marsupial devil. Its habits and family life. *Aust. Mus. Mag.* 10, 275–80.

Flynn, T. T. (1910). Contribution to a knowledge of the anatomy and development of the Marsupialia. I. The genitalia of *Sarcophilus satanicus* (♀). *Proc. Linn. Soc. NSW* 35, 873–87.

Forbes, D. K. and Tribe, D. E. (1969). Salivary glands of kangaroos. *Aust. J. Zool.* 17, 765–75.

Ford, E. (1934). A note on the sternal gland of *Myrmecobius. J. Anat.* 68, 346–9.

Fortney, J. A. (1973). Cytology of eccrine sweat glands in the opossum. *Am. J. Anat.* 136, 205–20.

Frith, H. J. and Calaby, J. H. (1969). *Kangaroos.* Cheshire, Melbourne.

Ganslosser, U. (1979). Soziale Interaktionen der Doriabaumkänguruhs (*Dendrolagus dorianus* Ramsay 1883). (Marsupialia, Macropodidae). *Z. Säugetierk.* 44, 1–18.

Gaughwin, M. D. (1979*a*). The occurrence of flehmen in a marsupial—the hairy-nosed wombat (*Lasiorhinus latifrons*). *Anim. Behav.* 27, 1063–5.

— (1979*b*). Ethology and social structure of the hairy-nosed wombat (*Lasiorhinus latifrons*). (Abstr.) *Bull. Aust. mammal. Soc.* 5(1), 289.

Germann-Meyer, U. (1974). *Verhaltens-studien am grauen Riesenkänguruh*, Macropus (Macropus) giganteus. Juris, Zürich.

Gordon, G. (1974). Movements and activity of the short-nosed bandicoot *Isoodon macrourus* Gould (Marsupialia). *Mammalia* 38, 405–31.

Grant, T. R. (1974). Observations of enclosed and free-ranging grey kangaroos *Macropus giganteus. Z. Säugetierk.* 39, 65–78.

Graziadei, P. P. C. (1971). The olfactory mucosa of vertebrates. In *Handbook of sensory physiology*, Vol. IV, Pt. 1, (ed. L. M. Beidler), pp. 27–58. Springer, Berlin.

— (1976). Functional anatomy of the mammalian chemoreceptor system. In *Chemical signals in vertebrates* (ed. D. Müller-Schwarze and M. M. Mozell). Plenum Press, New York.

Green, L. M. (1963). Distribution and comparative histology of cutaneous glands in certain marsupials. *Aust. J. Zool.* 11, 250–72.

Guiller, E. R. (1970). Observations on the Tasmanian devil, *Sarcophilus harrisii* (Marsupialia: Dasyuridae) I. Numbers, home range, movements and food in two populations. *Aust. J. Zool.* 18, 49–62.

Hediger, H. (1958). Verhalten der Beuteltiere (Marsupialia). *Handb. Zool.* 8:18, 10(9), 1–28.

Heinsohn, G. E. (1966). Ecology and reproduction of the Tasmanian bandicoots (*Perameles gunnii* and *Isoodon obesulus*). *Univ. Calif. Publs Zool.* 80, 1–96.

Hill, J. P. (1900), Contributions to the morphology and development of the female urogenital organs in the Marsupialia. II–V. *Proc. Linn. Soc. NSW* 25, 519–32.

— and Hill, W. C. O. (1955). The growth stages of the native cat (*Dasyurus viverrinus*) together with observations on the anatomy of the newborn young. *Trans. zool. Soc. Lond.* 28, 349–452.

How, R. A. (1978). Population strategies in four species of Australian possums. In *The ecology of arboreal folivores* (ed. G. Montgomery). *Proc. Nat. Zool. Park. Symp.* No. 1. Smithsonian Press, Washington, DC.

Hulbert, A. J. (1972). Growth and development of pouch-young in the rabbit-eared bandicoot, *Macrotis lagotis* (Peramelidae). *Aust. Mammal.* 1, 38–9.

Hunsaker, D. and Shupe, D. (1977). Behaviour of new world marsupials. In *The biology of marsupials* (ed. D. Hunsaker). Academic Press, New York.

Johnson, J. I. (1977). Central nervous system of marsupials. In *The biology of marsupials* (ed. D. Hunsaker). Academic Press, New York.

Johnson, K. A. (1977). Ecology and management of the red-necked pademelon, *Thylogale thetis*, on the Dorrigo Plateau of northern N.S.W. Unpublished Ph.D. thesis, University of New England, Armidale, Australia.

Johnson, P. M. (1980). Observations of the behaviour of the rufous rat-kangaroo, *Aepyprymnus rufescens* (Gray), in captivity. *Aust. wildl. Res.* 7, 347–57.

Kaneko, N., Debski, E. A., Wilson, M. C., and Whitten, W. K. (1980). Puberty acceleration in mice II. Evidence that the vomeronasal organ is a receptor for the primer pheromone in male mouse urine. *Biol. Reprod.* 22, 873–8.

Kaufmann, J. H. (1974a). Social ethology of the whiptail wallaby, *Macropus parryi*, in northeastern New South Wales. *Anim. Behav.* 22, 281–369.

— (1974b). The ecology and evolution of social organization in the kangaroo family (Macropodidae). *Am. Zool.* 14, 51–62.

— (1975). Field observations of the social behaviour of the eastern grey kangaroo, *Macropus giganteus. Anim. Behav.* 23, 214–21.

Kean, R. I. (1967). Behaviour and territorialism in *Trichosurus vulpecula* (Marsupialia). *Proc. NZ ecol. Soc.* 18, 42–7.

Kirsch, J. A. W. (1977a). The classification of marsupials. In *The biology of marsupials* (ed. D. Hunsaker). Academic Press, New York.

— (1977b). Notes on nomenclature. In *The biology of marsupials* (ed. D. Hunsaker). Academic Press, New York.

— and Calaby, J. H. (1977). The species of living marsupials: an annotated list. In *The biology of marsupials* (ed. B. Stonehouse and D. P. Gilmore). Macmillan, London.

— and Poole, W. E. (1972). Taxonomy and distribution of the grey kangaroos, *Macropus giganteus* Shaw and *Macropus fuliginosus* (Desmarest), and their subspecies (Marsupialia: Macropodidae). *Aust. J. Zool.* 20, 315–39.

Kitchener, D. J. (1973). Notes on home range and movement in two small macropods, the potoroo (*Potorous apicalis*) and the quokka (*Setonix brachyurus*). *Mammalia* 37, 231–40.

Kratzing, J. E. (1978). The olfactory apparatus of the bandicoot (*Isoodon macrourus*): fine structure and the presence of a septal olfactory organ. *J. Anat.* 125, 601–13.

La Follette, R. M. (1971). Agonistic behaviour and dominance in confined wallabies, *Wallabia rufogriseus frutica. Anim. Behav.* 19, 93–101.

Langworthy, O. R. (1928). The behaviour of pouch-young opossums correlated with the myelinization of tracts in the nervous system. *J. comp. Neurol.* 46, 201–40.

Lyne, A. G., Pilton, P. E., and Sharman, G. B. (1959). Oestrous cycle, gestation period and parturition in the marsupial *Trichosurus vulpecula. Nature, Lond.* 183, 622–3.

McCotter, R. (1912). The connection of the vomeronasal nerves with the accessory olfactory bulb in the opossum and other mammals. *Anat. Rec.* 6, 299–318.

McIlroy, J. C. (1973). Aspects of the ecology of the common wombat *Vombatus ursinus* (Shaw 1800). Unpublished Ph.D. thesis, Australian National University, Canberra, Australia.

McKay, G. M. (1974). *Planigale ingrami* in captivity. *Koolewong* 3, 4–5.

McManus, J. J. (1970). Behaviour of captive opossums, *Didelphis marsupialis virginiana. Am. Midl. Nat.* 84, 144–69.

Martin, W. (1836). Notes on the anatomy of the koala. *Proc. zool. Soc. Lond.* 109–13.

Masera, R. (1943). Sul'esistenza di un particolare organo olfativo nel setto nasale della cavia e di altri roditori. *Archo ital. Anat. Embriol.* 48, 157–212.

Maynes, G. (1973). Reproduction in the parma wallaby, *Macropus parma* Waterhouse. *Aust. J. Zool.* 21, 331–51.

— (1977). Distribution and aspects of the biology of the parma wallaby, *Macropus parma*, in New South Wales. *Aust. wildl. Res.* 4, 109–25.

Merchant, J. C. (1976). Breeding biology of the agile wallaby, *Macropus agilis* (Gould) (Marsupialia: Macropodidae), in captivity. *Aust. wildl. Res.* 3, 93–103.

— and Sharman, G. B. (1966). Observations on the attachment of marsupial pouch-young to the teats and on the rearing of pouch-young by foster-mothers of the same or different species. *Aust. J. Zool.* 14, 593–609.

Morton, S. R. (1978). An ecological study of *Smithopsis crassicaudata* (Marsupialia: Dasyuridae). II. Behaviour and social organization. *Aust. wild. Res.* 5, 163–82.

— and Burton, T. C. (1973). Observations on the behaviour of the macropodid marsupial *Thylogale billardieri* (Desmarest) in captivity. *Aust. Zool.* 18, 1–14.

Moulton, D. G. and Beidler, L. M. (1967). Structure and function in the peripheral olfactory system. *Physiol. Rev.* 47, 1–52.

Munhoz, C. O. G. and Merzel, J. (1967). Morphologic and histochemical study on perianal glands of the opossum (*Didelphis aurita*). *Acta anat.* 68, 258–71.

Mykytowycz, R. and Nay, T. (1964). Studies of the cutaneous glands and hair follicles of some species of Macropodidae. *CSIRO wildl. Res.* 9, 200–17.

Negus, V. (1958). *The comparative anatomy and physiology of the nose and paranasal sinuses.* Livingstone, Edinburgh.

Nicholls, E. M. and Rienits, K. G. (1971). Tryptophan derivatives and pigment in the hair of some Australian marsupials. *Int. J. Biochem.* 2, 593–603.

Ortmann, R. (1960). Die Analregion der Säugetiere. *Handb. Zool.* 8:26, 3(7), 1–68.

Packer, W. C. (1969). Observations on the behaviour of the marsupial *Setonix brachyurus* (Quoy and Gaimard) in an enclosure. *J. Mammal.* 50, 8–20.

Parsons, T. S. (1971). Anatomy of nasal structures from a comparative viewpoint. In *Handbook of sensory physiology* Vol. IV, Pt. 1, (ed. L. M. Beidler), pp. 1–26. Springer, Berlin.

Phillips, D. S. and Michels, K. M. (1964). Selective stimulation and electrophysiological responses of the olfactory bulb of the opossum. *Percept. Mot. Skills* 8, 63–9.

Pitelka, F. A. (1959). Numbers, breeding schedule, and territoriality in pectoral sandpipers of northern Alaska. *Condor* 61, 233–64.

Poole, W. E. (1975). Reproduction in two species of grey kangaroos, *Macropus giganteus* Shaw and *M. fuliginosus* (Desmarest). II. Gestation, parturition and pouch life. *Aust. J. Zool.* 23, 333–53.

Powers, J. B. and Winans, S. S. (1975). Vomeronasal organ: critical role in mediating sexual behavior of the male hamster. *Science, NY* 187, 961–3.

Reynolds, H. C. (1952). Studies on reproduction in the opossum (*Didelphis virginiana virginiana*). *Univ. Calif. Publs. Zool.* 52, 223–84.

Rigby, R. G. (1972). A study of the behaviour of caged *Antechinus stuartii*. *Z. Tierpsychol.* **31**, 15–25.

Robbins, M. and Russell, E. M. (1978). Observation on movements and feeding activity of the koala in a semi-natural situation. In *The koala: proceedings of the Taronga symposium on koala biology, management and medicine* (ed. T. J. Bergin). Zoological Parks Board of NSW, Sydney.

Robinson, P. T. (1978). Koala management and medicine at the San Diego Zoo. In *The koala: proceedings of the Taronga symposium on koala biology, management and medicine* (ed. T. J. Bergin). Zoological Parks Board of NSW, Sydney.

Röse, C. (1893). Über das Jacobson-Organ von Wombat und Opossum. *Anat. Anz.* **8**, 766–8.

Rotenberg, D. (1928). Notes on the male generative apparatus of *Tarsipes spenserae*. *J. Proc. R. Soc. West. Aust.* **15**, 9–17.

Russell, E. M. (1970). Observations on the behaviour of the red kangaroo (*Megaleia rufa*) in captivity. *Z. Tierpsychol.* **27**, 385–404.

— (1973). Mother–young relations and early behavioural development in the marsupials *Macropus eugenii* and *Megaleia rufa*. *Z. Tierpsychol.* **33**, 163–203.

— (1974). The biology of kangaroos (Marsupialia, Macropodidae). *Mammal. Rev.* **4**, 1–59.

— and Giles, D. C. (1974). Allogrooming in *Macropus eugenii*. *Aust. Mammal.* **1**, 387.

Russell, R. (1979). Interactions between various individuals of a fluffy glider (*Petaurus australis*) colony on the Atherton Tableland. *N. Qd Nat.* **45**, 4–5.

Scalia, F. and Winans, S. S. (1975). The differential projections of the olfactory bulb and accessory olfactory bulb in mammals. *J. comp. Neurol.* **161**, 31–57.

— — (1976). New perspectives on the morphology of the olfactory system: olfactory and vomeronasal pathways in mammals. In *Mammalian olfaction, reproductive processes, and behavior* (ed. R. L. Doty). Academic Press, New York.

Schaffer, J. (1940). *Die Hautdrüsenorgane der Säugetiere*. Urban and Schwarzenberg, Berlin.

— and Hamperl, H. (1926). Über Anal-und Circumanaldrüsen. 3. Mitteilung: Marsupialier. *Z. wiss. Zool.* **127**, 527–89.

Schultze-Westrum, T. (1965). Innerartliche Verstandigung durch düfte beim Gleitbeutler *Petaurus breviceps papuanus* Thomas (Marsupialia: Phalangeridae). *Z. vergl. Physiol.* **50**, 151–220.

Sharman, G. B. and Calaby, J. H. (1964). Reproductive behaviour in the red kangaroo, *Megaleia rufa*, in captivity. *CSIRO wildl. Res.* **9**, 58–85.

Smith, M. (1980a). Behaviour of the koala, *Phascolarctos cinereus* Goldfuss, in captivity. III. Vocalizations. *Aust. wildl. Res.* **7**, 13–34.

— (1980b). Behaviour of the koala *Phascolarctos cinereus* Goldfuss, in captivity. IV. Scent-marking. *Aust. wildl. Res.* **7**, 35–40.

Sonntag, C. F. (1921). The comparative anatomy of the Koala (*Phascolarctos cinereus*) and the vulpine phalanger (*Trichosurus vulpecula*). *Proc. zool. Soc. Lond.* **1921**, 547–77.

Sorenson, M. W. (1970). Observations on the behaviour of *Dasycercus cristicauda* and *Dasyuroides byrnei* in captivity. *J. Mammal.* **51**, 123–31.

Steinbrecht, R. A. (1969). Comparative morphology of olfactory receptors. In *Olfaction and taste III* (ed. C. Pfaffmann), pp. 3–21. Rockefeller University Press, New York.

Stodart, E. (1966*a*). Management and behaviour of breeding groups of the marsupial *Perameles nasuta* Geoffroy in captivity. *Aust. J. Zool.* 14, 611-23.
— (1966*b*). Observations on the behaviour of the marsupial *Bettongia lesueuri* (Quoy and Gaimard) in an enclosure. *CSIRO wildl. Res.* 11, 91-9.
Stoddart, D. M. (1980). Observations on the structure and function of cephalic skin glands in bandicoots (Marsupialia: Paramelidae). *Aust. J. Zool.* 28, 33-41.
Suckling, G. C. (1980). The effects of fragmentation and disturbance of forest on mammals in a region of Gippsland, Victoria. Unpublished Ph.D. thesis, Monash University, Victoria, Australia.
Sweet, G. (1907). The skin, hair and reproductive organs of *Notoryctes*. *Q. Jl. microsc. Sci.* 51, 325-44.
Thomson, J. A. and Owen, W. H. (1964). A field study of the Australian ringtail possum *Pseudocheirus peregrinus* (Marsupialia: Phalangeridae). *Ecol. Monogr.* 34, 27-52.
— Pears, F. N. (1962). The functions of the anal glands of the brushtail possum. *Vict. Nat.* 78, 306-8.
Tyndale-Biscoe, C. H. (1973). *Life of marsupials.* Edward Arnold, London.
— and Smith, R. F. C. (1969). Studies on the marsupial glider, *Schoinobates volans* (Kerr). II. Population structure and regulatory mechanisms. *J. anim. Ecol.* 38, 637-50.
Van den Broek, A. J. P. (1904). Über Rectaldrüsen weiblicher Beuteltiere. *Petrus Camper ned. Bijdr. Anat.* 2, 328-49.
— (1910). Untersuchungen über den Bau der mannlichen Geschlechtsorgane der Beuteltiere. *Gegenbaurs morph. Jb.* 41, 347-436.
Van Dyck, S. (1979). Behaviour in captive individuals of the dasyurid marsupial *Planigale maculata* (Gould 1851). *Mem. Qd Mus.* 19, 413-31.
Veselovsky, Z. (1969). Beitrag zur Kenntnis des Fortpflanzungsverhaltens der Känguruhs. *Zool. Garten (NF)* 37, 93-107.
Waite, E. R. (1894). Observations on *Dendrolagus bennettianus* De Vis. *Proc. Linn. Soc. NSW* 9, 571-82.
Watts, C. H. S. (1969). Distribution and habits of the rabbit-eared bandicoot. *Trans. R. Soc. S. Aust.* 93, 135-41.
Wells, R. T. (1978). Field observations of the hairy-nosed wombat *Lasiorhinus latifrons* (Owen). *Aust. wildl. Res.* 5, 299-303.
Wemmer, C. and Collins, L. (1978). Communication patterns in two phalangerid marsupials, the gray cuscus (*Phalanger gymnotis*) and the brush possum (*Trichosurus vulpecula*). *Säugetier. Mitt.* 26, 161-72.
Whitten, W. K. and Bronson, F. H. (1970). The role of pheromones in mammalian reproduction. In *Advances in chemoreception*, Vol. I—*Communication by chemical signals* (ed. J. W. Johnston, D. G. Moulton, and A. Turk), pp. 309-26. Appleton-Century-Crofts, New York.
Winter, J. W. (1977). The behaviour and social organization of the brush-tail possum (*Trichosurus vulpecula*: Kerr). Unpublished Ph.D. thesis, University of Queensland, Australia.
Wood-Jones, F. (1923). *The mammals of South Australia*, Part I. Government Printer, Adelaide.
— (1949). The study of a generalized marsupial *Dasycercus cristicauda* Krefft. *Trans. zool. Soc. Lond.* 26, 409-501.
Wooller, R. D., Renfree, M. B., Russell, E. M., Dunning, A., Green, S. W., and Duncan, P. (1981). A population study of the nectar-feeding marsupial *Tarsipes spencerae* (Marsupialia: Tarsipedidae). *J. Zool., Lond.* 195, 267-79.

Wünschmann, A. (1966). Einige Gefangenschaftsbeobachtungen an Breitstirn-Wombats (*Lashiorhinus latifrons* Owen 1845). *Z. Tierpsychol.* 23, 56–71.

Wysocki, C. J., Wellington, J. L., and Beauchamp, G. K. (1980). Access of urinary nonvolatiles to the mammalian vomeronasal organ. *Science, NY* 207, 781–3.

Young, J. A. and van Lennep, E. W. (1978). *The morphology of salivary glands.* Academic Press, London.

4 The primitive eutherians I: orders Insectivora, Macroscelidea, and Scandentia

DIETRICH V. HOLST

Introduction

The groups to be discussed here were, until quite recently, usually included in the order Insectivora (e.g. Heim de Balsac and Bourlière 1955). However, they do not represent a genuine phylogenetic unit, since their affinities are due essentially to the retention of basal mammalian characters from the insectivorous/carnivorous early stock. In 1866, Ernst Haeckel had already divided the Insectivora into the suborders Menotyphla (Macroscelididae and Tupaiidae) and Lipotyphla (Soricidae, Talpidae, Erinaceidae, and Centetidae) on the basis of the presence of a caecum in the digestive tract of the former and its absence in the latter. Although a single feature, such as the presence or absence of a caecum, is scarcely sufficient to establish phylogenetic relationships, the separation of the Menotyphla from the remaining insectivorous groups has subsequently proved to be well-founded.

Although there is, as yet, no definitive evidence that groups contained in Haeckel's 'Lipotyphla' represent a proper phylogenetic unit, they are now generally regarded as relatively closely related and it is usual to refer only these groups to the order Insectivora (Starck 1978). The group defined by Haeckel as Centetidae is treated here as the superfamily Tenrecoidea, including the three families Solenodontidae, Tenrecidae, and Potamogalidae. The moles, which have often been placed in their own family, Talpidae, are in fact so closely related to the Soricidae that the recent trend has been to include them in this latter family (for details, see for example Thenius and Hofer 1960; McKenna 1975; Starck 1978). Thus constituted, the order Insectivora now includes the following six families: Solenodontiade, Tenrecidae, Potamogalidae, Chrysochloridae, Erinaceidae, and Soricidae.

The two groups included by Haeckel in the Menotyphla (viz. Tupaiidae and Macroscelididae) are today generally regarded as taxonomically distinct from the Insectivora. The Maroscelididae (elephant-shrews) are highly specialized mammals which have occurred in Africa since the early Tertiary. In contrast to the typically nocturnal insectivores listed above, the elephant-shrews are large-eyed, diurnal forms with a relatively well-developed brain. They are quite distinct from all insectivores in ontogeny, placentation, and immunological properties, and the family Macroscelididae is accordingly placed in its own order Macroscelidea by several modern authorities (for summaries, see Patterson 1965; Heberer and Wendt 1972; Goodman 1975; Starck 1978; Luckett 1980).

According to the evidence now available, the Tupaiidae (tree-shrews) are related neither to the Insectivora nor to the Macroscelidea. Because of numerous shared morphological features first emphasized by Kaudern (1911), Carlsson (1922) and, particularly, Le Gros Clark (summary: 1959), the Tupaiidae were included by Simpson (1945) in the order primates in his classification of mammals. However, this inclusion has been increasingly questioned in recent years on the grounds of the methodology of primate phylogenetic reconstruction, since earlier approaches had confused primitive retentions and derived features, and on the basis of new evidence regarding cranial, dental, and cerebral morphology in addition to reproductive characters and biochemical properties (e.g. Hill 1953; van Valen 1965; Martin 1968*a*). As a result, tree-shrews are now generally placed in the separate order Scandentia (or Tupaioidea) (for summaries, see Butler 1972; Doyle 1974; Luckett and Szalay 1975; Luckett 1980).

Regardless of the systematic relationships of the Tupaiidae, there is no recent mammalian group which appears to be more similar to the ancestral primates. This resemblance is, doubtless, partly due to convergent adaptations for arboreal life, but it is primarily based on the retention of ancestral placental mammal features in both tree-shrews and primates. As Romer aptly pointed out in 1968: 'It may well be that in tree-shrews we see the most primitive of living placentals— forms not too distant from the common base of all eutherian stocks'. Given this general retention of primitive mammalian characters it would seem to be justifiable to deal with the following eight families contained in the orders Insectivora, Macroscelidea, and Scandentia all together in the following survey:

Order: Insectivora
    Family: Solenodontidae
        Tenrecidae (subfamilies: Tenrecinae, Oryzorictinae)
        Potamogalidae
        Chrysochloridae
        Erinaceidae (subfamilies: Echinosoricinae, Erinaceinae)
        Soricidae (subfamilies: Soricinae, Crocidurinae, Scutisorichinae, Uropsinnae, Desmaninae, Talpinae, Scalopinae, Condylurinae)
Order: Macroscelidea
    Family: Macroscelididae
Order: Scandentia
    Family: Tupaiidae (subfamilies: Ptilocercinae, Tupaiinae)

These eight families all represent relatively primitive descendants from the stock which gave rise to all the more advanced mammals. Although a knowledge of their behaviour and physiology is an essential prerequisite for an understanding of the emergence of more advanced mammals, we still know very little about these groups. For most species we have, at the most, only fragmentary observations of the behaviour of individuals kept in captivity, and, with few exceptions, virtually nothing is known of their ecology. There are two reasons for this state of affairs: most species are small-bodied and nocturnal in habits, and many of them are either fossorial or semiaquatic. Only the Macrosceliddae and the Tupaiinae are diurnal in habits, and it is only among the latter (along with the noctural Ptilocercinae) that varying degrees of arboreality are found.

In all of the species concerned, the sense of smell is undoubtedly of considerable importance, as is reflected anatomically by the well-developed olfactory bulbs. Further, it would seem that a vomeronasal organ (Jacobson's organ) is typically present (see Table 4.1). Efferents from this organ evidently have access, through the accessory olfactory bulb, to parts of the diencephalon and telencephalon which are not reached by axons arising from the olfactory mucosa (for summaries, see Scalia and Winans 1976; Graziadei 1977; Johns 1980).

Table 4.1. Occurrence of vomeronasal organ (Jacobson's organ) in members of the different families

Family	Species	References
Tenrecidae	*Setifer setosus*	57
	Echinops telfairi	57
Chrysochloridae	*Chrysochloris* sp	8
Erinaceidae	*Echinosorex gymnurus*	2, 8
	Hylomys suillus	2
	Erinaceus europaeus	2, 33, 59
Soricidae	*Sorex araneus*	2
	Neomys fodiens	2
	Crocidura russula	2
	Talpa europaea*	25
Macroscelididae	*Macroscelidides proboscideus*	7
Tupaiidae	*Ptilocercus lowii*	12
	Tupaia sp.	2, 8, 12
	Tupaia javanica	2
	Tupaia belangeri	42, 43

* In *Talpa* a well-developed vomeronasal organ is present only in embryonic stages; the organ is vestigial in adults. (See Table 4.3 for reference list.)

According to recent findings, the vomeronasal organ of various mammalian species is not only an extremely precise sense organ, but along with the olfactory system, is particularly important for the reception and interpretation of chemical signals from conspecifics, notably in connection with reproduction (e.g. Poduschka and Firbas 1968; Estes 1972; Powers and Winans 1975; Johns 1980; Meredith, Marques, O'Connell, and Stern 1980). It may also be stimulated especially by liquid-borne compounds of low volatility, which do not reach the olfactory epithelium (Wysocki, Wellington, and Beauchamp 1980). This would appear to be very significant, since many substances (e.g. sex steriods and some other chemical substances carried in urine), which could potentially communicate information on sexual condition and the like, are generally of low volatility.

In correspondence with the pronounced development of the olfactory and vomeronasal systems in the mammal groups considered here, scent substances are likely to play a major part in their behaviour, particularly since there is conspicuous development of various glands whose secretions are distributed in the animal's surroundings through scent-marking behaviour. In addition, chemical substances which are discharged in the urine, faeces, or saliva may also have important behavioural functions. Since most species belonging to the groups considered have a solitary life-style, it is likely that these chemical substances are of only minor significance with respect to communication between conspecifics and that they serve primarily as a source of important information for the marking animal itself (e.g. through 'signposting' of its home-range). However, the situation is obviously different during the limited periods over which conspecifics are obliged to live together (e.g. courtship and mating periods or during the rearing of young). At such times, chemical substances might not only serve as a means of facilitating encounters between conspecifics, but also permit identification of particular individuals and recognition of their sex and specific sexual state. In line with this, in many species the specialized cutaneous glands are notably well-developed and actively secreting at the appropriate times.

In the course of social interactions, be they aggressive or sexual in nature, auditory and tactile stimulation also have an important part to play (for summaries, see Herter 1968; Eisenberg and Gould 1970; Tembrock 1973; Poduschka 1977a). As a rule, in these predominantly nocturnal animals sight is far less developed than the senses of smell and hearing; consequently, visual signals are generally of little significance. However, there are some groups with well-developed vision, such as the diurnal Macroscelididae and Tupaiinae, which make extensive use of visual signals.

Scent substances: their origins and presumed functions in the individual families

Solenodontidae

There are two species of solenodons, one living on Cuba and the other on Hispaniola. They are nocturnal forms with body weights of up to 1 kg, and count among the largest of extant insectivores. In contrast to most other Insectivora, which lead solitary lives except during mating periods, solenodons would appear to have a different pattern of social organization. Thumb (cited by Mohr 1937) found up to eight animals at a time in nest cavities located in the wild, and such groups seemed to consist of families including two successive litters of young. No behavioural observations have been conducted on solenodons under natural conditions, but they have been maintained in captivity on a number of different occasions.

According to Mohr (1936), solenodons (*Solenodon paradoxus*) of both sexes possess glandular areas on the axilla, the flanks, and (probably) the abdomen, which begin to produce a greenish secretion when the animals reach sexual maturity at the age of about six months. The males kept by Mohr always exhibited a sticky deposit of this secretion in these body regions. In females, secretion was only observable at certain times and even then it was less obvious than in males; at other times, no sign of secretion would be found. Mohr explains this in the following way: the secretion of the flank glands plays a part in courtship and mating of solenodons; whereas males are able to mate at any time of the year, and therefore continuously produce secretion, females have limited periods of receptivity and the secretory activity of their glands is restricted to these times. Mohr maintained that his interpretation is supported by an experiment in which an adult male was introduced to an unfamiliar female. Both animals at once began to sniff at each other's axillae and flanks, prodding each other with their snouts in these areas of the body. The male repeatedly pushed his way beneath the female's body, but copulation did not take place. Mohr did not see any specific marking actions performed with the flanks of the body, but the male (and occasionally the female) would slide on his belly around and alongside his partner, pushing himself along with his forefeet and dragging his legs behind. Yet Eisenberg and Gould (1966) and Poduschka (1977a) reported that the solenodons maintained by them in captivity exhibited no marking behaviour with the flanks or any other body region, and that no secretion was found. To date, the body regions involved have not been subjected to histological examination, so no definitive conclusion can be reached regarding the presence of corresponding skinglands and their secretions.

In solenodons—as in all other Insectivora—there are well-developed proctodaeal glands which open into the anal canal, along with anal glands which incorporate both sebaceous and apocrine glands (Schaffer 1940; Ortmann 1960). The secretion from these glands may be discharged with the faeces, or it may be directly deposited on the substrate by means of the commonly observed behaviour pattern known as 'anal drag'. It seems unlikely, however, that this is genuine marking behaviour associated with some biological significance of the deposited scent substances in the context of the animals' social behaviour. As a rule the behaviour pattern concerned is performed either after defecation (Eisenberg and Gould 1966) or after urination (Poduschka 1977a), anywhere in the home-cage, and conspecifics do not respond to such 'anal marks', any more than they do to urine or faeces.

Tenrecidae

Tenrecs occur on Madagascar and some neighbouring islands. They have occupied a wide variety of ecological niches and adaptive radiation has produced a total of 10 genera comprising 28 species (Starck 1978). All tenrec species are either crepuscular or fully nocturnal in habits and most of them are apparently solitary as a rule, with the exception of the genus *Hemicentetes*, which exhibits a trend towards the formation of colonies (for details, see Eisenberg and Gould 1966, 1970). As yet, no field studies of tenrec behaviour have been conducted, though a number of different species have been maintained in captivity, in some cases leading to establishment of successful breeding. Nevertheless, our understanding of the morphology of possible scent glands and their potential significance in chemical communication is still extremely limited.

According to Eisenberg and Gould (1970), the various tenrec genera (*Tenrec, Hemicentetes, Setifer, Echinops, Microgale*) possess glandular areas in axillary, inguinal, cranial, ear, and caudal regions. The secretions of these glands are deposited in the animal's general surroundings and on conspecifics. Marking of objects (rubbing chin, cheeks, flanks, or abdomen) is rarely observed. The suspected scent glands are likely to be of significance in courtship behaviour. Following initial contact of their noses with the nose, ear, flank, and anogenital region of the opposite sex, males and females rub their sides or other regions against each other and very often crawl over and under their partners (Eisenberg and Gould 1970). All this suggests a combination of tactile communication with stimuli from potential glandular areas. It must be emphasized, however, that none of the potential scent-gland areas have, to date, been examined histologically, nor have specific secretions from these areas been described. This is also true of the sternal or ventral gland of *Setifer setosus* mentioned by Poduschka (1977a). When placed in unfamiliar surroundings, or during the presence of strange conspecifics, *Setifer setosus* will press these body areas on the substrate or on available objects with a conspicuous behaviour pattern: the forelegs are stretched out sidewards passively and the body is propelled forwards only by the hindlegs. No interpretation can yet be made of the function of this behaviour pattern, nor of the significance any chemical substances which may be deposited, particularly since conspecifics show no response to these apparent 'scent-marks'.

There are, apparently, no published descriptions of glands in the anal region of tenrecs, but it seems likely that they also possess proctodaeal and anal glands in common with the insectivores. At least, both males and females of all tenrec species so far maintained in captivity have exhibited marking with the perineal region (perineal drag) in unfamiliar surroundings and during interactions with conspecifics (Eisenberg and Gould 1970; Poduschka 1974a, 1977a).

All of the Tenrecinae so far studied in captivity (but not the Oryzorictinae) have been found to deposit faeces, and sometimes urine, at specific sites— usually close to retreats used for sleeping (Eisentraut 1955; Koch-Isenburg 1955; Herter 1962; Malzy 1965; Eisenberg and Gould 1970; Poduschka 1977a). With *Setifer*, the defecation site of the female was also used by her young (Poduschka 1977a).

The eye-gland secretion has been investigated in more detail in various tenrec species. Honegger and Noth (1966) were the first to describe, for the pigmy hedgehog tenrec (*Echinops telefairi*), the discharge of viscous white secretion from the angle of the eye, and to some extent from the nose as well, which could cover the entire eye with a white film. Eisenberg and Gould (1970) also saw such eye-gland secretion with males of other tenrec genera (*Setifer, Echinops, Microgale*), while Poduschka (1974a,b, 1977b) noted this secretion not only with male tenrecs but also in females. The secretion is produced from specialized eye-glands, which occur in addition to the usual lacrimal glands and which have been histologically identified (Cei 1946a, 1947; Poduschka 1974b). According to Poduschka (1974b) the secretion itself has a strong, persistent taste to humans. Honegger and Noth (1966), in agreement with Eisenberg and Gould (1970), interpret the eye-gland secretion as a component of male mating behaviour. But Poduschka (1974b, 1977a) rejects this narrow interpretation on the grounds that he observed production of the secretion in a wide variety of situations in both sexes of all the tenrecs he studied (*Setifer, Echinops, Microgale*). It was seen as a response to strange smells and novel tastes and occurred in connection with defensive behaviour, courtship, and even death. The secretion was commonly deposited on the surroundings with a wiping action made with the cheek or the eye region. Conspecifics responded to this secretion by becoming extremely excited and eventually rubbing off their own secretion as it exuded from their eyelids and nostrils.

Poduschka (1977a) suggests that the eye-gland secretion has multiple functions, such as warding off rivals, reducing insecurity in unfamiliar surroundings, and stimulating the female during courtship. He even discusses the possibility that it could act as a primer pheromone, inducing ovulation in the females (induced ovulation has been inferred for tenrecs by Eisenberg and Gould (1970)). Although these various functions are, indeed, conceivable, not one of them has been confirmed experimentally as yet.

The salivation of some tenrecs (*Setifer, Echinops*) in response to strong olfactory stimuli, which has been interpreted by Eibl-Eibesfeldt (1965) as a form of marking behaviour, should not be considered in this context, since it does not seem to have any relevance to active communication processes (Poduschka 1974c, 1977a; see also p. 142.

Potamogalidae

The otter-shrews (comprising three genera with one species each) are semi-aquatic forms whose distribution is limited to west and central Africa. Virtually nothing is known about their general behaviour or about the significance of chemical substances for their social behaviour, in particular. The only individual of this family so far kept in captivity (*Mesopotamogale ruwenzorii*; Rahm 1961) was nocturnal in habits and utilized a specific site in the cage for defecation, which could have some relevance to communication. According to Cei

(1946*a*), *Potamogale velox* possesses special glands in the eyelids which are presumably equivalent to those described for tenrecs. However, Rahm observed no secretion from these glands and nothing is known about their function.

Chrysochloridae

The golden moles comprise a close-knit family of highly specialized fossorial, blind insectivores inhabiting south and central Africa (six genera containing 18 species; Starck 1978). They possess large olfactory bulbs (Stephan and Bauchot (1960) and they also have well-developed proctodaeal and anal glands (Hamperl 1923, 1926; Ortmann 1960), so it seems possible that chemical substances play a part in their communicative behaviour. However, nothing seems to have been published about their behaviour.

Erinaceida

The hedgehogs are represented by two recent subfamilies, the Echinosoricinae of South-East Asia (four genera with one species each) and the Erinaceinae of Europe, Africa, and Asia (15 genera including about 40 species; Starck 1978).

Echinosoricinae   Scarcely anything is known about the behaviour of the terrestrial spiny hedgehogs. The only species for which limited behavioural observations have been conducted in captivity is the moonrat (*Echinosorex gymnurus*). Moonrats are apparently solitary in habits under natural conditions (Lim 1967). Medway (1969) has claimed that moonrats are active both by day and by night, but diurnal activity of any kind seems unlikely in view of the uniformly crepuscular or nocturnal habits of the other Insectivora, the more so as both Davis (1962) and Gould (1978) report a purely nocturnal pattern of activity for *Echinosorex*.

A very strong odour, reminiscent of ammonia (Davis 1962) or rotten garlic (Gould 1978) emanates from moonrats of both sexes, apparently deriving from the anal glands. If an animal is held by the tail, it will usually evert its anus to expose the two large lateral anal glands. These glands vary in size and colour, independently of the sex of the animals. According to Gould (1978), moonrats exhibit perineal dragging during exploratory behaviour, and anal eversion is sometimes visible in such situations. Secretion from the anal glands is evidently deposited during perineal dragging, since the entrance ways of nest-boxes and the wooden thresholds of doorways connecting adjacent cages were stained black and were impregnated with a strong smell. During social encounters, moonrats often show intensive sniffing of the anal regions of conspecifics and perineal dragging may also be seen on rare occasions. Although these observations provide some indication of a behavioural significance of the anal-gland secretion in moonrats, the extremely limited nature of the evidence permits no conclusions about possible functions. The same applies to the urination of subordinate animals during social encounters, which Gould (1978) interpreted as 'urine marking'. No response of conspecifics to the urine deposited was recorded and it is difficult to imagine what function it might serve. As a rule, moonrats do not deposit urine at specific sites, but they do exhibit defecation at a single location in each cage.

Erinaceinae   Hedgehogs are nocturnal or crepuscular animals which lead solitary lives for most of the year. Apart from the European hedgehog (*Erinaceus europaeus*) very little is known about their behaviour.

The European hedgehogs have very large home-ranges (males about 20 ha and females about 8 ha). Although hedgehogs will normally attack and try to drive away any conspecifics they encounter, there is a notable absence of territorial demarcation and associated behaviour patterns such as scent-marking. The home-ranges, correspondingly, overlap extensively and often completely, regardless of sex (Reeve 1980). Furthermore, individual scents in hedgehog nests will not repel conspecifics; a strange nest can be occupied at any time and the sex of the previous tenant is of no account (Poduschka 1977*a*).

Hedgehogs of both sexes possess relatively poorly developed anal glands and prominent proctodaeal glands (Hamperl 1923, 1926; Schaffer 1940; Ortmann 1960). Whether or not conspecifics are present, young hedgehogs will often press their faeces, and probably associated secretions from their anal region, onto vertical objects. However, it is doubtful whether this represents a marking process (perineal drag) with a defined function, since such behaviour is not observed with adults (Poduschka 1977*a*). In contrast to most other insectivores, hedgehogs do not make use of localized defecation sites, either in the wild or in captivity (Poduschka 1969).

Chemical signals are probably of considerable significance in hedgehog reproduction. With European hedgehogs, which are normally extremely aggressive to conspecifics, copulation is preceded by a drawn out courtship during which males often deposit scent marks and secretions on the substrate (Strieve 1948; Poduschka 1969, 1977*b*). In the process, the male's hindlegs are kept close together and the spine is arched upward convexly, while the partially protruding penis exudes a (sometimes whitish) secretion as the body is rocked gently from side to side. This secretion is probably produced by the accessory sex glands, which are extremely well-developed in hedgehogs (Ottow 1955), and it seems to be discharged in dilute form with urine. The smell of the secretion is clearly detectable to the human nose and it is completely different from the usual odour of the male's body and urine outside mating periods (Poduschka 1977*b*). The male typically deposits these scent marks behind the female (or on her, during actual mounting), while she allows herself to be driven around by the male on a zig-zag or circular course within a limited area of only 20–100 m² (even under natural conditions when there are no constraints on the area available). During such courtship behaviour, the female repeatedly passes over the male's previous scent marks so that she is subjected to continued olfactory stimulation. Poduschka (1977*b*) has concluded that the male's scent marks promote receptivity to mating in the female (with the scent acting as a 'releaser pheromone') and they might even exert a 'primer pheromone' effect by modifying the oestrous cycle. At times other than during mating, on the other hand, the urine of hedgehogs appears to carry no chemical signals of biological significance.

In contrast to *Erinaceus europaeus*, the courtship of *Hemiechinus auritus* is characterized by tactile interactions rather than by conspicuous scent-marking (Poduschka 1977*a*). The animals nestle up to one another and touch their heads against the partner's flanks in a manner closely corresponding to that seen in the courtship behaviour of tenrecs.

Just like tenrecs, European hedgehogs possess specialized eye-glands (Cei 1947), but their secretory activity is relatively limited (Poduschka 1977*b*). There is no indication of any special function of this secretion.

The conspicuous 'self-anointing' (spreading of saliva over the body), which is also found with the pigmy hedgehog tenrec (*Echinops telfairi*), was reported quite early for a variety of hedgehog genera (*Erinaceus, Paraechinus, Hemiechinus*; for summary, see Eisentraut 1953). This behaviour occurs as a response to olfactory and gustatory stimuli. Poduschka and Firbas (1968) interpret 'self-anointing' as the terminal event in a behavioural sequence through which hedgehogs identify strongly irritant new smells, notably stimuli encountered in their sexual behaviour, with the aid of their Jacobson's organs. It is suggested that smell or taste substances taken up by the snout or mouth are dissolved in the foamy saliva through chewing motions. Subsequently, probably via the nasopalatine duct, these substances would come into contact with the sensory epithelium of the Jacobson's organ and might then be identified. Thereafter the saliva is removed from the mouth with the tongue and wiped onto the animal's body. According to Poduschka and Firbas, it would seem unlikely that, in addition to this cleaning of the mouth after perception, self-anointing could serve other functions (e.g. self-marking with specific substances dissolved in the saliva).

Soricidae

The modern shrew and mole family, with eight subfamilies and 32 genera containing 280 species, is the best represented and most widely distributed group of insectivores. Their distribution is worldwide, with the exception of Australia, most of South America and the polar regions. Terrestrial, burrowing, climbing, and semiaquatic forms occur. Within the family Soricidae, there is a major division between the shrews (Soricinae, Crocidurinae, Scutisoricinae) and the moles (Uropsilinae, Desmaninae, Talpinae, Scalopinae, Condylurinae), and these two groups are often given the rank of separate families.

Shrews   There are virtually no reports of behavioural observations conducted on shrews under natural conditions. They are more or less nocturnal in habits and live predominantly in burrow systems or under dense cover. Apparently, species of the genera *Sorex* and *Neomys* are solitary in habits and conspecifics are chased away during encounters (Crowcroft 1955). By contrast, representatives of the genera *Crocidura, Suncus,* and *Cryptotis* are less aggressive towards conspecifics, and can be kept and bred in pairs or in groups in captivity (Gould 1969; Vogel 1969; Mock and Conaway 1976). Our knowledge of their social behaviour and of the significance of chemical signals is extremely fragmentary and stems almost entirely from observations conducted on animals maintained in captivity.

In all species so far investigated, covering a wide variety of genera (*Sorex, Neomys, Blarina, Crocidura, Suncus*), both sexes possess flank glands halfway between the fore- and hindlimbs at the level of the axilla. These glands, which are developed to different degrees in different species, can usually be recognized externally as small oval areas of moist hair. The flank glands were first described as early as 1815 by Geoffroy-Saint-Hillaire and subsequently they were subjected to histological examination in a variety of species (e.g. Johnsen 1914; Eadie 1938; Schaffer 1940; Pearson 1946). According to these studies, there are only minor differences between species in the structure of the glands, which always consist of an aggregation of large sweat glands and sebaceous glands. There is enormous variation in the secretory activity of the flank glands of adult animals, both in the wild and in captivity. As a rule, the activity of the glands is greater in males than in females. In addition, there seems to be a connection between flank-gland activity and reproductive condition in both sexes, though the findings are far from uniform. According to Johnsen (1914), Eadie (1938), and Hamilton (1940), the flank glands of males (*Sorex, Neomys, Blarina, Crocidura*) show a clear-cut pattern of size variation associated with the reproductive cycle; the gland components increase in size concurrently with the growth of the testes during the breeding period, indicating a corresponding increase in secretory activity. Pearson (1944), by contrast, noticed no difference in development of either sebaceous or sweat-gland tubules in the flank glands between breeding and non-breeding males, or even compared to castrates, in *Blarina* and *Sorex.* Further, injections of testosterone, oestrogen, or progesterone into normal or castrate males had no histologically noticeable effect on gland size or morphology. This finding is all the more surprising, since the flank glands of male *Suncus murinus* are clearly dependent upon the presence of sex hormones (Dryden and Conaway 1967). Their size and secretory activity decrease following castration, and this process can be reversed by administration of testosterone or progesterone, but not oestrogen.

Eadie (1938) reports that the flank glands of female *Blarina* show no variation in size with the reproductive cycle, whereas Johnsen (1914) and Pearson (1944) have both recorded a clear relationship between the secretory activity of the glands in females and their reproductive condition. Adult nulliparous or anoestrous females (as well as ovariectomized females) always have fully developed glands, while oestrus, pregnancy, and lactation are consistently accompanied by progressive reduction in function to a level where no further activity can be demonstrated. Injection of oestrogen, however, did not result in a corresponding decline of flank-gland development of females (one normal and one ovariectomized), even when it produced hypertrophy of the vagina to attain the full oestrous condition (Pearson 1944). According to Hamilton (1940), the flank glands of *Sorex fumeus* are most active in both sexes throughout the breeding season, only non-breeding adult females do not exhibit pronounced glands.

Shrews give off a strong odour, which has been claimed to be sufficiently repulsive to prevent certain carnivorous mammals from eating them, at least when alternative food is available (Eadie 1938). Wagner (1841) seems to have been the first to link this odour to the flank glands, and this assumption was implicitly or explicitly accepted by subsequent authors. But Dryden and Conaway showed in 1967 that this assumption is incorrect at least in the case of *Suncus murinus*. If the skin containing flank glands is removed from males of this species, they continue to produce the same strong odour. According to Dryden and Conaway, the odour is produced by concentrations of sweat glands on the throat (throat gland) and, especially, behind the ears (postauricular glands). The activity of these apocrine glands is dependent upon sex hormones. In both sexes, castration resulted within three weeks in sweat-gland atrophy and cessation of odour production. Testosterone or oestradiol implants brought about the recovery of the sweat glands and resumption of odour production by castrates of both sexes, as did progesterone in males (females were not tested). Dryden and Conaway state that the scent of these 'musk glands' can be discharged immediately following some kind of disturbance in such high concentration (particularly for males) that it can be detected by the human observer several metres away within the space of a minute. No special marking behaviour is observed in connection with such scent discharge. The 'musky' scent is also frequently discharged during marking of objects with the flank glands or other gland areas, and it may be detected on such 'marked' objects for days afterwards. The authors conclude that *Suncus* has the ability to discharge a highly volatile scent which is normally rapidly dissipated. This scent may be trapped by the oily secretions of the specialized sebaceous tubules of the flank glands (and of other, similar glands). The flank glands and their secretions, by this interpretation, act only as a carrier for the actual scent from the 'musk glands'. Through marking with their flank glands and other glandular areas, these shrews would accordingly deposit relatively persistent 'scent marks' on objects in an indirect fashion.

Kivett and Mock (1980) report that females of the lesser American short-tailed shrew (*Cryptotis parva*) also possess well-developed glands anterior to the ventral part of the aural cartilage, which exhibit apocrine secretion and are most active during pregnancy. These authors did not determine whether these auricular glands are also present in males and the extent to which they are involved in the production of the animals' body odour is unknown.

Both sexes of the greater American short-tailed shrew (*Blarina brevicauda*) possess, in addition to the flank glands, large ventral glands on the midline of the abdomen which, just like the flank glands, are composed of greatly enlarged sebaceous glands mixed with a relatively small proportion of sweat tubules (Coues 1896; Eadie 1938; Pearson 1944, 1946). According to Pearson, the ventral and flank glands increase or decrease in size in unison. Correspondingly, males show no noticeable variation in ventral gland size or activity throughout the year, whereas there is a marked decline in function of the ventral gland in pregnant females. In contrast to the situation reported for the flank glands, these observations agree completely with those of Eadie (1938). With the genus *Sorex*, Eadie (1938) did find normal sebaceous glands individually associated with hair follicles on the belly, just as elsewhere on the body, but there was no special ventral gland of the kind found in *Blarina*. No other genera have been investigated to date.

Males (but not females) of the common European white-toothed shrew (*Crocidura russula*) possess on the ventral aspect of the root of the tail a conspicuous subcaudal gland composed of greatly enlarged sebaceous and sweat glands, resembling the flank gland of this species (Niethammer 1962). It is therefore histologically quite different from the subcaudal glands of other soricids (*Desmana, Galemys*: Schaffer 1940). Growth of the subcaudal gland of male *Crocidura russula* begins with the attainment of sexual maturity. Related species of the same genus (*C. suavolens, C. leucodon*), along with *Sorex minutus*, lack this gland entirely.

All of the shrew species studied to date, including a wide range of genera (*Sorex, Neomys, Crocidura, Suncus*), exhibit well-developed proctodaeal and circumanal glands in both sexes (Hamperl 1923, 1926; Ortmann 1960). According to Balakrishnan and Alexander (1976, 1980), *Suncus* urinate and defecate following introduction into new cages, predominantly along the walls. Whilst urinating, the animals move forward dragging their hindquarters on the ground, such that urine marks can sometimes be seen as trails along the cage floor.

A specific pattern of defecation has been reported for captive *Crocidura* (Spannhof 1952; Herter 1957). The animals stand on their forepaws and deposit faecal pellets on the cage wall by pressing their hindquarters against it a few centimetres above ground level. According to these authors, similar observations have been made with other shrews as well, both in captivity and under natural conditions. However, there are no indications of possible functions for urine, faeces, or secretions from the anal region. Shrews do not seem to have localized defecation sites.

A white secretion that issues from the eyelids at times of excitation, comparable to that found among tenrecs, is known to be present in a number of shrews (*Crocidura, Sorex*: Poduschka 1977a). As yet, however, nothing is known regarding the possible function of this secretion.

Summarizing remarks on chemical substances and their biological relevance in shrews: Shrews overall possess relatively similar skin glands located in a wide variety of areas of the body (head, flanks, belly, tail root) and composed of greatly enlarged holocrine sebaceous glands together with tubular apocrine glands. It is, as yet, largely undecided whether these individual skin gland areas produce specific olfactorily active components or (only) serve as carriers for scent substances which originate elsewhere (e.g. from the urine or from the postauricular glands, as in *Suncus*). It is clear, however, that such glands are particularly suitable for the deposition of their secretions through specific behaviour patterns (marking behaviour) on environmental objects or on conspecifics. Such marking with the gland areas discussed here has indeed been described for a wide variety of shrew species. In addition, Balakrishnan and Alexander (1976) have demonstrated a dependency on sex hormones in the marking behaviour of male *Suncus murinus*: following castration, marking behaviour with the flank, throat and perineal glands decreased to a very low level within three to four weeks, and it proved possible to restore the original levels of marking behaviour by subsequent treatment with either androgen or oestradiol. This decrease in marking behaviour following castration matches the atrophy of sweat glands and cessation of odour production described by Dryden and Conaway (1967).

As yet, very little is known about the functions of the various skin glands or their secretions in any soricid species. The original suggestion that the skin-gland secretion of shrews, or their typical smell, acts to repel predators has been largely invalidated by numerous ecological findings and will not be discussed further here. Already in 1815, Geoffroy-Saint-Hillaire suggested a function of skin glands in reproductive behaviour on the basis of the relationship between the size and activity of skin glands with the sex and reproductive conditions of the animals: he proposed that male and female locate one another during the breeding season through odours produced by the scent glands. This interpretation was subsequently accepted, implicitly or explicitly, by a number of authors (e.g. Johnson 1914; Eadie 1938; Hamilton 1940; Crowcroft 1957; Niethammer 1962; Mock and Conaway 1976). In fact, in all of the genera so far investigated in detail (*Suncus, Blarina*: Gould 1969; *Cryptotis*: Kivett and Mock 1980), copulation is preceded by drawn-out courtship involving intensive reciprocal sniffing. During courtship, males repeatedly mark the substrate with their skin glands and, as a result, increasingly impregnate their own pelage with their secretion. It is thought that the strong smell given off in the process inhibits the, initially very aggressive, female from attacking the male and renders her receptive so that she will ultimately permit copulation. The authors do not discuss the possible significance of the female's smell.

Pearson (1946) was the first to suggest a totally different function for the skin-gland secretions. He proposed that in *Blarina* both sexes mark the tunnels making up their territories, thus informing other shrews of the occupation of the

tunnels and keeping them away except during the breeding season. With subterranean mammals of this kind, whose scent-mark odours are likely to persist
for relatively long periods in their tunnels, such a possibility is quite plausible,
particularly since shrews kept in captivity repeatedly mark objects in their
cages with their different skin glands as well as with urine and faeces (*Crocidura*:
Marlow 1955; *Suncus*: Balakrishnan and Alexander 1976, 1980). A direct
deterrent effect of the scent marks has not as yet been demonstrated, but both
male and female *Suncus* show clearly changed behaviour in cages containing
scent marks of strange conspecifics of either sex, compared with cages lacking
this scent (Balakrishnan and Alexander 1980): they exhibit a decrease in locomotion associated with an increased rate of sniffing, while their rubbing (marking) with flanks, throat, and perineum is significantly reduced, along with
urination and defecation. In addition, a special form of self-grooming, which is
normally seen with animals introduced into new enclosures, is inhibited.

On the basis of Pearson's (1946) report that the skin glands of oestrous
females are reduced in comparison to non-breeding females, several authors have
concluded that conspecifics would be encouraged to enter tunnel systems of
females only when their deterrent smell is 'switched off' during oestrus (Pearson
1946; Crowcroft 1957; Kivett and Mock 1980). It has been further suggested
that males would be more likely to penetrate into the tunnel systems of oestrous
females lacking a deterrent odour since they have considerably large home-
ranges (Blair 1940). However, it is difficult to understand why the skin-gland
regions of pregnant and lactating females, which no longer tolerate the presence
of adult conspecifics, are reduced even more than in oestrous females and
usually produce no more recognizable odour at all (Pearson 1946). Kivett and
Mock (1980) postulated that pregnant females, in contrast to oestrous females,
once again produce the deterrent smell signifying 'stay away, I am pregnant' and
thus avoid disturbance by males, but this interpretation was based on incorrect
citation of Pearson's (1946) data. Although the functions of the scent glands
of shrews therefore remain largely obscure, the interpretation first suggested
by Pearson would seem to me to be the most likely, namely that the typical
strong smell produced by shrews primarily serves to keep animals separate
except during the breeding season, rather than operating to bring male and
female together.

It is self-evident that a deterrent effect of shrew scent inevitably requires the
capacity for individual recognition. As yet, there is no evidence as to whether
the necessary information is provided by the same scent glands or derives from
some other source (e.g. from urine, which is regularly discharged during marking
with the scent glands). Similarly, it is not known whether oestrous females
produce additional scent substances carrying information relating to their
receptivity to mating (i.e. sexual attractants) though this seems quite probable.

Moles   Members of the mole subfamily are either fossorial forms living in subterranean burrow systems or semiaquatic forms. In line with this, only fragmentary information is available with respect to their ecology, behaviour, and communication systems. Most species seem to be typically solitary in habits and will drive off any conspecifics when encountered in their home-ranges or tunnel systems (Godfrey and Crowcroft 1960; Richard and Valette Viallard 1969). Exceptions are North American starnose moles (*Condylura cristata*: Hamilton 1931a) and the American shrew-moles (*Neurotrichus gibbsii*: Dalquest and Orcutt 1942), which are found in small groups and are thought by these authors to be to some extent colonial.

The visual sense has been partly reduced or completely abolished in all moles and their behaviour might therefore depend largely on tactile and acoustic stimuli. The olfactory bulbs are well-developed (Godet 1951; Stephan and Bauchot 1968), indicating that olfactory stimuli are also important in the lives of these mammals. Some authors, however, have presented behavioural evidence suggesting that the sense of smell is of only minor importance (*Talpa*: Godfrey and Crowcroft 1960; *Neurotrichus*: Dalquest and Orcutt 1942; *Condylura*: Hamilton 1931a).

It has not yet been determined whether an active Jacobson's organ is present in adult moles. Godet (1951) reported that it is clearly formed in the embryo of *Talpa*, but later degenerates and is functionless in adults. Regardless of the degree to which the chemical senses may be developed in moles, all of the species so far investigated have been found to possess more or less pronounced glandular areas, which may be of importance in the animals' behaviour. To date, however, there have been virtually no studies of possible patterns of marking with these glands or observations of the responses of conspecifics to the secretions of these glands and other chemical substances discharged by moles (e.g. in urine or faeces).

Eadie (1939) has described for males of *Parascalops breweri* a large glandular area extending from the lower jaw over the chest and down as far as the genital region. This gland is much less developed in females and is essentially confined to the underside of the jaw. The glands consist primarily of greatly enlarged and proliferated tubular sweat glands with a relatively small number of normal sebaceous glands individually associated with hair follicles. The gland exudes a yellow-brown secretion through which the entire ventral surface of the body can be discoloured, particularly in males during the breeding period. The secretion is also found with females, in fact, but it is much less abundant. In young animals, the gland begins to show secretory activity at the age of about six months. This ventral gland secretion is apparently responsible for the intensive smell which characterizes these animals, especially males during the breeding period, and it is thought to act as a sexual attractant. No corresponding gland area is known for other mole species.

In addition, males of *Parascalops* (females have not been examined) possess on each side of the genital region an anal (perineal gland) of combined holocrine-merocrine type which opens at the posterior border of the perineum (Eadie 1947). Anal glands have also been described for *Galemys* and *Talpa* (Hamperl 1923, 1926; Schaffer 1940; Ortmann 1960). In *Talpa*, they are extremely well-developed and have a special structural pattern: Each gland consists of three extensive sebaceous gland complexes combined with two aggregates of tubular glands which open to the exterior through three common ducts, forming a large, demarcated gland complex on either side of the body.

In contrast to the other moles, males of *Condylura cristata* in breeding condition (females have not been examined) possess an entire complex of skin glands in the inguinal/perineal region (Eadie 1948). Each male possesses an anal (perineal) gland with two structurally different components. The larger, more dorsal part is a holocrine gland 10 mm long and 6 mm dorsoventrally. The smaller, ventral part is a merocrine component, measuring up to 14 mm at its greatest width. Both components of this perineal gland open on to the perineal skin near the anterior lips of the rectal papilla. In addition, *Condylura* males possess two pairs of large holocrine glands of the tubulo-alveolar type, which have been termed inguinal glands I and II by Eadie (1948). The two components of inguinal gland I lie more or less in the angle formed between the perineal gland and the rectum and are separated from the latter by a thick layer of voluntary muscle (internal anal sphincter). Each component of this gland is provided with a single terminal duct associated with a large hair follicle and opening laterally on the rectal papilla, close to its base. The two components of inguinal gland II are located to the side of the proximal portion of the *glans penis* and prepuce and anterior to the urogenital papilla. Each component has a single duct opening on the surface of the skin lateral to the urogenital papilla. No such inguinal glands have been reported in other moles and they are definitely lacking in *Parascalops*, according to Eadie (1948).

Proctodaeal glands are present in all moles which have been studied to date (*Talpa, Galemys*: Hamperl 1923, 1926; Schaffer 1940; Ortmann 1960; *Condylura*: Eadie 1948). Their structure is comparable to that found in other insectivores. Prominent skin glands on the ventral surface of the base of the tail have been described for *Desmana moschata* (Brandt 1860—cited by Weber 1928; Brinkmann 1911). This subcaudal organ consists of a 3-4 cm long gland area composed of sebaceous glands. The individual gland tubules show multiple branching in some cases, producing a radiating, bush-like appearance. The glands open through pores on the individual scales on the tail base. In both sexes the tail gland exudes a viscous, oily secretion which has a strong musk-like smell, according to Schaffer (1940). It must be emphasized, however, that this description does not precisely characterize the smell, since almost all of the widely differing odours discussed here have been referred to as 'musky' by numerous different authors. This term is apparently used to connote a typical 'animal smell' without implying or demonstrating any connection with the specific quality of the various genuinely musk-odoured chemical substances.

The closely related species *Galemys pyrenaica* (sex unspecified) also possesses a subcaudal organ (Weber 1928; Schaffer 1940), but this differs in its internal structure. It consists exclusively of sebaceous glands which open into large vesicles from which excurrent ducts lead on to open out between the individual scales on the ventral surface of the tail. Given this difference in structure, it can be concluded that the tail gland of *Galemys* is not related to that found in *Desmana* (nor to that occurring in *Crocidura russula*—see p. 117). As with the tail gland of *Desmana*, the odour emanating from that of *Galemys* has been described as 'musk-like' (Richard and Valette Viallard 1969; Richard 1973).

A conspicuous phenomenon has been described for *Condylura* in this connection (Weber 1928; Hamilton 1931*b*). In animals of both sexes, with few exceptions, the tail periodically increases in girth, reaching a maximum at approximately the time of the breeding season. The tail may become swollen to such an extent that the scales covering its surface are pulled apart so that the flesh shows through and gives the tail a pinkish coloration. Swelling may be accompanied by production of a greasy secretion, which lends to the tail an appearance of having been rubbed in fat. Weber (1928) attributed this effect to subcutaneous deposition of strong-smelling fats, but there has been no histological or histochemical investigation to date which might confirm this.

Although moles do indeed possess a wide variety of glands producing clearly observable secretions—usually in connection with reproduction—their significance has not yet been established. Only in recent years have Pyrenean desmans (*Galemys pyrenaicus*) of both sexes been maintained for a substantial period of time in captivity (Richard and Valette Viallard 1969; Richard 1973). The observations conducted by these two authors provide clear indications of a possible role played by chemical scent substances in the social behaviour of this species. Whenever animals of opposite sex are placed together, faeces are deposited in specific areas, which are subjected to olfactory and tactile inspection. In addition, chemical scent marks are deposited by the subcaudal glands as black, glistening streaks on the substrate as the desman moves about. This is observed particularly frequently with males during the mating season in spring. The odour of a single scent mark from the subcaudal gland is barely perceptible to the human nose, but the nest and the immediately surrounding area become so heavily impregnated with the male's gland secretion that a strong 'gamey' smell can be detected by humans. The function of the odour remains unclear, however. For other species, we do not have even this limited amount of information.

Macroscelididae

The family Macroscelididae, containing five genera and approximately 18 species, is a highly specialized group of mammals which is limited to Africa and is first known from early Tertiary deposits in that continent (Starck 1978). In contrast to the Insectivora, they are large-eyed, predominantly diurnal and crepuscular forms with a well-developed brain. They possess a well-developed Jacobson's organ, which is quite different from that of other eutherians and is typically marsupial in almost every respect (Broom 1902). Under natural conditions, species of the genus *Macroscelides* apparently live as pairs, while *Elephantulus* species seem to live in large, loosely co-ordinated bands (Hoesch 1959).

Wagner (1841) described a subcaudal gland in *Macroscelides* which seemed to be responsible for the animal's typical smell. A similar gland is also found in *Rhynchocyon* (Carlsson 1910). The subcaudal gland consists of large tubular and alveolar glands occurring along the first third of the ventral surface of the tail, and its secretion collects in large storage sacs. From these sacs, excurrent ducts carry the secretion to openings located between the scales in the middle of the tail (Carlsson 1910; Weber 1928). This subcaudal organ, as a mixed type, is therefore clearly distinct from the gland found in the same region of the body in *Desmana* and *Galemys*, but resembles the above-mentioned tail-gland of *Crocidura russula* (see p. 117). The gland is apparently only present in sexually mature males, there being no trace of it in young males or in adult females (Peters 1852; Carlsson 1910).

According to Carlsson (1910), both *Macroscelides* and *Petrodromus* possess anal glands, but it is not clear what type of gland is involved. According to Hoesch (1959), faeces are never deposited by elephant-shrews in the course of movement around their home-ranges, but only at special defecation sites on the range borders.

No reports seem to be available of observations relating to marking behaviour involving skin-gland secretion, faeces, or urine, and nothing is known of the possible functions of chemical substances in the social behaviour of elephant-shrews.

Tupaiidae

The tree-shrews are divided into two subfamilies: the Ptilocercinae with one genus and the single species *Ptilocercus lowii*, and the Tupaiinae with five genera and about 18 species (Starck 1978). The classification within the latter group is, however, so uncertain that one cannot be at all sure that the nomenclature used by many authors is correct (for details, see Martin 1968*b*).

Ptilocercinae   Pen-tailed tree-shrews are small (body weight less than 80 g), nocturnal, arboreal mammals inhabiting Borneo and Malaya. As is the case with the other tree-shrews, they possess a Jacobson's organ which shows affinity with that of marsupials and differs markedly from that of insectivores in a number of characteristics (Clark 1926). They differ from the Tupaiinae in several primitive features (e.g. possession of scales on the tail) and, above all, in having a relatively poorly developed brain. *Ptilocercus* appears to live in small groups sharing a common sleeping nest under natural conditions (Lim 1967; Gould 1978). They have not, as yet, been studied in the wild with respect to their behaviour, but some behavioural data are available for a few animals of both sexes which Gould (1978) maintained in captivity for several months. According to Gould, pen-tailed tree-shrews, when faced with unfamiliar situations, frequently press their bodies against the substrate, particularly when climbing on flat, vertical surfaces. In addition, strange animals do show intensive reciprocal sniffing during encounters, concentrating especially on the anogenital region; but (with one exception) no actual rubbing behaviour with presumed skin glands has been described. Nevertheless, it should be pointed out that this observation is based only on six brief encounters involving a total of five females and one male.

The fur on the flanks and on the belly between the forelimbs in *Ptilocercus* is often stained orange, suggesting special glands located in this region, as is known for *Tupaia* (see p. 125). According to Gould, this colour (or any connected secretion) is unrelated to the animal's sex, age, or the time of the year.

Ptilocercus of either sex have a strong smell which Gould believes derives from the urine, but no actual marking with urine was described. According to Clark (1926), *Ptilocercus* males (females have not been studied) possess on either side of the anal canal a small gland composed of simple acini filled with homogeneous colloidal material. These anal glands open through small ducts onto the surface of the perineal region, lateral to the anal orifice. On two occasions, Gould (1978) observed perineal dragging, which may act to deposit the secretion of these glands on the substrate. The secretion, combined with faeces, may also be responsible for the dark discoloration of branches and nest entrances described by Gould. There are no other indications of the presence of skinglands or of scent-marking and its possible functions.

Tupaiinae   The remaining tree-shrews are distributed in South-East Asia and various regions of the Malay Archipelago. They are omnivorous, squirrel-like mammals with pointed snouts and either arboreal or terrestrial habits, depending on the species (Jenkins 1974). Their body weights lie between about 60 g (*Tupaia minor*) and more than 300 g (*Tupaia (Lyonogale) tana*). These purely diurnal tree-shrews differ from the predominantly nocturnal insectivores through their well-developed vision (e.g. retina composed primarily of cones: Samorajski, Ordy, and Keefe 1966; Tigges, Brooks, and Klee 1967) and the greater degree of development of the brain. However, despite this incipient dominance of the visual system, the sense of smell is well-developed (e.g. Meinel and Woehrmann-Repenning 1973; Campbell 1980). Under natural conditions, they live singly or in pairs in territories which they defend vigorously against conspecifics (Cantor 1846; von Holst, unpublished observations; Chorazyna and Kurup 1975; Kawamichi and Kawamichi 1979).

Studies of gland areas and possible scent-marking behaviour have only been conducted with species belonging to the genus *Tupaia*, notably with *Tupaia belangeri* (Lyon 1913), which has been referred to as *Tupaia glis* by a number of authors (for details, see Martin 1968*b*). Male *Tupaia* possess in their sternal region a prominent gland field which—at least in *Tupaia belangeri*—consists of greatly enlarged sebaceous glands together with apocrine glands. Secretion from this sternal gland begins when the animals reach sexual maturity at about eight weeks of age. Adults produce a varying amount of secretion, according to the individual, which leads to a yellow-brown discoloration of the chest region and to some extent of the belly as well (Sprankel 1961*a,b*; Vandenbergh 1963; Kaufmann 1965; Martin 1968*b*; Sorenson 1970; von Holst 1974; von Holst and Buergel-Goodwin 1975*a,b*; Richarz and Sprankel 1978; Kawamichi and Kawamichi 1979). Castration results, within a few days, in reduction of secretion and degeneration of the gland, while administration of testosterone can according to dosage) restore secretion to varying degrees in 1–2 weeks (von Holst, unpublished results). Males distribute the secretion of this gland in the environment by pressing the gland area on to objects or conspecifics and then rubbing to and fro (for references, see above). Following such marking, it is possible to identify on suitable objects an oily scent mark, which carries traces of the typical treeshrew odour. By means of this 'sternal marking' or 'chinning' in adult males—both under natural conditions and in captivity—the hairs in the gland region are worn away to expose the gland area itself (Martin 1968*b*; von Holst, unpublished results; Kawamichi and Kawamichi 1979). Following castration of males, however, the hair grows back again as a result of the marked decrease in marking activity. A corresponding sternal gland (also connected with chinning behaviour) is apparently also present in males of *Tupaia minor*, though in this species it has not been subjected to histological examination (von Holst, unpublished observations). Female *Tupaia belangeri* also possess a sternal gland, but they lack the proliferation of sebaceous glands which typifies the male gland (Sprankel 1961*b*; von Holst, unpublished results). The secretory activity of this gland is very limited in females, in captivity as well as in the wild (Kaufmann 1965; Martin 1968*b*; von Holst 1969; Kawamichi and Kawamichi 1979; for details, see also von Holst, p. 155 of this book), though it appears to be increased during pregnancy and lactation. Pregnant and lactating females can exhibit varying degrees of rust-brown discoloration of their normally whitish thoracic and abdominal fur, presumably because of oxidation of the secretion. Females also exhibit chinning in the same way as males, but they show this behaviour pattern far less frequently than males, which, at least in captivity, perform marking with their sternal glands several hundred times a day (for details, see p. 159). In line with this, the sternal gland region in females always retains its covering of hair.

Information content of sternal gland secretion In order to investigate the possible information content of the sternal gland secretion, it was rubbed off both male and female *Tupaia belangeri* with filter paper and the information content of the resulting 'scent marks' was tested with individuals of both sexes by means of conditioning experiments using a T-maze. Certainly, such tests cannot provide evidence of the biological significance of the information, but they do permit recognition of characteristics of the secretion which may possibly have some biological relevance, and which can be identified by the tree-shrews, independently of the animals' experience with the relevant chemical scent substances. These experiments have shown that the sternal-gland secretion conveys a variety of different kinds of information. Among other things, tree-shrews of either sex can discern from such scent marks the fact that they were made by conspecifics. Further, they can recognize the sex and sexual condition of the producer of the scent marks (e.g. juvenile male; fertile or castrated adult), along with its individual identity. Under constant laboratory conditions, the sternal-gland secretions of males are unique to each individual over periods of months and the scent marks are extremely durable. The information conveyed remains the same regardless of whether the scent marks are fresh or have been stored at room temperature for up to eight months. In other words, tree-shrews can recognize the individual producer of the scent even from a scent mark which is eight months old (von Holst and Lesk 1975). At this point, however, a special note must be made (for details, see p. 158). Although a secretion is quite definitely produced by the sternal-gland area, one can by no means exclude the possibility that the olfactory information demonstrated by the above tests is actually derived from some other source (e.g. from secretions of the abdominal or anal gland, from saliva or from urine). After all, tree-shrews regularly 'mark' their surroundings with a wide variety of body areas and excreta, such that scent substances of different origins are continually mixed together.

Male *Tupaia belangeri* also possess an abdominal gland, some 2 × 2 cm in size, which is located anterior to the penis and has basically the same structure as the sternal gland, except that it contains far fewer sebaceous glands (von Holst, unpublished results). This glandular area also produces a yellowish secretion which is, to the human nose, indistinguishable in smell from that produced by the sternal-gland region. The activity of the abdominal gland is similarly androgen-dependent. Males deposit the secretion of this gland by pressing it against longitudinal objects (such as downward sloping branches) or the substrate and pulling the body along by moving only the forelimbs. The hindlimbs are stretched out passively behind or (in the case of branches) allowed to hang downwards. This behaviour pattern is generally observed less frequently than chinning (Sprankel 1961a; Kaufmann 1965; Martin 1968b; Richarz and Sprankel 1978; Kawamichi and Kawamichi 1979; von Holst, unpublished results—for details, see p. 160). As with the sternal gland, the occurrence of abdominal marking or 'sledging' (Martin 1968b) is correlated with the development of a bare patch of skin, which recovers its hair following castration of males and the consequent reduction in their sledging activity.

With females, there is no histological evidence of a special abdominal gland area and there is no external sign of glandular secretion. Nevertheless, females do regularly exhibit 'sledging', though—once again—this behaviour is less prevalent than in males (Martin 1968*b*; see von Holst, p. 160 of this book).

In all animal species, urine contains substances derived from a wide variety of metabolic processes, notably hormones and their metabolites. It is therefore possible that urine itself or the products of specific glands discharged with the urine might convey information about an individual and thus perform a social function. For example, tree-shrews of both sexes are able to distinguish male from female urine and to tell urine from other chemical substances (unpublished results of conditioning experiments in a T-maze; other potential kinds of information have not yet been investigated). It is particularly likely that urine plays some communicative role when—as in the case with tree-shrews—urine is discharged through ritualized behaviour patterns in specific situations. In addition to normal urination, in which the animals empty their bladders without any special behaviour pattern and without moving away, male tree-shrews exhibit two forms of urine marking: 'dropwise urination' and 'urine trippling', with the latter occurring only rarely (Sprankel 1961*a*; Kaufmann 1965; Martin 1968*b*; Sorensen 1970; Richarz and Sprankel 1978; von Holst, unpublished results; for details, see p. 176).

Dropwise urination is an extremely common behaviour pattern of males. Females also show this pattern, but less frequently. During this urine-marking activity, the tree-shrew's body is hunched in a characteristic manner such that the haunches are brought forward and the penis (or the prominent clitoris in females) is lowered to contact the substrate. Droplets of urine are then expelled, while the animal moves forward with a shuffling gait, leaving a trail of urine droplets behind. In some circumstances, an individual can deposit 1–2 ml of urine within a few minutes in a small area by repeatedly moving to and fro, with the result that the urine is distributed even further on the tree-shrew's feet. In large enclosures, branches marked in this manner (especially by males) become, in the course of a few months, covered with a thick, sticky deposit which may hang down as large orange-coloured flanges beneath the branches. This substance has a strong smell of meat-broth (probably due to the presence of oxybutyric acid) which clearly distinguishes it from the smell of fresh urine.

More rarely, a behaviour pattern termed urine trippling (Urintrippeln: Sprankel 1961*a*) is exhibited by male tree-shrews, though it has never been seen with females. It would seem to me that this is in fact a more intensive, modified form of dropwise urination, since it usually occurs as a sequel to this form of urine marking. During urine trippling, the male moves forward only very slowly with his snout dipped to sniff at the ground and simultaneously making scratching movements with his hindlegs, with the result that the urine deposited is effectively spread over a patch of the substrate. This scratching increases in intensity to lead to a dance performed on the spot with the hindlegs, while the forelegs are kept unmoved.

According to Ortmann (1960), male tree-shrews (species unknown) possess two small glands in the anal region, and our own investigations have shown that these are lacking in female *Tupaia belangeri* (von Holst, unpublished data). Ortmann (1960) further states that *Tupaia* does not possess a proctodaeal gland, but Hamperl (1926) interpreted a gland he found in a male tree-shrew as a rudimentary proctodaeal gland. There is some evidence suggesting that anal gland secretions may act as scent marks, since males exhibit 'marking' with this region with considerable frequency (Sprankel 1961a; Martin 1968b; von Holst, unpublished data). Following defecation and at other times as well, male tree-shrews will drop their haunches and drag the perineum along the surfaces of objects. As a result, cages which have been occupied for some time, always have a dark, strong-smelling coating over certain branches and specific areas of the floor. This coating is probably a combined product containing not only secretion from the anal gland, but also urine, the secretion of other skin glands and possibly faeces as well.

In addition, isolated tree-shrews, pairs or even large family groups use specific sites on branches as defecation places. As a rule, all of the animals in one cage will use one and the same defecation site (Sprankel 1961a; Martin 1968b; Sorenson 1970; Richarz and Spankel 1978; von Holst, unpublished results—for details, see p. 166). These defecation sites are precisely maintained for months on end, such that high piles of faeces may build up on the ground below (reaching up to 70 cm in height in our enclosures). Use of a single defecation site of this kind is only lacking when animals are not yet accustomed to their cage or when there is social stress among the cage occupants, as with incompatible pairs showing mutual aggression or when young animals in a family group reach sexual maturity (for details, see von Holst 1974; see also p. 166).

Tree-shrews, particularly males, not only show the above forms of marking behaviour in response to new objects, but also exhibit frequent and intensive licking and smearing of saliva over the object. This behaviour can be so pronounced that branches and even entire cage-walls (in small cages) may become moist with saliva. However, in contrast to the other forms of marking behaviour described, such spreading of saliva is not regularly observed and it is not exhibited by all animals. It is conceivable that, as has been inferred for the Tenrecidae (Poduschka 1974b) and for the Erinaceinae (Eisentraut 1953; Poduschka and Firbas 1968), this behaviour permits close monitoring of chemical substances with the nose and Jacobson's organ, particularly since tree-shrews have a well-developed Jacobson's organ. But it is also possible that the saliva contains scent substances which are deposited on the animal's surroundings as 'scent marks'.

In some situations, notably in unfamiliar surroundings and during social confrontations, male *Tupaia belangeri* can exhibit abrupt discharge of a white secretion from the eyes, and sometimes from the nose as well. Within the space of a few seconds, this secretion can cover both eyes with a viscous white coating which is then shaken off or wiped away on suitable objects. In addition to tear glands and Meibomian (tarsal) glands, tree-shrews of both sexes have a histologically identifiable, large glandular complex which might be responsible for the production of this secretion (Cei 1947; von Holst, unpublished data). Despite the fact that the glands are present in both sexes, the secretion has only been observable externally in a number of male *Tupaia belangeri*, which exhibited regular signs of secretion. Such eye-gland secretion has never been observed in females.

Presumed functions of chemical substances in tree-shrews' social behaviour
Although a variety of tree-shrew species have been maintained in captivity from quite an early date, giving rise to numerous accounts of their behaviour under captive conditions, only one proper field-study has been published as yet (*Tupaia glis*: Kawamichi and Kawamichi 1979). This single study has, however, provided confirmation of interpretations derived from studies in captivity relating both to the social behaviour of tree-shrews and to the significance of chemical signals within the social context. The following account provides a short survey of the importance of scent-marks to tree-shrews, paying particular attention to the results of this field study. A more comprehensive account covering not only the social situations in which olfactory signals may be important in communication, but also the endogenous (hormonal) and exogenous (olfactory) conditions controlling the elicitation of marking behaviour, is given elsewhere (see p. 176).

According to Kawamichi and Kawamichi (1979), free-living *Tupaia glis* live in (permanent) pairs, as was first suggested by Martin (1968*b*) for captive *Tupaia belangeri*. Although the partners of each pair exhibit solitary ranging, they occupy the same territory, which they defend vigorously against conspecifics of their own sex. As a rule, the territories of different pairs overlap only to a slight degree. Young animals apparently live with their parents within the parental territory, thus forming family groups. After the attainment of sexual maturity, young animals are no longer to be found in the territories of resident pairs and they lead a nomadic, peripheral existence. These observations are in close accord with laboratory reports (especially of Martin (1968*b*) and von Holst (1969, 1974)).

As has been mentioned above, the members of each pair (together with any immature young) occupy virtually the same territory, with almost complete overlap between their individual ranges, despite the fact that the partners of a pair show solitary ranging. The coincidence of home-range boundaries between a male and a female ranging separately through a dense forest area with poor visibility obviously requires some effective characterization of the territory, and particularly of its boundaries. Such characterization is probably achieved predominantly by means of scent marks, particularly since marking with both sternal and abdominal glands counted among the commonest behaviour patterns observed under natural conditions by Kawamichi and Kawamichi. Marking behaviour was particularly commonly observed in the vicinity of territorial boundaries. In overlap areas, prominent places were sometimes used as common marking sites by different individuals from neighbouring territories. When animals came to such a marking site, they sniffed at it and very often marked it afterwards. On the basis of their field observations, Kawamichi and Kawamichi regard marking (with both sternal and abdominal glands) as playing a major part in territorial behaviour, though this does not in fact explain its function. As has been noted by Martin (1968*b*), among others, with respect to tree-shrews, marking of a territory could conceivably serve a number of functions, such as: to reassure the individual within its own home-range; to deter conspecific rivals from entering the marked area, or to demoralize them (thus contributing to the victory of the territory owner); or to label pathways within the home area, thus enabling the individual (or other members of its group) to achieve better orientation within the area normally inhabited. Scent marks could assume any of these functions, or even all of them combined, with the same result: the territory owner gains an advantage over any intruder. Such advantage is, in fact, evident under natural conditions. In all but one of the 34 aggressive chases observed by Kawamichi and Kawamichi, a resident chased another tree-shrew of the same sex from the area normally occupied by the chaser (about 80 per cent of the fights involved males). In addition, almost all of the chase sequences observed occurred on territorial boundaries.

It is, of course, possible that territorial demarcation might be achieved through scent-marking by both members of a pair, particularly since females do exhibit marking behaviour as well, both in the laboratory and under natural conditions (although at lower frequencies than males). However, numerous laboratory test results indicate that this is unlikely. Whereas the marking behaviour of females does not seem to be aggressively motivated (for details, see p. 161), males typically exhibit intensive marking of their surroundings with sternal, abdominal, and perineal regions, as well as with urine, before, during and (particularly) after aggressive interactions. Further, the laboratory males mark cage boundaries separating them from neighbouring rivals several hundred times a day, even without direct confrontation. Regardless of whether females also take part in the demarcation of a territory against neighbouring animals, the characterization of territorial area (amounting to about 10 000 m^2 under natural conditions) requires considerable durability of the scent marks, or of their information content (e.g. concerning the sex, sexual condition, and identity of the scent-mark donor). In our laboratory, 2,5-dimethylpyrazine has been identified as the primary stimulus for eliciting aggression by conspecifics and for 'territorial' marking with the various skin glands and with urine (von Stralendorff 1982*a*). This substance is extremely volatile in its pure form and in isolation its effect on the behaviour of conspecifics lasts only a few minutes. However, when it is present as a component of marking with urine and with the predominantly lipid-based secretions of the sternal and abdominal glands (lacking nearly completely in females), its effect is far more persistent. For example, it takes 24 hours for the effect in eliciting marking behaviour from conspecifics to fall to 50 per cent of the value found with fresh scent marks, and it takes 48 hours for the effect to disappear completely. In contrast to this effect on marking behaviour, the identity of an individual can still be determined from scent marks even after a period of months (as has been shown through conditioning experiments, see p. 210). The scent marks therefore contain at least two different kinds of information with quite different patterns of persistence. Under laboratory conditions, a marked area can apparently still be recognized individually after months have elapsed, whereas the territorial information relating to the rival or his sexual condition is relatively short-lived.

We have not observed any deterrent effect of the scent marks deposited by an unfamiliar rival. Instead, such scent marks are immediately sought out, intensively sniffed and then 'overmarked'. Nevertheless, one cannot exclude the possibility (and some laboratory results point in this direction, see p. 161) that scent marks may have a deterrent effect when a specific individual (or its scent) becomes recognized as dominant through fighting in a given area. We also have some evidence (as yet unpublished) that even scent-marking by a strange conspecific, if sufficiently concentrated, can exert an inhibitory effect on the marking and other behaviour of a rival. This could provide the basis for the superiority of territory owners over intruders: within a given range, the odour of an experimental cage previously scent-marked by a male occupant will stimulate subsequent marking activity by a male conspecific placed in the cage, and the intensity of his marking will increase with the duration of prior occupation by the 'scent donor'. However, when the duration of prior occupancy exceeds a certain level, the effect is reversed and the scent marks (or, more exactly, the concentration of the scent deposited by the donor) will inhibit marking by the individual subsequently tested in the cage. The same result is obtained by using synthetic 2,5-dimethylpyrazine; as its concentration increases, the odour of this substance progressively stimulates marking by male tree-shrews, but if the concentration exceeds a certain level, an inhibitory effect is exerted on marking behaviour (for details, see also p. 199).

In summary, both field and laboratory studies have provided strong evidence of the importance of scent glands and associated marking behaviour of males in the context of territorial behaviour. Scent marks may serve to label a territory (especially its borders), thus reassuring the territory owner as well as unsettling any intruder, with the combined effect that the territory owner's chances of victory are enhanced, as has been observed under natural conditions. Since the female of a pair occupies the same area as the male in the wild, she must be able to recognize individually her partner's scent marks and adjust her behaviour accordingly.

As yet, no information is available from the field regarding observations of scent-marking by tree-shrews in other social situations or with scent substances from other sources (e.g. urine, faeces). In captivity, however, in addition to the prominent border (territorial) marking performed by males, tree-shrews of both sexes also exhibit regular marking of the rest of their enclosures and will mark newly-introduced objects not bearing the odour of conspecifics (Sprankel 1961*a*; Kaufmann 1965; Martin 1968*b*; Sorenson 1970; Richarz and Sprankel 1978). Nevertheless, although these authors have equated this behaviour with territorial marking, in so far as any interpretation has been suggested, we regard this as likely to be incorrect (von Holst and Buergel-Goodwin 1975*a,b*). On the basis of ontogenetic and endocrinological studies, we favour the interpretation that such marking behaviour is differently motivated and serves for characterization of the home-range by both members of a pair (for details, see p. 161), though a functional separation between characterization of a range and territorial demarcation is admittedly difficult to construe.

Finally, in captivity one can observe (albeit relatively rarely) reciprocal marking between the members of a pair and marking of offspring by their parents (Martin 1968*b*; von Holst 1969, 1974; von Holst and Buergel-Goodwin 1975*b*; Richarz and Sprankel 1978). Martin (1968*b*) has suggested that this behaviour reinforces the pair-bond and we would also subscribe to this interpretation, although no functional explanation can as yet be given.

Other details of the various patterns of marking behaviour and their possible functions, along with the possible significance of olfactory stimuli in direct interactions between individuals (particularly during courtships and rearing of offspring) are discussed fully elsewhere (von Holst, see p. 158).

Comparative remarks on scent substances and their presumed functions within the groups considered

As was pointed out on page 106, the eight families considered in this account do not constitute a proper phylogenetic unit; instead they represent very ancient groups which had, to a large extent, already diverged along different evolutionary pathways in the Mesozoic (Thenius and Hofer 1960), undergoing adaptation to a wide variety of habitats and life-styles. It is therefore hardly surprising that the scent glands found in these different families are extremely diverse. In fact, histological and histochemical investigations of presumed scent glands have been few and far between, so that the tabular survey (Table 4.2) is more effective in highlighting the gaps in our knowledge than in providing a basis for morphological comparison. But even the scanty data so far available bear witness to the great diversity in structure of glands among closely related species of a single family or even within a single genus, as in the development of the subcaudal gland in *Crocidura russula*, which is lacking in closely related species of the same genus.

We know even less about the significance of these scent glands in the behaviour of these species. The available data are predominantly derived from individual studies conducted to varying depths on the behaviour of animals kept in captivity. Quite often, the sex of the animals studied is not even mentioned and the social contexts in which the scent substances are used remain unspecified. Detailed observations on specific marking activity involving gland areas, on the hormonal control of this behaviour, and on its relationship to exogenous stimuli have only been conducted with a few species (e.g. *Suncus* and *Tupaia* species; see also Table 4.3).

Sources of potential scent substances

The chemical substances mentioned here, which have potential relevance in intraspecific communication, can in principle be classified somewhat arbitrarily into three groups on the basis of their sources of production:

Faeces, including secretions from proctodaeal and/or anal glands   Faeces provide an obvious source of odours which could have communicative properties in mammals, as was first pointed out by Hediger (1944). However, it is as yet by no means clear how far faeces do actually contain specific information of biological relevance. Simply on the basis of differential environmental factors (such as variations in diet), faeces doubtless might exhibit recognizable differences from one animal to another. But it is likely that secretions from glandular areas of the proctodaeum and perineum are of particular significance. These secretions, which may be deposited alone or with the faeces, might exhibit recognizable differences due to metabolic variations between sexes and among individuals. Proctodaeal and/or anal glands have been identified in representatives of all the families dealt with here, with the exception of tenrecs and otter-shrews, which have not yet been examined (Table 4.2). The function of these glands may depend upon gonadal hormones, as has been shown for the anal glands in adult male *Tupaia belangeri*: their size and secretory activity decrease within a few days of surgical castration or under conditions of pronounced social stress, which inhibit gonadal functions (von Holst, unpublished data).

Faeces (together with the secretions of any associated glands) are deposited by species of all families so far examined either at specific defecation sites (e.g. tenrecs, otter-shrews, hedgehogs, elephant-shrews, and tree-shrews), or spread on objects in their living accommodation, most notably on the walls of their cages (e.g. shrews). Further, representatives of all families investigated are known to exhibit a specific form of marking behaviour, perineal drag, by means of which secretions from glandular areas of the perineal region (perhaps mixed with the faeces) can be deposited on the substrate as lasting scent marks. Although specific responses of conspecifics to faecal marks have not been described for any species, with the exception of *Galemys pyrenaicus*, the use of a common defecation site by different individuals in a family group (as in tenrecs or tree-shrews) as least suggests that such sites serve some biological function which remains to be identified.

Table 4.2. Scent glands of presumed relevance to social behaviour in members of different families. Entries have been restricted to glands which have been identified on morphological/histological grounds (M) or whose presence is suggested by pronounced secretion from specific regions of the body (S). Because of the generally limited nature of available data, no details can be provided on the types of glands involved.

The table specifies which of the sexes possess a particular gland. If it is definitely known that only one sex possesses the gland, that is all that is indicated. Where only one sex has been investigated to date, the opposite sex is marked with a '?'. In cases where both sexes have been investigated without yielding any evidence of a particular gland, this is indicated by '—'. Finally, where a gland has been identified but the sex of the animal was not specified, this is indicated by '+'. For references, see numbers in Table 4.3.

Genera investigated	Scent glands with presumed relevance to social behaviour							
	Sternal or ventral glands	Flank glands	Inguinal or abdominal glands	Aural glands	Eye-lid glands	Caudal or subcaudal glands	Anal glands	Proctodaeal glands
Solenodon	♂♀ S 47	♂♀ S 47			♂?? M.S. 10,11, 55		♂?? M.S. 49,63	♂?? M 49,63
Hemicentetes								
Setifer					♂♀ M.S. 22,38, 55			
Echinops					♂♀ M.S. 11,22, 55			
Oryzorictes					+ M 11			
Microgale					♂♀ M.S. 11,22, 55			
Geogale					+ M 11			
Potamogale					+ M 10			
Chrysochloris					—		♂?♀ M 31,32, 49	♂?♀ M 31,32, 49

Echinosorex							
Erinaceus			♂♀M.S. 11,57		♂♀M.S. 27 ♂♀M.S. 31,32, 49,64	♂♀M	31,32, 49,64
Sorex	– 16	♂♀M.S. 1,24,30, 39,64	♂♀S 57	♂M.S.* 16,48, 69	31,32, 49	♂♀M	31,32, 49
Neomys		♂♀M.S. 1,39,64	57		49	♂♀M	31,32
Blarina	♂♀M.S. 14,16, 50,51	♂♀M.S. 14,16, 50,51			31,32, 49	♂♀M	31,32
Crocidura		♂♀M.S. 1,39,44, 48	♂♀S 57	♂M.S.† 48	49	♂♀M	49
Suncus	♂♀M.S. 3,4,15	♂♀M.S. 3,4,15					
Cryptotis	♂♀M.S. 15						
Desmana	♂?♀M 41						
Galemys			+ M.S. 5,6,9, 64		31,32, 49,64	+ M	31,32, 49,64
Talpa			+ M.S. 63,69	♂♀M	31,32, 49,64	+ M	31,32, 49,64
Parascalops	♂♀M.S. 17				18,19	♂♀M	31,32, 49,64
Condylura	– 19 ♂?M 19		♂?♀S* 29,69	♂?♀M	19	♂?♀M	19,32
Macroscelides			♂M 9,52, 64,68				
Elephantulus			♂M 9		49	+ M	49
Rhynchocyon			♂M 9		49	+ M	49
Ptilocercus	♂♀S 27						
Tupaia	♂♀M.S. 37,40, 46,67 37	♂M.S. 37,40, 46,66 37	♂♀M.S. 37,11	♂M 37	12 32	♂?♀M ♂M	32,49

*No defined glandular region present; swelling of the entire tail.
†Only in *Crocidura russula*; not in *C. leucodon* or *C. suaveolens*.

Table 4.3. Secretions/excretions (or body regions) used in substrate marking (in familiar or unfamiliar areas) as well as in partner marking (including wrestling and grooming) during encounters between males and more or less receptive females. Body regions subjected to especially intensive investigation during courtship are mentioned along with the presumed social organization and the natural history of the species described so far.

Genera investigated	Social organization	Natural history	Marking behaviours observed in			Chemical scent substances of relevance during encounters between males and receptive females				Elimination	
			Familiar area	Unfamiliar area or familiar area with rival present	References	Partner marking	Regions especially investigated	Substrate marking	References	Specific sites	References
Solenodon	family group	nocturnal, terrestrial	PD		21,57	RV,RF,W	F,H		47	–	21,57
Tenrec	solitary	nocturnal, terrestrial		PD,RF	22	G	N,H,A,F		22,28	D,U	22,35
Hemicentetes	family group	crepuscular, terrestrial		PD,RV,RF, U,D	22,28	SE,W	N,H,V, B,A	PD	22,55	D,U	22,35
Setifer	solitary	nocturnal, semiarborial	PD,RE,U	PD,RV,RH, RE	22,55,56	RE	N,B,A,H	RV,RE,RH, D,U	22,54,55	D,U	22,57
Echinops	solitary	nocturnal, semiarboreal	RE	PD,RE,U,D	20,22,55	RF	F,H,N		22,38,55	D,U	22
Microgale	solitary (pair)	nocturnal, semifossorial	RE	PD,RE,UM	22,55	RF,W,G	N,A,H,F	U,D	22,55	U?	22
Limnogale	?	nocturnal, aquatic								D	28,45
Potamogale	solitary	?, fossorial		U,D	60					D	60
Echinosorex	solitary	nocturnal, terrestrial	PD	PD,UM	27	W	A,N		27	D	27
Erinaceus	solitary	nocturnal, terrestrial	UM,FM		53,55	U	B,A	U	58	–	53,57
Hemiechinus	solitary?	nocturnal, terrestrial				RH,W			57		
Sorex	solitary	polyphasic, subterranean	FM		34,57		A,H		13		
Blarina	solitary	polyphasic, subterranean						RV,RF,RA	26		
Crocidura	pair	polyphasic, subterranean	FM,RL		44,57,65		A,F		48	D	34
Suncus	pair	polyphasic, subterranean	RF,RV,UM FM,PD		3,4,15		N	RV,RF,RA	22,26	D	22

Genus	Social structure	Activity pattern, habitat	Marking / rubbing	Ref.	Other behaviour	Grooming (regions)	Ref.	Urine marking	Ref.	Elimination	Ref.
Cryptotis	pair (group?)	polyphasic, subterranean	RC	61,62						D	61
Galemys	solitary	nocturnal, aquatic								D	17
Parascalops	solitary	polyphasic, fossorial									
Macroscelides	pair	diurnal, terrestrial			F,RC	A,H,F	41			D	36
Elephantulus	group	diurnal, terrestrial				F	61			D	36
Ptilocercus	group	nocturnal, arboreal	RV,W		RV,RH,RE,A,H,B	A,N	27				
Tupaia	pair	diurnal, semiarboreal	RD,RV,UM	PD,RV,U 27; PD,RV,RE,UM 37,40,46,66	RV,W; W,G	A,H,B		UM,RV,RA	37,46,66	D	37,46,66

Abbreviations:

Behaviour:
FM: faecal marking
G: grooming
PD: perineal drag
RA: rubbing (Marking) with anogenital (cloacal) region
RC: rubbing with caudal region
RE: rubbing with eye region
RF: rubbing with flanks
RH: rubbing with different head regions (except the eye region)
RV: rubbing with ventral region
UM: urine marking
W: wrestling

Body regions investigated:
A: anogenital region
B: body general
F: flanks
H: head
N: nose

Elimination:
D: defecation
U: urination

References for Tables 4.1–4.3.

1. Ärnbäck-Christie-Linde (1907).
2. Ärnbäck-Christie-Linde (1914).
3. Balakrishnan and Alexander (1976).
4. Balakrishnan and Alexander (1980).
5. Brandt (1860).
6. Brinkmann (1911).
7. Broom (1902).
8. Broom (1915).
9. Carlsson (1910).
10. Cei (1946a,b).
11. Cei (1947).
12. Clark (1926).
13. Crowcroft (1957).
14. Coues (1896).
15. Dryden and Conaway (1967).
16. Eadie (1938).
17. Eadie (1939).
18. Eadie (1947).
19. Eadie (1948).
20. Eibl-Eibesfeldt (1965).
21. Eisenberg and Gould (1966).
22. Eisenberg and Gould (1970).
23. Eisentraut (1955).
24. Geoffroy-Saint-Hillaire (1815).
25. Godet (1951).
26. Gould (1969).
27. Gould (1978).
28. Gould and Eisenberg (1966).
29. Hamilton (1931a).
30. Hamilton (1940).
31. Hampel (1923).
32. Hampel (1926).
33. Harvey (1882).
34. Herter (1957).
35. Herter (1962).
36. Hoesch (1959).
37. Holst [Chapter 5 in this book] (1966).
38. Honegger and Noth (1966).
39. Johnsen (1914).
40. Kawamichi and Kawamichi (1979).
41. Kivett and Mock (1980).
42. Kolnberger (1971).
43. Kolnberger and Altner (1971).
44. Marlow (1955).
45. Malzy (1965).
46. Martin (1968b).
47. Mohr (1936).
48. Niethammer (1962).
49. Ortmann (1960).
50. Pearson (1944).
51. Pearson (1946).
52. Peters (1852).
53. Poduschka (1969).
54. Poduschka (1974a).
55. Poduschka (1974b).
56. Poduschka (1974c).
57. Poduschka (1977a).
58. Poduschka (1977b).
59. Poduschka and Firbas (1968).
60. Rahm (1961).
61. Richard (1973).
62. Richard and Valette Viallard (1969).
63. Richarz and Sprankel (1978).
64. Schaffer (1940).
65. Spannhof (1952).
66. Sprankel (1961a).
67. Sprankel (1961b).
68. Wagner (1841).
69. Weber (1928).

Urine, including secretions from accessory glands of the genital tract   Urine is likely to convey information on the sex and sexual conditions of an individual as a general rule, as has been demonstrated with conditioning experiments in the specific case of *Tupaia belangeri*. In males of this species, the chemical basis for this information is 2,5-dimethylpyrazine. With the European hedgehog, urine discharged by a male during the rutting season can be clearly distinguished from other urine by the human nose even at a considerable distance away. It has not yet been established whether, in the families considered here, urine also carries information regarding individual identity, but this would seem to be a reasonable assumption.

Tenrecs deposit urine, just like faeces, at specific sites, while representatives of other families (Erinaceidae, Soricidae, Tupaiidae) distribute their urine, often with specific patterns of marking behaviour, throughout their cages and particularly at cage boundaries. With all species so far investigated in detail, urine-marking is particularly prevalent with males during periods of mating activity. Further, in those species which have been studied in this respect (males of *Suncus* and *Tupaia* species), the marking behaviour responsible for the distribution of urine has been found to be androgen-dependent.

Particularly when discharged in the process of courtship, urine can also contain secretions from accessory sex glands of the genital tract and thus carry stimuli which are relevant for sexual behaviour. This is, at least, the case with hedgehogs and tree-shrews, which exhibit production of secretions from the genital tract during courtship. The same could also be true for other species, since many of them exhibit during courtship intensive interaction between the mating partners involving crawling above and below the other's body. In the course of this behaviour, substances from the genital tract might be deposited, though this is not directly apparent to the observer. A specific response to secretions from the genital tract alone (without the presence of any mating partner) has so far been observed only with *Tupaia belangeri* (see p. 169). In this species, secretions from the genital tract immediately bring about high sexual arousal in animals of the opposite sex. Such arousal is accompanied by specific vocalization known as 'rhythmic clicking' ('Kieseln' of Sprankel 1961*a*), which our observations have shown to be consistently elicited by both sexes during courtship.

Secretions from specific glandular areas of the skin, such as flank, sternal, ventral abdominal, inguinal, caudal, aural, and eyelid glands   Specific glandular regions of the skin have been described for members of all eight families. In many cases, secretions from these potential scent glands are deposited on the substrate or on conspecifics with highly ritualized marking behaviour. In addition to lachrymal and Meibomian glands, members of most families possess special eye glands producing a white secretion (particularly evident with tenrecs). In addition, with the exception of the hedgehogs and otter-shrews, all species possess histologically identified or at least inferred (in the case of tenrecs) glandular regions in a variety of specific regions of the body surface. This is particularly prevalent in shrews, which possess not only lateral (flank) glands, but also potential scent glands on the head, belly, and caudal areas, according to species. As a rule, these glands are better developed in males than in females, or they may even be completely lacking in females, as is the case for example, with the caudal glandular areas of *Sorex, Crocidura*, and the Macroscelididae. In both sexes, the activity of such skin glands is evidently dependent upon sexual condition, beginning with puberty and showing seasonal fluctuations in species with limited breeding periods. However, whereas the glandular activity of males generally shows a positive correlation with sexual condition (though for some contradictory results on shrews see p. 115), the situation with females is far from uniform. With female shrews, the activity of their normally prominent lateral (aural and ventral) glands shows a progressive decrease, accompanied by a reduction in intensity of their body odour, with oestrus, pregnancy, and lactation. By contrast, in female tree-shrews the activity of the otherwise quiescent sternal gland appears to increase during pregnancy to reach a maximum at about the time of birth of her offspring (for details, see p. 169).

Virtually nothing is known as yet about the information content of the secretions from these various body regions. Conditioning experiments have shown that the secretion from tree-shrews of either sex can carry a great deal of information, characterizing the species, the sex, sexual condition, and individual identity of the animals concerned. But it remains to be determined how far these various possible items of information are actually utilized by conspecifics. No comparable studies have so far been conducted with other species.

To date it remains completely unknown whether different skin glands of a given individual can also produce and discharge secretions with differing information content. It is, in any case, certain that scent marks made with specific glandular areas need not contain only chemical substances from those areas, since all of the species dealt with here deposit substances of very different origins on the substrate in the course of marking (e.g. in male tree-shrews, urine in addition to the secretions of sternal and abdominal glands and those from the perineal region). The substances concerned thus become mixed not only on the substrate, but also on the body regions used for marking. Similarly, scents from different sources may be spread over an individual's body (or even over the bodies of several animals) during self-grooming (vs. allogrooming) and solitary (vs. communal) occupation of nest-boxes for resting and sleeping. It is even possible that regions with especially pronounced secretory activity serve predominantly as carriers for highly volatile scent substances from other glandular areas (as proposed for *Suncus* by Dryden and Conaway 1967) or from urine (as in *Tupaia belangeri*—von Holst; for details, see p. 155). Further, behavioural data for the groups considered here provide no evidence for differing information content (vs. differential functions) of scent marks according to their origins. In situations which elicit marking activity (whether in home-cages, in unfamiliar surroundings, or during encounters with conspecifics), all species consistently exhibit marking with all available products (e.g. male *Suncus* mark with urine, faeces, ventral glands, and lateral glands). But regardless of the actual origin of scent substances of communicative relevance, all species so far studied have been found to distribute secretions from their different glandular areas on the general surroundings and on conspecifics by repeated marking behaviour.

The hormonal control of specific patterns of marking behaviour has so far only been investigated with respect to *Suncus* and *Tupaia*. With these two genera, it has been shown that gonadal hormones influence marking with urine, with the perineum and with cutaneous scent glands in males (and, in the case of *Tupaia*, in females as well; von Holst, unpublished preliminary results).

Presumed functions of the scent substances

Any attempt to analyse the significance of olfactory signals in mammalian behaviour is faced with the difficulty that the possibilities are not limited to urine, faeces, and saliva. Even in species without well-defined glandular skin areas, the entire body surface (including the soles of the feet) is covered with sebaceous and tubular glands whose secretions might have some communicative function. The presence of specific areas with pronounced concentrations of sebaceous and/or tubular glands simply underlines the significance of the relevant scent substances in the behaviour of the species concerned. This is particularly evident when the secretions from these skin areas can be deposited in the environment as more or less persistent scent marks and when the secretory activity of these glandular regions—along with the frequency of marking behaviour used to distribute their secretions—is correlated with specific functional systems (e.g. reproduction).

The release of potential scent substances, wherever they may be produced, can be achieved in various ways. In fact, the manner in which scent is deposited may to some extent permit conclusions regarding the significance of the relevant substances for the behaviour of the animals concerned (for more detailed information on this aspect, see for example Mykytowycz 1970; Eisenberg and Kleiman 1972; Bronson 1976; Petrovich and Hess 1978; Brown 1979).

Discharge of scent from the animals' general body surface   In the first place, scent is doubtless continuously discharged into the air to a greater or lesser extent from the general body surface and from specialized skin glands. According to prevailing climatic conditions, the scent would spread to varying degrees through the surrounding environment. Thus, in a manner comparable to visual signals in birds, individuals can receive information even from some distance away not only regarding the sex and sexual state of conspecifics, but also (under certain circumstances) regarding individual identity, as has been described for the common shrew by Crowcroft (1955, p. 74):

Before entering a rival's nest or cover box a shrew always 'pointed' and sniffed tentatively; if the resident was present, or had been a few seconds earlier, a retreat followed without further investigation. If the nest approached was its own and a stranger had entered in its absence, 'pointing' was followed by a precipituous rush, and the intruder was driven out.

It should be emphasized in this connection that even human beings can, with certain species, relatively easily recognize from some distance away not only the species itself but also the sex on the basis of body scents. With tree-shrews, it is even possible to discriminate (lactating) females with young from those without young.

Accordingly, members of species with a well-developed olfactory sense should be able to obtain, without direct bodily contact, all necessary advance information to adjust their behaviour appropriately to the situation. Depending on the sex and the sexual state of the conspecific and, possibly, depending upon previous experiences, such behaviour could be indifference to the conspecific, active avoidance, attack, or a sexual approach. Correspondingly, fighting between animals of the same sex usually takes place—in the species considered here—at once and without previous direct sniffing taking place. Exceptions occur in situations where the scent given off from an individual's body is perhaps too weak to permit clear-cut olfactory recognition. This is, for example, the case with tree-shrews of both sexes at the time when they are just reaching sexual maturity. Whereas adult tree-shrews usually attack adult conspecifics of the same sex without any preliminary (direct) sniffing of the body, young animals are repeatedly sniffed intensively and in a drawn-out fashion. Such sniffing is directed for a more or less extensive period of time at the sternal gland and at the anogenital region, before the investigating adult either loses interest in the young animal or attacks (see also p. 165).

The situation is different when animals of opposite sex are involved, particularly during mating seasons. In all of the species so far studied (see Table 4.3), such encounters are characterized by intensive reciprocal sniffing, licking, nibbling, and snout-thrusting of various areas of the body, most notably the genital region. However, even in this situation it is doubtless the case that the individuals concerned might have obtained olfactory information regarding the sex of the partner prior to any body contact. It is, of course, conceivable that the resulting intensive stimulation, especially of the genital region, might promote the discharge of secretions from the genital tract carrying additional information. Nevertheless, this behaviour is probably concerned only in a minor way with recognition, rather the continual and persistent stimulation by tactile and chemical stimuli might enhance receptivity to mating and, in females, might even induce ovulation (see also p. 168).

Overall, in my opinion, the essential function of the various scent glands resides in the broadcasting of an unequivocally identifiable body odour carrying a wide range of information at high concentration, which facilitates prediction of the likely behaviour of conspecifics during encounters and thus helps to avoid unnecessary or dangerous aggressive interactions. All other aspects, such as the deposition of persistent scent marks containing specific information, are probably subordinate to this primary function.

Passive discharge of scent marks   Chemical substances are continuously deposited on the surrounding substrate in a passive fashion, as is the case (for example) with substances from the soles of the feet or from subcaudal, ventral, or flank glands. Such 'scent marks' are subsequently discernable to varying degrees both to the individual itself and to conspecifics. Such scent deposition is likely to be especially prevalent with species which live in relatively closed tunnel systems and are repeatedly brought into direct contact with the surrounding walls (e.g. various species of the family Soricidae). Passively deposited scent marks of this kind might not only permit the animals concerned to orient themselves more effectively in their habitual ranging areas, but also provide conspecifics with information concerning the occupant of a range. Hence, along with actively deposited scent marks, such passively deposited chemical substances could be important in both territorial and sexual behaviour.

Active object marking   Active scent-marking has been documented for members of all families considered here. Such marking leads to intensive contamination of specific points or areas in the environment with the secretions of specialized gland regions, with faeces, or with urine. As has been shown, this often involves ritualized behaviour patterns (marking behaviour) such as ventral, flank, or abdominal rubbing; urine marking; or perineal drag. As a rule, members of the species discussed here mark with a balanced combination of various secretions and excreta (Table 4.3). Without going into possible differences between scent marks from different sources, scent substances deposited in this way, especially in association with the contribution of lipid-rich secretions (which are the dominant feature of most cutaneous glands), might be relatively durable. As a result, they would be particularly suited to the individual characterization of a given range area. In line with this, members of all species investigated so far in captivity mark their home-cages repeatedly every day and show particularly intensive marking of any introduced novel objects, or of the substrate of any unfamiliar cage whilst engaged in exploratory behaviour (Table 4.3).

The presence of such scent marks is recognized by conspecifics, as is evident from changes in their own marking and other behaviour. Whereas tree-shrews of both sexes respond to the presence of conspecific odours by marking more frequently than in a cage lacking such odours or bearing their own scent marks (for details, see p. 190), musk shrews of either sex exhibit inhibition of their marking behaviour in cages previously scented by male or female conspecifics (see p. 119).

Such observations suggest that scent marks (be they actively or passively deposited on the substrate) serve for the characterization of a home area, especially since all of the species discussed live (as far as is known) under natural conditions either in home-ranges of limited size (e.g. tenrecs, hedgehogs, and shrews), or in territories with well-defined borders (e.g. tree-shrews).

As has already been mentioned, scent-marking of an area can potentially serve a number of functions such as the marking of pathways within the home-range (enhancing the orientation of the range owner), the reassurance of the owner within its own range, or the deterrence of conspecific rivals from entering the marked area. Although the use of pathways is known from a number of different species under natural conditions, as yet there are no findings either from the field or from the laboratory which document the significance of scent marks in this context. The same is essentially true of the potentially reassuring effect of an animal's own scent marks. For virtually all of the species considered here, there is no evidence as to whether the presence of an animal's own scent marks actually represents (for example) an advantage during aggressive encounters. However, field observations on *Tupaia glis* do provide some evidence for such an effect (in combination with a possible deterrent effect exerted upon strange conspecifics): in this species, territory owners of either sex have been found to emerge victorious from encounters with conspecifics almost without exception. Nevertheless, it must be emphasized that under laboratory conditions the presence of an individual tree-shrew's own scent marks is by no means sufficient to guarantee a decisive advantage in a fight with a conspecific.

Finally, the scent marks of individuals may also exert a deterrent effect which excludes conspecifics from a given area. This possibility was first suggested for shrews by Pearson (1946): shrews of either sex apparently mark their territorial boundaries with secretions from their skin glands (notably the flank glands), thus informing other shrews that the area is occupied and deterring them from encroaching, except during the breeding season, when (in females, at least) the secretory activity of the presumed scent glands decreases (see p. 115). Observations conducted on *Suncus* (already mentioned on p. 119) are in accordance with this hypothesis. However, field data either fail to support this interpretation or even contradict it. For example, short-tailed shrews (*Blarina brevicauda*) of either sex have separate home-ranges, which are in each case covered virtually completely by the occupant in the course of a few nights. In most cases, however, the home-ranges are found to overlap extensively between neighbouring *Blarina*, regardless of their sex (Hamilton 1931*b*; Blair 1940). Further, the few field studies conducted on species of other groups (with the exception of the tree-shrews) have provided no indication of demarcated territories in the sense defined by Burt (1943), with exclusive occupation by the territory owner or group of owners. Thus, both European hedgehogs (Reeve 1980) as well as various mole species (Eadie 1938; Dalquest and Orcutt 1942; Godfrey and Crowcroft 1960; Richard and Valette Viallard 1969) lead solitary lives in home-ranges or tunnel systems which typically exhibit considerable overlap with neighbours. Encounters between conspecifics, regardless of their sex, can either be completely free of aggression, with the animals showing mutual avoidance, or lead to serious fights which result in one of the combatants being driven from the area or tunnel system. In captivity, where no such escape is possible, such encounters can lead to the death of the subordinate animals (e.g. with some tenrec species, hedgehogs, shrews, moles, and tree-shrews). It is, indeed, conceivable that such a fight may have the result that the ranging area of the victor is subsequently avoided by the loser and that the victor's scent may thus acquire a deterrent property. Nevertheless, in my opinion the evidence available to date generally contradicts the assumption of a deterrent effect of scent marks.

Scent marks do generally transmit information about the animal that deposited them, however. As a result, they permit the marking individual to achieve better orientation within its own home-range, and they also provide conspecifics with an advance indication, prior to any confrontation, of the occupation of the range by a conspecific which might be encountered in the course of any encroachment. Such information is probably of particular significance with respect to reproduction, since individuals of most of the species concerned lead solitary lives for most of the year and only seek out a mating partner during a limited mating season or period. The males typically play the active role in this. During mating periods, they generally show not only a considerable increase in their ranging activity, but also in some cases congregate in tunnel systems which are for the rest of the year occupied only by solitary females (e.g. shrews, moles, hedgehogs; references cited above). Correspondingly, species with restricted mating periods also show enhanced activity of their scent glands at such times.

Partner marking   Active deposition of scent substances on a partner (partner marking) is especially common among species living in pairs (or groups), as is the case with tree-shrews. In those species which lead solitary lives for most of the year, such as the majority of insectivore species, both active and passive partner marking is prominent during sexual behaviour. In the first place, the scent given off by the female's body surface and the scent marks she deposits doubtless assist the male in locating her. The subsequent encounter between male and female is then accompanied by intensive reciprocal sniffing and licking of various body areas in all species which normally lead solitary lives and in all cases where the sexes are kept apart in captivity. During such mutual investigation, particular interest is shown in the specific glandular areas of the body surface and in the anal (or cloacal) regions (see Table 4.3). Active marking of the partner has indeed been observed in various cases (e.g. in solenodons, in some tenrec species and in tree-shrews), but it does not represent a major component of courtship. However, there is a considerable degree of passive exchange of scent substances in all species. This occurs to some extent because during courtship the animals scent-mark the substrate, as is found with female tenrecs or male hedgehogs (which mark with urine) and with shrews, moles or tree-shrews (which mark with skin-gland secretions). As a consequence, the surrounding air is heavily scented with the chemical substances involved and the animals themselves pick up the scent substances directly by moving over them or during further marking behaviour. This is particularly pronounced with the European hedgehog in which—as has already been mentioned—the female is driven around by the male in a very limited area for a period of hours, such that she crosses his scent marks again and again. In addition, all species regularly exhibit intensive and long-lasting body contact during courtship in the form of reciprocal crawling over and crawling under, grooming, and copulation attempts made by the male. In this way, scent substances are also exchanged between the mating partners and spread over the body surface. Nevertheless, although courtship thus involves intensive olfactory stimulation of both partners, it is far from easy to decide whether the scent substances concerned serve any function other than the recognition of sex between partners and what such a function might be.

It is important to note at this point that insectivores, elephant-shrews and tree-shrews uniformly lack an oestrous cycle characterized by spontaneous ovulation with immediate female sexual responsiveness, as is found in most of the mammals. Instead, they apparently exhibit induced ovulation (tenrecs: Eisenberg and Gould 1970; hedgehogs: Poduschka 1977*b*; shrews: Pearson 1944; Crowcroft 1957; Mock and Conaway 1976; elephant-shrews: Tripp 1971; tree-shrews: Martin 1968*b*; von Holst, unpublished data). Although various species exhibit a post-partum oestrus which is usually fertile if a male is present (e.g. apparently all shrews as well as *Tupaia*: Brambell and Hall 1936; Hamilton 1949; Price 1953; Crowcroft 1955; Jameson 1955; Johnston and Rudd 1957; Martin 1968*b*; Dryden 1969; Hellwing 1971; Fons 1973; Mock and Conaway 1976; von Holst, unpublished data), it is usually the case (and always for females without young) that the activity of the male during courtship is apparently necessary to bring the female into a fully receptive state (references as cited above for the different families). There is considerable variation from species to species in the length of time that a male must be in contact with the female before she becomes sexually receptive (e.g. from less than an hour in *Tupaia* to several hours in hedgehogs and up to one or two days in some shrews such as *Cryptotis*).

There can be great variations in the importance of chemical substances in this context. With the European hedgehog, it has been suggested by Poduschka (1977*b*) that the female is kept within a limited area scented by the male during courtship. But scent marks may also inhibit the female, initially unresponsive and aggressive towards the male, from actually attacking her mating partner, as seems to be the case with the skin-gland secretions of shrews (Gould 1969). Poduschka's (1977*b*) suggestion that scent substances alone may be sufficient to induce ovulation in hedgehogs is, however, less convincing. It would seem far more likely that induction of ovulation depends primarily on tactile stimulation taking place during courtship and mating, as has been shown for several shrew species. On the other hand, it seems extremely probable that chemical substances produced by the female—most notably from the genital tract—would contain information regarding her receptive state. In other words, such substances (via the nose in general or the Jacobson's organ in particular) could inform the male during courtship of the female's readiness to mate and might even act as a releaser for mating behaviour.

Partner marking outside the context of mating has so far been described only for tree-shrews of both sexes. Such marking is an essential component of the behaviour of a well-adjusted and regularly breeding pair and it is doubtless of biological relevance, though its exact function remains unclear. The same applies to the marking of recently weaned young tree-shrews by their parents (for details, see p. 169).

Chemical substances are also highly likely to be of significance in the behavioural relationship between the mother and her offspring during the nest-phase, as has been demonstrated for rodents (e.g. Brown 1979). However, nothing is known about the importance of scent substances in this context among insectivores and elephant-shrews. With tree-shrews, it seems that substances produced (either exclusively or in greatly increased quantities) by the mother at the time of birth are deposited on the infants and protect them from being devoured by conspecifics (von Holst 1969; von Stralendorff 1982b). As yet, the origin of these substances remains obscure (for details, see p. 172).

Conclusions

In summary, it may be concluded that scent substances probably play a major role in the behaviour of members of the families considered here. But the functions of these substances are, as with species belonging to most other mammalian orders, largely unknown. The smell of an individual undoubtedly carries general information identifying the species, sex, sexual state, and individual identity, thus fulfilling the essential preconditions of any interaction between conspecifics. Further, in most species chemical substances are likely to be deposited as lasting scent marks carrying information concerning a home-range owner even when the latter is not physically present, thus assisting in the avoidance of unnecessary or dangerous aggressive interactions. Whereas the marking behaviour of tree-shrews probably has (among other things) central significance in territorial demarcation, it would appear that in insectivores the significance of chemical signals is greatest in the context of reproduction. Such signals facilitate encounters between mating partners, inhibit their aggressive tendencies during courtship and have an overall stimulatory effect on receptivity to mating. The same might hold true for elephant-shrews as well.

Thus, in comparison to other mammal species, the groups considered here (with the exception of the tree-shrews) appear to possess a relatively primitive chemical communication system whose operation is largely confined to a single functional context. However, this would not seem to be an original ancestral condition attributable to the phylogenetically ancient character of these groups, since even marsupials possess highly developed chemical communication systems (see Chapter 3). As yet, of course, one cannot exclude the possibility that the conclusions reached here are incomplete since the necessary data on the lives of the species involved (especially under natural conditions) are largely lacking. Yet it would seem more likely that these species, which live largely solitary lives except during mating periods, do not actually require a more elaborate chemical communication system. The information transmitted regarding sex and sexual state is adequate to permit avoidance of unnecessary social interactions and to facilitate encounters between mating partners. During direct encounters, on the other hand, it is likely that—as with various other mammalian orders—other kinds of signals (e.g. acoustic, tactile and, to some extent, visual signals) are brought into play.

Acknowledgements

I would like to thank Dr R. D. Martin for translating the manuscript and for his valuable comments. The work described is supported by the Deutsche Forschungsgemeinschaft.

References

Ärnbäck-Christie-Linde, A. (1907). Der Bau der Soriciden and ihre Beziehungen zu anderen Säugetieren. In *Morphologisches Jahrbuch* (ed. G. Ruge) pp. 463–514. Engelmann, Leipzig.
— (1914). On the cartilago palatina and the organ of Jacobson in some mammals. *Morph. Jb.* **48**, 343–64.
Balakrishnan, M. and Alexander, K. M. (1976). Hormonal control of scent marking in the Indian musk shrew, *Suncus murinus viridescens* (Blyth). *Horm. Behav.* **7**, 431–9.
— — (1980). A study on scent marking and its olfactory inhibition in the Indian musk shrew, *Suncus murinus viridescens* (Blyth). *Bonn. zool. Beitr.* **31**, 2–13.
Blair, W. F. (1940). Notes on home ranges and populations of the short-tailed shrew. *Ecology* **21**, 284–8.
Brambell, F. W. R. and Hall, K. (1936). Reproduction in the lesser shrew (*Sorex minutus* Linnaeus). *Proc. zool. Soc. Lond.* **1936**, 957–69.
Brandt, J. F. (1860). Observations nouvelles sur l'anatomie du musc. (*Moschus moschiferus* L.). *Bull. sc. Acad. imp. sc. St. Petersbourg*, 549.
Brinkmann, A. (1911). Die Hautdrüsen der Säugetiere (Bau und Sekretionsverhältnisse). *Ergebn. Anat. EntwGesch.* **20**, 1173–231.
Bronson, F. H. (1976). Urine marking in mice: causes and effects. In *Mammalian olfaction, reproductive processes, and behavior* (ed. R. L. Doty) pp. 119–41. Academic Press, New York.
Broom, R. (1902). On the organ of Jacobson in the elephant-shrew (*Macroscelides proboscideus*). *Proc. zool. Soc. Lond.* **1902**, 224–8.

— (1915). On the organ of Jacobson and its relations in the 'insectivora'. Part I. *Tupaia* and *Gymnura. Proc. zool. Soc. Lond.* 12, 157–62.

Brown, K. (1979). Chemical communication between animals. In *Chemical influences on behaviour* (ed. K. Brown and S. J. Cooper) pp. 599–649. Academic Press, London.

Burt, W. H. (1943). Territoriality and home range concepts as applied to mammals. *J. Mammal.* 24, 346–52.

Butler, P. M. (1972). The problem of insectivore classification. In *Studies in vertebrate evolution* (ed. K. A. Joysey and T. S. Kemp) pp. 253–65. Oliver and Boyd, Edinburgh.

Campbell, C. B. G. (1980). The nervous system of the Tupaiidae: its bearing on phyletic relationships. In *Comparative biology and evolutionary relationships of tree shrews* (ed. W. P. Luckett) pp. 219–42. Plenum Press, New York.

Cantor, T. (1846). Catalogue of mammals inhabiting the Malayan peninsula and islands. *J. Asiat. Soc. Beng.* 15, 171–203; 241–79.

Carlsson, A. (1910). Die Macroscelididae und ihre Beziehungen zu den übrigen Insectivoren. *Zool. Jb.* 349–400.

— (1922). Über die Tupaiidae und ihre Beziehungen zu den Insectivora und den Prosimiae. *Acta zool.* 3, 227–70.

Cei, G. M. (1946*a*). Morfologia degli organi della vista negli insettivori. I. Centetidi e potamogalidi (*Hemicentes semispinosus* e *Potamogale velox*). *Arch. ital. Anat. Embriol.* 52, 1–17.

— (1946*b*). Morfologia degli organi della vista negli insettivori. II. Sorecidi e talpidi. *Arch. ital. Anat. Embriol.* 52, 18–42.

— (1947). La différenciation de la grande nyctitante et sa signification phyletique chez les insectivores et les rongeurs. *Mammalia* 11, 69–110.

Chorazyna, H. and Kurup, G. U. (1975). Observations on the ecology and behaviour of *Anathana ellioti* in the wild. In *Contemporary primatology* pp. 342–4. Karger, Basel.

Clark, W. E. Le Gros (1926). On the anatomy of the pen-tailed tree-shrew (*Ptilocerus lowii*). *Proc. zool. Soc. Lond.* 4, 1179–309.

— (1959). *The antecedents of man.* Edinburgh University Press.

Coues, E. (1896). Three subcutaneous glandular areas of *Blarina brevicauda. Science, NY* 3, 779–80.

Crowcroft, P. (1955). Notes on the behaviour of shrews. *Behaviour* 8, 61–81.

— (1957). *The life of the shrew.* Reinhardt, London.

Dalquest, W. W. and Orcutt, D. R. (1942). The biology of the least shrew-mole, *Neurotrichus gibbsi minor. Am. Midl. Nat.* 27, 387–401.

Davis, D. D. (1962). Mammals of the lowland rain-forest of North Borneo. *Bull. nat. Mus. Singapore* 31, 1–129.

Doyle, G. A. (1974). Behavior of prosimians. In *Behavior of nonhuman primates*, Vol. 5 (ed. A. M. Schrier and F. Stollnitz) pp. 155–353. Academic Press, New York.

Dryden, G. L. (1969). Reproduction in *Suncus murinus. J. Reprod. Fert.* Suppl. 6, 377–96.

— and Conaway, C. H. (1967). The origin and hormonal control of scent production in *Suncus murinus. J. Mammal.* 48, 420–8.

Eadie, W. R. (1938). The dermal glands of shrews. *J. Mammal.* 19, 171–4.

— (1939). A contribution to the biology of *Parascalops breweri. J. Mammal.* 20, 150–73.

— (1947). The accessory reproductive glands of *Parascalops* with notes on homologies. *Anat. Rec.* 97, 239–51.

— (1948). The male accessory reproductive glands of *Condylura* with notes on a unique prostatic secretion. *Anat. Rec.* **101**, 59–79.

Eibl-Eibesfeldt, I. (1965). Das Duftmarkieren des Igeltanrek (*Echinops telfairi* Martin). *Z. Tierpsychol.* **22**, 810–12.

Eisenberg, J. F. and Gould, E. (1966). The behavior of *Solenodon paradoxus* in captivity with comments on the behavior of other insectivora. *Zoologica* **51**, 49–58.

— — (1970). The tenrecs: a study in mammalian behavior and evolution. *Smithson. Contrib. Zool.* **27**, 1–137.

— and Kleiman, D. G. (1972). Olfactory communication in mammals. *A. Rev. Ecol. Syst.* **3**, 1–32.

Eisentraut, M. (1953). Vergleichende Beobachtungen über das Selbstbespeicheln bei Igeln. *Z. Tierpsychol.* **22**, 810–12.

— (1955). A propos de la température de quelques mammifères de type primitif. *Mammalia* **19**, 437–43.

Estes, R. D. (1972). The role of the vomeronasal organ in mammalian reproduction. *Mammalia* **36**, 315–41.

Fons, R. (1973). Modalites de la reproduction et developpement postnatal en captivite chez *Suncus etruscus* (Savi, 1822). *Mammalia* **37**, 288–324.

Geoffroy-Saint-Hillaire, M. (1815). Mémoire sur les glandes odoriférantes des musaraignes. *Mém. mus. hist. nat.* **6**, 299–311.

Godet, R. (1951). Contribution à l'éthologie de la taupe (*Talpa europaea* L.). *Bull. Soc. zool. Fr.* **76**, 107–28.

Godfrey, G. K. and Crowcroft, P. (1960). *The life of the mole* (Talpa europea *Linnaeus*), Museum Press, London.

Goodman, M. (1975). Protein sequence and immunological specificity. In *Phylogeny of the primates* (ed. W. P. Luckett and F. S. Szalay) pp. 219–48. Plenum Press, New York.

Gould, E. (1969). Communication in three genera of shrews (Soricidae): *Suncus, Blarina* and *Cryptotis. Commun. Behav. Biol.* **3A**, 11–31.

— (1978). The behavior of the Moonrat, *Echinosorex gymnurus* (Erinaceidae) and the Pentail Shrew, *Ptilocercus lowi* (Tupaiidae) with comments on the behavior of other Insectivora. *Z. Tierpsychol.* **48**, 1–27.

— and Eisenberg, J. F. (1966). Notes on the biology of the Tenrecidae. *J. Mammal.* **47**, 660–86.

Graziadei, P. P. C. (1977). Functional anatomy of the mammalian chemoreceptor system. In *Chemical signals in vertebrates* (ed. D. Müller-Schwarze and M. M. Mozell) pp. 435–54. Plenum Press, New York.

Haeckel, E. (1866). *Generelle Morphologie der Organismen*. Band. II: *Allgemeine Entwicklungsgeschichte der Organismen*. Reimer, Berlin.

Hamilton, W. J. Jr (1931*a*). Habits of the star-nosed mole, *Condylura cristata. J. Mammal.* **12**, 345–55.

— (1931*b*). Habits of the short-tailed shrew, *Blarina brevicaudata* (Say). *Ohio J. Sci.* **31**, 97–106.

— (1940). The biology of the smocky shrew (*Sorex fumeus fumeus* Miller). *Zoologica* **25**, 473–92.

— (1949). The reproductive rates of some small mammals. *J. Mammal.* **30**, 257–60.

Hamperl, H. (1923). Zur Kenntnis in der Analgegend bei Insektivoren vorkommenden Drüsen. *Anat. Anz.*, Ergänzungsheft **57**, 233–42.

— (1926). Über Anal- und Circumanaldrüsen. 4. Mitteilung: Insektivoren. *Z. wiss. Zool.* **127**, 570–89.

Harvey, R. T. (1882). Note on the organ of Jacobson. *Q. Jl. Microsc. Sci.* **22**, 50–2.

Heberer, G. and Wendt, H. (eds.) (1972). *Entwicklungsgeschichte der Lebewesen.* Kindler, Zürich.

Hediger, H. (1944). Die Bedeutung von Miktion und Defäkation bei Wildtieren. *Schweiz. Z. Psychol.* **3**, 170–82.

Heim de Balsac, H. and Bourlière, F. (1955). Ordre de Insectivores: systématique. In *Traité de zoologie. Anatomie, systématique, biologie*, XVII, 2. (ed. P. Grassé) pp. 1653–1712. Mason, Paris.

Hellwing, S. (1971). Maintenance and reproduction in the white toothed shrew, *Crocidura russula monacha* Thomas, in captivity. *Z. Säugetierk.* **36**, 103–13.

Herter, K. (1957). Das Verhalten der Insektivoren. *Handb. Zool.* **8** (10), 1–15.

— (1962). Über die Borstenigel von Madagaskar (Tenrecinae). *Sber. Ges. naturf. Freunde Berl.* **2**, 5–37.

— (1968). Die Insektenesser. In *Grzimeks Tierleben*, 10. Band. *Säugetiere* 1, pp. 169–232. Kindler, Zürich.

Hill. W. C. O. (1953). *Primates: comparative anatomy and taxonomy*, Vol. I. *Strepshirhini.* Edinburgh University Press.

Hoesch, W. (1959). Elefanten-Spitzmäuse in Freiheit und in Gefangenschaft. *Natur Volk* **89**, 53–9.

Holst, D. v. (1969). Sozialer Stress bei Tupajas (*Tupaia belangeri*). Die Aktivierung des sympathischen Nervensystems und ihre Beziehung zu hormonal ausgelösten ethologischen und physiologischen Veränderungen. *Z. vergl. Physiol.* **63**, 1–58.

— (1974). Social stress in the tree-shrew: its causes and physiological and ethological consequences. In *Prosimian biology* (ed. R. D. Martin, G. A. Doyle, and A. C. Walker) pp. 389–411. Duckworth, London.

— and Buergel-Goodwin, U. (1975a). The influence of sex hormones on chinning by male *Tupaia belangeri. J. comp. Physiol.* **103**, 123–51.

— — (1975b). Chinning by male *Tupaia belangeri*: the effects of scent marks of conspecifics and of other species. *J. comp. Physiol.* **103**, 153–71.

— and Lesk, S. (1975). Über den Informationsinhalt des Sternaldrüsensekretes männlicher und weiblicher *Tupaia belangeri. J. comp. Physiol.* **103**, 173–88.

Honegger, R. E. and Noth, W. (1966). Beobachtungen bei der Aufzucht von Igeltanreks *Echinops telfairi* Martin. *Zool. Beitr.* **12**, 191–218.

Jameson, E. W. Jr (1955). Observations on the biology of *Sorex trowbridgei* in the Sierra Nevada, California. *J. Mammal.* **36**, 339–45.

Jenkins, F. A. Jr (1974). Tree shrew locomotion and the origins of primate arborealism. In *Primate locomotion* (ed. F. A. Jenkins Jr) pp. 85–115. Academic Press, New York.

Johns, M. A. (1980). The role of the vomeronasal system in mammalian reproductive physiology. In *Chemical signals: vertebrates and aquatic invertebrates* (ed. D. Müller-Schwarze and R. M. Silverstein) pp. 341–64. Plenum Press, New York.

Johnsen, S. (1914). Über die Seitendrüsen der Soriciden. *Anat. Anz.* **46**, 139–49.

Johnston, R. F. and Rudd, R. L. (1957). Breeding of the salt marsh shrew. *J. Mammal.* **38**, 157–63.

Kaudern, W. (1911). Studien über die männlichen Geschlechtsorgane von Insectivoren und Lemuriden. *Zoo. Jahrb.* **31**, 1–106.

Kaufmann, J. H. (1965). Studies on the behavior of captive tree shrews (*Tupaia glis*). *Folia primat.* **3**, 50–74.

Kawamichi, T. and Kawamichi, M. (1979). Spatial organization and territory of tree shrews (*Tupaia glis*). *Anim. Behav.* 27, 381-93.

Kivett, V. K. and Mock, O. B. (1980). Reproductive behavior in the least shrew (*Cryptotis parva*) with special reference to the aural glandular region of the female. *Am. Midl. Nat.* 103, 339-45.

Koch-Isenburg, L. (1955). Madagassische Wirbeltiere. *Umschau* 55, 335-8.

Kolnberger, I. (1971). Vergleichende Untersuchungen am Riechepithel, insbesondere des Jacobsonschen Organs von Amphibien, Reptilien und Säugetieren. *Z. Zellforsch.* 122, 53-67.

— and Altner, H. (1971). Ciliary-structure precursor bodies as stable constituents in the sensory cells of the vomero-nasal organ of reptiles and mammals. *Z. Zellforsch.* 118, 254-62.

Lim, B. L. (1967). Note on the food habits of *Ptilocercus lowii* Gray (Pentail tree-shrew) and *Echinosorex gymnurus* (Raffles) (Moonrat) in Malaya with remarks on 'ecological labelling' by parasite patterns. *J. Zool., Lond.* 152, 375-9.

Luckett, W. P. (1980). The suggested evolutionary relationships and classification of tree shrews. In *Comparative biology and evolutionary relationships of tree shrews* (ed. W. P. Luckett) pp. 3-31. Plenum Press, New York.

— and Szalay, F. S. (eds.) (1975). *Phylogeny of the primates. A multidisciplinary approach.* Plenum Press, New York.

Lyon, M. W. (1913). Treeshrews: an account of the mammalian family Tupaiidae. *Proc. U.S. natn. Mus.* 45, 1-188.

McKenna, M. C. (1975). Toward a phylogenetic classification of the mammalia. In *Phylogeny of the primates* (ed. W. P. Luckett and F. S. Szalay) pp. 21-46. Plenum Press, New York.

Malzy, P. (1965). Un mammifere aquatique de Madagascar: le limnogale. *Mammalia* 29, 400-12.

Marlow, J. G. (1955). Observations on the Herero musk shrew, *Crocidura flavescens herero* St. Leger, in captivity. *Proc. zool. Soc. Lond.* 124, 803-8.

Martin, R. D. (1968a). Towards a new definition of primates. *Man* 3, 377-401.

— (1968b). Reproduction and ontogeny in tree shrews (*Tupaia belangeri*) with reference to their general behaviour and taxonomic relationships. *Z. Tierpsychol.* 25, 409-95; 505-32.

Medway, L. (1969). *The wild mammals of Malaya and offshore islands including Singapore.* Oxford University Press, London.

Meinel, W. and Woehrmann-Reppening, A. (1973). Zur Morphologie und Histologie des Geruchsorgans von *Tupaia glis* (Diard 1820). *Folia primat.* 20, 294-311.

Meredith, M., Marques, D. M., O'Connell, R. J., and Stern, F. L. (1980). Vomeronasal pump: significance for male hamster sexual behavior. *Science, NY* 207, 1224-6.

Mock, O. B. and Conaway, C. H. (1976). Reproduction of the least shrew (*Cryptotis parva*) in captivity. In *The laboratory animal in the study of reproduction* (ed. T. Antikatzides, S. Erichsen, and A. Spiegel) pp. 59-74. Fischer, Stuttgart.

Mohr, E. (1936). Biologische Beobachtungen an *Solenodon paradoxus* Brandt in Gefangenschaft. II. *Zool. Anz.* 116, 65-76.

— (1937). Biologische Beobachtungen an *Solenodon paradoxus* Brandt in Gefangenschaft. III. *Zool. Anz.* 117, 233-41.

Mykytowycz, R. (1970). The role of skin glands in mammalian communication. In *Advances in chemoreception*, Vol. I. *Communication by chemical signals* (ed. J. W. Johnston Jr, D. G. Moulton, and A. Turk) pp. 327–60. Appleton-Century-Crofts, New York.

Niethammer, G. (1962). Die (bisher unbekannte) Schwanzdrüse der Hausspitzmaus, *Crocidura russula* (Hermann, 1780). *Z. Säugetierk.* **27**, 228–34.

Ortmann, R. (1960). Die Analregion der Säugetiere. *Handb. Zool.* **8**:26, 3(7),

Ottow, B. (1955). *Biologische Anatomie der Genitalorgane und der Fortpflanzung der Säugetiere.* Fischer, Jena.

Patterson, B. (1965). The fossil elephant shrews (family Macroscelididae). *Bull. Mus. comp. Zool.* **133**, 295–385.

Pearson, O. P. (1944). Reproduction in the shrew (*Blarina brevicauda* Say). *Am. J. Anat.* **75**, 39–93.

— (1946). Scent glands of the short-tailed shrew. *Anat. Rec.* **94**, 615–29.

Peters, W. C. H. (1852). *Naturwissenschaftliche Reise nach Mossambique. Zoologie. I. Säugetiere.* Reimer, Berlin.

Petrovich, S. B. and Hess, E. H. (1978). An introduction to animal communication. In *Nonverbal behavior and communication* (ed. A. W. Siegman and S. Feldstein) pp. 17-53. Erlbaum, Hillsdale, NJ.

Poduschka, W. (1969). Ergänzungen zum Wissen über Erinaceus e. roumanicus und kritische Überlegungen zur bisherigen Literatur über europäische Igel. *Z. Tierpsychol.* **26**, 761–804.

— (1974a). Das Paarungsverhalten des grossen Igel-Tenrek (*Setifer setosus*, Froriep, 1806) und die Frage des phylogentischen Alters einiger Paarungseinzelheiten. *Z. Tierpsychol.* **34**, 345–58.

— (1974b). Augendrüsensekretionen bei den Tenreciden *Setifer setosus* (Froriep 1806), *Echinops telfairi* (Martin 1838), *Microgale dobsoni* (Thomas 1918) und *Microgale talazaci* (Thomas 1918). *Z. Tierpsychol.* **35**, 303–19.

— (1974c). Fortpflanzungseigenheiten und Jungenaufzucht des grossen Igel-Tenrek *Setifer setosus* (Froriep 1806). *Zool. Anz.* **193**, 145–80.

— (1977a). Insectivore communication. In *How animals communicate* (ed. T. A. Sebeok) pp. 600–33. Indiana University Press, London.

— (1977b). Das Paarungsvorspiel des osteuropäischen Igels (*Erinaceus e. roumanicus*) und theoretische Überlegungen zum Problem männlicher Sexualhormone. *Zool. Anz.* **199**, 187–208.

— and Firbas, W. (1968). Das Selbstbespeicheln des Igels, *Erinaceus europaeus* Linné, 1758, steht in Beziehung zur Funktion des Jacobsonschen Organes. *Z. Säugetierk.* **33**, 160–72.

Powers, J. B. and Winans, S. S. (1975). Vomeronasal organ: critical role in mediating sexual behavior of the male hamster. *Science, NY* **187**, 961–3.

Price, M. (1953). The reproductive cycle of the water shrew *Neomys fodiens bicolor* Shaw. *Proc. zool. Soc., Lond.* **123**, 599–621.

Rahm, U. (1961). Beobachtungen an dem ersten in Gefangenschaft gehaltenen Mesopotamogale ruwenzorii (Mammalia-Insectivora). *Revue Suisse Zool.* **68**, 73–90.

Reeve, N. J. (1980). A simple and cheap radio tracking system for use on hedgehogs. In *A handbook on biotelemetry and radio tracking* (ed. C. J. Amlaner and D. W. Macdonald) pp. 169–73. Pergamon Press, Oxford.

Richard, P. B. (1973). Le desman des pyrenees (*Galemys pyrenaicus*). Mode de vie. Univers sensoriel. *Mammalia* **37**, 1–16.

— and Valette Viallard, A. (1969). Le desman des Pyrenées (*Galemys pyrenaicus*): premières notes sur sa biologie. *Terre Vie* **116**, 225–45.

Richarz, K. and Sprankel, H. (1978). Daten zum Territorial-, Sexual- und Sozialverhalten von *Tupa glis* Diard, 1820. *Z. Säugetierk.* 43, 336–56.

Romer, A. S. (1968). *Notes and comments on vertebrate paleontology*. University of Chicago Press.

Samorajski, T., Ordy, J., and Keefe, J. (1966). Structural organization of the retina in the tree shrew (*Tupaia glis*). *J. cell. Biol.* 28, 489–504.

Scalia, F. and Winans, S. S. (1976). New perspectives on the morphology of the olfactory system: olfactory and vomeronasal pathways in mammals. In *Mammalian olfaction, reproductive processes, and behavior* (ed. R. L. Doty) pp. 7–28. Academic Press, New York.

Schaffer, J. (1940). *Die Hautdrüsenorgane der Säugetiere mit besonderer Berücksichtigung ihres histologischen Aufbaues und Bemerkungen über die Proktodäaldrüsen*. Urban and Schwarzenberg, Berlin.

Simons, E. L. (1962). Fossil evidence relating to the early evolution of primate behavior. *Ann. NY Acad. Sci.* 102, 282–94.

Simpson, G. G. (1945). The principles of classification and a classification of mammals. *Bull. Am. Mus. nat. Hist.* 85, 1–350.

Sorenson, M. W. (1970). Behavior of tree shrews. In *Primate behavior* (ed. L. A. Rosenblum) pp. 141–93. Academic Press, New York.

Spannhof, L. (1952). Spitzmäuse. *Neue Brehm Büch.* 48, 1–44.

Sprankel, H. (1961*a*). Über Verhaltensweisen und Zucht von *Tupaia glis* (Diard, 1820) in Gefangenschaft. *Z. wiss. Zool.* 165, 186–220.

— (1961*b*). Histologie und biologische Bedeutung eines jugulosternalen Duftdrüsenfeldes bei *Tupaia glis* (Diard 1820). *Verh. d. zool. Ges.* 198–206.

Starck, D. (1978). *Vergleichende Anatomie der Wirbeltiere auf evolutionsbiologischer Grundlage. Band 1: Theoretische Grundlagen. Stammesgeschichte und Systematik unter Berücksichtigung der niederen Chordata*. Springer, Berlin.

Stephan, H. and Bauchot, R. (1960). Le cerveau de *Chlorotalpa stuhlmani* (Matschie) 1894 et de *Chrysochloris asiatica* (Linné) 1758 (Insectivora, Chrysochloridae). *Mammalia* 24, 495–510.

— — (1968). Vergleichende Volumenuntersuchungen an Gehirnen europäischer Maulwürfe (Talpidae). *Z. Hirnforsch.* 10, 247–58.

Stieve, H. (1948). Fortpflanzungsbiologie des Igels. *Verh. dt. zool. Ges.* 253–6.

Stralendorff, F. v. (1982*a*). A behaviorally relevant component of the scent signals of male *Tupaia belangeri*: 2,5-dimethylpyrazine. *Behav. Ecol. Sociobiol.* 11, 101–7.

— (1982*b*). Maternal odor substances protect newborn tree shrews from cannibalism. *Naturw.* 11, 553–4.

Tembrock, G. (1973). Land mammals. In *Animal communication* (ed. T. A. Sebeok) pp. 338–404. Indiana University Press, Bloomington.

Thenius, E. and Hofer, H. (1960). *Stammesgeschichte der Säugetiere. Eine Übersicht über Tatsachen und Probleme der Evolution der Säugetiere*. Springer, Berlin.

Tigges, J., Brooks, B. A., and Klee, M. R. (1967). ERG recordings of a primate pure cone retina (*Tupaia glis*). *Vision Res.* 7, 553–63.

Tripp, H. R. H. (1971). Reproduction in elephant-shrews (Macroscelididae) with special reference to ovulation and implantation. *J. Reprod. Fert.* 26, 149–59.

Vandenbergh, J. G. (1963). Feeding, activity and social behaviour of the tree shrew, *Tupaia glis*, in a large outdoor enclosure. *Folia primat.* 1, 199–207.

Valen, L. van (1965). Tree shrews, primates, and fossils. *Evolution* 19, 137–51.

Vogel, P. (1969). Beobachtungen zum intraspezifischen Verhalten der Haus-spitzmaus (*Crocidura russula* Hermann, 1870). *Revue suisse zool.* **76**, 1079–86.

Wagner, J. A. (1841). *Die Säugetiere in Abbildungen nach der Natur mit Be-schreibungen*, Supplement 2. Voss'sche Buchhandlung., Erlangen.

Weber, M. (1928). *Die Säugetiere. Einführung in die Anatomie und Systematik der recenten und fossilen Mammalia.* Gustav Fischer, Jena.

Wysocki, C. J., Wellington, J. L., and Beauchamp, G. K. (1980). Access of urin-ary nonvolatiles to the mammalian vomeronasal organ. *Science, NY* **207**, 781–3.

5 The primitive eutherians II: a case study of the tree shrew, *Tupaia belangeri*

DIETRICH V. HOLST

Introduction

Communication is a basic feature of all animal life. It is an indispensable component of any interaction between individuals of the same species, determining social structure and the content of social organization and thus the pattern of distribution of animals within their habitats.

It would seem to be perfectly clear what communication means in this context, although numerous authors have pointed out how difficult it is to draw a sharp boundary between that which is termed communication and that which is not (see, for example, Smith 1969; Burghardt 1970; MacKay 1972; Otte 1974; von Glasersfeld 1977; Petrovich and Hess 1978). Certainly, communication is not a purely mechanical phenomenon in that the amount of energy involved in the sender's transmission does not determine the receiver's response. Instead, in communication there is always a transfer of 'information' and it is in this respect that communication processes differ from chemical or mechanical interactions. To be effective in terms of communication, a signal—in the case of olfaction a mixture of ephemeral substances—must be both received and understood; that is, it must be decoded. By this means the receiver extracts information (a 'message'), for instance regarding the identity and sexual state of the individual whose scent has been received. The term 'information' is here used in the everyday sense, as a message which removes uncertainty. As the next step, the receiver must interpret the message; in other words, the appropriate meaning must be selected. In the process of decoding, all that is required is selection of the message which has been phylogenetically or ontogenetically linked to the signal. In interpretation, by contrast, the possible meanings are not pre-established in a conventionally fixed list; rather the receiver must take into account the context, that is aspects of his own age, sex, and state as well as aspects of the sender's state. This means that the same chemical substances may have quite different effects, depending upon the receiver's state and experience or upon the situation in which the substances are encountered. For example, a male *Tupaia* which has been surgically castrated or subjected to extreme stress over a period of several days (and thus rendered sterile) will not be attacked by strange conspecifics because the resulting withdrawal of androgens suppresses the synthesis and secretion of the (aggression-eliciting) scent typical of the intact male. Yet the lack of the typical male smell will not prevent a male con-

This work is dedicated to my esteemed teacher Professor Dr H. Autrum, on occasion of his 75th birthday.

specific from attacking if it has become acquainted with the (sterile) male as a rival in previous fights (nine experiments).

Context-dependent interpretation of a chemical signal is therefore an essential feature of communication in mammals, setting it apart from most forms of communication by means of pheromones in insects.* In this, a receiver must take into account not only its own state and that of the sender, but above all an explicit or implicit hypothesis as to why the message was sent. This concept of 'intent', 'purpose', or 'motivation' of the sender is essential for the definition of communication, as has been particularly pointed out by Burghardt (1970) and von Glasersfeld (1977). 'Intent' in this context means that it is of advantage (at least statistically) for the sender or its group to have its message received and understood—that means it is of adaptive value.

In the case of olfactory signals, it is especially difficult to recognize either the intent (motivation) of the sender or the adaptive significance of the signal to the sender or its group, since in mammals olfactory signals (in contrast to acoustic or tactile signals) are generally not produced over a short time span within a specific situation. Instead, the bodies of mammals continuously discharge a bewildering variety of chemical signals of widely varying origins. These signals, just like certain morphological features, characterize the sex, condition, and identity of the sender. The continuously emitted scent renders the behaviour of the sender to some extent more predictable during social interactions and is therefore of advantage to the receiver of the signal. But natural selection will have operated only if the response to the message conveyed meets the needs of the sender even more than those of the receiver. As a rule, however, the function (selective advantage) of such passive and continuously produced scents cannot be assessed, since in a social interaction an individual will always emit in addition a variety of visual, acoustic or tactile signals which combine with any chemical signals to determine the behaviour of the individuals involved.

Further, an individual's scent is not only present and operative during direct interactions between conspecifics; it can also be carried over long distances by air movements before exerting its effect upon another animal. Under more-or-less crowded laboratory conditions, scents dispersed in this way can of course have drastic physiological effects, particularly in rodents (e.g. on growth and maturation of young and on adrenocortical activity and reproduction of adults: Whitten and Bronson 1970; Macrides 1976; Loyber, Perassi, and Lecuona 1976; Wuensch 1979; Cowley 1980; Herreid and Schlenker 1980; Milligan 1980; Johns 1980; Vandenbergh 1980). However, it would seem unlikely that this is the actual function for which these scents were selected, since under natural

---

* For this reason, and in agreement with several other authors (e.g. Beauchamp, Doty, Moulton, and Mugford 1976; Bronson 1976; Brown 1979; Goldfoot 1981), I consider use of the term 'pheromone' for known or suspected chemical compounds in mammals to be unhelpful (or even misleading, due to the implied existence of an inherited specific predisposition to respond in a stereotyped manner), except where those compounds regularly release a defined response which is not context-dependent.

conditions the scent of an individual in the air is dissipated more or less rapidly. It seems more probable that these are side-effects which are not normally of significance (for a critical review of the differences between evolved functions and incidental effects, see Williams 1966; Eisenberg and Kleiman 1972; Otte 1974).

In addition to the continuous emission of total body odour, most mammals have the ability to make special scent marks on their surroundings. Tree-shrews, for instance, possess on certain parts of their bodies pronounced concentrations of sebaceous glands and scent glands which can be regarded as marking glands (sternal, abdominal, and anal glands). Tree-shrews regularly distribute the secretions of these glands in their surroundings by means of certain ritualized behaviour patterns, collectively referred to as 'marking behaviour' (sternal marking or 'chinning'; abdominal marking or 'sledging'; perineal dragging). Scent marks laid down in this fashion can persist for very long periods. For example, male tree-shrews can detect and discriminate sternal gland scent marks made by different individuals even when the marks are more than four months old (von Holst and Lesk 1975). Certain biologically relevant substances can also be contained in urine, faeces, and even saliva or secretions from eye-lid glands. These can be perceived by conspecifics either through the general olfactory epithelium lining the nasal cavity or through the special vomeronasal (Jacobson's) organs, which are well-developed in tree-shrews as well as in insectivores and elephant-shrews. Taste receptors can also play a role in the perception of such substances, since tree-shrews regularly deposit secretions and excretions on conspecifics as well as on their physical surroundings, sometimes with conspicuous marking or other behaviour patterns, such as 'dropwise urination' or 'mouth-licking' (see p. 173). Often these chemical substances were subsequently not only sniffed at but also intensively licked by conspecifics, thus activating olfactory as well as taste receptors.

The development of specific marking behaviour patterns, through which secretions or excretory products from a variety of sources are systematically deposited on surrounding objects or even on conspecifics, naturally indicates that the resulting scent marks have a communicatory function in the social behaviour of tree-shrews, and of other mammals too. However, although scent marking is known to occur in most mammalian species, we still have very little precise information on its function. Ralls (1971), in her review of this subject, sees marking as an integral part of agonistic behaviour and infers that the underlying motivation is intolerance of conspecifics on the part of dominant animals. Bronson (1976), on the other hand, suggests that marking is a general reaction to novelty, either physical or biological. Eisenberg and Kleiman (1972), agreeing with Johnson (1973), conclude instead that there is probably no single motivational state characteristic of all marking and that scent marks may be used for any of the following functions:

as a deterrent, or substitute for aggression, warning conspecifics away from an occupied territory;

as a sexual attractant or stimulant;

as a system labelling the habitat for an animal's own use or orientation, or to maintain a sense of familiarity with an area;

as an indicator of individual identity, perhaps including information regarding sexual status, age, dominance, etc.;

as an alarm signal to conspecifics.

Although all of these functions are plausible, and to some extent also supported by various findings, not one of them has received conclusive experimental backing as yet. In my view, this can only be possible when the behaviourally relevant substances have been chemically identified and isolated in a pure form. Through precise manipulation (introduction or removal) of such purified marking substances, their effects on the behaviour and physiological state of conspecifics in complex social groups can be analysed to reveal their functions and adaptive value for individuals. As yet, we are far from achieving this level of understanding.

The following text considers in some detail a number of investigations of tree-shrews (*Tupaia belangeri*) which help to clarify the significance of olfactory signals in the behaviour of this species. These investigations involved the following methodological approaches:

I. Behavioural observations of tree-shrews kept as pairs or family groups in relatively spacious cages. These studies were designed to identify situations and functional systems within which scent-marking is involved, in order to permit recognition of the motivational factors underlying marking behaviour. In parallel, various investigations were conducted into the responses of tree-shrews to conspecifics in a variety of physiological states.

II. Standardized experiments designed to reveal the information content of tree-shrew scent marks, the olfactory signals responsible for the elicitation of marking behaviour (along with their hormonal control and chemical nature), and the endogenous mechanisms responsible for the occurrence of marking behaviour.

Behavioural observations

Materials and methods

The investigations were conducted in a large room containing three cages (floor area approx. 7 m^2; height approx. 2 m) arranged side-by-side. The fronts of the cages and the intervening partitions were made of cage-wire while the remaining walls were brick-built. All cages were equipped with branches, several nest-boxes, and a floor covering of wood shavings. The animals were maintained under an artificial light regime with alternating 12-hour periods of light and dark; the light phase (= activity period) was from 8 a.m. to 8 p.m. Room temperature was maintained relatively constant at 25–28 °C and relative humidity fluctuated between 45 and 70 per cent. Food and water were always provided *ad libitum.* The observer (von Holst) kept his work desk in the room so that any behavioural peculiarities of the tree-shrews could be noted on a daily basis. According to the behaviour of interest at any given time, daily ethograms were compiled by selecting two animals on each occasion, observing them without interruption for the entire 12-hour activity period, and recording all clearly defined behaviour patterns on an event recorder (Metrawatt: Miniscript) for later evaluation. In all, eight adult males and eight adult females were observed for periods of up to three years, during which time between three and 30 daily ethograms per individual were compiled. In addition, according to particular conditions, daily ethograms ($n = 38$) were recorded for a total of seven young animals living together with their parents in the cages. These behavioural records will only be discussed here to the extent that they throw light on marking behaviour. Of the behaviour patterns which are relevant in this context, it was possible to quantify sternal marking, abdominal marking, mouth-licking, and defecation. Other marking processes were not unequivocally identifiable in all cases, so they were not included in quantitative evaluation. Under the large cage conditions it was impossible to identify eye-gland secretions or saliva-deposition on objects since the average distance of the observer from the animals was more than 5 m.

Results

General findings When accustomed to their cage conditions, adult tree-shrews of both sexes exhibit chinning (sternal marking) and sledging (abdominal marking) as their most conspicuous marking patterns. Under stable conditions, marking activity (number of marking actions per observation day) remains extraordinarily constant over periods running into years, though there are considerable differences between individuals in the values recorded (Table 5.1). Males mark considerably more than females and in both sexes marking with the sternal gland (chinning) predominates.

Both chinning and sledging (along with dropwise urination and perineal marking) generally occur in the same context. However, there is no significant relationship between the frequencies of these different kinds of marking. This is true whether one compares the overall average values from one animal to another, whether one compares the individual daily values from a single individual with one another, or whether the frequencies of occurrence of chinning and sledging of an individual within a single day are compared (time interval of 60, 30, or 10 minutes; correlation coefficients range from $r = 0.0$ to $r = 0.4$).

Although the marking behaviour of an individual can occur at different times on different days, such that on a given day more than 25 per cent of all marking events may take place within one hour, on average tree-shrews will exhibit a relatively uniform rate of marking over the day. In other words, there is no recognizable daily rhythm in marking activity. For five male tree-shrews, evaluation of a total of 15 observation days gave a figure of 8.33 ± 0.36 chinnings per hour, expressed as a percentage of total daily marking activity.

Table 5.1. Marking activity of male and female tree-shrews under stable conditions

	Marking activity of paired Tupaias		
	Days	Chinning per day	Sledging per day
Pair No.	n	M ± SE	M ± SE
♂BB	24	405.9 ± 9.4	21.0 ± 1.4
♂BP	12	252.1 ± 25.8	107.2 ± 7.9
♂VT	20	208.2 ± 18.4	11.9 ± 9.2
♂VT	3	185.7 ± 14.5	74.0 ± 10.0
♂CT	3	134.0 ± 34.4	105.0 ± 19.9
♂CK	3	97.7 ± 14.2	58.7 ± 15.3
♀BB	24	16.6 ± 1.6	0.1 ± 0.1
♀BP	12	32.4 ± 3.4	21.5 ± 4.4
♀VT	20	19.0 ± 4.4	7.8 ± 2.2
♀CT	3	12.2 ± 3.1	8.1 ± 2.0
♀PV	3	10.3 ± 4.2	13.0 ± 3.6
♀CK	3	62.3 ± 30.6	38.0 ± 17.2

In young animals, marking with the sternal and abdominal areas commences with the attainment of sexual maturity. Within four weeks, marking frequencies increase to reach typical adult levels. The first onset of marking was recorded between 46 and 81 days of age in 14 males, and between 55 and 72 days of age with 11 females.

As has already been mentioned, under stable conditions the marking activity of adult tree-shrews of both sexes remains constant for periods of several years. However, social conflict has an immediate influence on marking behaviour of subordinate animals. Tree-shrews of both sexes, when in familiar surroundings, will attack strange conspecifics and the latter are normally completely subjugated within a few minutes. From the time of subjugation onwards, the subordinate animal, when housed in a large cage, will hide in a relatively secluded place which it leaves only to eat and to drink. On subsequent days, fights between the victor and the loser are very rare, but the subordinate animal nevertheless shows drastic weight-loss and in males (females have not yet been investigated) gonadal hormone secretion falls to levels typical of castrates. The subordinate male will become sterile and die within 20 days if not removed from this situation in time (for details see von Holst 1969, 1972, 1974, 1977). Even in otherwise peaceable family groups, the presence of sexually mature offspring will lead sooner or later (after three to 14 months) to fighting between animals of the same sex, which similarly results in the subjugation of one of the two combatants.

Whereas the marking activity of dominant animals does not seem to show any overall change (six experiments; 14 observation days), sternal gland secretion is arrested within two to four days in subordinate males and from the time of their subjugation onwards they show no signs of marking behaviour whatsoever, be it chinning, sledging, perineal dragging, or dropwise urination. The same applies to females. However, if the dominant conspecific is removed, marking behaviour re-appears and returns to its previous level over a period of eight to 10 days (Table 5.2).

Table 5.2. Marking activity of one male and one female tree-shrew within six months prior to defeat by a conspecific, within 10 days after defeat, and within 12 months following removal of the dominant animal (n = number of observation days per period)

Marking (n/day)	Animal no.	Before defeat		After defeat		After removal of winner	
		n	M ± SE	n	M ± SE	n	M ± SE
Chinning	♂P	6	162.3 ± 17.3	4	0.0 ± 0.0	7	161.3 ± 11.5
	♀B	12	32.4 ± 3.4	3	0.0 ± 0.0	7	37.7 ± 5.8
Sledging	♂P	6	91.0 ± 10.8	4	0.0 ± 0.0	7	41.4 ± 11.0
	♀B	12	21.5 ± 4.4	3	0.0 ± 0.0	7	22.7 ± 4.9

The context of marking   Three different 'marking situations' can be distinguished on the basis of our observations of both male and female tree-shrews in large cages.

Border (territorial) marking: Tree-shrews of both sexes will vigorously attack strange conspecifics of the same sex both under natural conditions (Cantor 1846; von Holst, unpublished observations; Kawamichi and Kawamichi 1979) and in the laboratory, provided that they are in familiar surroundings. Even when animals are separated in individual cages by cage-wire partitions they will engage in daily fights of varying degrees of severity. Before, during, and (above all) following such fights, males regularly mark their surroundings very intensively with their sternal and abdominal glands as well as with urine. Moreover, they regularly mark the boundaries shared with animals occupying neighbouring cages, even if the occupants themselves are not actually present at the time. As a result, branches and areas of the floor at cage boundaries become covered with a thick, sticky conglomerate of urine and marking gland secretions which may hang down as orange coloured flanges beneath branches.

Such marking at boundaries, or in association with aggressive confrontations, accounts for approximately 80 per cent of all marking activities in males (chinning, sledging, perineal, and urine marking). Marking at boundaries is, as a rule, accompanied by a high degree of sympathetic nervous activity, which can be easily recognized in tree-shrews from the ruffled appearance of their tails (for details, see von Holst 1969, 1974).

If a male has more than one neighbour separated by a cage boundary, as was the case with the male occupying the middle of our three cages, the relative frequencies of marking of the different boundaries reflect the relative levels of aggressive interaction with the neighbouring males. For example, male BB, when occupying the middle cage, exhibited 90 per cent of all aggressive encounters with male A and only 10 per cent with male C. Correspondingly, BB performed 80 per cent of all boundary marking with sternal and abdominal glands on the border with A's cage (marked area 0.2 m^2; 24 observation days). By contrast, when male BP occupied the middle cage, he exhibited only 30 per cent of his aggressive interactions with male A and the rest with male C. In line with this, 60 per cent of all chinning and 90 per cent of all sledging was performed by male BP on the border with C's cage (12 observation days). In both cases the differences were highly significant ($p < 0.001$).

Females also fight with neighbours of the same sex, though not as often as males. However, no marking by females was observed during aggressive interactions or on cage boundaries.

Familiarization marking: In familiar surroundings any previously encountered objects which do not bear the scent of conspecifics (e.g. clean feeding bowls, branches, or sleeping-boxes), together with any novel but non-alarming items, are usually marked immediately by tree-shrews of both sexes with a combination of sternal and abdominal gland secretions as well as with urine. In addition, both males and females mark their enclosures repeatedly every day with secretions and excreta, in the absence of any stimulation from neighbouring conspecifics (and even when the latter are removed completely). In contrast to the 'aggressive' marking included in the first category, these scent marks are not deposited specifically on boundaries, but are distributed uniformly over the entire cage. This form of marking occurs, as a rule, in the absence of any apparent excitation and in the absence of any recognizable releasing situation. Such marking accounts for about 10 to 20 per cent of all daily marking with the sternal and abdominal glands in males and for about 50 per cent of marking with these glands in females. Marking with urine or with the perineal glands is rarely observed in this context and therefore no quantitative assessment is possible.

In our view, marking of this kind can be interpreted as a form of familiarization of the individual with its entire living space by impregnation with its own odour. This interpretation receives some support from the fact that both the marking frequency of male tree-shrews and the pattern of distribution of marking behaviour in the cage change in the presence of sexually mature offspring.

Effects of the presence of young animals on their parents' marking behaviour
From the time that they leave the nest, at the age of about 30 days, to the time they reach sexual maturity, young tree-shrews of both sexes are marked by both their parents with sternal and abdominal gland secretions (see p. 169). However, the total number of marking actions performed on the young is so limited that the marking activity of the males (in contrast to that of the females —see p. 167) is not significantly modified. The sternal glands of males begin to produce secretion at an average age of 60 days, at which time marking and copulation also make their appearance. At the same age, young females also become sexually attractive to their fathers (and any other males that are present); they are mounted at this time and usually prove to be fertile. Even after attainment of sexual maturity, months may pass without any form of agonistic interaction between parents and their offspring, whereas such sexually mature young animals are immediately attacked by strange conspecifics (see p. 165). When female offspring reach sexual maturity, the marking behaviour of their parents is not noticeably modified. However, when young males become sexually mature, the father's marking activity with sternal and abdominal glands increases to a greater or lesser degree, whereas the mother's marking activity is not appreciably altered (Table 5.3).

Table 5.3. The influence exerted by sexual maturity of young males on the marking activity of their parents.

Marking activity of two pairs of Tupaia before and after puberty of two male young living with their parents

	Before puberty of the young				After puberty of the young			
	Chinning/day		Sledging/day		Chinning/day		Sledging/day	
Pair No.	n	M ± SE	n	M ± SE	n	M ± SE	n	M ± SE
♂P1	3	170.0 ± 8.1	3	78.3 ± 11.2	6	317.7 ± 20.0	6	120.0 ± 10.1
♀P1	3	49.3 ± 7.6	3	24.0 ± 13.4	6	34.3 ± 4.5	6	13.2 ± 3.7
♂P2	9	379.6 ± 19.3	9	19.3 ± 1.5	10	427.6 ± 15.8	10	23.2 ± 2.8
♀P2	9	17.1 ± 2.3	9	0.2 ± 0.2	10	16.4 ± 2.1	10	0.0 ± 0.0

Even more conspicuous is the change that takes place in the spatial distribution of the adult male's scent-marking behaviour. (Observations were conducted on three different pairs with offspring, though it was only possible to record detailed quantitative information with one pair.) Prior to attainment of sexual maturity by his two male offspring, the father (just like the males of pairs without young) performed 90 per cent of all his marking with sternal and abdominal glands on the cage boundaries separating him from his two neighbours (marked area < 0.4 m²), only 5 per cent of his marking with these glands was conducted in the remaining cage area of 7 m². However, as soon as the young males reached sexual maturity the figure for marking in the remaining cage area increased to more than 20 per cent. Following the castration or removal of the young males, marking of the general cage area fell to a low level once again, but there was another significant increase when testosterone was administered to the castrated young males (25 mg Testoviron-Depot (Schering) i.m.) (Fig. 5.1).

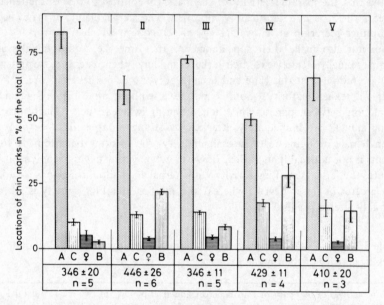

Fig. 5.1. Relationship between the chinning behaviour of an adult male and the sexual state of two male offspring (total observation period: 214 days).
  I: Prior to attainment of sexual maturity by the young males.
 II: Following attainment of sexual maturity by the young males.
III: After castration of the young males.
IV: Following testosterone treatment of the castrate males.
 V: After removal of the young males.
For each experimental period, the average marking activity level (mean ± SE) is given along with the number of observation days (n). Distribution of marking activity is indicated as follows:
A: Boundary with cage A.
C: Boundary with cage C.
B: Remaining area of home cage.
♀: Subject marking of the female by the male.

Hence the presence of sexually mature young males (or of their scent marks) leads to a redistribution of the adult male's scent-marking. He increases his own marking of the general cage area which is occupied and marked by young males.

But, in contrast to marking of the cage boundary, which—as has already been mentioned—often takes place as a direct response to contact with a neighbour, increased marking of the general cage area is not associated with direct inter-action with the young males or with their scent marks. However, such increased marking is associated with the male's raised level of arousal, which is possibly also responsible for the increase in the total level of marking activity (for details, see von Holst 1973, 1974).

Response of strange conspecifics to young animals before and after puberty
In parallel to the investigation of the distribution of the adult male's scent-marking in relation to the sexual condition of his offspring, studies were also conducted on the responses of strange conspecifics to the young. To this end, the young were placed singly and at irregular intervals in the cage of an isolated, strange adult male for a period of 10 minutes (age of the young animals at the first experimental introduction = 59 days). With one young male (SW), the sternal gland first exhibited signs of secretory activity at the age of 74 days, whereas with the second male (SM) secretory activity was first seen at 80 days. Simul-taneously, both males exhibited descent of the testes into the scrotum; they also began to mark their home-cage with their sternal and abdominal glands and chasing and mounting of the mother was observed. In other words, they had reached sexual maturity. As has already been noted, no aggressive interactions between the young and their parents occurred throughout the entire period of investigation.

Before they reached sexual maturity, both young males were immediately sought out by any strange male after introduction to his cage. He exhibited regular and intensive sniffing, for minutes at a time, of their anal regions and (to some degree) of their sternal regions. On each occasion, the strange male marked them once or twice with his sternal gland. Subsequently, however, the strange male would lose interest in the immature young (two experiments per young male). By contrast, once the young males had reached sexual maturity, accompanied by the onset of sternal-gland secretion (at 74 and 80 days respect-ively), a strange male would briefly sniff at their anal region, then at the sternal area and immediately attack so violently that the young males had to be re-moved (three experiments per young male).

Both young males were castrated at the age of 123 days, at which time they had fully developed testes and sperm were present in the epididymes. By eight days after castration, sternal-gland secretion was no longer discernible with either animal. When subsequently introduced to strange males, these castrate males were no longer attacked, though the stranger would sniff at the anal, sternal, and abdominal areas throughout the period of experimental intro-duction (two experiments per young male). Thirty days after castration, both young males were injected with testosterone. Their sternal glands soon began to show signs of secretory activity (after two days for male SW and after eight days for male SM). Eight days after the hormone treatment, both young males were introduced in turn to separate strange males. Both were immediately attacked so violently that they had to be removed immediately (one experiment per young male). Thus, in contrast to the situation within the family group, young males are immediately attacked by strange males as soon as they reach sexual maturity (or as soon as testosterone is administered to castrate males). Whereas attacks were preceded in all cases by relatively drawn-out sniffing of the anal and sternal regions when males had just attained sexual maturity, subsequently attacks would occur without any direct olfactory investigation, as is the case with adult males in general. Presumably, this is because such older individuals can be recognized as 'rivals' by conspecifics from some distance away.

Similar observations were conducted with two young males and two young females which, from the time they emerged from the nest onwards, were alternately placed every two to four days for 10 minutes at a time in cages containing single adult males or females or in an enclosure occupied by an adult pair (total of 54 experiments). All of the young animals were left largely unheeded by strange conspecifics, following a preliminary intensive bout of sniffing, as long as they were sexually immature. But both males were attacked after olfactory investigation of their anogenital and sternal gland regions, once their testes descended and the onset of sternal gland secretion occurred (at the age of 64 and 68 days, respectively). Adult females showed no interest in the young males after they had attained sexual maturity, but they attacked the two young females (from the age of 60 and 61 days, respectively) after drawn-out olfactory investigation of the anogenital region and the rest of the body, lasting several minutes. At about the same time, the young females became sexually attractive to males in their home-cage and they were seen copulating, so it can be concluded that they had reached sexual maturity.

Defecation   Isolated tree-shrews, pairs, and even large family groups usually defecate at one particular spot in their home-cage, usually selecting an elevated branch for this purpose. This defecation site is usually utilized by all cage occupants in such a closely co-ordinated fashion that a pinnacle of faeces forms on the ground beneath the branch (rising up to a height of up to 70 cm in our large enclosures). Further, if the branch originally used for defecation is moved to another spot and replaced by a fresh branch, it is the previous defecation site rather than the original branch which is used. In our investigations, the defecation site was usually located close to the centre of the cage.

Although all tree-shrews in a cage, both parents and offspring, normally use one and the same defecation site, faeces may be deposited at several different sites or even at random throughout the cage if the occupants are subject to persistent disturbance. This is the case if the animals are transferred to a new cage or if there are frequent aggressive interactions between the members of a pair or between parents and offspring. Although the presence or absence of a communal defecation site is therefore an extremely sensitive index of the degree of social tension existing between individuals in pairs or family groups of tree-shrews, we have as yet no evidence of a clear function for the faecal deposits. It is possible that tree-shrew faeces are impregnated with a special scent by means of the anal glands and hence serve in marking of the home-range, either for advertisement or for familiarization. However, such a function would seem to be questionable since tree-shrews generally show no response to fresh or old faeces of strange conspecifics.

Subject marking—*Marking of adults* Pair-living males mark their females several times a day with their sternal and abdominal glands. Such marking some-times occurs during occasional encounters between members of the pair, but it predominantly takes place prior to resting in close contact. The male will approach the resting female and mark her with his sternal gland. Subsequently, he will almost always slide over the female so that his entire abdominal and genital area is dragged across her body. Such sliding over the female's body may be repeated a number of times before the male settles down with her. This behaviour of the male doubtless leads to intensive exchange of scents with the female. But it should be noted in this context that paired tree-shrews, and indeed all members of a family group, will always sleep together in a single nest-box every night. This alone ensures extensive exchange of scents between individuals. Whereas this form of marking behaviour is always incorporated in the behavioural reper-toire of the males of well-adjusted, regularly breeding pairs, and comprises (for both marking behaviours) about 5 per cent of total marking activity (five pairs; 54 observation days; see the example given in Fig. 5.1), it has never been observed with pairs showing antagonism between male and female.

Females can also mark their males in a corresponding fashion, though there is enormous variation from one observation day to another in the level of such marking (mean value = about 50 per cent of total daily marking activity; range 0–80 per cent). As yet there is no explanation for this great variability.

When young animals are present in the cage, the male will mark them from the day they leave the nest onwards. Regardless of the sex of the offspring, he marks them at about the same frequency as he marks the female (four pairs with offspring; 17 observation days). When young males attain sexual maturity, they are no longer marked by their father, whereas he will continue to mark female offspring regularly even after they are sexually mature. However, the total amount of marking behaviour which the male performs on young animals is so small in comparison to boundary marking that only a negligible increase occurs in the male's overall marking activity. By contrast, females mark their young so frequently in the first few days following their emergence from the nest (up to 100 times a day or more) that her total marking activity is significantly in-creased. Females cease to mark their young in this way by about eight days after their emergence from the nest.

The male marks the female particularly frequently and intensively prior to, and sometimes after, copulation. In tree-shrew pairs, copulation generally only occurs during the post-partum oestrus of the female, that is to say approximately every 45 days, after the birth of each litter. It is the male that normally plays the active part during courtship. He sniffs and licks at the female's anogenital region, repeatedly nibbles cautiously at her neck and ears, and marks her intensively with his sternal gland. Following this, the female will usually adopt a presentation posture and copulation takes place. If the female is not (yet) receptive, the excitation of the male rises continuously. Then he will slide his body over hers, push beneath her while lying on his back, and rub against her side by side. The male continues to rub against the female in this way, with intermittent licking of her anogenital region and attempted mounts, until the female permits copulation to take place. When copulating, the male ejaculates after 10–50 thrusts. Subsequently he licks his penis and may repeatedly break off from this to mark the female with his sternal gland. Copulation, interrupted by brief pauses, can take place up to 100 times in a 2–3 hour period, after which the female rejects any further mounting attempts by the male. In the course of such a day when she is in oestrus, a female may be marked as much as 150 times with the male's sternal, abdominal, and anogenital areas.

In about 20 per cent of all cases, usually at the beginning of a copulatory period, the female took the initiative in mating. In a typical situation, the female would approach the male—usually while he was resting on a branch or nest-box—to sniff his sternal region and then make repeated mild jabs at his sternal gland with her snout. During such olfactory investigation of the male's sternal gland, the female may also mark him with her own sternal gland. Subsequently, the female will become increasingly aroused and will slide her body over the male repeatedly, pass beneath him on her back (just as the male might do) and then stand in front of the male with her tail raised. At this point, the male normally mounts the female, but if he does not do so the female resumes her sliding above and below the male's body for up to 10 minutes until mating takes place.

Thus, the most conspicuous feature of courtship in both male and female tree-shrews is this intensive 'wrestling' which will usually lead quite rapidly to sexual arousal of the partner. Of course, such behaviour—just like subject-marking taking place when the female is not in oestrus—doubtless leads to impregnation of the partner with skin-gland secretions. However, it does not follow that these scent substances, which (just like body contact alone) are also present during other forms of subject-marking, are responsible for the resulting sexual arousal. It seems far more likely, on the basis of observations of tame animals, that during such 'wrestling' the active partner (be it the male or the female) produces specific scent substances from the genital tract, which are rubbed on the mate and hence elicit sexual arousal.

Two hand-reared tame females consistently showed the behaviour described above at specific times, by responding to the human hand. They repeatedly rubbed themselves above and below the hand with ever-increasing signs of sexual arousal, producing a specific mating call (rhythmic clicking) audible only at close quarters and attempting to elicit mating through adoption of the copulatory posture (a similar observation has been described by Wharton (1950) for *Urogale everetti*). Within a short space of time (30–120 s) a white fluid, with no odour recognizable to the human nose, was discharged from the vagina and within a few minutes the observer's hand was entirely covered with this secretion. When the secretion was presented to tame males, it immediately elicited intense sexual arousal with continuous production of the mating call and attempts to mount the proffered hand. A corresponding discharge of secretion from the penis also took place with wrestling males, though no clear-cut response of females to this secretion could be identified. Anoestrous females showed no interest at all in the secretion, whereas oestrous females responded with the same pronounced sexual response (lordosis, mating call, etc.) with or without the male secretion. This does not preclude the possibility that this secretion has a sexually stimulating effect; it was simply impossible to demonstrate such an effect with tree-shrews kept in isolation.

Olfactory protection of the neonates   Female tree-shrews when paired with a male typically produce litters of offspring (1–4 infants) every 45 days. The infants are suckled immediately after birth and take in up to 50 per cent of their body weight in milk. Subsequently, the mother leaves the nest containing the infants and then visits them only once every 48 hours for 5–10 minutes at a time to suckle. No other conspecifics will normally enter the nest containing the young, even if they have previously used it as a sleeping-site. The following observations suggest that this effect can be attributed to scent substances produced by the female following birth.

As a general rule, females without young exhibit little or no signs of secretion in the sternal gland area. By contrast, in every single female examined directly after birth (37 births involving 14 different females) both the sternal gland region and the abdominal area were found to be moist. The secretion, in contrast to that of males, is watery and colourless. As yet, however, it has not been possible to establish whether this fluid is actually secreted by the sternal gland. Nevertheless, the histological appearance of the sternal glands of females sacrificed immediately after the birth of their young is suggestive of a mild increase in secretory activity (the lumina of the glands are approximately 20–50 per cent larger than in control females; $n = 10$). The secretion changes colour in the course of 24 hours, such that the entire ventral, thoracic, and abdominal region of the female changes from a yellowish white to a conspicuous dark rust hue. Young tree-shrews and their mothers have a strong smell detectable by the human nose which clearly distinguishes them from mothers without young. Further, tree-shrews of both sexes rapidly learned to distinguish (in a T-maze) filter paper which had been rubbed on the sternal gland region of a female just after birth, or on the body surface of the young themselves, from filter paper which had been rubbed on the sternal gland region of a female lacking young (see also p. 172). It should be noted, though, that there were three regularly breeding females which never exhibited this dark discoloration of the ventral body area. In addition, isolated females (particularly older individuals) were repeatedly found to exhibit rust-brown discoloration of the thoracic and abdominal regions.

If adult females are subjected to stress for a certain minimal period every day (at least two hours) by some kind of disturbance (fighting with the male; presence of sexually mature daughters; noise; unfamiliar human visitors), an interesting phenomenon is observed. Although such females will continue to produce young at regular intervals and will usually suckle them, within a few hours of birth one of the cage occupants (often the mother herself) will enter the juvenile nest, carry out the young and devour them just like some kind of animal prey (for details, see von Holst 1969, 1973, 1974). In such cases, there was usually no observable dampness of the ventral glandular area or discoloration of the ventral fur (90 per cent of all observations conducted on 118 litters produced by 27 different females). This suggests that females normally produce a scent substance which protects the young from conspecifics, and that there is a qualitative change in scent production when females are disturbed over long periods, with the result that the protective effect is lacking. In order to examine this question in more detail, a variety of experiments were conducted on young tree-shrews at the age of 1–4 days. Unless otherwise stated, the offspring were believed to be marked with the protective scent, since any remaining litter-mates were left unharmed by the parents. In the experiments, the infant tree-shrews were presented to single or grouped strange conspecifics in the home-cages of the latter.

A nest-box containing one infant tree-shrew was placed in a cage containing four adult tree-shrews. All of the cage occupants immediately approached the nest-box and the males marked it intensively with urine, with the sternal gland and with the abdominal gland. (Approximately 170 chinnings were performed by the two adult males within the first 30 minutes.) Subsequently, the adult tree-shrews gradually lost interest in the nest-box. Not one of them entered the box in a 24-hour period and it was therefore removed from the cage.

In two experiments, a nest-box containing two infants was placed in a cage occupied by six adult tree-shrews along with an identical nest-box used by the parents for sleeping. As in the previous case, all of the cage occupants immediately investigated both boxes and the two males marked them both, though predominantly marking the box containing the infants (197 against 162 chinnings respectively within the first 30 minutes). The parental sleeping-box was repeatedly entered by all of the cage occupants, whereas not one of them entered the infants' nest-box. In one instance, one of the infants was transferred from the original nest-box to the parental sleeping-box after one hour had elapsed. Although the cage occupants continued to enter the sleeping-box both infants were still unharmed after 24 hours and were therefore removed.

A neonate which was still partially covered with its embryonic membranes was immediately visited by the cage occupants. A female initially ate the embryonic membranes and remnants of the placenta, which were still attached to the umbilical cord, before marking the embryo more than 100 times with her sternal gland in an uninterrupted sequence. Yet, accompanied by her five cage-mates, she subsequently devoured the infant entirely within the space of three minutes. It is not known for certain whether this infant had been marked with the maternal secretion prior to the test. A litter-mate which remained in the original nest-box was not eaten by the parents.

Two infants, which had possibly been left unmarked (a third litter-mate had already been eaten by the parents) were transferred to a cage containing four adult tree-shrews after one infant had been previously rubbed on the dis-coloured ventral abdominal region of a lactating female. Both infants were immediately visited by the cage occupants. The untreated infant was first of all marked intensively more than 20 times by the two males, using the sternal gland, but after two minutes all four cage-mates joined in devouring the infant. The marked infant, by contrast, was left untouched by the cage occupants and was removed unharmed after 24 hours.

Three infants in a litter were carefully washed with soap and water. One of them was then rubbed on the sternal region of an adult male, one was rubbed on the corresponding region of a pregnant female, and the third was coated with paraffin oil. Subsequently, all three infants were laid side-by-side in a cage containing six adult tree-shrews. All of the cage occupants immediately approached the infants. The four males present at once showed very intensive marking of the infant impregnated with adult male secretion. The two females also marked this infant a number of times, but they showed far more interest in marking (160 times within the first 10 minutes) the infant marked with pregnant female scent, using both chinning and sledging. The adult tree-shrews showed no interest in the infant which had been rubbed with paraffin oil, but they began to eat it after eight minutes. This infant had been completely devoured within 24 hours; the two infants impregnated with tree-shrew scents were both still alive, but their tails and extremities had been bitten.

Two four-day-old infants which were obviously marked with the protective scent were placed in a cage containing an adult pair of tree-shrews after one of them had been washed, as described above, while the other had been left un-treated. Both infants were marked to the same extent by the adult male and female, with both the sternal gland and the abdominal gland (male: 31 chin-nings; female: 17 chinnings, within the first three minutes). Subsequently, however, the washed infant was devoured by the male of the pair, whereas the untreated infant was removed unharmed from the cage two hours later.

In two other experiments, a number of mealworms and two young mice in each case were rubbed on neonates so that reciprocal transfer of scent occurred. When the mealworms and mice were presented to two males and two females in a cage, they were immediately inspected, but then left untouched despite the occurrence of pronounced bite intention movements (two of the tree-shrews repeatedly made biting motions in the air alongside the prey). Untreated mealworms which were labelled and placed between the treated prey were immediately eaten.

Further tests were conducted with infants taken from females which had already eaten a litter-mate or which, on the basis of information concerning the degree of stress experienced by the mothers, could be predicted to devour their offspring. In all cases, strange conspecifics, usually following a bout of marking, were found to devour the infants within the first hour after presentation (15 experiments with 22 neonates).

Without going into these findings in detail, it seems reasonable to conclude that, at about the time of birth of their offspring, females produce scent substances through which infants are protected from conspecifics. It must, however, be noted that tree-shrew mothers have never been seen actively marking their offspring either immediately following birth (five observations) or during subsequent suckling visits. However, there is such intimate body contact between a female and her young during suckling that any scent substances present would inevitably be transferred from one to the other. As the above observations indicated, though, any (strange) infant found outside a nest is regularly marked with the sternal and abdominal regions by adult females, particularly when the latter are pregnant or lactating. It should also be remembered that females exhibit extremely intensive marking of their own offspring during the first eight days after their emergence from the nest. The scent substances produced by females obviously operate in an unspecific manner; the secretion from a strange female will protect the neonate just as effectively as the mother's own scent. In line with this, it is perfectly possible to exchange single infants or even entire litters from different females without any negative influence on the rearing of the young (five experiments). This is true even if the infants interchanged are of different ages. For example, no disruption of rearing occurred when single infants were exchanged between two litters of two young each, where one litter was 17 days old and the other was only two days old.

Lately it has been possible to obtain a 'musky' scent substance (by extracting nest material from successfully raised litters with acetone) which is found lacking in nest material from females whose young had been cannibalized after birth. When this extract is applied on neonates of rats or Mongolian gerbils it repels tree-shrews of both sexes and thus protects this 'prey' effectively from being eaten. The chemical analysis is as yet not resolved; the data, however, point to a similarity with phenylacetic acid (v. Stralendorff 1982a).

Hence, female scent subtances, whose origin is unknown to date, have the function of protecting infants from conspecifics. If this protection is lacking, as is always the case (for instance) with any female whose cage is also occupied by her sexually mature daughters, the infants are sought out and eaten like some kind of animal prey.

Mouth-licking   A conspicuous behaviour pattern which is likely to involve the transfer of scent or taste substances between individuals is that described by Martin (1968) as 'mouth-licking'. Pair-living adult tree-shrews exhibit mouth-licking 2–20 times a day, with a total duration of up to 10 minutes per day. The initiative is usually taken by an adult of either sex which 'invites' the partner to perform mouth-licking. The initiator approaches the partner, jabs at the angle of the latter's mouth with its snout, and then holds its head in front of the partner's mouth as saliva begins to dribble forth. As a rule, the partner immediately responds to this 'invitation' and licks at the proffered saliva. However, if the presentation gesture does not immediately elicit this response the initiator repeatedly jabs at the partner's mouth with its snout, often holding the partner's head in its fore-paws so that it cannot be turned away. A mouth-licking invitation is in rare cases completely rejected when the partner repeatedly turns its head away or even runs off.

Mouth-licking is one of the first behaviour patterns exhibited by tree-shrew neonates. Directly after birth, and in many cases before they have ever been suckled, neonate tree-shrews scramble up towards the mother's head and lick saliva flowing from her mouth (five observations). Such mouth-licking by the infants is repeated at all subsequent suckling visits made by the mother, and the mother herself usually elicits this behaviour by thrusting her snout at the off-spring and then holding her mouth, with its dribbling saliva, directly in front of their snouts.

The young first leave the nest, for a period of about 20–30 minutes, when they are approximately 30 days old. There is a rapid increase in the time spent away from the nest each day until, at an age of about 40 days, they cease altogether to visit the nest during the daytime. Subsequently, all rest periods occur at specific resting sites, often in conjunction with the parents and other cage occupants. As soon as the young tree-shrews make their first contact with the father following their emergence from the nest, he remains with them almost incessantly. As has already been mentioned, the father regularly marks the off-spring with his sternal and abdominal glands. He will also exhibit occasional nibbling at the neck region of the young, which is reminiscent of his behaviour towards the female during mating. However, the most conspicuous feature of his behaviour is his persistent invitation for mouth-licking. He presents his head to the young again and again for mouth-licking, to the extent that the young spend 5–10 minutes of their first day outside the nest in licking the father's mouth (32 records). Females, on the other hand, show considerably less interest in the young. Although they do mark the emergent offspring frequently, mouth-licking occurs only rarely and is of very brief duration (1-3 times on the first day of emergence; maximal total duration = 50 s). In subsequent days, mouth-licking between the mother and her infants shows very little change, whereas mouth-licking directed at the father by them increases daily to reach a total duration of up to 20 minutes as the young spend more and more time outside the nest. However, although the father exhibits a continuing high level of motivation for mouth-licking, the young show decreasing responsiveness to his invitations as they grow older, avoiding these by turning away or moving off.

As soon as young tree-shrews reach sexual maturity, mouth-licking between individuals of the same sex ceases completely. Even if a sexually mature young male attempts to join in mouth-licking of his father at the same time as the mother, the former will avoid this by repeatedly turning his head away or by moving off. Similarly, I have never seen mouth-licking performed between members of pairs which exhibited some degree of agonistic interaction. The cessation of mouth-licking between a male and his sons is accompanied by the above-mentioned increase in his overall marking activity and the redistribution of scent marking characterizing the attainment of sexual maturity of the young males. Correspondingly, when young females reach sexual maturity, they cease to exhibit mouth-licking with their mothers.

Although mouth-licking is clearly a sensitive indicator of the presence or absence of social tensions between adult tree-shrews, its function is as yet obscure. Whereas it is conceivable that this behaviour might guarantee the transfer of nutrients or symbiotic bacteria to young animals, this function would seem to be unlikely among adults. With the latter, however, mouth-licking may represent a ritualized form of the juvenile feeding pattern through which biologically relevant scent or taste substances are exchanged.

Conclusions

These observations conducted on tree-shrews provided with the opportunity for social contact with conspecifics have revealed situations in which marking behaviour predominates. These, in turn, indicate functional systems in which marking might be important and suggest possible motivational factors under-lying marking behaviour. The following three marking situations can be dis-tinguished for males:

1. *Border (territorial) marking:* this is the commonest context for marking. It occurs during fights and on the boundaries separating rival neighbours. Mark-ing is performed with urine as well as with the sternal, abdominal, and anal gland fields.

2. *Marking for familiarization of an occupied area:* this accounts for about 10 per cent of all marking activity. It increases when scent marks of fertile young males are present in the cage. Once again, marking involves urine as well as the sternal, abdominal, and anal glands. It is possible that the well defined defecation site also has a function in this context.

3. *Marking for the reinforcement of pair and group bonds*: this accounts for less than 10 per cent of total marking activity. Such marking is regularly performed on females of all ages and on sexually immature males, using the sternal and abdominal glands. Mouth-licking, which occurs in parallel with such marking, might also serve to reinforce pair and group bonds.

It would seem that the behaviour of sliding above and below the female's body during mating represents a special case. This, possibly in association with secretions from accessory sex glands in the genital tract, can evoke strong sexual arousal in a previously unreceptive female.

Female tree-shrews also exhibit marking in three different situations or functional contexts:

1. *Marking for familiarization of an occupied area:* in absolute terms such marking occurs about as frequently as in males, but since females (without young) exhibit considerably less marking activity overall than do males, it represents about 50 per cent of female marking activity. Females also mark with a combination of urine and rubbing of the sternal, abdominal, and anal regions, though they do not possess such well-developed glandular fields as found in males. Specific defecation sites are typically utilized in common by all the animals occupying a given area and might therefore have a marking function. 'Territorial marking' has not been observed with females.

2. *Marking for the reinforcement of pair and group bonds:* except when young have just emerged from the nest, this form of marking accounts for the remaining 50 per cent of female marking behaviour. As with males, it involves the use of the sternal and abdominal regions, and mouth-licking may similarly play a role in pair and group bonding.

As suggested for male tree-shrews, it is likely that sliding above and below the male's body prior to copulation represents a special case in which the sexual arousal of the male is enhanced through secretions from the female's genital tract.

3. *Marking of the young for protection against conspecifics:* the evidence collected to date suggests that the mother tree-shrew produces protective scent substances. However, in contrast to their behaviour towards young which have just left the nest, mothers have not been observed actively marking infants following birth, either with the sternal region or with any other area of the body. Therefore, it cannot be assumed that there is a specific motivational factor underlying the distribution of this productive scent on the neonates.

The observations so far conducted also reveal the following overall trends: marking behaviour (with the exception of mouth-licking and defecation at special sites) first appears with the attainment of sexual maturity and then increases over a period of about four weeks to reach a level which is subsequently maintained more or less constant. Males mark more than females overall, but the quantitiative difference is largely attributable to territorial marking, which is exclusively performed by the male. In addition there are considerable differences between animals in levels of marking activity. If adult animals are subjected to stress through subordination by a dominant conspecific followed by the continued presence of the latter, or if their gonadal functions are otherwise suppressed, marking activity is at once eliminated, though the process is reversible.

Standardized experiments

Introduction

As has been demonstrated by the findings to date, both marking with the sternal (and other) glands as well as with urine occur regularly in a wide variety of situations and functional systems in tree-shrews. Through a number of different behaviour patterns, secretions and excretory products from several different sources can be selectively deposited in the general environment or on conspecifics. Nevertheless, it is generally not possible to conclude that different patterns of marking behaviour and the substances involved (e.g. secretions of sternal, abdominal, or anal glands and urine) necessarily have different functions. It is indeed likely that some of the different kinds of scent mark do contain different categories of information, but tree-shrews in the course of their marking behaviour generally deposit secretions and excretory products from a variety of sources together in the environment. For instance, adult males will mark any object bearing the scent of strange conspecifics with their sternal, abdominal, and anal glands as well as with urine. In addition, saliva and sometimes secretion from the eye-lid glands can be smeared across the object. In the process, the various scent substances produced by the marking tree-shrew are deposited not only on the object but also over the animal's own body. Similarly, during communal resting in a nest-box or during auto- and allogrooming there is continual thorough exchange of scent substances both between the different individuals composing a group and between the various body areas of each single individual. Thus, even if a tree-shrew marks an object with, say, its sternal gland alone, the deposited scent mark may also contain substances deriving from other glands or from the urine. For example, the urine of male tree-shrews—which may be deposited in a specific pattern of urine marking (dropwise urination)—contains a chemical substance (2,5-dimethylpyrazine; see p. 199) which, according to its concentration, will elicit a greater or lesser degree of marking with the sternal and abdominal glands in male conspecifics. However, the same substance is also present in urine which the male discharges in the course of 'routine' urination (and which can be obtained by direct puncture of the bladder) and it is similarly detectable in quite high concentration in the vapour given off from a male into the surrounding air. So far, it has not been possible to determine whether this substance is also produced by any of the marking glands or whether it only reaches the fur through the transfer of urine. It is, of course, also possible that the glands of the sternal and abdominal fields in tree-shrews, which are particularly rich in fat content, largely operate to produce a carrier for this extremely volatile substance. In any event, differences in marking behaviour do not necessarily reflect differences in information-content and there is therefore no evidence for differential functions. Further, when the same marking patterns, such as chinning, sledging, and drop-wise urination (as in the above example) occur in different functional contexts, this must indicate that distinct motivational factors may underlie such patterns according to context.

One level of study of these different motivational factors is represented by analysis of the form or frequency of the marking behaviour in standardized test situations. Marking behaviour is particularly suitable for such analysis, since it is possible to investigate the responses of tree-shrews in different physiological states to scent substances, in the absence of any disruptive influence arising from conspecifics themselves. Tree-shrews respond to the presence or absence of scent marks from conspecifics or from other species with quite specific, quantifiable marking behaviour.

Materials and methods

The experiments were conducted with healthy tree-shrews (180 males, 18 females) of different ages which had been captured as juveniles or adults in Thailand or were born in our Institute's colony. All animals were kept in individual cages under constant ambient conditions (light–dark cycle of 12:12 hours; temperature $= 25 \pm 1\ ^\circ C$; relative humidity $= 50 - 60$ per cent) and had an *ad libitum* supply of water and food (Altromin pellets for tree-shrews).

The experiments were carried out in a separate room which could be viewed without disturbance to the animals from an observation chamber. The experimental cages each had a floor area of 70×50 cm and a height of 50 cm. The floors and roofs of the cages were made of wire mesh, the front walls of transparent Plexiglas, and the remaining walls and a rectangular rod fixed inside the cage consisted of opaque plastic material. On one side of each cage was a circular opening over which the sleeping-box from the animal's home-cage could be fastened from the outside. The animals were thus able to move directly from the sleeping-box (transport-box) into the experimental arena (Fig. 5.2).

Depending on the experimental situation, the experimental cage could be presented clean ('unscented') or intentionally contaminated with animal scent marks ('scented'). Unscented cages were soaked in a tub containing detergent, scrubbed thoroughly with a brush and finally rinsed with tap water before drying. Scented cages were clean cages that had been occupied by a so-called 'scent-donor', so that they had been contaminated with scent marks such as urine, faeces, skin gland secretions, and saliva.

For an actual experiment, a tree-shrew was brought into the experimental room in its transport box, which was then fastened to the experimental cage. The experiment began as soon as the animal entered the cage and (unless otherwise stated) lasted for 10 minutes. The animal was subsequently returned to its home-cage. During the experimental period a record was kept of the number of marking events involving the gland fields (sternal, abdominal, and anal), urination, and defecation.

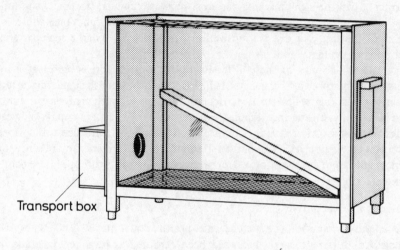

Fig. 5.2. Experimental cage.

The following account will largely concentrate on chinning, as changes in the internal state of any tree-shrew as well as changes in its surroundings generally had similar qualitative and quantitative effects on all patterns of marking behaviour, and since marking with the sternal gland was far more common than other types of marking, making it more suitable for quantitative analysis.

A marking action was recorded as one chinning event if an animal unambiguously pressed its sternal gland region against an object such as the rod, wall or floor of the cage and performed one to-and-fro motion. The number of such chinning actions performed during a 10-minute period differed considerably from one animal to another (as did other forms of marking behaviour), but remained essentially constant for a given individual in repeated experiments. The average number of chinning events per 10-minute experiment for each animal is referred to as its chinning value; it is a measure of the mean chinning activity of that animal in a single specific situation. The chinning values given below are always accompanied by their standard errors (= standard deviations of their means).

Results

General results for the marking behaviour of fertile males   If an adult tree-shrew is placed in an experimental cage, it will usually begin to explore and to mark it at once. If the experimental cage contains the scent marks of strange conspecifics, the marked areas are always sniffed intensively and then overmarked with urine, the sternal gland, the abdominal gland, and (more rarely) the anal gland. Males will often exhibit pronounced production of saliva, which dribbles from the mouth and is spread across the foreign scent mark with wiping motions of the head. Within a few seconds of entering the experimental cage, about 20 per cent of all males would regularly produce a copious discharge of eye-gland secretion, which could cover the eye completely with a white film or dribble out of the nose. This secretion was then wiped on objects or scattered with a shaking movement of the head. The great majority of males never showed such eye-gland secretion, however.

The introduced tree-shrews would mark not only the scent marks of 'scent-donors', but also the entire remaining area of the cage. In the course of this, independently of the condition of the cage, particular individuals exhibited specific preferences for marking sites (e.g. especially heavy marking of the rod, the floor, or one of the cage walls). No habituation to the scent marks of a specific scent-donor was observed. Even if a tree-shrew is daily placed in a cage previously marked by the same scent-donor in a series of experiments lasting for several weeks, its marking behaviour in response to the scent marks does not perceptibly change.

Marking activity in such experiments is highest in the first few minutes; approximately 40 per cent of all scent marks are deposited in the first three minutes of a 10-minute experimental period. Subsequently, marking activity progressively decreases along with exploratory behaviour (see Fig. 5.3).

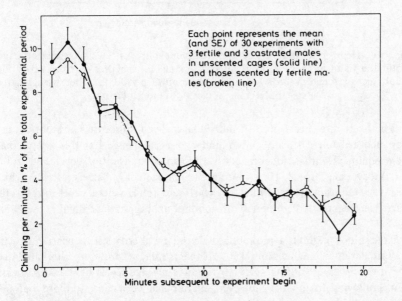

Fig. 5.3. Decrease in marking activity (chinnings/min) with the passage of time (duration of experiment: 20 minutes).

Whereas the animals remain continuously active in the first minutes of the experiment, they subsequently begin to exhibit self-grooming and rest-periods with increasing frequency. In unscented experimental cages, rest-periods can account for up to 70 per cent of each minute interval, while in cages which bear the scent of conspecifics the value is considerably smaller, at about 25 per cent (see Fig. 5.4).

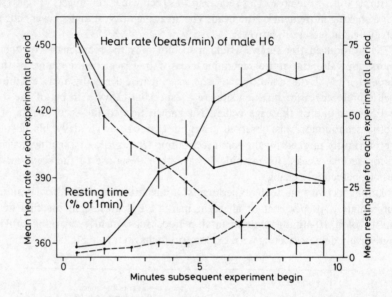

Fig. 5.4. Heart rate and resting time of a male tree-shrew in unscented cages (solid line) and in cages scented by fertile male conspecifics (broken line). The graph gives average values for each one-minute period together with vertical bars (unilateral) indicating standard errors.

The heart rate (beats/min) of individual males in unscented or scented cages was recorded telemetrically in a number of experiments. For this, the animals were equipped with a subcutaneously implanted transmitter developed in our laboratory (size: 15 × 10 × 4 mm; weight including battery: 1.5 g) which permitted continuous recording of heart rate (and, where necessary, of the electrocardiogram) for a period of about four months (for details, see Stöhr 1978).

Tree-shrews exhibit a pronounced day/night alternation in heart rate with nocturnal values ranging from 150–210 beats/min, according to the individual, and diurnal values (during activity) reaching average levels of 330–420 beats/min. Following disturbance, however, the frequency can climb to more than 660 beats/min within a few seconds; it may return equally quickly to the lower value existing previously (hundreds of observations with 21 different animals).

As a control experiment, single tree-shrews were shut into their sleeping-box (transport-box) and removed, but instead of being carried to an experimental cage they were returned to the home-cage. Although the heart rate of such control animals did increase to 420–540 beats/min during transport and directly after emerging from the sleeping-box, in less than a minute the heart rate had returned to the lower initial values (84 experiments with four males). On the other hand, when the sleeping-box was fastened to an experimental cage, the heart rate (increased by handling and transport) only decreased very slowly after the tree-shrews had entered the cage. Even after the elapse of the full 10-minute period of the experiment, the heart rate had never returned to the lower value prevailing before the experiment or following transfer back to the home-cage, despite the fact that all of the animals were completely familiar with the experimental situation (252 experiments with four males).

At the beginning of the experiments, all of the tree-shrews had high heart rates at comparable levels, regardless of the particular condition of the experimental cage. Contrary to expectation, the heart rates always declined faster in cages previously marked by fertile male or female conspecifics, and reached a lower level, than in unscented experimental cages. Nevertheless, scented cages were marked far more than unscented cages. For example, male H6 marked unscented cages an average of 26.8 ± 1.4 times with his sternal gland (26 experiments), while cages previously marked by a fertile male were marked an average of 51.1 ± 2.1 times (23 experiments). Prior to the beginning of the experiments (directly before handling produced a rise to over 420 beats/min), his average heart rate was 341 ± 7 beats/min. After 10 minutes in the experimental cage, the value descended to an average level of 388 ± 8 beats per minute if the cage was unscented, whereas it descended to 361 ± 10 beats/min if the cage was scented (Fig. 5.4). This difference is admittedly rather small but it is nevertheless statistically significant ($p \ll 0.05$), as in all the other experiments ($n = 203$). The more pronounced decrease in heart rate in scented cages is not the outcome of decreased locomotor activity in the animals concerned, since tree-shrews introduced to unscented cages are always considerably less active in locomotor terms, exhibit less marking behaviour and rest for a greater proportion of time (see Fig. 5.4). Further, there is no discernible difference in heart rate of the animals between periods of rest, self-grooming exploration, general locomotor activity or marking activity (for details, see Stöhr 1982).

Overall, these experiments have established the following points: the scent of a strange male or female conspecific does not evoke greater arousal (viz. higher heart rates) at the beginning of the experiment than does a situation in which no conspecific scent is present. As the experimental period progresses, arousal decreases faster and reaches a lower level in a scented cage, despite the fact that the tree-shrew in the cage is more active and marks more frequently than in the absence of conspecific scent. It might therefore be inferred that the overall decrease in arousal with the passage of time and the difference between scented and unscented cages could be a function of the test tree-shrews' own scent concentration: the longer an animal occupies the experimental cage, the more it invests it with its own scent marks, and the process is considerably accelerated in scented cages. This inference is further supported by the fact that male tree-shrews introduced to experimental cages already containing their own scent marks from previous experiments consistently show less marking activity than in unscented cages (for details, see p. 205). However, such an explanation conflicts with findings that the heart rate of an animal in an experimental cage already bearing its own scent marks does not show a different pattern from that found with a cage scent-marked by a strange conspecific (70 experiments with four males). As yet, it has not been possible to resolve this enigma.

Influence of the male's hormonal status on its marking behaviour   Ontogeny of marking activity: Male tree-shrews leave the nest at an age of about 30 days and reach sexual maturity at 60–80 days. As mentioned previously, puberty is indicated externally by a darkening of the scrotal skin and by descent of the testes. Simultaneously, the sternal and abdominal gland areas, which were previously inconspicuous, begin to assume a yellowish discoloration owing to the onset of secretory activity and marking of the surroundings begins.

In order to follow the ontogeny of marking in detail, five males were separated from their parents at an age of 33–35 days, placed in individual standard home-cages and tested for 10-minute periods in unscented and scented cages on alternate days. Unless otherwise stated, scent-donors in this and subsequent experiments were always strange fertile male conspecifics.

No animals showed any sort of marking behaviour prior to puberty; urine was discharged in the normal way without any sign of special urine-marking activity. With the onset of puberty, the tree-shrews began to mark with their sternal, abdominal, and anal glands as well as with urine (dropwise urination). The frequency of marking increased within a few days in all cases to reach a level that remained approximately constant for about 40 days. Then, at an age of roughly 100 days, marking activity with both glands and urine increased once again to attain a new plateau which remained unchanged thereafter. This second increase in marking activity coincides with the end of the growth phrase (Fig. 5.5).

Especially interesting is the response of the animals to the scent marks of fertile adult male conspecifics. For the first 30 days after the onset of marking activity, no animal showed different marking levels for scented and unscented experimental cages. The differences characteristic of adult tree-shrews did not appear until they were about three months old, at which age the scent marks of fertile male conspecifics, in comparison to unscented cages, began to exert a clear stimulatory influence on marking behaviour (chinning, sledging, and urine marking). This phenomenon cannot be ascribed to an inability of the young, prior to three months of age, to recognize the scent of conspecifics, since at all ages they were able to discriminate between pieces of unscented filter paper and paper bearing scent marks (urine or skin-gland secretions) of male conspecifics. After briefly sniffing at unscented paper, they showed no further interest in it, whereas they repeatedly sniffed at the scent-marked paper and sometimes bit and shook it in an apparently excited state (10 experiments).

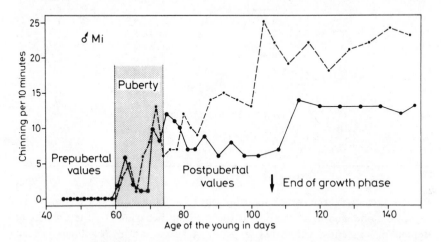

Fig. 5.5. Ontogeny of chinning in a male tree-shrew measured in unscented cages (solid line) and in cages scented by fertile males (broken line).

Marking behaviour in unscented and scented cages must therefore be controlled by two different central nervous sub-systems. Whereas one of these sub-systems matures at puberty, or at that time becomes accessible to stimulation by the increased concentration of circulating androgens, the sub-system responsible for marking in a cage scented by a fertile male conspecific does not mature until several weeks later.

Marking activity of adult males   The marking activity of adult males varies greatly from one individual to another. In unscented experimental cages, fertile males mark with their sternal glands at average rates of less than one to more than 30 times per 10-minute period, depending on the individual. The scent of fertile male conspecifics always raises the chinning rate (as well as other forms of marking behaviour), but the increase varies from individual to individual, from less than 50 per cent to more than 600 per cent (see Table 5.4). To take one example, the chinning value of male A in scented cages is somewhat higher than that of male 435, whereas in unscented cages male A marks only 20 per cent as often as male 435.

The concentrations of testosterone (and of dihydroxytestosterone) in the serum vary between 10 and 80 ng/ml depending on the individual (blood samples

Table 5.4. Chinning values of fertile male tree-shrews in unscented cages and in cages scented by fertile males

♂♂ No.	Chinning per 10 min				Increase in % of A
	A: Cage unscented		B: Cage scented		
	n	M ± SE	*n*	M ± SE	
9	27	34.2 ± 2.8	36	46.3 ± 3.4	35
76	12	30.4 ± 3.3	8	38.0 ± 5.5	25
435	10	25.4 ± 1.8	10	36.9 ± 2.6	45
J4	6	22.7 ± 3.8	7	50.6 ± 3.6	123
10	11	16.1 ± 1.8	12	32.9 ± 2.9	104
35	14	12.2 ± 2.3	11	20.8 ± 2.7	70
597	13	10.1 ± 1.0	18	24.8 ± 0.7	145
Ba	25	8.5 ± 1.0	17	34.7 ± 2.2	308
A	14	5.1 ± 1.3	28	37.8 ± 5.5	641
51	10	2.8 ± 0.7	20	21.1 ± 1.6	654
Mu	11	1.6 ± 0.4	10	5.3 ± 0.7	231
14	29	0.3 ± 0.1	38	0.8 ± 0.2	167

always taken two hours prior to lights on), but there is no correlation at all between blood androgen levels and the marking behaviour of the animals in scented or unscented cages (two experimental series with 26 animals yielded correlation coefficient values of less than 0.1). Hence, there must be variation between individuals in their sensitivity to circulating androgens.

Castration of fertile males   In order to determine the influence of androgens of gonadal origin on marking activity, 15 adult males with different initial marking values were castrated. In general, marking activity decreased following castration, but the degree and pattern of this decrease varied greatly from one animal to another. In some individuals, marking activity was reduced to the low post-castration level within 24 hours; in others, the reduction occurred progressively over a period lasting as long as four weeks (see Fig. 5.6 for examples). Yet, as expected, the serum testosterone concentrations of all individuals fell to less than 1.0 ng/m on the day following castration.

In general, castration of fertile males typically leads to a reduction in marking activity, especially pronounced with chinning, sledging, and dropwise urination. Anal marking also decreases, but because of the low overall level of marking activity involving the anal glands this effect is not very drastic. Although pre-pubertal tree-shrews show no marking behaviour at all, adult castrates all retain a certain degree of 'residual marking activity' which varies from 0.0 to 7.8 chinnings per 10 minutes in unscented cages and from 2.4 to 15.4 chinnings in scented cages (see examples, Table 5.5). Because individual variation is so great, some castrates may even mark more frequently than certain fertile males.

Under both experimental conditions (scent of fertile males present or absent

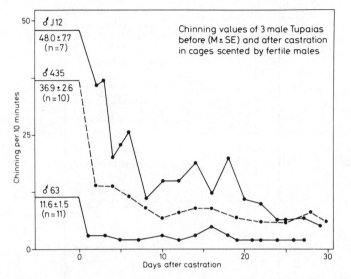

Fig. 5.6. Decrease in chinning activity of male tree-shrews following castration.

in the experimental cage), the marking activity of castrates is positively cor-related with the level of activity measured prior to castration in the same type of experimental set-up ($r \gg 0.9$; $p \ll 0.001$). Residual marking activity (in per cent of the precastration values), on the other hand, varies appreciably both between individuals and also for single individuals in comparisons of unscented and scented cages ($r \ll 0.1$). This shows that castration has differential effects on the marking behaviour of individuals in unscented cages as opposed to scented cages.

Influence of testosterone on marking activity of fertile and castrated males
If fertile male tree-shrews are injected with testosterone (2-25 mg testosterone propionate; Schering), this has no discernible effect on their marking behaviour, even if the testosterone levels measured in the blood are elevated several times above their normal values (22 experiments with five males). However, when castrates are injected with testosterone within nine months of castration, marking activity in both scented and unscented cages increased to levels which are maintained for 4 -20 days (according to the dose injected) before falling back to pre-injection values. The response to injection was always fully estab-lished 24 hours later (the earliest time of testing), as is shown in Fig. 5.7. As a rule, however, it proved to be impossible to restore marking activity to pre-castration levels even through repeated hormone treatment (11 experiments with eight animals; 203 data points). Injection of a testosterone depot (Testo-viron-Depot; Schering) similarly led to an increase in marking activity in castrate males, but the effect usually emerged somewhat more slowly in both scented and unscented cages and persisted for a number of weeks (Fig. 5.7).

Table 5.5. Marking behaviour of male tree-shrews before and 4–8 weeks after castration

| | ♂♂ No. | Marking behaviour per 10 minutes of fertile male Tupaias | | | | Cage scented by fertile males | | | |
| | | Experimental cage unscented | | | | | | | |
		Exp. n	Chinning M ± SE	Sledging M ± SE	Anal drags M ± SE	Exp. n	Chinning M ± SE	Sledging M ± SE	Anal drags M ± SE
Before castration	F43	8	27.8 ± 1.0	3.5 ± 1.0	0.7 ± 0.2	6	44.2 ± 4.7	9.8 ± 1.3	0.7 ± 0.3
	B38	44	18.8 ± 1.2	2.2 ± 0.3	0.8 ± 0.2	38	46.4 ± 1.6	6.5 ± 0.7	1.6 ± 0.3
	Mi	10	11.5 ± 0.8	4.5 ± 0.8	1.5 ± 0.5	7	19.0 ± 0.6	6.7 ± 0.9	3.4 ± 1.2
	J12	7	9.0 ± 2.3	1.7 ± 0.7	0.0 ± 0.0	7	48.0 ± 7.7	13.1 ± 1.9	0.7 ± 0.2
	B51	34	0.8 ± 0.2	0.2 ± 0.1	0.5 ± 0.2	37	13.0 ± 1.7	0.9 ± 0.4	1.1 ± 0.2
After castration	F43	13	6.6 ± 1.5	2.7 ± 1.0	0.9 ± 0.3	10	15.0 ± 1.8	3.2 ± 1.1	0.8 ± 0.3
	B38	10	3.3 ± 0.8	0.0 ± 0.0	1.3 ± 0.5	11	11.6 ± 2.5	4.2 ± 1.1	2.5 ± 0.5
	Mi	9	0.9 ± 0.3	0.7 ± 0.3	0.4 ± 0.3	10	6.1 ± 0.9	2.8 ± 0.7	0.2 ± 0.1
	J12	12	3.6 ± 0.6	0.3 ± 0.3	0.3 ± 0.3	17	10.7 ± 1.2	3.5 ± 0.6	1.4 ± 0.4
	B51	10	0.0 ± 0.0	0.0 ± 0.0	0.0 ± 0.0	14	5.3 ± 0.6	0.2 ± 0.1	0.6 ± 0.2

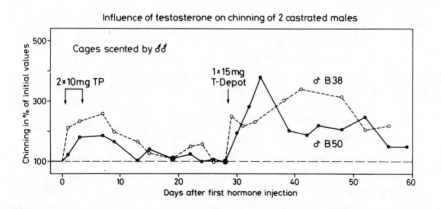

Fig. 5.7. Influence of testosterone propionate (TP) and testosterone depot (T-Depot) injections on chinning values of two castrate males (B38; B50) in cages scented by fertile males.

When castrates are injected with testosterone propionate at doses between 0.5 and 2.0 mg twice weekly, their marking activity increases in both scented and unscented cages in rough linear relationship to the logarithm of the dose. But higher testosterone doses have no additional influence on marking values even when they result in serum testosterone levels several times higher than those present prior to castration (> 250 ng/m) (see Fig. 5.8).

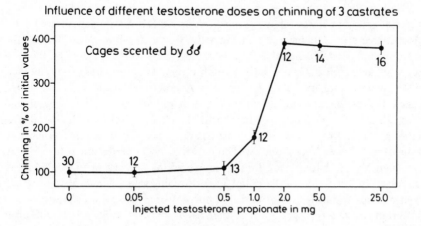

Fig. 5.8. Chinning activity (M ± SE) of castrated males as a function of testosterone propionate injection dose. For each point, the number of experiments is indicated.

The response to testosterone treatment differed from one individual to another, showing a correlation with the marking levels prevailing both prior to castration and following castration but prior to treatment ($r \gg 0.9$; $p \ll 0.001$; 25 animals). Nevertheless, although the absolute marking values differ widely between individuals for a given hormone replacement dose, the percentage changes (with respect to initial values) are virtually the same and are entirely reproducible. The behaviour of any particular individual thus quite precisely reflects its hormonal state, but comparison of individuals is of little value unless their differential sensitivities to androgens are already known.

Influence of ACTH on marking activity of castrated males   The results thus far indicate a clear dependence of male tree-shrew marking behaviour on blood androgen levels. One might therefore ask why the marking activity of males following castration does not fall to zero—the prepubertal level. It is possible that androgenic substances of adrenal origin could be responsible. The production of such adrenal androgens has been demonstrated with tree-shrews in our laboratory, but it was not known whether they would be physiologically effective under normal conditions. To test this possibility, three castrate males were given a daily intramuscular injection of adrenocorticotrophic hormone (20 I.U. of ACTH; Schering) on 12 consecutive days. In tree-shrews, this dosage results in a maximal rise in blood cortisol levels within 30 minutes, followed by a decline to initial levels within six hours. Our experiments to date with seven animals indicate that the adrenal androgens show a similar pattern of change; on average, they increase within 30 minutes from the basal value of less than 1.0 ng/ml serum to reach a level of about 3.5 ng/ml and then decrease again slowly to return to the basal value.

Marking behaviour of the castrates was measured in unscented and scented cages at different times following injection of ACTH. All animals showed a distinct increase in chinning and sledging activity after injection of ACTH (for the relationship between chinning and sledging in unscented cages, $r = 0.86$; in scented cages, $r = 0.73$; $p < 0.05$ in both cases). The responses were, however, different both qualitatively and quantitatively between scented and unscented cages. In unscented cages, chinning (and sledging) activity reached a maximum after 30 minutes post-injection and steadily declined to return to its initial level after a further 4.5 hours. In contrast, marking activity in scented cages continued to increase for about two hours after ACTH injection before slowly returning to the initial level. In addition, marking activity in unscented cages increased to several times the initial levels attaining values approximately 80 per cent of those resulting from maximal stimulation of marking through testosterone injection. In scented cages, on the other hand, marking activity 1–2 hours after ACTH injection was barely 50 per cent higher than the initial level ($p \approx 0.05$). In short, marking is rapidly and strongly affected by activation of the adrenal cortex, but the pattern of response differs between scented and unscented cages (Fig. 5.9).

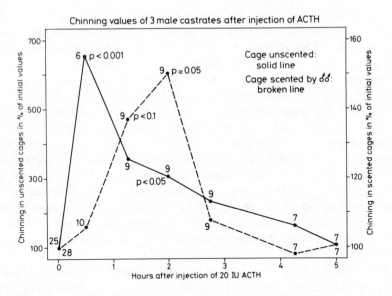

Fig. 5.9. Influence of ACTH injection on chinning activity of three castrate males. For each point, the number of experiments is indicated and (where relevant) the significance levels of differences from initial values are shown.

Influence of female sex hormones on marking activity of fertile and castrate males   In order to investigate the effects of female sex hormones on marking activity, three fertile and three castrate males were given intramuscular injections of 0.5 mg oestrogen (oestradiol benzoate: Vitis) twice a week. Some weeks later, the same animals were injected in the same way with 1.0 mg doses of progesterone (Proluton: Schering). Contrary to expectation, the animals responded in identical fashion to the two different female hormones, so the results have been combined (Table 5.6). Fertile male tree-shrews exhibit no

Table 5.6. Influence of female sex hormones on chinning values of fertile and castrate males. Post-treatment values which differ significantly from basal values are indicated ($* = p < 0.05$; $** = p < 0.01$).

| | Chinning values: changes in % of basal values | | | | |
| | A: Cage unscented | | B: Cage scented by ♂♂ | | Difference between A and B |
Experimental animals	n	M ± SE	n	M ± SE	
Castrated males	37	719 ± 245**	37	2 ± 11	$p < 0.01$
Fertile males	32	24 ± 19	32	−24 ± 9*	$p < 0.05$

change in chinning activity (or in other marking behaviours) in unscented cages following injection of female sex hormones, whereas castrate males respond with increases of about 700 per cent. This finding can be explained as follows: the central nervous sub-systems responsible for the performance of chinning behaviour in unscented cages and for the release of gonadotrophins have androgen receptors which respond equally well to oestrogen, progesterone, or testosterone. The female sex hormones thus on the one hand inhibit the release of gonadotrophins—and hence the production and release of androgens by the testes—while on the other hand they stimulate the central nervous sub-systems responsible for the performance of marking behaviour in unscented experimental cages. As a result they greatly increase the chinning activity of castrate males. In fertile males, testicular activity (and thus stimulation by androgens of the central nervous sub-systems responsible for the evocation of marking behaviour) decreases, but this effect is compensated by the stimulatory effect of female sex hormones. Hence, there is no overall change in the marking activity of fertile males.

In scented cages, the situation is entirely different. Whereas the chinning values of the castrates do not change, in comparison to pre-treatment levels, those of fertile males actually decrease by about 25 per cent. Assuming that the androgen receptors responsible for the performance of chinning behaviour do not respond at all to the two female hormones in this situation, it is relatively easy to explain this difference in behaviour. In castrates, marking activity would not change after administration of the hormones. In fertile males, by contrast, marking levels would decline because the lowered blood androgen concentration (resulting from reduced secretion of gonadotrophins by the hypothalamus) is not compensated by a stimulating effect of the female sex hormones on marking behaviour.

The evidence therefore suggests that marking behaviour in unscented and scented cages is controlled by two different central nervous sub-systems which respond differentially to hormones. Whereas the sub-system responsible for marking of unscented cages responds to oestrogen, progesterone, and adrenal androgens almost as strongly as to testosterone, the sub-system responsible for marking in scented cages responds (for those hormones so far investigated) not at all to female hormones, only weakly to adrenal androgens, and strongly only to testosterone.

Influence of scent marks on marking behaviour of males—Scent marks of male conspecifics   Influence of duration of scent-marking by the scent donor: As has been demonstrated above, the scent marks of fertile male tree-shrews have a pronounced stimulating effect upon the marking of adult male conspecifics, regardless of their hormonal status. Overall, the degree of stimulation of chinning (and other marking behaviour) by male tree-shrews increases with the duration of prior occupation (namely scent-marking) of the cage by a fertile male conspecific (Fig. 5.10).

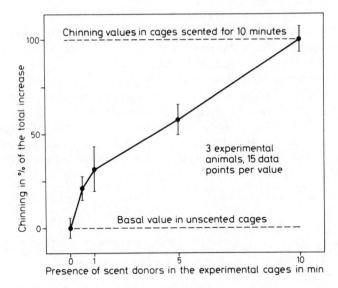

Fig. 5.10. Relationship between the chinning activity (mean ± SE) of male tree-shrews and the duration of prior occupation of the cage by a scent-donor.

Prior occupation of the experimental cage for a period of 30 seconds by the scent-donor is already enough to exert a significant stimulatory effect compared to the unscented condition ($p \approx 0.02$). This effect is even observable when the scent-donor has not directly marked the cage with urine, faeces, or cutaneous-gland secretions during the 30-second occupation period. In other words, the scent substances given off by the animal's body surface and by its fur are in themselves sufficient to have this clear-cut effect. As the duration of prior occupation of the experimental cage by the scent-donor increases (accompanied by a corresponding increase in scent-marking of the cage) the stimulatory effect of the cage, or rather of the scent marks it contains, also increases. However, the results obtained to date indicate that occupation beyond 5–15 minutes (according to the particular individual concerned) does not lead to any further increase in the marking activity of a tree-shrew subsequently tested in the cage. In fact, longer prior occupation of the cage actually leads to depression of the test animal's marking activity to levels corresponding to those found with unscented experimental cages. Thus, scent concentrations in excess of a certain level inhibit marking by an animal subsequently tested in the cage. This finding could be of great significance in the context of territorial behaviour, but it will not be considered any further here as investigations are still in progress.

Persistence of fertile male scent marks: In order to test the effectiveness of scent marks of differing degrees of recency on the marking behaviour of male conspecifics, scent-donors were placed for 10-minute periods in cages which were subsequently tested with expermental animals at various times ranging from immediately after occupation to 10 days later. In the intervening period between occupation and testing, the experimental cages were kept in a separate, well-aired room which (as far as could be determined) did not carry tree-shrew scent itself. The effectiveness of the scent marks of fertile males as stimulators of chinning (and other marking) by male conspecifics decreases progressively over 48 hours to reach a low baseline level which is subsequently maintained. Tested males marked even 10-day-old scented cages at a higher average level than they did unscented cages, though the difference is relatively small and not statistically significant (Fig. 5.11).

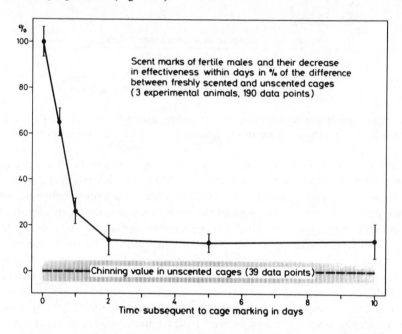

Fig. 5.11. Persistence of substances eliciting marking from male conspecifics in scented cages. Standard errors are indicated by vertical bars for individual points and by the hatched area for the baseline value in unscented cages.

Differential responsiveness of individuals: The scent marks of fertile male tree-shrews typically stimulate the marking activity of adult male conspecifics (chinning, sledging, dropwise urination, and—to a small extent—anal marking). However, there are pronounced individual differences among both fertile and castrated males with respect to their responses to scent marks. Thus, in comparison with values recorded in unscented cages, chinning and sledging activity of different individuals can increase by between 50 and 600 per cent (Tables 5.4

and 5.9). Since the experiments were usually conducted using a different male as scent-donor each time, these differences cannot be ascribed to differential effectiveness of the scent marks from different animals. Rather they must reflect individual characteristics of the individual animals themselves. In other words, tree-shrews not only exhibit different levels of marking activity in unscented cages; they also show independent differences from one another in the strength of their response (in terms of increased marking behaviour) to the presence of scent marks deposited by strange conspecifics.

Differential effectiveness of the scent marks of different fertile males: As a general rule, the scent marks of fertile males exert a stronger stimulatory influence on the marking behaviour of other males than the scent marks of castrate males or unscented experimental cages (see p. 198). Yet there can be considerable differences between fertile males in the effectiveness of their scent marks (findings from more than 60 different experimental animals). For example, male 51 showed the following different values: in an experimental cage previously scent-marked by male 53, he marked 14.3 ± 3.3 times with his sternal gland in 10 minutes ($n = 5$), whereas in a cage previously occupied by male Mu he marked 29.4 ± 3.2 times ($n = 10$). In cages previously marked by castrated males, he marked 5.4 ± 2.1 times ($n = 5$) and in unscented cages he marked only 2.8 ± 0.7 times ($n = 10$). There was no detectable meaningful relationship between the marking behaviour of tested animals and the numbers of times scent-donors had marked with sternal gland, abdominal gland, perineal gland, urine, or faeces, nor with the overall number of scent marks from all sources.

The differential effectiveness of male scent marks was investigated in more detail in a series of experiments lasting approximately two years. The experimental animals were 11 healthy adult males which exhibited extreme differences in body weight, appearance, locomotor activity, and development of the sternal gland (Table 5.7). All of these animals were weighed once a month and four separate blood samples were taken from each of them (by extraction from the caudal vein) in the course of the investigation in order to permit measurement of testosterone levels and various other parameters. (All blood samples were taken two hours before lights on.) During the investigation, each male was also subjected six times to a procedure in which the sternal gland region was rubbed with a sheet of filter paper of known weight, which was subsequently weighed to determine the quantity of secretion thus removed. Further, at each monthly weighing the coloration of the sternal gland area was recorded For each male, its levels of marking activity in unscented cages and in cages scent-marked in turn by all other males of the experimental group were determined. Each male acted as a scent-donor approximately 3–4 times for every other male. The chinning activity of each test animal in cages scented by each of the different scent-donors was calculated as a percentage of its mean chinning

Table 5.7. Body weight, quantity of secretion and coloration of the sternal gland, chinning values in unscented and scented cages, stimulation of scent marks, and testosterone levels of 11 fertile male tree-shrews. For details, see text

| δδ No. | Body weight (g) | | Sternal gland | | | Chinning per 10 minutes | | | | Stimulation of scent marks in % | | Testosterone (ng/ml serum) | |
| | | | Secretion (mg) | | | Cage unscented | | Cage scented | | | | | |
	n	M ± SE	n	M ± SE	Coloration	n	M ± SE	n	M ± SE	n	M ± SE	n	M ± SE
32	24	253.0 ± 3.2	6	10.5 ± 0.6	ye/br	28	0.9 ± 0.4	38	11.4 ± 1.0	34	77 ± 9	4	19.1 ± 1.6
7	24	235.7 ± 1.6	6	2.3 ± 0.2	ye	28	2.5 ± 0.6	25	5.2 ± 0.7	34	118 ± 15	4	53.1 ± 12.0
12	24	224.3 ± 1.6	6	7.7 ± 1.1	ye	28	12.0 ± 1.2	37	21.1 ± 1.7	32	113 ± 15	4	41.8 ± 9.0
10	24	207.1 ± 1.6	6	10.8 ± 0.8	ye	29	8.1 ± 1.0	36	22.6 ± 1.4	32	122 ± 15	4	49.3 ± 6.8
9	24	201.5 ± 1.1	6	8.3 ± 0.4	ye	27	34.2 ± 2.8	36	46.3 ± 3.4	33	126 ± 10	4	39.6 ± 1.2
13	24	194.7 ± 1.5	6	12.9 ± 0.6	ye/br	25	11.2 ± 1.2	33	15.3 ± 1.0	37	95 ± 11	4	26.8 ± 10.8
26	24	181.5 ± 1.7	6	5.4 ± 1.2	br	29	4.9 ± 0.5	38	7.9 ± 0.6	36	71 ± 7	4	26.9 ± 8.6
3	24	181.0 ± 1.7	6	5.3 ± 0.3	ye	27	15.9 ± 2.1	33	25.8 ± 2.0	32	104 ± 9	4	33.6 ± 12.0
4	24	174.1 ± 2.1	6	7.6 ± 0.4	ye	28	18.2 ± 0.9	28	29.3 ± 1.3	46	103 ± 9	4	15.6 ± 6.4
14	24	148.5 ± 1.1	6	3.0 ± 0.3	ye/br	29	0.3 ± 0.1	38	0.8 ± 0.2	22	77 ± 9	4	31.7 ± 7.9
8	24	138.9 ± 1.2	6	5.3 ± 0.5	br	27	12.5 ± 2.1	33	22.8 ± 1.8	36	88 ± 9	4	59.7 ± 12.9

activity in all scented cages taken together. This percentage 'stimulation value' thus indicates the extent to which scent marks of any given individual stimulate the marking behaviour of a conspecific in comparison to the average stimulatory effect of all animals considered together. The stimulation values of the scent marks of the individual tree-shrews were relatively consistent with respect to all of the tested conspecifics. There were certain individuals whose scent marks elicited only a small amount of marking activity in all conspecifics tested, whereas scent marks of other individuals stimulated a high level of marking activity across the board. Accordingly, in Table 5.7 the stimulatory effect of each individual on the remaining 10 conspecifics is collectively given as 'stimulation of scent marks'. As the results show (Table 5.7), the average stimulatory effectiveness of the scent marks of different individuals varies between 70 and 130 per cent, with extreme values being significantly different from one another. This means that there are qualitative differences between the scent marks of fertile males. These differences are perceived by conspecifics in a generally comparable fashion and they respond in a way which is, on average, similar for all individuals according to the quality of the scent marks (and probably according to the quantity of scent substances contained in the scent marks). This response is reflected in a greater or lesser number of marking actions conducted with the sternal (and abdominal) gland.

One reservation must, however, be made at this point. There were in some cases, notably in the intermediate range, conflicting results. For example, the scent marks of male 4 had the strongest effect on the marking activity of male 9, whereas for male 8 they had the weakest stimulating effect. It is, of course, conceivable that these conflicting results can be attributed to the small number of experiments (only four per animal), but one cannot exclude the possibility that the general stimulatory effect of the scent marks is overlain by individual responses to particular scent mixtures.

Preliminary results suggest that the effectiveness of the scent marks of fertile males can be modified by alteration of their testosterone levels. Two males which exhibited quite high levels of marking activity, but whose scent marks exerted only a relatively mild effect on the chinning activity of conspecifics, were each given an injection of testosterone propionate (10 mg; Schering). Within 24 hours, the injections led to an increase of about 100 per cent in the effectiveness of their scent marks, to reach levels more typical of other males (Fig. 5.12). By contrast, results obtained to date indicate that the injection of testosterone has no discernible effect on the effectiveness of the scent marks of males which already exhibit relatively high stimulatory values (five experiments).

Thus, although the effectiveness of scent marks of fertile males is (up to a certain limit) dependent upon the circulating testosterone levels, just as with castrate males (see p. 185), there is no significant correlation between the stimulatory effect of the scent marks of the 11 experimental animals (Table

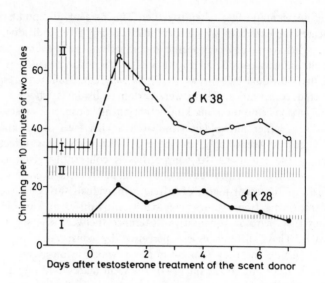

Fig. 5.12. Chinning activity of two males (K28; K38) before and after testosterone treatment of their fertile male scent-donors.
I: Mean ± SE of the male prior to testosterone treatment of the scent-donor (12 experiments in each case).
II: Mean ± SE of the male in cages scent-marked by other fertile males with relatively high stimulation values (20 experiments in each case).

5.11) and their testosterone levels ($r = 0.45$). It is possible that this poor correlation is partially due to the fact that single hormone measurements do not give an accurate picture of the total amount of testosterone produced by individual males. However, it seems more likely that the organs or structures which produce the scent substances stimulating marking behaviour by conspecifics respond differently to the testosterone levels in the blood in different males. In accordance with this, there is no significant correlation between testosterone levels and the amount of sternal gland secretion ($r = -0.33$), or between testosterone levels and marking with the sternal gland (and other forms of marking) in either unscented or scented experimental cages ($r = 0.17$ and 0.07, respectively).

The quantity of secretion obtained by rubbing of the sternal gland area remained relatively constant for each of the 11 experimental animals, but there was wide variation (between about 2 and 13 mg) from one animal to another. Indeed, more than 20 mg of secretion has been obtained from other fertile males by rubbing of the sternal gland. The amount of secretion obtained by rubbing of the sternal gland of the male showing the lowest level of secretory activity (2.3 mg average from male 7) was still notably greater than the quantity obtained from castrate males (0.5–1.1 mg; 10 experimental animals).

There is no relationship between the extent of sternal-gland secretion of a particular male and its level of marking activity or the stimulatory effect of its scent marks. However, there is a striking correlation between the externally visible coloration of the sternal gland regions of the 11 experimental animals and the stimulatory effects of their scent marks. All males whose scent marks stimulated marking in conspecifics by more than 100 per cent possessed (regardless of the quantity of secretion obtained by rubbing) a bright yellow coloration in the sternal-gland region, while in males with a lesser stimulatory effect the colour ranged from yellow-brown to brown. Despite the fact that our results to date indicate that the stimulatory effect of conspecific scent is not exclusively, or even predominantly, dependent upon chemical substances contained in the sternal-gland secretion, the colour of the sternal-gland area—in contrast to all other parameters measures—is the only reliable indicator of the effectiveness of an animal's scent marks. Since chemical substances (e.g. 2,5-dimethylpyrazine) are present in the male's body scent in general, and in his urine in particular, it is possible that these substances are also produced and discharged by the sternal gland. Thus, at least with respect to the stimulation of marking activity in conspecifics, urine and sternal-gland secretion might contain the same information. On the other hand, it is also possible that the sternal gland only acts as a carrier for scents originally discharged in the urine. In both cases, however, the composition of the sternal-gland secretion must change according to the concentration of the scent substances such that it decomposes or undergoes oxidation at varying rates and thus becomes either bright yellow or brown in colour.

Taken as a whole, these findings permit the following conclusions: Scents from different males have different degrees of stimulatory effect on the marking activity of male conspecifics. The effectiveness of the scent marks of fertile males with weak stimulatory properties can be markedly enhanced within a few hours by administration of testosterone. There is no (significant) relationship between the stimulatory effect of a given individual's scent marks and his body weight, his own marking activity, his circulating testosterone level, or the quantity of sternal-gland secretion he produces. By contrast, the coloration of the sternal gland area does provide an index of the effectiveness of a male's scent marks. Animals whose scent marks have a strong stimulatory effect always have bright yellow sternal gland areas, whereas those with weak stimulatory properties have yellow-brown to dark brown sternal areas.

Sexual status of male scent-donors: The scent marks of fertile adult males stimulate marking activity in male conspecifics to varying degrees, but overall these differences are relatively limited, since they amount to departures of no more than ±30 per cent from the average stimulatory effect determined for fertile males in general. By contrast, there are great variations in effectiveness in stimulating marking behaviour by male conspecifics, according to the sexual status of the scent-donor. When compared with the marking activity elicited in unscented experimental cages, the scent marks of sexually immature tree-shrews have no significant stimulatory effect on the marking behaviour of adult conspecifics (30 experiments with nine prepubertal animals; see also Fig. 5.14). With the onset of puberty—and always a few days before the appearance of any sternal-gland secretion or chinning behaviour—the effectiveness of the young male's scent marks rapidly increases to attain adult values within a few days. The stimulatory effect must, therefore, be exerted by substances derived (probably) from the male's urine (for further details see: von Holst and Buergel-Goodwin 1975*b*).

Following castration, the stimulatory effectiveness of the scent marks of adult male tree-shrews with respect to the marking behaviour of male conspecifics decreases over the course of about 10 days, to reach values for test animals that are on average only about twice as high as those recorded with unscented cages (18 animals; for examples, see Fig. 5.15). However, there are great individual differences in the response to scent marks of castrated male conspecifics. Whereas some individuals marked no more frequently in cages scent-marked by castrates, other males showed marking frequencies in response to the scent marks of castrate males up to three times higher than those recorded in unscented cages. However, despite this great variation in response of individual males to the scent marks of castrated conspecifics, the chinning values of test males in the presence of scent marks of fertile males are always considerably higher (see Table 5.8 for examples).

Table 5.8. Chinning values of fertile male tree-shrews in unscented cages and in cages scented by either castrated or fertile male conspecifics

	Chinning per 10 minutes					
	Cage unscented		Cage scented by castrates		Cage scented by fertile males	
♂♂ No.	*n*	M ± SE	*n*	M ± SE	*n*	M ± SE
45	21	14.5 ± 1.3	7	15.3 ± 1.5	20	28.3 ± 2.0
497	13	10.1 ± 1.0	11	14.2 ± 1.2	18	24.8 ± 0.7
Ba	25	8.5 ± 1.0	9	20.3 ± 2.0	17	34.7 ± 2.2
A	14	5.1 ± 1.3	14	13.4 ± 1.4	28	37.8 ± 5.5
455	23	4.4 ± 0.5	4	5.5 ± 0.6	22	11.9 ± 0.7
43	5	4.2 ± 1.6	10	12.5 ± 1.8	20	28.7 ± 2.0
51	10	2.8 ± 0.7	5	5.4 ± 2.1	20	21.1 ± 1.6
315	10	2.7 ± 0.7	9	8.8 ± 1.6	35	14.8 ± 0.9
Mug	11	1.6 ± 0.4	4	1.7 ± 0.2	10	5.3 ± 0.7

The effectiveness of the scent marks of castrated males ($n = 18$) increases following injection of testosterone (testosterone propionate; Schering) over the entire range of testosterone doses tested (0.05–25.0 mg per animal), reaching maximum values within 20 hours (the minimum period examined prior to testing). The increase in chinning (and sledging) values of the test animals is related in approximately linear fashion to the dose injected into the scent-donor. In other words, the test animal can quantitatively detect the androgen level of the scent-donor, or rather the concentration of androgen-dependent substances contained in its scent marks, and can respond to this with a corresponding amount of marking activity (Fig. 5.13). This finding is all the more surprising

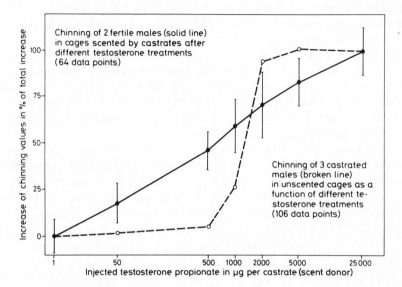

Fig. 5.13. Influence of different doses of testosterone on the chinning activity of castrated males in unscented cages and the effects exerted by the resulting scent marks on the marking activity of fertile males.

in that the chinning activity of castrates injected with testosterone increases as a function of dosage only over the initial part of the dosage range (from 0.5–2.0 mg testosterone). Higher doses of testosterone produce no further increases in chinning activity (Fig. 5.13). Therefore, the increased effectiveness of the castrate male's scent marks at higher testosterone doses is due not to an increase in the number of marking actions, but to some differential property of the marks themselves.

2,5-Dimethylpyrazine as a substance eliciting male marking behaviour: A number of male tree-shrews were provided with closeable glass nest-boxes which they used for their period of night-time rest. Each of these all-glass nest-boxes was closed off at night and supplied with purified artificial air which passed over a charcoal filter after leaving the box. The scent substances concentrated on the filter in a given time-span (e.g. an entire night) were subsequently separated using capillary column chromatography (for details, see von Stralendorff 1982*b*).

In this way, scent profiles for individual animals could be obtained, and these proved to be relatively constant for months on end provided that conditions were left unchanged. However, there were considerable differences from individual to individual in a wide variety of components. In addition, one can identify general differences between the sexes, particularly with respect to 2,5-dimethylpyrazine. Whereas this substance is virtually unidentifiable in the female scent profile (< 0.17 per cent of the total scent material), it is present in high concentration in the scent of fertile males (range 10–30 per cent of total scent material) and the stimulatory effect of scent marks is found to correlate with the concentration of 2,5-dimethylpyrazine. By contrast, this component of scent material is no longer identifiable with castrate males (< 1.0 per cent), though it can be made to re-appear within 24 hours through administration of testosterone, reaching levels reflecting the dose given. The effectiveness of the scent marks of castrate males treated in this way, as measured through the stimulation of marking activity in male conspecifics, also increases in line with testosterone dosage (Fig. 5.14).

Experimental cages which had been previously marked by castrated males were sprayed with an aqueous solution of synthetic 2,5-dimethylpyrazine. When male tree-shrews were subsequently tested with these cages, their marking activity was stimulated according to the concentration of dimethylpyrazine utilized. Cages which had been sprayed with a concentration of this substance corresponding to that in a cage scent-marked by a fertile male showed the same effectiveness in stimulation of marking activity as cages actually marked by fertile males.

Thus 2,5-dimethylpyrazine (or probably the corresponding dihydroxy-dimethylpyrazine which could not yet be identified due to technical reasons) elicits a dose-specific response from male conspecifics, namely marking behaviour. Accordingly, it would seem justifiable to regard it as a genuine pheromone following the original definition of Karlson and Lüscher (1959) (for details, see von Stralendorff 1982*b*).

2,5-Dimethylpyrazine is always present at high concentrations in the urine deposited by fertile males during marking. However, it is evidently not produced by androgen-dependent accessory· glands of the male genital tract, since it is already present in urine contained in the bladder. As yet, its site of origin is unclear. Similarly, it is still not known whether dimethylpyrazine is also produced by marking gland areas (e.g. the sternal gland). It is certain, though, that in the course of marking this substance does appear in the sternal region and on the body fur in general. As a result, dimethylpyrazine is continuously discharged from the animal's surface into the environment even without any specific marking behaviour. Further, along with other scent substances, it can be directly deposited on particular objects as a persistent scent mark during chinning.

Influence of female sex hormones on the scent marks of males: In order to assess the influence of female sex hormones on the effectiveness of scent marks of male tree-shrews, four fertile and four castrated males were each injected intramuscularly with 0.5 mg oestrogen (oestradiol benzoate: Vitis), followed by 1 mg progesterone (Proluton: Schering) eight weeks later. The effectiveness of the scent marks of these treated animals, with respect to the marking activity of fertile males, was determined 1-3 days after each hormone injection. Since there were no significant differences between the effects of the two hormones, at the doses given, the values obtained have been grouped together.

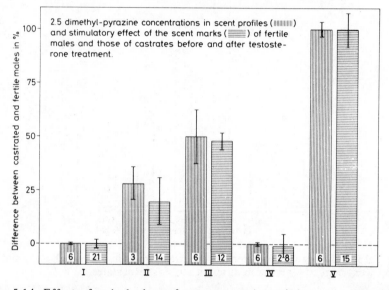

Fig. 5.14. Effect of a single dose of testosterone depot (10 mg per animal) on 2,5-dimethylpyrazine concentrations in scent profiles of three castrate males and on the stimulatory effect of the scent marks of three identically treated castrates on the chinning behaviour of three fertile test males. Values are given as a percentage of the differences between the values of the castrate males prior to treatment (I) and the values of three fertile males (V), and standard errors along with the numbers of experiments are indicated in each case. Dimethyl-pyrazine concentrations were determined by enclosing each experimental animal once a week in an all-glass nest-box for an entire 12-hour night period. The scent substances given off during this period were collected on a charcoal filter and subsequently, using capillary column chromatography, the relative proportions of 2,5-dimethylpyrazine in comparison to the total amount of scenting substances were measured. For untreated castrate males (I), the proportion of dimethylpyrazine was < 0.1 per cent; for fertile males (V) it was 25.4 per cent.
Key:
  I: Values of castrates before testosterone injection.
 II: Values obtained within one week after hormone treatment.
III: Values determined 2-3 weeks after treatment.
 IV: Values determined 4-5 weeks after treatment.
  V: Values of fertile males.

Injection of the female sex hormones resulted in a significant ($p < 0.01$) decrease in the effectiveness of the scent marks of both fertile and castrated males by 15-30 per cent (Fig. 5.15).

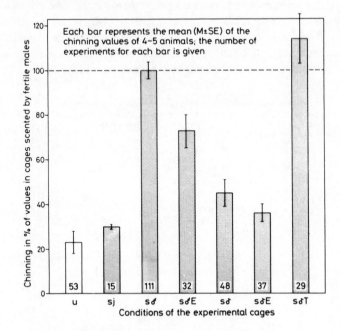

Fig. 5.15. Chinning values of fertile male tree-shrews in unscented cages and in cages scent-marked by males in different physiological states.
Key:
u   = unscented.
sj   = scent-marked by (sexually immature) juveniles.
s♂   = scent-marked by fertile males.
s♂E = scent-marked by fertile males after injection of female sex hormones.
s♂   = scent-marked by castrated males.
s♂E = scent-marked by castrated males after injection of female sex hormones.
s♂T = scent-marked by castrated males after injection with 25 mg testosterone propionate.

In the case of fertile males, this decrease in the effectiveness of scent marks may well depend primarily upon decreased production and release of androgens by the testes, since female sex hormones might occupy the 'negative-feedback receptors' located in the hypothalamus and thus inhibit release of gonadotrophins. However, it is difficult to understand why the effectiveness of the scent marks of the castrated males also decreased after injection of female sex hormones, especially since their own chinning activity in unscented cages increased by about 700 per cent (see p. 189). It is possible that the effectiveness of the scent marks of castrated males, as is the case for fertile males, depends upon the presence of androgen-dependent substances in their secretions and excretions, the formation of which is stimulated by adrenal androgens. The female sex hormones injected could compete with these androgens, which are in any case present at very low concentrations (< 1 ng/ml serum), at the peripheral receptors and thus prevent them from having any effect, while the female sex hormones themselves produce little or no enhancement of scent-substance synthesis and discharge.

In summary, the results of injection of female sex hormones provide clear evidence for a difference in response to hormones between the central nervous sub-systems controlling marking behaviour (at least in unscented cages) and the peripheral systems associated with production of scent substances. Whereas the marking activity of castrate males in unscented cages is strongly stimulated by female sex hormones, formation of the scent substances which stimulate marking activity in other individuals decreases.

Scent marks of stressed males: Further experiments were conducted using as scent-donors males which had been subordinated by a fertile male conspecific and then subjected to his continued presence. This is a very stressful situation for subordinate tree-shrews, leading to death within a few days (for details, see von Holst 1972, 1974, 1977). All such stressed animals moved about in the experimental cage and urinated, but they made no marking actions with their sternal, abdominal, or anal glands. In contrast to the situation with surgically castrated males, the scent marks of these stressed males did not significantly stimulate the marking activity of fertile males in comparison with the effects of unscented cages (26 experiments with five stressed males).

Effectiveness of the scent marks of female conspecifics   The scent marks of female tree-shrews, and particularly their urine deposits, have a more or less stimulatory effect upon the marking behaviour of male conspecifics (132 experiments with 12 females; see Table 5.9 for examples). This effect is, on average, approximately the same as that of castrated males which have not been treated with hormones (see also Table 5.8).

In view of the fact that the stimulatory effects of the scent marks made by male tree-shrews depend upon their circulating androgen levels, it seemed possible that the effectiveness of female scent marks depends upon the presence of androgens, notably those produced by the adrenal cortex, just like that of the marks made by castrate males. To clarify this point, five females were injected with high doses of testosterone (10 mg testosterone propionate; Schering). Contrary to expectation, however, the scent marks of the females did not exhibit an increased stimulatory effect following hormone treatment; virtually no change was detectable (Table 5.10). Thus, in contrast to the situation with male scent marks, the effectiveness of female scent marks in eliciting the chinning behaviour of males cannot be ascribed to testosterone metabolites or to androgen-dependent secretions and excretions. Instead, it is probably dependent upon substances whose production is under the control of female sex hormones. Preliminary results from experiments conducted with five females support this interpretation in that injection of oestrogen (0.5 mg oestradiol benzoate per female) gave rise to a significant increase ($p < 0.001$) in the effectiveness of female scent marks (Table 5.10), but further corroboration is required.

Table 5.9. Marking activity of five fertile male tree-shrews in unscented cages and in cages scent-marked by fertile male or female conspecifics

		Marking behaviour of fertile male Tupaias					
		Cage unscented		Cage scented by ♂♂		Cage scented by ♀♀	
	♂♂ No.	n	M ± SE	n	M ± SE	n	M ± SE
Chinning	3	15	22.8 ± 2.3	11	38.3 ± 1.7	13	14.3 ± 2.5
per 10 min	13	11	18.0 ± 1.7	11	22.0 ± 1.5	13	15.4 ± 0.8
	8	8	6.6 ± 1.5	10	29.9 ± 2.0	12	10.9 ± 2.0
	A	14	5.1 ± 1.3	8	37.8 ± 5.5	8	15.3 ± 0.7
	43	5	4.2 ± 1.6	20	28.7 ± 2.0	8	13.3 ± 0.7
Sledging	3	15	1.3 ± 0.3	11	3.5 ± 0.7	13	0.6 ± 0.3
per 10 min	13	11	3.0 ± 0.6	11	3.9 ± 0.3	13	3.2 ± 0.8
	8	10	0.0 ± 0.0	10	1.1 ± 0.4	12	0.0 ± 0.0
	A	14	1.3 ± 0.4	8	10.8 ± 2.0	8	3.8 ± 0.9
	43	5	2.6 ± 1.1	20	6.1 ± 0.6	10	5.6 ± 1.1
Anal drags	3	15	0.7 ± 0.2	11	0.6 ± 0.3	13	0.2 ± 0.1
per 10 min	13	11	1.5 ± 0.5	11	0.9 ± 0.3	13	0.6 ± 0.2
	8	10	0.7 ± 0.3	10	1.0 ± 0.3	12	0.6 ± 0.2
	A	14	0.2 ± 0.1	8	0.6 ± 0.2	8	0.5 ± 0.3
	43	5	4.2 ± 1.5	20	3.7 ± 0.5	10	3.5 ± 0.9

Table 5.10. Influence of testosterone and oestrogen on the effectiveness of scent marks of five fertile females in stimulating marking behaviour of fertile male tree-shrews

	Chinning values of five fertile males in cages scented by female Tupaias (in % of basal values)	
Female scent-donors	n	M ± SE
Untreated	72	100.0 ± 5.6
After injection of testosterone	91	105.3 ± 6.7
After injection of oestrogen	40	143.9 ± 7.8

Influence of an animal's own scent marks   To study the influence of an animal's own scent marks on its marking behaviour, individual fertile or castrate males were left for 10 minutes in an unscented cage or cage scented by a fertile male, during which time they deposited their own scent marks. They were then removed and replaced in the same cage about three hours later. Control animals which were placed in an unscented experimental cage three hours later marked 99.7 ± 3.3 per cent as often with their sternal glands as they had marked during the first test (80 experiments with four fertile and four castrate males). Correspondingly, control animals which were tested twice with an intervening period of three hours in experimental cages marked by the same scent-donors marked 96.9 ± 2.5 per cent as often in the second test as in the first (50 experiments with four fertile and four castrate males). Thus, repeat of an experiment after three hours does not in itself lead to a reduction in marking activity.

In cages containing their own scent marks, however, all animals marked far less frequently ($p < 0.001$) than in unscented cages (Fig. 5.16; for examples, see Table 5.11). This demonstrates that the absence of personal scent in unscented cages is an important stimulus governing the release of marking behaviour. Similarly, animals placed in cages previously marked by a strange male conspecific, but also bearing their own scent marks, marked considerably less than they did in cages bearing only the scent marks of the strange conspecific (Fig. 5.16).

Table 5.11. Chinning activity of males and females in unscented cages and in cages bearing their own prior scent marks

Experimental animals	No.	Chinning per 10 min in cages				Difference (II in % of I)	
		I unscented		II with own scent			
		n	M ± SE	n	M ± SE	%	p
Fertile males	76	8	30.7 ± 3.2	8	20.5 ± 2.1	67	< 0.05
	10	8	16.3 ± 1.8	8	8.9 ± 1.3	55	< 0.01
	32	5	2.8 ± 1.8	5	0.0 ± 0.0	0	< 0.001
Castrated males	43	11	25.2 ± 2.6	11	15.8 ± 1.5	63	< 0.01
	36	10	5.2 ± 1.2	10	2.0 ± 0.6	38	< 0.05
	24	10	4.2 ± 0.8	10	1.7 ± 0.3	40	< 0.01
Fertile females	1	10	7.7 ± 0.9	10	7.8 ± 0.8	101	n.s.
	2	10	5.9 ± 1.3	10	8.3 ± 1.3	141	n.s.
	7	10	0.7 ± 0.3	10	0.8 ± 0.3	114	n.s.

Surprisingly, this effect of an animal's own scent marks is identifiable not only when an experimental animal has overmarked the scent marks of a strange scent-donor (i.e. has occupied the cage subsequently), but also when the experimental animal has marked an unscented cage first and is tested three hours after a strange male has overmarked his original scent marks (preliminary experiments with four experimental males). This indicates that the presence of an animal's own scent marks modifies the scent of a strange conspecific, or the olfactory environment of the test animal in general, such that the effectiveness of the strange conspecific's stimuli, as releasers of marking behaviour, is reduced to a greater or lesser extent.

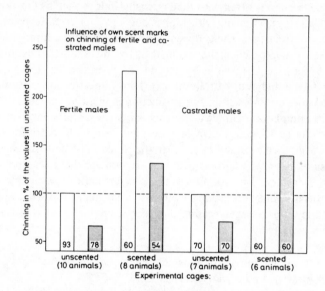

Fig. 5.16. Influence of own scent marks (dark bars) on chinning activity of fertile and castrate males in cages with or without prior scent marks of strange fertile male conspecifics. The number of experiments is indicated within each bar. All differences were highly significant ($p < 0.001$).

Effectiveness of the scent marks of males of other species   Male tree-shrews were placed in experimental cages which had previously been occupied for a period of 10 minutes by a male of one of the following species: laboratory rabbit, laboratory rat, laboratory mouse, lesser bushbaby (*Galago senegalensis*), pigmy tree-shrew (*Tupaia minor*), or which had been previously sprayed with fresh human male urine.

Despite the fact that in all cases the experimental cages bore smells which were clearly detectable to the human nose, and the test tree-shrews sniffed at the scent marks (particularly urine) with great intensity, marking activity in these cages did not exceed the level found with unscented experimental cages (41 experiments with four test animals). Hence the scent marks of males of foreign species—including one belonging to the same genus (viz. *Tupaia minor*)—do not stimulate marking activity, in contrast to the scent marks of conspecifics.

Marking behaviour of fertile female tree-shrews General observations: As mentioned initially, the adult females of mated pairs kept in large cages mark a number of times a day with the sternal gland regions, though the secretory activity of the sternal gland is very limited in comparison to that of the males. On average, rubbing with filter-paper on the sternal gland region led to the collection of only 0.41 ± 0.01 mg of secretion from isolated (non-pregnant) females (30 experiments with 10 females). This is only 50 per cent of the quantity obtained from castrate males (0.81 ± 0.12 mg; 25 experiments with 10 castrates).

In experimental cages, females always mark with their sternal gland regions and they also mark with abdominal and anal regions, just like males, although they do not possess specialized glands in these latter two areas, in contrast to males (see p. 125). Females also mark particularly intensively with urine, which is usually deposited in a dropwise fashion, as is the case with males. Finally, females may repeatedly trample and roll in the urine so that their entire body becomes completely impregnated with urine within a few minutes. As a result, any subsequent marking with the sternal or abdominal areas will spread urine in the surroundings. Such behaviour is particularly predominant in cages previously scent-marked by fertile males. Females also exhibit intensive salivation, just like males, as a regular phenomenon, but eye-gland secretion was never noticed with females, although the eye-gland itself is histologically comparable in males and females.

Detailed studies have so far been conducted on the marking behaviour of 12 females which were maintained in individual cages in a room completely isolated from any male tree-shrews. All other conditions were the same as for experiments conducted with males (see p. 176). Investigations of the hormonal control of female marking behaviour are currently in progress, but as yet only preliminary results are available with respect to the influence exerted by the scent marks of conspecifics on female marking activity.

Influence of scent marks on female marking behaviour   All females exhibit relatively constant levels of marking behaviour in unscented cages over long periods of testing. However, just as is the case with males, there are great individual differences in marking activity (for examples, see Table 5.12).

In contrast to the findings with males, however, a female's own scent marks have no influence on her subsequent marking behaviour. Marking with sternal, abdominal, and anal regions and with urine is just as frequent in cages previously marked with a female's own scent marks as in unscented experimental cages (120 experiments with six females; see Table 5.11 for examples).

If an experimental cage had been previously scent-marked by a strange female conspecific, the tested females ($n = 12$) did show an overall increase in marking with the sternal, abdominal, and perineal glands and with urine. The increase recorded, however, is relatively limited, particularly in comparison with the effect of scent marks of fertile males (see also Table 5.12 (females) and Table 5.9 (males)). By contrast, the presence of scent marks of fertile males led to a dramatic increase in the marking activity of females ($n = 10$) with sternal area, abdominal area, and urine. In this context, females sometimes marked far more frequently with their sternal glands than ever recorded with males. Particularly conspicuous is intensive urination which, as already noted, leads to the distribution of urine throughout the cage as a result of marking with the sternal and abdominal areas. On the other hand, anal marking does not change in comparison with other experimental situations (Table 5.12).

Table 5.12. Marking activity levels of fertile female tree-shrews in unscented cages and cages scent-marked by fertile male or female conspecifics

	♀♀ No.	Marking behaviour of fertile female Tupaias					
		Cage unscented		Cage scented by ♀♀		Cage scented by ♂♂	
		n	M ± SE	n	M ± SE	n	M ± SE
Chinning per 10 min	B72	33	18.1 ± 3.1	32	22.9 ± 2.3	25	110.2 ± 7.7
	B71	33	5.6 ± 0.5	32	7.7 ± 0.7	22	85.2 ± 10.7
	B67	34	3.1 ± 0.6	31	3.4 ± 0.6	24	127.9 ± 12.8
	B87	25	1.1 ± 0.3	37	5.0 ± 1.3	25	87.5 ± 10.6
	B94	27	0.3 ± 0.1	27	1.6 ± 0.2	19	33.5 ± 8.3
Sledging per 10 min	B72	33	10.7 ± 1.4	32	14.4 ± 1.1	25	22.8 ± 2.8
	B71	33	1.5 ± 0.4	32	1.4 ± 0.4	22	8.6 ± 1.8
	B67	34	4.8 ± 0.8	31	5.4 ± 1.0	24	16.3 ± 2.1
	B87	25	2.5 ± 0.7	37	6.8 ± 0.9	25	24.2 ± 2.5
	B94	27	0.6 ± 0.3	27	0.9 ± 0.3	19	13.6 ± 3.8
Anal drags per 10 min	B72	33	4.4 ± 0.4	32	5.1 ± 0.7	25	4.1 ± 0.1
	B71	33	1.6 ± 0.3	32	1.9 ± 0.4	22	2.1 ± 0.4
	B67	34	1.4 ± 0.3	31	1.4 ± 0.3	24	0.9 ± 0.3
	B87	25	1.2 ± 0.2	37	2.4 ± 0.4	25	3.5 ± 1.4
	B94	27	0.4 ± 0.1	27	1.7 ± 0.6	19	1.5 ± 0.5

General conclusions

Marking behaviour and various effects of scent marks of tree-shrews

As has been shown, marking behaviour commences in tree-shrews of both sexes with the attainment of puberty, indicating that gonadal hormones exert a controlling influence. In line with this, castration of males (comparable ovariectomy of females has not yet been conducted) leads to a reduction in marking activity which can be restored in a dose-dependent fashion by administration of testosterone (over a specific range of dosage). Male tree-shrews respond in an extremely sensitive fashion to changes in their circulating blood hormone levels. The degree

of marking activity accordingly provides a quantitative reflection of the hormonal status of any given individual. Nevertheless, it is not possible to relate the
great differences in marking activity among intact animals to corresponding differences in the production and release of sex hormones. Instead, different individuals exhibit differential responsiveness to circulating hormone levels in
the systems responsible for the performance of marking behaviour. Under our
conditions of maintenance and testing, fertile males always exhibit a maximal
level of marking activity which cannot be further increased by the administration of exogenous testosterone. Evidence obtained to date suggests that this
maximal level of marking activity may be established during ontogeny (though
the participation of genetic factors may, of course, play a part). For example,
three males which were exposed to major confrontations with conspecifics at
the time of attaining puberty continued to show significantly lower marking
values than their sheltered male siblings even when maintained separately and
free of conflict as adults.

One striking feature is the context-dependent variation in the response of
males to modification of their hormonal status. With respect to almost every
parameter examined, marking behaviour by adult male tree-shrews differs
fundamentally between unscented cages and cages previously scent-marked by
male conspecifics. There are differences in the baseline rates of marking, depending on the situation, and the levels are affected differentially, rapidly and
strongly by activation of the adrenal cortex. Changes in the opposite direction
are brought about by administration of female sex hormones. The nature of
these differential effects indicates that marking activities in cages scent-marked
by males and in unscented cages are controlled by two different central nevous
sub-systems which differ in the characteristics of their hormone receptors.
Whereas the sub-system responsible for marking in unscented cages responds to
female sex hormones and to stimulation of the adrenal cortex by ACTH almost
as strongly as to testosterone, the sub-system responsible for marking in cages
scent-marked by male conspecifics responds not at all to female sex hormones,
only weakly to adrenocortical activation, and strongly only to testosterone (for
those hormones investigated). In this context, it is noteworthy that only testosterone stimulated the production and release of those male marking substances
which elicit marking behaviour of male conspecifics.

The marking experiments conducted demonstrate that tree-shrews of both
sexes can detect whether scent marks of strange animals were made by conspecifics, and they can also distinguish the sex of them. In addition, the sexual
or androgenic state of male scent-donors (females have yet to be examined) can
also be discerned by males and females. In all cases, the animals respond to any
given test situation by performing a relatively fixed number of marking actions.
Information relating to the sex and sexual state of a conspecific can be obtained
from the urine alone, but conditioning experiments have shown that the same
information is also carried in scent marks obtained by rubbing of a tree-shrew's

sternal or abdominal gland area. In males, the chemical basis of this information is provided by 2,5-dimethylpyrazine. While lacking in females, it is not only the predominant component, in quantitative terms, of the body odour and the urine of fertile males; it is also capable of eliciting marking responses from both male and female tree-shrews, when sprayed within an experimental cage in a chemically pure form. (This, however, does not exclude the possibility that other, as yet unidentified substances could also transmit information regarding the sexual state of males.)

The effectiveness of male scent marks in eliciting marking behaviour of male conspecifics is quite limited. After only 48 hours, no further significant effect can be detected. As yet, it cannot be decided whether this is because the scent marks are relatively volatile and therefore rapidly become attenuated, or whether tree-shrews are able to recognize the age (and thus perhaps the biological relevance) of scent marks so that they respond accordingly. In any case, male tree-shrews must have the capacity not only to produce large quantities of scent substances, but also to perform a great deal of marking activity, in order to ensure continuous deposition of adequate scent marks throughout their home area and particularly on its boundaries. This proves to be the case, since male tree-shrews in the laboratory mark several hundred times a day with their sternal-gland regions, and this is accompanied by frequent urine deposition. Since observation conducted under natural conditions (Kawamichi and Kawamichi 1979) indicate that tree-shrews have relatively small territories and exhibit frequent marking behaviour (see also p. 161), effective marking of the home-range, and particularly of its borders, with these scent marks is perfectly possible.

In addition to the relatively short-lived information regarding the sexual state of males, tree-shrew scent marks from animals of both sexes contain substances permitting individual recognition (urine and sternal-gland secretion have been investigated in this respect to date). It is not yet known what chemical substances provide the information necessary for such individual recognition. It is, indeed, true that the scent-profiles of individual tree-shrews (established by gas chromatography) show pronounced differences which persist for months on end (for details, see von Stralendorff 1982*b*), such that the relative contributions of these components could provide a basis for individual recognition. However, there are a number of problems with this interpretation. For example, tree-shrews can be trained to respond to the sternal-gland secretion obtained by rubbing from a particular individual, and then (without any prior training with such old scent marks) they can recognize the secretion as belonging to the same animal even when the scent mark is months old, despite the fact that the scent-profile must inevitably be changed because of the differential volatility of the various components. It is conceivable that the persistence of an individual scent-profile is enhanced by the lipids which represent 99 per cent of male sternal-gland secretion (primarily wax and sterol esters: von Stralendorff 1977;

see also Regnier and Goodwin 1977). But this conflicts with the results obtained with marking experiments involving a tree-shrew's own scent marks. It is, of course, instructive that an adult tree-shrew, placed in an experimental cage containing its own scent from a previous experiment, can recognize its own scent mark (leading, in the case of males, to reduced marking activity). However, it is difficult to understand how a male can recognize his own scent in a mixture incorporating scent marks from a strange male, and accordingly perform less marking. The individual scent components present at different concentrations in the scent-profiles of two different individuals must surely combine to form a third scent-profile distinct from either of the original patterns.

The scent marks of strange fertile males have no repellent effect on male conspecifics. In fact, in choice experiments, such scent marks are preferred to all other conspecific scent marks or those of foreign species. In common with the scent marks of female conspecifics, those of fertile male conspecifics have no excitatory effect. Indeed, in marking experiments the heart rate of the tested animal returns more rapidly to the baseline value when conspecific scent marks are present than when the cage is free of tree-shrew odour. By contrast, scent marks (or the general scent) of known male conspecifics can give rise to varying degrees of arousal when they have acquired 'social relevance' for an individual through prior interactions. On the one hand, this is indicated by observations of tree-shrews in large cages, where contact with neighbouring animals or their scent marks is accompanied by high levels of arousal (sympathetic activation is reflected by tail-ruffling; for details, see von Holst 1969). In addition it is confirmed by studies of sniffing frequency, which provide a quantitative measure of an animal's central nervous arousal, according to our previous findings (for details, see von Holst and Kolb 1976). In comparison to the effects of scent marks prior to confrontation or the influence of an unfamiliar conspecific, even a one-minute confrontation of an experimental animal with a rival through a wire-mesh partition will lead to dramatic increase in sniffing frequency in response to the latter's scent marks. This drastic increase can still be recorded several weeks after the short confrontation. Hence, a tree-shrew is not only able to recognize the scent of a conspecific some time after initial exposure (periods of up to six weeks were investigated), but will also respond to this scent with central nervous arousal.

If a subordinate male lives together with the dominant individual in a large cage (see p. 161), it will cease to exhibit marking behaviour completely, in contrast to unstressed castrate males. Particularly with experiments lasting many days, this effect is partially due to inhibition of androgen production by the subordinate, resulting from its subjection. The primary influence, however, is exerted by the dominant male's presence, since the subordinate's marking activity ceases immediately following its subjection (in contrast to the situation with surgically castrated males). Further, preliminary experiments ($n = 3$) have shown that a single subjection of an otherwise isolated male suffices to suppress its marking behaviour in a cage previously marked by the dominant male, whereas the scent marks of other (strange) males exert a stimulatory effect on his marking activity in the usual way.

Hence, the scent marks of strange conspecifics elicit marking behaviour by males, while those of known dominant conspecifics inhibit marking activity. Findings to date indicate that the effectiveness of the scent marks of known subordinate males is not detectably different from that of unknown fertile males, as long as the subordinate is not so severely stressed that he is 'psychologically castrated' and has no stimulatory effect on any males (known or unknown) because of the complete absence of 2,5-dimethylpyrazine.

The scent marks of female conspecifics also elicit marking behaviour from male tree-shrews, but this effect is not due to androgen-dependent substances. Instead, the effectiveness of female scent marks seems to be augmented by oestrogens. Conditioning experiments have shown that tree-shrews of both sexes can distinguish females from males when tested with either urine or scent substances obtained by rubbing of the sternal gland area. Accordingly, the gas chromatographic scent-profiles of males and females (obtained either from total body scent or from urine) differ conspicuously in a number of constituents. However, the constituents concerned have not yet been identified, nor has their effect on male marking behaviour been investigated.

Contrary to the situation with male tree-shrews, the hormonal control of marking behaviour in females has not yet been studied. The scent marks of females, unlike those of males, do not evoke any appreciable marking response from tree-shrews of the same sex, despite the fact that females will vigorously attack one another. It would therefore seem that there is no aggressive motivation underlying female marking behaviour. On the other hand, the scent marks of fertile males elicit an intense marking response from any female kept in isolation. As in males, this response is elicited by 2,5-dimethylpyrazine probably the corresponding dihydroxydimethylpyrazine).

Possible functions of scent-marking in tree-shrews

These results of experiments on marking behaviour in standardized test situations evidently cannot provide a general explanation of the functions of scent marks and marking activity in tree-shrews. However, when combined with observations conducted on tree-shrews in more natural conditions, they do permit certain inferences concerning the complexes of behaviour patterns of which marking forms a part. As mentioned at the beginning, there are several distinct situations—defined in terms of the their temporal correlation with certain behaviour patterns—in which marking by males and females with chemical substances of differing origins is a regular occurrence. Furthermore, it is possible to distinguish by means of specific experiments corresponding scent-mark situations which elicit marking behaviour by tree-shrews. Finally, there are in males two operationally distinct central nervous sub-systems which govern marking behaviour in different scent-mark contexts. All of these findings, taken together, permit us to draw some conclusions regarding possible functions of marking behaviour in tree-shrews, which could be:

1. To generate familiarity of both males and females with their home area. The secretions of the different cutaneous glands and urine provide for this function. Familiarization marking is a routine occurrence in familiar enclosures lacking the scent of strange conspecifics and it is also elicited in experimental cages lacking tree-shrew scent. In males such marking is governed by a central nervous sub-system which differs in its ontogeny and its responsiveness to hormones from that governing other forms of marking behaviour. This sub-system responds to male and female sex hormones in the same way.

2. To strengthen the pair-bond between males and females. This function is also served by the secretions of the various cutaneous glands and by urine. It is a routine component of encounters between the members of a pair and forms part of their sexual behaviour. In experimental cages, males and females respond to the scent marks of fertile conspecifics of the opposite sex.

3. To augment the sexual arousal of a partner of the opposite sex. This probably involves substances produced by the genital tract, which are spread over the body of an unreceptive partner during the invitation behaviour known as 'wrestling' (sliding above and below the partner's body). This form of behaviour was commonly observed with mated pairs kept in large cages. It is likely that the same substances (in addition to secretions of cutaneous scent glands and urine) are discharged by isolated females when placed in cages previously scent-marked by males.

4. Protection of neonates by the mother. As has been mentioned before (p. 169), the substance responsible for the protection of the infants from cannibalism by conspecifics has a close similarity to phenylacetic acid. The source of this substance has not yet been established, though there is some evidence that the female's urine is involved.

5. To mark the boundaries of a territory by males and, in the context of appropriate dominance relationships, perhaps to deter subordinate rivals. While lacking in females, 'territorial marking' is the predominant form of male marking and this is the only context in which males have been seen to exhibit marking behaviour in the wild (Kawamichi and Kawamichi 1979). Once again, secretions of the various cutaneous scent glands and urine seem to serve this function in males. Boundary marking always occurs during disputes between male conspecifics and it is a routine phenomenon on the shared borders of cages separating adult males. In experimental cages, such marking is elicited to varying degrees by the scent of male conspecifics, depending on their sexual state (mediated through the production and release of 2,5-dimethylpyrazine). It is governed by a central nervous sub-system in the male which differs from that controlling familiarization marking in its ontogeny and in its responsiveness to hormones; it responds very specifically to testosterone.

In three of the five possible functions listed (familiarization; pair-bonding; boundary marking), the same secretions of cutaneous scent glands and urine are employed by males, and the use of these must therefore be governed by different context-dependent motivational factors. Marking behaviour of females with these secretory and excretory products is generally relatively limited, and the essential function is probably sexual (in combination with the effects of vaginal secretions). In other words, marking of males by females is primarily sexually motivated.

No explanation has yet been obtained for the function of saliva spreading and eye-gland secretion, or for the intensive mouth-licking.

A number of authors have suggested that marking behaviour might serve more than one function, and the results of the present experiments support this view. Yet, both for tree-shrews and for other mammalian species, it remains to be seen just how these potential functions are achieved. To put it another way, further research is required to determine the adaptive significance of such communicatory behaviour (scent-marking) in these different contexts.

Acknowledgements

I should like to thank Dr R. D. Martin for translating the manuscript and for his valuable comments. The work described is supported by the Deutsche Forschungsgemeinschaft.

References

Beauchamp, G. K., Doty, R. L., Moulton, D. G., and Mugford R. A. (1976). The pheromone concept in mammalian chemical communication: a critique. In *Mammalian olfaction, reproductive processes and behavior* (ed. R. L. Doty) pp. 143-60. Academic Press, New York.

Bronson, F. H. (1976). Urine marking mice: causes and effects. In *Mammalian olfaction, reproductive processes and behavior* (ed. R. L. Doty) pp. 119-41. Academic Press, New York.

Brown, K. (1979). Chemical communication between animals. In *Chemical influences on behaviour* (ed. K. Brown and S. J. Cooper) pp. 599-649. Academic Press, London.

Burghardt, G. M. (1970). Defining 'communication'. In *Advances in chemoreception*, Vol. I *Communication by chemical signals* (ed. J. W. Johnston Jr, D. G. Moulton, and A. Turk) pp. 5-18. Appleton-Century-Crofts, New York.

Cantor, T. (1846). Catalogue of mammals inhabiting the Malayan peninsula and its islands. *J. Asiat. Soc. Beng.* **15**, 171-203, 241-79.

Cowley, J. J. (1980). Growth and malturation in mice (*Mus musculus*). *Symp. zool. Soc. Lond.* **45**, 213-50.

Eisenberg, J. R. and Kleiman, D. G. (1972). Olfactory communication in mammals. *A. Rev. Ecol. Syst.* **3**, 1-32.

Glasersfeld, E. v. (1977). Linguistic communication: theory and definition. In *Language learning by a chimpanzee. The Lana project* (ed. D. M. Rumbaugh) pp. 55-71. Academic Press, New York.

Goldfoot, D. A. (1981). Olfaction, sexual behaviour, and the pheromone hypothesis in Rhesus monkeys: a critique. *Am. Zool.* 21, 153–64.

Herreid II, D. F. and Schlenker, E. H. (1980). Energetics of mice in stable and unstable social conditions: evidence of an air-borne factor affecting metabolism. *Anim. Behav.* 28, 20–8.

Holst, D. v. (1969). Sozialer Stress bei Tupajas (*Tupaia belangeri*). Die Aktivierung des sympathischen Nervensystems und ihre Beziehung zu hormonal ausgelösten ethologischen und physiologischen Veränderungen. *Z. vergl. Physiol.* 63, 1–58.

— (1972). Renal failure as the cause of death in *Tupaia belangeri* exposed to persistent social stress. *J. comp. Physiol.* 78, 236–73.

— (1973). Sozialverhalten und sozialer Stress bei Tupajas. *Umschau* 73, 8–12.

— (1974). Social stress in the tree-shrew: its causes and physiological and ethological consequences. In *Prosimian biology* (ed. R. D. Martin, G. A. Doyle, and A. C. Walker) pp. 389–411. Duckworth, London.

— (1977). Social stress in tree-shrews: problems, results, and goals. *J. comp. Physiol.* 120, 71–86.

— and Buergel-Goodwin, U. (1975a). The influence of sex hormones on chinning by male *Tupaia belangeri*. *J. comp. Physiol.* 103, 123–51.

— — (1975b). Chinning by male *Tupaia belangeri*: the effects of scent marks of conspecifics and of other species. *J. comp. Physiol.* 103, 153–71.

— and Kolb, H. (1976). Sniffing frequency of *Tupaia belangeri*: a measure of central nervous activity (arousal). *J. comp. Physiol.* 105, 243–57.

— and Lesk, S. (1975). Über den Informationsinhalt des Sternaldrüsensekretes männlicher und weiblicher *Tupaia belangeri*. *J. comp. Physiol.* 103, 173–88.

Johns, M. A. (1980). The role of the vomeronasal system in mammalian reproductive physiology. In *Chemical signals: vertebrates and aquatic invertebrates* (ed. D. Müller-Schwarze and R. M. Silverstein) pp. 341–64. Plenum Press, New York.

Johnson, R. P. (1973). Scent marking in mammals. *Anim. Behav.* 21, 521–35.

Karlson, P. and Lüscher, M. (1959). 'Pheromones': a new term for a class of biologically active substances. *Nature, Lond.* 183, 55–6.

Kawamichi, T. and Kawamichi, M. (1979). Spatial organization and territory of tree-shrews (*Tupaia glis*). *Anim. Behav.* 27, 381–93.

Loyber, I., Perassi, N. I., and Lecuona, F. A. (1976). Exteroceptive olfactory stimuli: their influence on plasma corticosterone in rats. *Physiol. Behav.* 17, 153–4.

MacKay, D. M. (1972). Formal analysis of communicative processes. In *Nonverbal communication* (ed. R. A. Hinde), pp. 3–25. Cambridge University Press.

Macrides, F. (1976). Olfactory influences on neuroendocrine function in mammals. In *Mammalian olfaction, reproductive processes, and behavior* (ed. R. L. Doty) pp. 29–65. Academic Press, New York.

Martin, R. D. (1968). Reproduction and ontogeny in tree-shrews (*Tupaia belangeri*) with reference to their general behaviour and taxonomic relationships. *Z. Tierpsychol.* 25, 409–95; 505–32.

Milligan, S. R. (1980). Pheromones and rodent reproductive physiology. *Symp. zool. Soc. Lond.* 45, 251–75.

Otte, D. (1974). Effects and functions in the evolution of signaling systems. *A. Rev. Ecol. Syst.* 5, 385–417.

Petrovich, S. B. and Hess, E. H. (1978). An introduction to animal communication. In *Nonverbal behavior and communication* (ed. A. W. Siegman and S. Feldstein) pp. 17–53. Erlbaum, Hillsdale, NJ.

Ralls, K. (1971). Mammalian scent marking. *Science, NY* 171, 443–9.

Regnier, F. E. and Goodwin, M. (1977). On the chemical and environmental modulation of pheromone release from vertebrate scent marks. In *Chemical signals in vertebrates* (ed. D. Müller-Schwarze and M. M. Mozell) pp. 115–33. Plenum Press, New York.

Smith, W. J. (1969). Messages of vertebrate communication. *Science, NY* 165, 145–50.

Stöhr, W. (1978). *Radio-telemetrische Messungen der Herzfrequenz bei* Tupaia belangeri. Diplomarbeit, München.

Stöhr, W. (1982). Telemetrische Langzeituntersuchungen der Herzfrequenz von *Tupaia belangeri*: Basalwerte sowie phasische und tonische Reaktionen auf nichtsoziale und soziale Belastungen. Dissertation, Bayreuth.

Stralendorff, F. v. (1977). Untersuchung zur intraspezifischen olfaktorischen Kommunikation bei männlichen *Tupaia belangeri*: Charakterisierung des Sternaldrüsensekretes. Dissertation, München.

— (1982*a*). Maternal odour substances protect newborn tree shrews from cannibalism. *Naturw.* 69, 553–4.

— (1982*b*). A behaviorally relevant component of the scent signals of male *Tupaia belangeri*: 2,5-dimethylpyrazine. *Behav. Ecol. Sociobiol.* 11, 101–7.

Vandenbergh, J. G. (1980). The influence of pheromones on puberty in rodents. In *Chemical signals, vertebrates and aquatic invertebrates* (ed. D. Müller-Schwarze and R. M. Silverstein) pp. 229–41. Plenum Press, New York.

Wharton, C. H. (1950). Notes on the Philippine tree shrew, *Urogale everetti* Thomas. *J. Mammal.* 31, 352–4.

Whitten, W. K. and Bronson, F. H. (1970). The role of pheromones in mammalian reproduction. In *Advances in chemoreception*, Vol. I *Communication by chemical signals* (ed. J. W. Johnston Jr, D. G. Moulton, and A. Turk) pp. 309–25. Appleton-Century-Crofts, New York.

Williams, G. C. (1966). *Adaptation and natural selection.* Princeton University Press.

Wuensch, K. L. (1979). Adrenal hypertrophy in mice following exposure to crowded males' odors. *Behav. neural Biol.* 27, 222–6.

6 The bats: order Chiroptera

UWE SCHMIDT

Introduction

The bats (order Chiroptera) are unique among the mammals because of their capacity for true flight. Although they are the second largest mammalian order —with more than 800 species, a number exceeded only by the rodents—they have received intensive scientific study in only a few respects. We have very detailed knowledge of their highly specialized auditory orientation system, echolocation, but only isolated aspects of their olfactory and visual abilities have been examined. Nor have there been many publications concerning their fairly complex social behaviour.

Despite the great uniformity in structure of these animals, they exhibit considerable differences in behaviour and way of life. There is a distinction between the two suborders Megachiroptera and Microchiroptera with regard to mode of orientation; all the microchiropterans have an ultrasonic echolocation system, whereas *Rousettus* is the only megachiropteran genus with a sonar system. Moreover, even closely related species can differ fundamentally in their habits. For example, certain neotropical leaf-nosed bats (Phyllostomatidae) feed on insects, whereas others are exclusively frugivorous and the vampire bats specialize in blood. Hand in hand with feeding habits go physiological and ethological adaptations that can extend to specializations in social behaviour.

The features of the olfactory system are highly correlated with feeding behaviour. Various observers have noted that olfactory impressions assist a number of bats in finding and selecting their food. The most dependent on olfactory stimuli while in search of food are the frugivorous megachiropterans (Ratcliffe 1931; van der Pijl 1936, Möhres and Kulzer 1956; Kulzer 1969) and the fruit- and nectar-eating microchiropterans (Mann 1951a; Vogel 1958). But prey-detection by the sense of smell has also been discussed as a possibility in the case of the sanguivorous vampire bats (Ruschi 1952; Mann 1960) and even various insectivorous vespertilionids (Dijkgraaf 1957; Kolb 1961; Roer 1969). Some bats have actually been reported (Dunning 1968) to detect by smell that an insect is poisonous, and thus avoid it.

It is doubtful that olfaction plays a role in long-distance orientation (Mueller 1969), but there is a possibility that some bats localize their daytime resting places by smell (Dijkgraaf 1946). In his study of the occupation of nest-boxes by bats, Henze (1966) found that in successive years various individuals sought out the same boxes, and he interpreted this as evidence for olfactory localization. It has been suggested that the intense odour of species of *Tadarida* helps them to find their daytime retreats (Kingdon 1974); Kulzer (1969) showed that the sense of smell of the Egyptian fruit bat (*Rousettus aegyptiacus*) is crucial for localization of its resting place at close range.

Very few papers contain any comments on the significance of olfaction in social behaviour. And most of these remarks are based on chance observations or are simply speculation; not a single publication describes an experimental approach to this question.

Anatomy and physiology of olfaction

Olfactory structures in the brain

The olfactory bulb is developed to widely differing degrees in the various bat species. Volumetric comparisons of structures in the brains of 18 bat species (Stephan and Pirlot 1970) showed that the primary olfactory region is smallest in the insectivorous Rhinolophidae, Hipposideridae, and Vespertilionidae, as well as the fish-eating bat *Noctilio leporinus* (regression index 21 to 42). The blood-, nectar-, and fruit-eating leaf-nosed bats (Phyllostomatidae) occupy an intermediate position (regression index 51 to 87), whereas the largest olfactory bulb is found in the frugivorous megachiropterans (regression index 78 to 114). This trend was confirmed by Bhatnagar and Kallen (1974*a*). In frugivorous bats they found bulbs of over 2 mm diameter, whereas the bulbs of insectivorous species had diameters of 2 mm or less.

A similar correlation between feeding habits and development of the olfactory structures in the brain was found by Mann (1960). In comparing the vespertilionid *Histiotus montanus* (insectivore) and the phyllostomatids *Desmodus rotundus* (sanguivore) and *Phyllostomus hastatus* (omnivore), he found by far the smallest olfactory bulb in the insectivorous bat.

Structure of the nose

As early as 1902 Grosser published detailed descriptions of the nasal morphology of eight European bat species, but no other species were studied in this regard until quite recently (Kolb and Pisker 1964; Kolb 1971; Schmidt and Greenhall 1971; Gothe 1973; Bhatnagar and Kallen 1974*b*; Kämper and Schmidt 1977).

Although the number of conchae in the nose is fairly constant (there are 3–4 endoturbinates and two ectoturbinates), the structural features of the nasal cavity differ in the various families. In the Phyllostomatidae the conchae are connected to the lateral wall and the floor of the cavity by a complicated system of struts; the voluminous ectoturbinates of the Vespertilionidae are not connected to the floor of the cavity, but are attached only in the posterior region, to the lateral wall. In the Rhinolophidae the conchae are thin, rounded ridges braced by long struts against the lateral air-space of the nose wall or the floor of the nose. The tendency to enlarge the olfactory area by infolding of the conchae is also much more pronounced in some bats than in others. The most complicated structures are exhibited by the ethmoturbinates of the frugivorous and sanguivorous phyllostomatids; in the insectivorous Rhinolophidae there is no enlargement of the olfactory surface, the turbinates leaving large open cavities within the nose. It may be that this modification represents an adaptation for emitting echolocation sounds through the nose.

The few quantitative studies of the olfactory mucosa show that the insect-eating species usually have a relatively small olfactory region with few sense cells, whereas both the area of the olfactory region and the number of receptors are considerably greater in some of the fruit-, nectar- and blood-eating bats (Table 6.1). For example, the vampire bat (*Desmodus rotundus*), with its c. 2×10^7 receptors, has almost 40 times as many olfactory cells as the insectivorous lesser horseshoe bat (*Rhinolophus hipposideros*), which has only 5.4×10^5 receptors (Kolb and Pisker 1964; Grosse-Braukmann and Schmidt, in preparation). But there are exceptions to this feeding-correlated trend; for example, some of the insectivorous vespertilionids (*Myotis myotis, Nyctalus noctula*) have well-developed olfactory organs, with c. 5×10^6 receptors.

Table 6.1. The development of the olfactory epithelium in different bat species

Species	Olfactory epithelium in both nasal cavities (mm^2)	Number of receptors	Food	Reference
Rhinolophus hipposideros	27	5.4×10^5	I	1
Plecotus auritus	35	1.1×10^6	I	1
Myotis lucifugus	36	–	I	2
Myotis blythi	68	–	I	2
Rhinolophus ferrumequinum	72	1.9×10^6	I	1
Rhinolophus bocharius	76	–	I	2
Macrotus waterhousii	94	6×10^6	I, F	3
Leptonycteris nivalis	124	9×10^6	I, N	3
Nyctalus noctula	176	5.3×10^6	I	1
Myotis myotis	188	5.5×10^6	I	1
Artibeus jamaicensis	232	–	F	2
Desmodus rotundus	308	2×10^7	B	3

Food: B—blood; F—fruit; I—insects; N—nectar.

References:
1—Kolb (1971); 2—Bhatnagar and Kallen (1975); 3—Grosse-Brauckmann and Schmidt (in preparation).

The external nares of most bat species show little differentiation. But in two groups, the Murininae (family Vespertilionidae) and the Nyctimeninae (family Pteropidae), tubular nostrils have evolved. In *Nyctimene* such a tube consists of a complicated scroll of skin and reaches a length of over 5 mm. The function of this conspicuous structure is not clear. Stoddart (1980) suggests that the tubes allow the bat to utilize tropotactic orientation to odour while in flight.

Although only a few species have been studied so far, the variations in morphology and histological structure suggest that the different bat species may differ profoundly in olfactory ability and in the uses to which this ability is put.

Vomeronasal organ

Although there is as yet no experimental evidence as to the function of the vomeronasal organ, the neural link between this accessory olfactory organ and the limbic system suggests that it is involved in social and reproductive behaviour. Among the bats, the accessory olfactory system is developed to quite different extents. In all of the Megachiroptera examined so far it is absent, but it has been shown to exist in members of various microchiropteran families (Grosser 1902; Cooper and Bhatnagar 1976; Kämper and Schmidt 1977).

The vomeronasal organ is most conspicuous in the Phyllostomatidae; all of these have a well-developed vomeronasal complex (Jacobson's organ and accessory olfactory bulb). A rudimentary vomeronasal organ is present in some bat species (e.g. *Rhinopoma, Rhinolophus, Hipposideros,* and *Megaderma*), but it appears unlikely to be functional because there is no accessory olfactory bulb. It should be noted, however, that in all these species except *Rhinolophus* a vomeronasal nerve and paravomeronasal ganglion have been found (Cooper and Bhatnagar 1976). Most of the vespertilionids lack the accessory olfactory organ altogether; only *Miniopterus* has a well-developed vomeronasal complex (Broom 1895). None of the Molossidae have as yet been found to have a vomeronasal organ (Humphrey 1936; Mann 1961; Schneider 1957).

In contrast to the normal olfactory organ, the degree of development of the vomeronasal organ is not correlated with feeding habits. Among both fruit-eating (Megachiroptera) and insectivorous species (many vespertilionids) the accessory olfactory organ may be lacking, though it is present in other species of similar habits. For example, all the frugivorous phyllostomatids and many insectivorous bats (*Miniopterus, Macrotus*) have a well-developed vomero-nasal complex. As far as the possibility of correlation with social behaviour patterns is concerned, no relevant data are available.

Olfactory ability of bats

The microchiropterans, which depend chiefly on echolocation for orientation, are often classified as microsmatic mammals. In the case of certain species, the Rhinolophidae in particular, the anatomical structure of the nose and the development of the olfactory bulb suggest that this assignment is justified. But the anatomy of the olfactory system of most microchiropteran species indicates that they have considerable olfactory sensitivity.

A number of field observations have shown that the olfactory system can perform quite well in localizing food. For example the South American leaf-nosed bat *Phyllostomus hastatus* can find small pieces of banana hidden under leaves (Mann 1951*b*), and the vampire bat, *Desmodus rotundus*, is thought to be capable of using its sense of smell to locate prey animals in a barn (Mann 1960) and to identify blood (Ruschi 1951*a*). It may be that olfaction is also involved in choice of prey animals by vampire bats. Goodwin and Greenhall (1961), for instance, found that of a group of people sleeping in the same room a particular individual was bitten every night, even though he changed the place where he slept.

Indications of the olfactory acuity of different bat species are provided by experiments in which the animals are required to find hidden food or respond to insects held out to them. For example, Möhres and Kulzer (1956) trained Egyptian fruit bats (*Rousettus aegyptiacus*) to take bananas only from a white box, and to ignore a black box in which the food was inaccessible to them. Once the bats had become familiar with this problem, the usual feeding place was suddenly left empty and only the black box was baited. The animals turned unhesitatingly to the previously negative training site, and paid no attention to the white box to which they had been trained, but which now no longer smelled of banana. These results show that while the Egyptian fruit bat takes visual features into account while searching for food, the dominant role is played by the sense of smell. Because the animals studied were able while flying to localize as little as 50 mg of banana, and could distinguish artificial banana flavour on ether from natural banana aroma, it can be assumed that the sense of smell of *Rousettus aegyptiacus* is considerably superior to that of man in both sensitivity and the ability to discriminate.

Certain insectivorous bats have also been found to use olfactory stimuli to find and choose prey. Dijkgraaf (1957) showed that both *Myotis emarginatus* and *Plecotus auritus* could be stimulated to search for food when a mealworm was placed 10 to 15 mm in front of the nose. These bats could also distinguish a mealworm from a piece of wood of the same size, by their smell. In his study of olfaction in various European vespertilionids, Kolb (1961) found that non-flying insects hidden on the ground under leaves could be detected by olfaction. The greatest distance at which a long-eared bat (*Plecotus auritus*) could localize dead moths was 10–20 cm. The vespertilionid species studied by Kolb were also able to discriminate between unpalatable potato beetles and the dung beetles they preferred. Kolb (1973, 1976) suggested that the mouse-eared bat (*Myotis myotis*) sends out special sounds that aid in the olfactory localization of insects by loosening odour molecules from the prey and whirling them into the air. The frequency of these sounds, 16–30 kHz, is below the frequency range of echolocation sounds.

Olfactory threshold measurements for defined odour molecules have as yet been made with very few species of bat. Schmidt and Greenhall (1971) set up a training programme in which the locomotor responses of vampire bats (*Desmodus rotundus*) could be used to establish the detection threshold for butyric acid. In these experiments the olfactory sensitivity of the bat was higher by about a factor of 10 than that of man. These bats could also be trained to distinguish among various aliphatic carboxylic acids (Schmidt 1973). Olfactory thresholds for a number of synthetic odour substances were measured with a conditioning method for three phyllostomatid species (*Phyllostomus discolor*, *Artibeus lituratus*, and *Desmodus rotundus*) as well as the greater mouse-eared bat (*Myotis myotis*) (Schmidt 1975; Obst and Schmidt 1976; Schmidt and Schmidt 1978). Here changes in respiratory and heart rates were taken as the criteria for the detection of an odour substance. Whereas the thresholds found for the three phyllostomatids were very near one another, *Myotis* was distinctly less sensitive to all the substances tested. Its thresholds were higher than those of the vampire bat by factors of 2 to 10^7. Comparison with rat and human shows that the olfactory thresholds of the phyllostomatids to aliphatic carboxylic acids are of the same order of magnitude. However, different mammalian species appear to have very different odour spectra. For example, the olfactory thresholds of the vampire bat to certain alcohols and aldehydes are lower than those of humans, whereas the situation is reversed in the case of fatty acids. Moreover, among the four bat species studied here the relative sensitivities to the different odour substances were not the same.

Olfactory thresholds to synthetic odour substances, however, give nothing like a complete picture of the olfactory abilities of a species. It cannot be ruled out that the olfactory sensitivity of bats to biologically important odour substances such as pheromones or food odours is much greater than to the more or less randomly selected substances used in experiments. Nevertheless, the physiological experiments reveal that all the bat species so far examined have an entirely competent olfactory organ, capable of a performance not inferior to that of other small mammals known to depend heavily on olfaction. Thus all the conditions for participation of the sense of smell in social communication are met.

Glands

Skin glands take many different forms in Chiroptera. The more conspicuous of these are mentioned in all the systematic works (e.g. Goodwin and Greenhall 1961; Husson 1962; Harrison 1964; Villa 1966; Kingdon 1974; Walker 1975), and some of the glands have been described by Schaffer (1940) and Grassé (1955, 1967). The secretion of the glands in the skin is frequently perceived as having a penetrating smell and is considered to be responsible for the specific odour of many bats.

Members of the Megachiroptera often have large glandular fields in the shoulder and neck region. In species of *Pteropus*, for example, an oval field of glands lies on the back of the neck toward the side; it is highly developed in the males but only slightly so in females (Nelson 1965; Neuweiler 1969). These neck glands consist of numerous holocrine sebaceous-gland tubules, each of which opens into a bed of hairs. During the rutting season these glands in the male release a secretion with a strong musky smell, sometimes in such large amounts that it soaks the neck fur. Similar neck glands are found in the genera *Rousettus* and *Eidolon*.

Male epauletted fruit bats (Epomorphorinae) bear an especially striking structure—pockets of skin at the side of the neck (Noack 1899). Within each of these pockets is a tuft of white hairs that can be pushed out or withdrawn entirely. The extent to which these pockets function as glands is unclear. Püscher (1972), who examined them histologically in several species of *Epomorphorus*, could find no special glands. The skin of the pockets differs little from the general body skin; indeed, in the pocket skin the sebaceous glands are distinctly reduced.

Neck glands are relatively rare in the Microchiroptera. A few Phyllostomatidae, such as the wrinkled-faced bat (*Centurio senex*), the false vampire bat *Chrotopterus*, and the American epauletted bat *Sturnira*, have glands in the neck region. Some species of *Tadarida* (Molossidae) also have tufts of glandular hairs behind their ears (Walker 1975).

Tadarida laticaudata has glands in the anal region (Stubbe 1969). These consist of a dense knot of apocrine tubules that surround the proctodeum, and of a holocrine glandular apparatus that secretes into five storage chambers. The apocrine glands and the ducts from the secretion reservoirs open at the anus in a single plane. Large anal glands with openings to the right and left of the penis are found in male fisherman bats (*Noctilio leporinus*). The extremely penetrating odour of the fisherman bats is ascribed to the secretion produced by these glands (Ruschi 1951*b*).

Many bat species have glands in the throat or chest region. Males of the species *Molossus ater* and *Molossus molossus* are distinguished by a large, unpaired throat gland with both apocrine and holocrine elements, surrounded by a thin connective-tissue capsule. The ducts from the gland tubules open into a small sac of skin, the external opening of which is formed by a transverse slit in the throat region (Werner and Lay 1963; Stubbe 1969). Similar throat sacs are found in the Southeast Asian naked bat *Cheiromeles* (Harrison and Davies 1949), the genera *Eumops* and *Otomops*, and males of species of *Phyllostomus* (the throat glands of females are much reduced). In most species of *Taphozous* (Emballonuridae) there is a large gland between the mandibles, with ducts that open on two papillae just behind a throat pouch with an opening that faces forward (Schaffer 1940). In this case, gland and throat pouch are well developed in both sexes (Kingdon 1974). Male *Saccopteryx bilineata* also have a throat gland (Starck 1958).

The wing membranes of some species also contain pouches of skin that are frequently regarded as glandular organs. The ghost bat, *Diclidurus albus*, has a pocket in the tail membrane (Ruschi 1953), but the wing pockets typical of many Emballonuridae are in the propatagium. These small pockets at the humeroradial angle, which are an important characteristic for species diagnosis, open on to the dorsal surface of the wing membrane. In *Saccopteryx* they exhibit a distinct sexual dimorphism (Starck 1958), being considerably longer and thicker in the male than in the female (Fig. 6.1). The slit-like opening of the pockets can be opened and closed by striated muscles. On the floor of the pocket there project 8–14 comb-like ridges. A surprising finding is that the wall of the fore-arm sac is entirely free of glands of any kind. However, these pockets

Fig. 6.1. A male *Saccopteryx bilineata* (*right*) salting to an intruder in his territory. (After Bradbury and Emmons 1974.)

frequently contain a large amount of granular matter that gives off a penetrating aroma. Because the much-folded epidermis of the pockets exhibits a peculiar loosening of the horny layer, Starck (1958) proposed that this region represents a special form of holocrine secretory mechanism, a 'holocrine surface gland' (Schaffer 1940). Whereas *Saccopteryx bilineata* has no special glands in its wings, the fish-eating vespertilionid *Pizonyx vivesi* has a glandular mass in the middle of the forearm (Walker 1975). Glands producing a strong-smelling secretion can also be present in the mouths of some bats. For example, the vampire bat *Diaemus youngi* has two cup-shaped glands in the oral cavity, the stinking contents of which are expelled through the mouth with an audible hiss when the bat is disturbed (Goodwin and Greenhall 1961).

All the bat species studied so far have glands in the facial region. In some families, such as the Vespertilionidae, these are located at the side of the nose (Werner and Dalquest 1952; Walton and Siegel 1966; Grassé 1967), and in others the nasal leaves and folds that give many bat species their characteristic appearance bear glands (Werner, Dalquest, and Roberts 1950; Dalquest and Werner 1954; Werner and Rutherford 1979). In hibernating species such as the Rhinolophidae, the size of the glands is reduced during hibernation (Grassé 1955); in some species they are especially prominent during the reproductive season (Dwyer 1966).

The facial glands seem usually to be identical in males and females; only those of the funnel-eared bats (*Natalus*) exhibit sexual dimorphism. Here the male has a large apocrine glandular mass under the skin of the forehead, the natalid organ, which makes a bulge in the frontal region of the head (Dalquest 1950).

Whereas the facial glands of closely related genera are quite similar in structure, there are considerable differences among the various families. Some families (Emballonuridae, Rhinolophidae, Natalidae, Vespertilionidae) have sweat glands, whereas there are none in the Phyllostomatidae, Chilonycteridae, and Molossidae. Sebaceous glands, on the other hand, are present in all families. They are especially numerous in the face region of the vespertilionids, where they may form enormous pits that open to the surface of the skin by way of short, broad canals. In the Phyllostomatidae the sebaceous glands are scattered over the nose leaf. Only in *Desmodus* is the nose ridge composed almost entirely of loose connective tissue, within which there is practically no glandular tissue (Dalquest and Werner 1954).

The examples listed here do not exhaust all of the known varieties of glands. The great diversity in skin glands among the different bat species suggests that there are pronounced differences in the relevance of the odours to social behaviour. In species sexually dimorphic in this regard it appears likely that odour substances play a role in reproductive behaviour. However, the absence of conspicuous glandular regions does not exclude the possibility that, for example, individuals can recognize one another by smell owing to the ubiquitous small facial glands.

The role of olfaction in social behaviour

Megachiroptera

The main sense by which all fruit bats conduct their search for food is olfaction. The advanced development of the olfactory system associated with this function facilitates the acquisition of social behaviour patterns influenced by olfactory signals. The presence of large shoulder glands in *Pteropus*, and their increased secretion during the rut, imply that the odour of this secretion is involved in sexual behaviour.

The most thoroughly investigated are the three sympatric Australian species

Pteropus poliocephalus, P. gouldi, and *P. scapulatus* (Ratcliffe 1932; Nelson 1964, 1965). These fruit bats live in colonies that comprise many individuals, spending the day in the branches of large trees. In the course of the year they migrate between winter camps and summer camps. Whereas the sexes live separately in the winter camps, in the summer camps the young are born and raised and mating occurs. Accordingly, the most conspicuous social interactions are observable in the summer camps. Behaviour patterns influenced by olfaction appear predominantly in two functional categories: (i) in the mother–offspring relationship, and (ii) in the mating season.

Between September and December, when the young are born and being raised, the females tend strongly to keep apart from the males. The most important social function of the sense of smell during this time is the identification of individual young by their mothers. During the first months of life the young are left behind in the home trees when the mothers fly out at night to look for food. When the adults return in the morning, they fly around the trees and give loud search cries, which are answered by the young animals with location calls. After some time the mother lands next to one of the young and smells its chest region. She accepts only her own offspring, by pulling back her wings and letting it climb onto her chest. Strange young are pushed away with thumbs and wings. Until the young are about three weeks old they cannot recognize their own mothers; they try to cling to any mature animal. Nelson (1964) proposed that the mothers can remember the position in the home-tree where they have left their young, and thus are usually able to find their own offspring quickly.

Localization of the young involves an interplay between auditory and olfactory sensations. As laboratory experiments have shown, *Pteropus* mothers whose young have been isolated in gauze sacks are attracted by the location calls of all young animals. But individual identification is an olfactory process; only after the females have smelled the gauze sacks do they remain near their own offspring, or fly away if the young bat is not their own (Nelson 1965).

The mating season of *Pteropus poliocephalus* begins in January. At this time the males set up territories and mark them by rubbing their shoulder glands over the twigs. The majority of the colony now has formed into family groups consisting of a male, a female, and her young. A few males gather several females about themselves, but in this case there are no young. The males of these adult groups also mark more frequently than those of the family groups. Two sites within the territory are always marked—one is the resting place of the male, the other is the place where the female hangs. The marked regions and the surrounding twigs are defended against intruders by males and females alike. The members of a group appear to recognize one another by smell. They often sniff each other in the shoulder region, particularly when an intruder has been driven away. When a female returns to the territory the male always moves along the branch toward her, briefly sniffs her shoulder region, and rapidly returns to his place. Finally, an intense sniffing of the female's scapula region, followed by licking of the genitalia, can always be observed prior to copulation.

Not all *Pteropus* species seem to exhibit marking behaviour. Neuweiler (1969) observed *P. giganteus* in India and saw no evidence of marking, although in other respects the behaviour patterns of these fruit bats during the reproductive season closely resemble those of *P. poliocephalus. P. giganteus* also sniffs the scapula region frequently, especially before copulation, and the mothers also recognize their young by their smell. According to Neuweiler, the secretion of the neck gland could function in the olfactory identification of the tree where the animals rest, which by being impregnated with the scent would acquire a familiar aura of home for the whole colony. The possibility of synchronization of readiness to reproduce by the odour of this secretion has also been discussed.

Social behaviour patterns based on olfaction have rarely been observed in other species of fruit bat. Kulzer (1958) made a laboratory study of the mother–young relationship of *Rousettus aegyptiacus.* His isolation experiments showed unequivocally that the young are recognized by the mother, as well as the mother by the young, on the basis of smell. Sniffing of the neck and the genital region prior to copulation appears to be a widespread behaviour pattern among fruit bats, as observations of *Eidolon helvum* have shown (Kulzer 1969). Opinions differ as to the significance of the shoulder pockets of the epauletted fruit bats, with their conspicuous tufts of hair. Whereas Kingdon (1974) proposed that their action as a visual signal is accompanied by an olfactory signal, other authors offer no evidence of glandular function (Wickler and Seibt 1976; Püscher 1972). Wickler and Seibt always observed a brief nose contact when members of a colony of *Epomorphorus wahlbergi* encountered one another; there was no sniffing of other parts of the body.

The few studies of social behaviour of the Megachiroptera indicate that an olfactory component is probably necessary for individual recognition in all species. As to the function of the widespread neck glands, except in the case of *Pteropus poliocephalus*, nothing can be said. The extent to which the secretion of these glands also serves to mark the territories of other species remains to be shown by further investigations.

Microchiroptera

In all the microchiropterans, the auditory sense is by far the most important for orientation as well as for social communication. For communication, all the species so far studied use either the echolocation sound itself or special communication sounds. By comparison, the other sensory modalities are of little significance. The great variability in the development of the olfactory organ within this group, and the great diversity in skin glands, suggest that olfaction has very disparate roles in social interactions within the different species.

Olfaction has been mentioned in the literature most frequently in the context of the mother–young relationship. Especially among the Vespertilionidae, it appears to be the rule that each mother suckles only her own offspring (Pearson, Koford, and Pearson 1952; Davis 1969; Dwyer 1970). Because the young of most species remain behind in the daytime shelter while the adults hunt at night, individual recognition must be possible (Davis, Barbour, and Hassell 1968). As Brown (1976) showed for *Antrozous pallidus*, the mothers are capable of identifying their young by the isolation calls. The individual characteristics of the sound, however, seem to be relatively ill-defined in the very young, because the isolation calls often attract strange mothers. It can be assumed that an animal's own young are identified by smell, for each offspring is smelled before being allowed to cling to the mother. Evidence for individual olfactory identification is given by the observation that a slightly lethargic mother immediately became attentive and emitted social sounds when her offspring was brought near her. Other nearby females showed no reaction at all.

Female *Myotis nigricans* (Wilson 1971) and *Myotis myotis* (Kolb 1950, 1957, 1972) have also been observed, when searching for their young, to sniff various other young bats before accepting their own and letting it suckle. That the young can in turn identify their mothers by smell has been shown by Kolb (1977). He put two *Myotis* mothers into a lethargic state by cooling them, so that they could give no sounds, and watched to see which female the young would suckle. Although neither female made any sort of defence movement, each young bat took only the teat of its own mother.

As to the function of olfaction in behavioural situations other than the mutual recognition of mother and young, little is known about the vespertilionids. Thomas, Fenton, and Barclay (1979) observed *Myotis lucifugus*, a species in which 2–11 individuals form small, labile groups, and noted that each animal joining such a group makes brief nose contact with the others in it. Such nose contact also occurred between males and females prior to copulation. Similar behaviour patterns are also found among the Phyllostomatidae. Vampire bats sniff each other at each meeting (Schmidt and Manske 1973). In a colony of these bats the individuals know one another (Schmidt and van de Flierdt 1973; Sailler and Schmidt 1978), and it can be assumed that olfactory as well as auditory factors are involved in this recognition. Porter (1978) described a very odd behaviour of *Carollia perpicillata*. Two males, one of which had a harem, would frequently hang motionless opposite one another for several minutes at a time, with their heads turned toward one another and their tongues flicking repeatedly in and out of the mouth. From this 'nosing' a ritualized combat could develop, in which the opponents struck at each other with their wings. In view of the fact that *Carollia* has a well-developed vomeronasal organ, the tongue movement could represent stimulation of this organ. The synchronization of reproduction in phyllostomatids has also been thought to stand under olfactory influence. For example, Rasweiler (1975) has proposed that in *Glossophaga soricina* the timing of ovulation is determined by pheromones produced by the stud male.

Among the microchiropteran species with highly developed specific scent glands are the Molossidae. The males and females of these colonial species differ considerably in body odour, at least during a few months of the year. For example, male *Tadaria brasiliensis* exude a strong sweetish scent from October to December, whereas the females smell slightly musky. During the summer months this sex-specific scent is not detectable (Herreid 1960). Some Molossidae have conspicuous throat glands. Male *Molossus molossus* use the secretion of these glands to mark members of the group and certain sites in the cage (Häussler *et al.*, 1981). A male that I kept isolated for a long time in the laboratory marked a young female held out toward him so intensely that the female was drenched in whitish secretion. The penetrating musky smell filled the whole room. When marking, the males rub their throats rhythmically over the neck and back fur of the partner. This marking behaviour is exhibited with females in the group, and subdominant males are marked in the same way whenever a posture of humility is adopted by the subordinate animal following an aggressive inter-action (Häussler, in preparation).

Among the molossids, the sense of smell is responsible for recognition of the mother by the young as it is in the vespertilionids. Kulzer (1962) studied the responses of a young *Tadarida condylura* to the orientation sounds of various bat species. The young animal responded to the echolocation sounds of three *Tadarida* species by emitting isolation sounds and running toward the sound source; the sounds of bats of other genera elicited no response. At close range, however, the young bat would respond only to its mother. If a number of individuals were hidden in gauze sacks, so that the young bat could smell them, it sniffed at all the sacks but only when it found that containing its own mother did it attempt to climb onto the sack or creep under it. Kulzer concluded from his experiments that the orientation sounds stimulate a young animal separated from its mother to look for her and send out isolation sounds, but that the identification of the mother is done by olfactory means. This behaviour is very probably not a uniform characteristic of all *Tadarida* species. In *Tadarida brasi-liensis*, a species that lives in huge colonies, no individual relationship between mother and child is discernible; the mothers suckle all the young indiscriminately (Davis *et al.* 1962). Accordingly, individual recognition appears unlikely in this species.

The most thorough study of microchiropteran social behaviour was done on the emballonurid *Saccopteryx bilineata* (Bradbury and Emmons 1974). These small South American bats spend the daylight hours in light but protected places —for example, in the niches among the buttress roots of large trees. Here the males gather harems and defend them against intruders. As part of this territorial defense there appears, in addition to vocalization, a form of behaviour that Bradbury and Emmons call 'salting'. In a typical case of salting, a male approaches his opponent and extends the folded wings toward him (Fig. 6.1). At this point the wing gland in the propatagium opens and the wing, raised above the ground, is moved rapidly back and forth 4-25 times. This wing movement can cover an angle of more than 90°. Often the two males salt one another alternately, several times in succession. The same behaviour pattern is also directed toward the females in the male's own harem and toward those in the neighbouring harem. Because a moist secretion becomes visible in the opened wing pockets during salting, it can be assumed that an olfactory signal is transmitted during the wing movement. It may be that the wing glands also go into action when the bats return in the morning from their feeding flight. The males are always the first to return to their territories. When a female returns shortly thereafter, the male flies up, swoops in a tight curve and hovers a few centimetres in front of the female. In so doing he emits brief chirps and opens the wing pockets. As yet the function of the odour of the wing-gland secretion during this behaviour remains a mystery. The closely related *Saccopteryx leptura*, in which the wing glands are similarly developed, has not been observed to engage in wing-shaking or any sort of marking behaviour.

Studies of bat social behaviour have been begun only recently. It is not yet possible to estimate the extent to which olfactory signals may be involved. The few results so far available, however, indicate that in at least a few species the odours produced by the skin glands can be employed in social communication. The oft-quoted assumption that bats are microsmatic and communicate exclusively by way of the auditory system certainly applies to only a few species.

References

Bhatnager, K. P. and Kallen, F. C. (1974*a*). Cribriform plate of ethmoid, olfactory bulb and olfactory acuity in forty species of bats. *J. Morph.* **142**, 71-90.
— — (1974*b*). Morphology of the nasal cavities and associated structures in *Artibeus jamaicensis* and *Myotis lucifugus*. *Am. J. Anat.* **139**, 167-90.
— — (1975). Quantitative observations on the nasal epithelia and olfactory innervation in bats. *Acta anat.* **91**, 272-82.
Bradbury, J. W. and Emmons, L. H. (1974). Social organization of some Trinidad bats. *Z. Tierpsychol.* **36**, 137-83.
Broom, R. (1895). On the organ of Jacobson in an Australian bat (Miniopterus). *Proc. Linn. Soc. NSW* **10**, 571-5.
Brown, P. (1976). Vocal communication in the pallid bat, *Antrozous pallidus*. *Z. Tierpsychol.* **41**, 34-54.
Cooper, J. G. and Bhatnagar, K. P. (1976). Comparative anatomy of the vomeronasal organ complex in bats. *J. Anat.* **122**, 571-601.
Dalquest, W. W. (1950). The genera of the chiropteran family Natalidae. *J. Mammal.* **31**, 436-43.
— and Werner, H. J. (1954). Histological aspects of the faces of North American bats. *J. Mammal.* **35**, 147-60.
Davis, R. B. (1969). Growth and development of young pallid bats, *Antrozous pallidus*. *J. Mammal.* **50**, 729-36.
– – Herreid, C. F., and Short, H. L. (1962). Mexican free-tailed bats in Texas. *Ecol. Monogr.* **32**, 311-46.

Davis, W. H., Barbour, R. W., and Hassell, M. D. (1968). Colonial behavior of *Eptesicus fuscus. J. Mammal.* **49**, 44–50.

Dijkgraaf, S. (1946). Die Sinneswelt der Fledermäuse. *Experientia* **2**, 438–48.

— (1957). Sinnesphysiologische Beobachtungen an Fledermäusen. *Acta physiol. pharmac. néerl.* **6**, 675–84.

Dunning, D. C. (1968). Warning sounds of moths. *Z. Tierpsychol.* **25**, 129–38.

Dwyer, P. D. (1966). Observations on *Chalinolobus dwyeri* (Chiroptera: Vespertilionidae) in Australia. *J. Mammal.* **37**, 716–18.

— (1970). Social organization in the bat *Myotis adversus. Science, NY* **168**, 1006–8.

Goodwin, G. G. and Greenhall, A. M. (1961). A review of the bats of Trinidad and Tobago. *Bull. Am. Mus. nat. Hist.* **122**, 189–302.

Gothe, J. (1973). Beiträge zu einer funktionellen Morphologie der Regio intermedia in der Nasenhöhle einheimischer Chiropteren (Vespertilionidae). *Z. wiss. Zool.* **185**, 222–34.

Grassé, P.-P. (1955). Traité de Zoologie. Anatomie, Systématique, Biologie, **17**, Pt. 2. *Mammifères*, pp. 1729–1853. Libraires de l'Académie de Médicine, Masson et Cie, Paris.

— (1967). Traité de Zoologie. Anatomie, Systématique, Biologie, **16**, Pt. 1. *Mammifères: téguments et squelette*, pp. 1–233. Libraires de l'Académie de Médicine, Masson et Cie, Paris.

Grosser, O. (1902). Zur Anatomie der Nasenhöhle und des Rachens der einheimischen Chiropteren. *Morphol. Jb.* **29**, 1–77.

Harrison, B. A. and Davies, D. V. (1949). A note on some epithelial structures in Microchiroptera. *Proc. zool. Soc. Lond.* **119**, 351–7.

Harrison, D. L. (1964). *The mammals of Arabia.* Ernest Benn, London.

Häussler, U., Möller, E., and Schmidt, U. (1981). Zur Haltung und Jugendentwicklung von *Molossus molossus* (Chiroptera). *Z. Sängetierkunde* **46**, 337–51.

Henze, O. (1966). Riechen Fledermäuse die Rastplätze ihrer Artgenossen vom Herbstzug des Jahres zuvor? *Myotis* **4**, 23–4.

Herreid, C. F. (1960). Comments on the odors of bats. *J. Mammal.* **41**, 396.

Humphrey, T. (1936). The telencephalon of the bat. I. The non-cortical nuclear masses and certain pertinent fiber connections. *J. comp. Neurol.* **65**, 603–711.

Husson, A. M. (1962). The bats of Suriname. *Zool. Verh.* No. 58. Brill, Leiden.

Kämper, R. and Schmidt, U. (1977). Die Morphologie der Nasenhöhle bei einigen neotropischen Chiropteren. *Zoomorph.* **87**, 3–19.

Kingdon, J. (1974). *East African mammals*, Vol. 2. Academic Press, London.

Kolb, A. (1950). Beiträge zur Biologie einheimischer Fledermäuse. *Zool. Jahrbücher, Abt. f. Syst., Okol. u. Geogr. d. Tiere* **78**, 547–71.

— (1957). Aus einer Wochenstube des Mausohrs, Myotis m. myotis. *Säugetierkdl. Mitt.* **5**, 10.

— (1961). Sinnesleistungen einheimischer Fledermäuse bei der Nahrungssuche und Nahrungsauswahl auf dem Boden und in der Luft. *Z. vergl. Physiol.* **44**, 550–64.

— (1971). Licht- und elektronenmikroskopische Untersuchungen der Nasenhöhle und des Riechepithels einiger Fledermausarten. *Z. Säugetierkunde* **36**, 202–13.

— (1972). Die Geburt einer Fledermaus. *Image* **49**, 5–13.

— (1973). Riechverhalten und Riechlaute der Mausohrfledermaus *Myotis myotis. Z. Säugetierkunde* **38**, 277–84.

— (1976). Funktion und Wirkungsweise der Riechlaute der Mausohrfledermaus, *Myotis myotis. Z. Säugetierkunde* **41**, 226–36.

— (1977). Wie erkennen sich Mutter und Junges des Mausohrs, *Myotis myotis*, bei der Rückkehr vom Jagdflug wieder? *Z. Tierpsychol.* 44, 423-31.

— and Pisker, W. (1964). Über das Riechepithel einiger einheimischer Fledermäuse. *Z. Zellforsch.* 63, 673-81.

Kulzer, E. (1958). Untersuchungen über die Biologie von Flughunden der Gattung *Rousettus* Gray. *Z. Morph. Ökol. Tiere* 47, 374-402.

— (1962). Über die Jugendentwicklung der Angola-Bulldogfledermaus Tadarida (Mops) condylura (A. Smith, 1833) (Mollossidae). *Säugetierkdl. Mitt.* 10, 116-24.

— (1969). Das Verhalten von *Eidolon helvum* (Kerr) in Gefangenschaft. *Z. Säugetierkunde* 34, 129-48.

Jones, C. (1967). Growth, development and wing loading in the evening bat, *Nycticeius humeralis* (Rafinesque). *J. Mammal.* 48, 1-19.

Mann, G. (1951a). Biologia del vampiro. *Biologica, Santiago* 12/13, 1-19.

— (1951b). *Esquema ecológico de selva, sabana y cordillera en Bolivia.* Publ. Universidad de Chile, Santiago.

— (1960). Neurobiologia de *Desmodus rotundus. Invest. Zool. Chil.* 6, 79-99.

— (1961). Bulbus olfactorius accessorius in Chiroptera. *J. comp. Neurol.* 116, 135-44.

Möhres, F. P. and Kulzer, E. (1956). Über die Orientierung der Flughunde (Chiroptera—Pteropodidae). *Z. vergl. Physiol.* 38, 1-29.

Mueller, H. C. (1969). Homing and distance—orientation in bats. *Z. Tierpsychol.* 23, 403-21.

Nelson, J. E. (1964). Vocal communication in Australian flying foxes (Peteropodidae, Megachiroptera). *Z. Tierpsychol.* 21, 857-70.

— (1965). Behaviour of Australian Pteropodidae (Megachiroptera). *Anim. Behav.* 13, 544-57.

Neuweiler, G. (1969). Verhaltensbeobachtungen an einer indischen Flughundkolonie (*Pteropus g. giganteus* Brünn). *Z. Tierpsychol.* 26, 166-99.

Noack, T. (1899). Beiträge zur Kenntnis der Säugetierfauna von Süd- und Südwestafrika. *Zool. Jb. Abt. Syst.* 4, 94-261.

Obst, Ch. and Schmidt, U. (1976). Untersuchungen zum Riechvermögen von *Myotis myotis* (Chiroptera). *Z. Säugetierkunde* 41, 101-8.

Pearson, O. P., Koford, M. R., and Pearson, A. K. (1952). Reproduction of the lump-nosed bat (*Corynorhinus rafinesquei*) in California. *J. Mammal.* 33, 273-320.

Pijl, L. van der (1936). Fledermäuse und Blumen. *Flora* 131, 1-40.

Porter, F. L. (1978). Roosting patterns and social behavior in captive *Carollia perspicillata. J. Mammal.* 59, 627-30.

Püscher, H. (1972). Über die Schultertaschen von Epomophorus (Empomophorini, Pteropodidae, Megachiroptera, Mammalia). *Z. Säugetierkunde* 37, 154-61.

Rasweiler, J. J. (1975). Maintaining and breeding neotropical frugivorous, nectarivorous and pollenivorous bats. *Zoo Yearbook* 15, 18-30.

Ratcliffe, F. N. (1931). The flying fox (*Pteropus*) in Australia. *Aust. Commonw. Counc. Sci. Ind. Res. Bull.* 53, 6.

— (1932). Notes on the fruit bats (*Pteropus* spp.) of Australia. *J. anim. Ecol.* 1, 32-57.

Roer, H. (1969). Zur Ernährungsbiologie von Plecotus auritus (L.) (Mamm. Chiroptera). *Bonner Zool. Beitr.* 20, 378-83.

Ruschi, A. (1951a). Morcegos do estado do Espirito Santo. Familia Desmodontidae. *Boln. Mus. biol. Prof. Mello-Leitão. Zool.* 2, 1-7.

— (1951b). Morcegos do estado do Espirito Santo. Familia Noctilionidae. *Boln. Mus. biol. Prof. Mello-Leitão. Zool.* 7, 1-7.

— (1952). Morcegos do estado do Espirito Santo—IX a. Familia Emballonuridae. *Boln. Mus. biol. Prof. Mello-Leitão, Zool.* 10, 1-16.

— (1953). Morcegos do estado do Espirito Santo—X. Familia Emballonuridae. *Boln. Mus. Biol. Prof. Mello-Leitão. Zool.* 12, 1-9.

Sailler, H. and Schmidt, U. (1978). Die sozialen Laute der Gemeinen Vampirfledermaus *Desmodus rotundus* bei Konfrontation am Futterplatz unter experimentellen Bedingungen. *Z. Säugetierkunde* 43, 249-61.

Schaffer, J. (1940). *Die Hautdrüsenorgane der Säugetiere.* Urban and Schwarzenberg, Berlin.

Schmidt, U. (1973). Olfactory threshold and odour discrimination of the vampire bat (*Desmodus rotundus*). *Period. Biol.* 75, 89-92.

— (1975). Vergleichende Riechschwellenbestimmungen bei neotropischen Chiropteren (*Desmodus rotundus, Artibeus lituratus, Phyllostomus discolor*). *Z. Säugetierkunde* 40, 269-98.

— and Greenhall, A. M. (1971). Untersuchungen zur geruchlichen Orientierung der Vampirfledermäuse (*Desmodus rotundus*). *Z. vergl. Physiol.* 74, 217-26.

— and Manske, U. (1973). Die Jugendentwicklung der Vampirfledermäuse (*Desmodus rotundus*). *Z. Säugetierkunde* 38, 14-33.

— and Schmidt, Ch. (1978). Olfactory thresholds in four microchiropteran bat species. *Proc. 4th Int. Bat Res. Conf.*, pp. 7-13. Kenya Literature Bureau.

— and Van de Flierdt, K. (1973). Innerartliche Aggression bei Vampirfledermäusen (*Desmodus rotundus*) amd Futterplatz. *Z. Tierpsychol.* 32, 139-46.

Schneider, R. (1957). Morphologische Untersuchungen am Gehirn der Chiropteren (Mammalia). *Abh. Senckenb. Naturf. Ges.* 495, 1-92.

Starck, D. (1958). Beitrag zur Kenntnis der Armtaschen und anderer Hautdrüsenorgane von *Saccopteryx bilineata* Temminck 1838 (Chiroptera, Emballonuridae). *Morphol. Jb.* 99, 3-25.

Stephan, H. and Pirlot, P. (1970). Volumetric comparisons of brain structures in bats. *Z. Zool. Syst. Evolutionsforsch.* 8, 200-36.

Stoddart, D. M. (1980). *The ecology of vertebrate olfaction.* Chapman and Hall, London.

Stubbe, M. (1969). Zur Biologie mexikanischer Fledermäuse. *Zool. Anz.* 183, 317-26.

Thomas, D. W., Fenton, M. B., and Barclay, R. M. R. (1979). Social behaviour of the little brown bat, *Myotis lucifugus.* I. Mating behavior. *Behav. Ecol. Sociobiol.* 6, 129-36.

Villa, B. (1966). *Los murcielagos de Mexico.* Univ. Nac. Anton. de Mexico.

Vogel, S. (1958). Fledermausblumen in Südamerika. *Oesterr. bot. Z.* 104, 491-530.

Walker, E. P. (1975). *Mammals of the world,* Vol. I. The Johns Hopkins University Press, Baltimore.

Walton, D. W. and Siegel, N. J. (1966). The histology of the pararhinal glands of the pallid bat, *Antrozous pallidus. J. Mammal.* 47, 357-60.

Werner, H. J. and Dalquest, W. W. (1952). Facial glands of the tree bats, *Lasiurus* and *Dasypterus. J. Mammal.* 33, 77-80.

— — and Roberts, J. H. (1950). Histological aspects of the glands of the bat, *Tadarida cynocephala* (Le Conte). *J. Mammal.* 31, 395-9.

— — and Lay, D. M. (1963). Morphologic aspects of the chest gland of the bat, *Molossus ater. J. Mammal.* 44, 552-5.

— and Rutherford, K. (1979). Histological aspects of the facial glands of the bat, *Monophyllis redmani portoricensis*. *J. Mammal.* **60**, 229.

Wickler, W. and Seibt, U. (1976). Field studies on the African fruit bat *Epomorphorus wahlbergi* (Sundevall), with special reference to male calling. *Z. Tierpsychol.* **40**, 345–76.

Wilson, D. E. (1971). Ecology of *Myotis nigricans* (Mammalis: Chiroptera) on Barro Colorado Island, Panama Canal Zone. *J. Zool.* **163**, 1–13.

7 The primitive ungulates: orders Tubulidentata, Proboscidea, and Hyracoidea

RICHARD E. BROWN

Tubulidentata

The aardvark (*Orycteropus afer*), the only living species of Tubulidentata, has evolved from the protoungulates. Although once classified with the Edentata, it is now considered a separate order (see Melton 1976). Found only in Africa south of the Sahara, the aardvark is almost hairless with a long tapering snout, long ears, and short powerful legs which it uses to dig into termite hills for food or into the ground to form a sleeping burrow and to escape predators. The mouth and tongue are elongated and large salivary glands are present (Melton 1976). The teeth are unlike those of other mammals, having no enamel but dentine surrounding tubular pulp cavities (hence the name tubulidentata). These teeth grow continuously and are used for chewing termites and ants before swallowing them (Eisenberg 1981).

Aardvarks are solitary and nocturnal and have a home range which is determined by the number of termite nests available. The termite hills in the home range are visited regularly and the aardvark travels between them and its burrow on regular paths (Grzimek 1975). The sexes spend most of the year in separate burrows. In the breeding season, which occurs during the first rainy season (April-May), pairs may gambol about together and share a burrow (Kingdon 1971). After a gestation period of about seven months a single young is born (rarely two) during the second rains (October-November) (Melton 1976).

The young stays in its mother's burrow until it is two weeks old when it will follow the mother on her feeding trips. The bright white patch on the tip of the female's tail has been suggested as a possible signal which may assist the young in following its mother (Kingdon 1971), but since vision is considered of low importance in Tubulidentata (Delany and Happold 1979) and since these foraging trips occur at night, this seems unlikely. The young continues to share its mother's burrow until it is six months of age when it digs its own burrow. Even then it may continue to forage with its mother until one year of age (Kingdon 1971; Grzimek 1975; Melton 1976). This suggests that there is a strong mother-young bond (Eisenberg 1981), but the role of olfaction in this bond is unknown.

The aardvark has large olfactory bulbs and an immense surface area of olfactory epithelium indicating a highly developed sense of smell (Kingdon 1971). According to Delany and Happold (1979), olfactory communication is very important in the Tubulidentata, but little is known about the types of information conveyed.

Kingdon (1971) states that there are glands on the elbows and hips but no histological examination has been reported on these glands. Both male and female aardvarks have anal glands, glands on the soles of the feet and large preputial or clitorial glands (Pocock 1916, 1924; Schaffer 1940) (see Fig. 7.1). According to Pocock (1916, p. 742) these 'inguinal' glands are filled with a yellow secretion having a strong odour. Pocock suggests that the secretion of these glands may be used for 'protection' like the anal glands of *Mephitis* or used to enable animals to find one another.

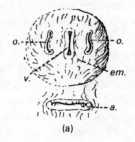

(a)

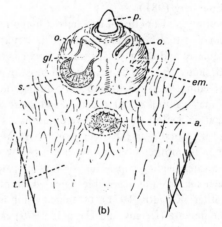

(b)

Fig. 7.1. Clitoral glands of the female *Orycteropus* (a) and preputial glands of the male (b). Abbreviations indicate the genital eminence (em), orifices (o) of the glands (gl) and the layer of secretory cells (s), the penis (p), vagina (v), anus (a), and base of the tail (t). (Reprinted from Pocock (1916) with permission from the Zoological Society of London.)

It seems likely that olfactory communication plays an important role in sexual behaviour, mother–young interactions and possibly home-range and burrow identification in aardvarks, but these have yet to be systematically investigated (Melton 1976).

Proboscidea

The order Proboscidea has two living species: the Asian elephant (*Elaphas maximus*) and the African elephant (*Loxodonta africana*), and each has at least two subspecies (Kingdon 1979; Eisenberg 1981). All elephants are vegetarian, eating grasses, roots, bark, and some fruit. Elephants live in groups which range in size from 8 to 20 animals up to herds of 100 animals. These groups are of two kinds: family groups consisting of females and their young up to 15 years of age and all-male groups consisting of adult bulls and young males. Adult males are often solitary. Descriptions of ecology and social behaviour are available for both *Elaphas* (Eisenberg, McKay, and Jainudeen 1971; McKay 1973; Kurt 1974; Eisenberg 1981) and *Loxodonta* (Buss and Smith 1966; Sikes 1971; Douglas-Hamilton and Douglas-Hamilton 1975; Laws, Parker, and Johnstone 1975; Kingdon 1979).

The most prominent feature of the elephant, the trunk, is used for tactile communication, object manipulation, and drinking, as well as a sense organ (Kingdon 1979). As the elephant moves, and even when it is still, the trunk is in motion, sampling the olfactory environment. It may be extended ahead of the animal or moved from side to side, apparently testing for scent. When faeces or urine of other elephants or the excreta of other animals are encountered, these are investigated with the trunk (McKay 1973).

Olfactory signals play an important role in mutual recognition and communication between elephants who greet each other by touching their trunks to the ear, mouth, eyes, temporal gland, tail, anus, feet, and genitalia of other animals. Both *Elaphas* and *Loxodonta* have a vomeronasal organ (Eales 1926; McKay 1973) and the trunk is used to bring odorous substances, particularly urine, into contact with this organ by placing the trunk into the mouth (McKay 1973; Grzimek 1975).

Aside from urine and faeces elephants produce odorous secretions from their temporal glands, tarsal or Meibomian glands, and interdigital glands (Schaffer 1940; McKay 1973). Most experimental work on olfactory communication in elephants has focused on the temporal gland. Both *Elaphus* and *Loxodonta* have temporal glands (Schaffer 1940) (see Plates 7.1 and 7.2) and cave drawings suggest that the woolly mammoths (*Mammuthus primigenius*) also had temporal glands (Pocock 1916) (see Fig. 7.2). The temporal gland is a large multi-lobed sac with an orifice half-way between the eye and the ear (Pocock 1916; Adams, Garcia, and Foote 1978). In both *Elaphas* and *Loxodonta* the temporal glands have the same histological structure of apocrine glands separated into tubules surrounded by epithelial muscle cells (Plate 7.3) (Estes and Buss 1976; Fernando, Jayasinghe, and Panabokke 1963; Schneider 1956; Short, Mann, and Hay 1967). While the histological structure of the temporal gland is similar in the two species the stimuli which cause it to secrete temporin (the name given to the secretion) appear to differ and the secretion of temporin may serve different functions in *Elaphas* and *Loxodonta* (Sikes 1971).

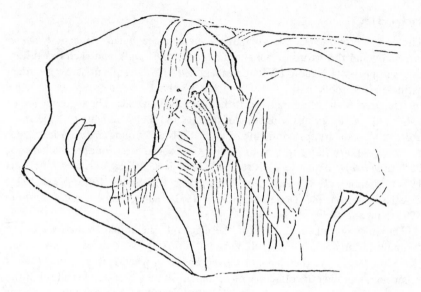

Fig. 7.2. Cave-drawing of a mammoth showing the secretion from the temporal gland. (Reprinted from Pocock (1916) with permission from the Zoological Society of London.)

Elaphas

In male *Elaphas* the secretion of the temporal gland begins at about 10 years of age and is seen most regularly in males over 25 years of age (Jainudeen, McKay, and Eisenberg (1972). The temporal gland of the female does not secrete at all and 'appears to be a vestigial organ' (Eisenberg, McKay, and Jainudeen 1971, p. 217).

The temporal gland of male *Elaphas* secretes most profusely during the period of 'musth'. Musth is characterized by an increase in aggression, an increase in temporal gland size and secretion, and an increase in 'disobedience' (in domestic elephants). During musth, bull elephants 'are too dangerous to handle so they have to be chained up and hand fed' (Jainudeen, Katongole, and Short 1972, p. 99). Musth may occur at any month of the year (Kurt 1974) but is most common in the rainy season and is correlated with large increases in testosterone levels (Jainudeen *et al.* 1972). While sexual activity occurs during musth, bull elephants also mate when not in musth.

The musth period lasts from one to 34 days and males can come into musth more than once a year (Jainudeen *et al.* 1972; Kurt 1974). Males in musth pay more attention to females, are more aggressive toward other males and mark trees, mud wallows, and waterholes with the temporal gland secretion (Eisenberg *et al.* 1971; Kurt 1974). Although a number of males may come into musth at the same time, the stronger male chases the others away from female herds and stays in musth while the other males may leave the area and stay in musth or come out of musth and stay near the herds (Kurt 1974). This suggests that subordinate males may have lower testosterone levels than dominant males (Kurt 1974). Musth in bull *Elaphas* thus appears to function as a rut period in which males become dominant, scent mark and mate.

During musth the female investigates the male's temporal gland with her trunk and may use these secretions to identify the age, sex, and the reproductive (rut) status of the male (Eisenberg *et al*. 1971; see Plate 7.1).

During musth, male elephants dribble urine and pay particular attention to the urine and faeces of other *Elaphas* (Eisenberg *et al*. 1971; Jainudeen *et al*. 1972). Male *Elaphas* also have preputial glands which may produce specific odours. Females investigate the preputial areas of males and also investigate their mouth area (Eisenberg *et al*. 1971) suggesting they may produce odorous secretions from the oral lips or in the saliva.

Female *Elaphas* appear to have a change in urogenital odours as they come into oestrus, and secretions from the clitoral glands or in the urine may be involved. Male *Elaphas* investigate the urogenital area of females very frequently before mating and show a urine-testing or flehmen reaction, placing the trunk in female urine and then bringing this in contact with the roof of the mouth, possibly to contact the Jacobson's organ (Eisenberg *et al*. 1971). The urine of females just coming into oestrus is tested more than that of females in oestrus, but as the female goes out of oestrus there is a large increase in the frequency of urine testing (Eisenberg *et al*. 1971).

Loxodonta

The African elephant *Loxodonta* possesses temporal glands which are histologically identical to those of the Asian elephant (Estes and Buss 1976; Schneider 1956; Sikes 1971). A biochemical analysis of the components of the temporal gland of *Loxodonta* indicates high protein and cholesterol levels (Buss, Rasmussen, and Smuts 1976) and analysis with gas-liquid chromatography and mass spectrometry has identified some of the major components as phenol and *m*- and *p*-cresol (Adams *et al*. 1978).

Unlike *Elaphas*, the secretion of the temporal gland of *Loxodonta* appears to be unrelated to musth. The temporal gland secretes in both juvenile and adult *Loxodonta* of both sexes, including pregnant females, and there are no sex differences in the size or weight of the temporal gland, nor is the temporal gland related to reproductive activity (Short, Mann, and Hay 1967). The size of the temporal gland does, however, appear to be related to the age of the elephant, being larger in older animals (Adams 1974).

There are few reports of musth in *Loxodonta*. Poole and Moss (1981), however, have described musth in *Loxodonta* and conclude that it is similar to musth in *Elaphas*. Male *Loxodonta* which are older than 30 years of age are seen to dribble urine, have copious secretions from the temporal gland, show an increase in aggression and spend more time near female herds during musth. High levels of aggression during musth are associated with the increased temporal gland secretion. Like *Elaphas*, male *Loxodonta* may have more than one musth period each year and the duration is variable, from one to 103 days. Male *Loxodonta* in musth also appear to have higher dominance ranks and to be more sexually active than males not in musth (Poole and Moss 1981).

Loxodonta rub the temporal gland on trees, termite mounds, and on the ground and Sikes (1971, p. 37) suggests that the resulting scent marks may function in territorial behaviour. When a strange elephant approaches a group, the temporal gland of the dominant group member begins to secrete, and aggressive behaviour may be followed by tree marking (Sikes 1971). Elephants may 'attack' trees with the scent of a foreign elephant.

Since the cholesterol levels of the temporal glands vary between individual elephants, Buss *et al.* (1976) suggest that the temporal gland secretion may be used for recognition of individual elephants. Elephants 'caress' each other with their trunks and rub their faces together and in doing so may be investigating each other's temporal gland odour (Sikes 1971). Young elephants secrete temporin during sexual and aggressive play (Sikes 1971).

Loxodonta also secrete from the temporal gland when stressed or frightened, so the temporal gland secretion may serve to indicate that the elephant is disturbed or excited (Adams 1974; Buss *et al.* 1976; Sikes 1971) (see Plate 7.2). Adams *et al.* (1978), however, reported that when the temporal glands secreted on some elephants during stressful situations, the presence of the secretion did not effect the behaviour of the nearby elephants.

Loxodonta have three pairs of salivary glands (parotid, submaxillary, and sublingual) and have mucous glands in the oral angle (Sikes 1971). Elephants investigate each other's mouths with the trunk and may be testing secretions from these glands. *Loxodonta* also have small Harderian glands (Sikes 1971), but there is no evidence that the secretion from these glands has a communicative function.

Faeces and urine may be used for olfactory communication in *Loxodonta*. Females in oestrus urinate frequently and males test the urine using a 'flehmen'-like behaviour (Sikes 1971). This urine testing may involve touching the urine to the Jacobson's organ.

Hyracoidea

The Hyraxes are small animals which look like rodents but are closely related to the elephants and Sirenia. They are vegetarians and live in large groups in the savanna and arid zones of Africa and the Middle East. The three living genera of hyraxes include the rock hyrax (*Procavia*); Bruce's Hyrax (*Heterohyrax*); and the tree hyrax (*Dendrohyrax*) (Walker 1975; Kingdon 1971).

Both *Procavia* and *Heterohyrax* live in burrows in rocky outcrops and are diurnal, having peaks of feeding early in the morning and late in the evening. They live in colonies ranging in size from 6 to 50 individuals in *Procavia* and up to 100 individuals in *Heterohyrax*. Colonies appear to be made up of small family groups, each with a single dominant male who protects the females and their young (Coe 1962; Kingdon 1971). *Dendrohyrax* differ from the other hyraxes in that they are mainly, but not exclusively arboreal, nocturnal, and less gregarious (Walker 1975; Kingdon 1971; Grzimek 1975). Whereas *Procavia* are territorial and very aggressive toward non-colony members, they share feeding grounds with *Heterohyrax* without inter-specific aggression (Kingdon 1971; Hoeck 1975). Auditory communication appears to be more important for hyraxes than olfaction (Delany and Happold 1979), but the hyraxes have a number of potential odour sources.

Both males and females of all three genera of hyraxes have a multi-lobed dorsal apocrine gland which is covered by tufts of distinctively coloured hair. These tufts may be black, pale yellow, or white, depending on the species (Schaffer 1940; Sale 1970) (see Plate 7.4). When the hyrax is angry or frightened the hairs are erected and the gland exposed (Walker 1975, p. 1326–30). According to Grzimek (1975, p. 515) 'if the animals get highly excited, the hairs unfold like a flower, exposing the naked glandular area' (see Fig. 7.3). The dorsal gland becomes prominent during the breeding season and in threat displays (Grzimek 1975, p. 515; Sale 1970), and Pocock (1916) suggested that the coloured hairs around the gland are used as social signals and raised, like the tails of some deer, when the animal is frightened. According to Sale (1970) alarm is signified when the hairs are raised 45° and threat when they are raised 90°.

The dorsal gland of *Dendrohyrax terricola* has been described by Mollison (1905), and Lederer (1950) has identified alcohols, fatty acids, and sterols in the secretion of this gland. The size and secretory rate of the dorsal gland depends on the age and sexual maturity of the animals; juveniles having small, inactive glands and sexually mature, sexually active adults having the largest glands (Sale 1970).

Although it does not appear to be used for scent-marking, the dorsal gland may provide a species-typical odour, a rut odour, a sex odour, and an individual odour. Rutting males appear to be able to identify oestrous females by the odours of their dorsal glands (Sale 1970) and female rock hyraxes (*Procavia capensis*) can discriminate between sexually active and inactive male hyraxes by the odours of their dorsal glands and by their urine odours (Stoddart and Fairall 1981). The dorsal gland may also be used for mother-pup identification (Sale 1965) and individuals sniff each others' dorsal glands when they meet, suggesting that it is used for individual identification (Sale 1970).

Members of the *Heterohyrax* genus also have abundant sudoriferous glands on the soles of the feet (Walker 1975, p. 1330). Whether or not these glands serve a communicative function seems to be unknown.

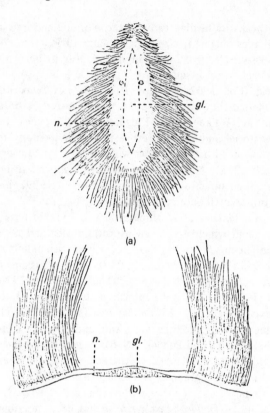

Fig. 7.3. The dorsal gland (gl) of *Dendrohyrax dorsalis* with the hairs raised. The gland is shown from above (a) and from the side (b). *n.* = naked skin. Reprinted from Pocock (1916) with permission from the Zoological Society of London.)

References

Adams, J. (1974). The behavioral significance of the temporal gland in the African elephant (*Loxodonta africana*). Paper presented at the annual meeting of The Animal Behavior Society.
— Garcia III, A., and Foote, C. S. (1978). Some chemical constituents of the secretion from the temporal gland of the African elephant (*Loxodonta africana*). *J. chem. Ecol.* **4**, 17–25.
Buss, I. O. and Smith, N. S. (1966). Observations on reproduction and breeding behavior of the African elephant. *J. Wildl. Manag.* **30**, 375–88.
Buss, I. O., Rasmussen, L. E., and Smuts, G. L. (1976). The role of stress and individual recognition in the function of the African elephant's temporal gland. *Mammalia* **40**, 437–51.
Coe, M. J. (1962). Notes on the habits of the Mount Kenya hyrax (*Procavia Johnstoni Mackinderi* Thomas). *Proc. zool. Soc., Lond.* **138**, 639–44.
Delany, M. J. and Happold, D. C. D. (1979). *Ecology of African mammals.* Longman, New York.

Douglas-Hamilton, I. and Douglas-Hamilton, O. (1975). *Among the elephants.* Collins and Harvill Press, London.

Eales, N. B. (1926). The anatomy of the head of a foetal African elephant (*L. africana*). *Trans. R. Soc. Edinb.* 54, 491–546.

Eisenberg, J. F. (1981). *The mammalian radiations.* University of Chicago Press, Chicago.

— McKay, G. M., and Jainudeen, M. R. (1971). Reproductive behavior of the Asiatic elephant (*Elaphas maximus maximulus* L.). *Behaviour* 38, 193–225.

Estes, J. A., and Buss, I. O. (1976). Microanatomical structure and development of the African elephant's temporal gland. *Mammalia* 40, 429–36.

Fernando, S. D. A., Jayasinghe, J. B., and Panabokke, R. G. (1963). A study of the temporal gland of an Asiatic elephant (*Elaphas maximus*). *Ceylon vet. J.* 11, 108–11.

Grzimek, B. (1975). *Grzimek's animal life encyclopedia,* Vol. 12. Van Nostrand Reinhold, New York.

Hoeck, H. N. (1975). Differential feeding behaviour of the sympatric hyrax *Procavia johnstoni* and *Heterohyrax brucei. Oecologia* 22, 15–47.

Jainudeen, M. R., Katongole, C. B. and Short, R. V. (1972). Plasma testosterone levels in relation to musth and sexual activity in the male Asiatic elephant, *Elaphas maximus. J. Reprod. Fert.* 29, 99–103.

— McKay, G. M., and Eisenberg, J. F. (1972). Observations on musth in the domesticated Asiatic elephant (*Elaphas maximus*). *Mammalia* 36, 247–61.

Kingdon, J. (1971). *East African mammals,* Vol. 1. Academic Press, London.

— (1979). *East African mammals,* Vol. 3B. Academic Press, London.

Kurt, F. (1974). Remarks on the social structure and ecology of the Ceylon elephant in the Yala National Park. In *The behaviour of ungulates and its relation to management* (ed. V. Geist and F. Walther) Vol. 2, pp. 618–34. IUCN Publications, Morges, Switzerland.

Laws, R. M., Parker, I. S. C., and Johnstone, R. C. B. (1975). *Elephants and their habitats.* Clarendon Press, Oxford.

Lederer, E. (1950). Odeurs et parfums des animaux. *Fortschr. Chem. org. Nat. Stoffe* 6, 87–153.

McKay, G. M. (1973). Behavior and ecology of the Asiatic elephant in south-eastern Ceylon. *Smithsonian Contrib. Zool.* 125, 1–113.

Melton, D. A. (1976). The biology of aardvark (*Tubulidentata: orycteropodidae*). *Mammal Rev.* 6, 75–88.

Mollison, T. (1905). Die Reickendrüse von *Dendrohyrax terricola. Gegenbaurs morph. Jb.* 34, 240–5.

Pocock, R. I. (1916). Scent glands in mammals. *Proc. zool. Soc. Lond.* 84, 742–55.

— (1924). Some external characters of *Orycteropus afer. Proc. zool. Soc. Lond.* 92, 697–706.

Poole, J. H. and Moss, C. J. (1981). Musth in the African elephant, *Loxodonta africana. Nature, Lond.* 292, 830–1.

Sale, J. B. (1965). Observations on parturition and related phenomena in the hyrax (Procaviidae). *Acta trop.* 22, 37–54.

— (1970). Unusual external adaptations in the rock hyrax. *Zool. Afr.* 5, 101–13.

Schaffer, J. (1940). *Die Hautdrüsenorgane der Säugetiere.* Urban & Schwarzenberg, Berlin.

Schneider, R. (1956). Untersuchungen über den Feinbau der Schläfendrüse beim afrikanischen und indischen Elefanten, *Loxodonta africana* Curier und *Elephas maximus* Linnaeus. *Acta anat.* 28, 203–312.

Short, R. V., Mann, T., and Hay, M. F. (1967). Male reproductive organs of the African elephant, *Loxodonta africana*. *J. Reprod. Fert.* 13, 517–36.

Sikes, S. K. (1971). *The natural history of the African elephant.* Weidenfeld and Nicolson, London.

Stoddart, D. M. and Fairall, N. (1981). Electrocardiographic technique for studying olfactory response in the rock hyrax *Procavia capensis* L. (Mammalia: Hyracoidea). *J. chem. Ecol.* 7, 257–63.

Walker, E. P. (1975). *Mammals of the world*, 3rd edn. The Johns Hopkins Press, Baltimore.

8 The rodents I: effects of odours on reproductive physiology (primer effects)

RICHARD E. BROWN

The effects of social odours on reproductive physiology are referred to as primer effects (Bronson and Whitten 1968; see Introduction). Through their action on the neuroendocrine system, odours can affect three of the most important aspects of rodent social behaviour: mate selection, reproduction, and parent–offspring interactions. Odours alone can produce the physiological changes underlying many of these behaviours (known primer effects) while others appear to require additional tactile, auditory, or visual stimulation (suspected primer effects, see Table 8.1).

There have been many reviews in the last 20 years which have covered some aspects of the primer effects listed in Table 8.1: Parkes and Bruce (1961); Whitten (1966, 1969); Bronson (1968, 1971, 1976b, 1979); Bruce (1969);

Table 8.1. Known and suspected olfactory primer effects in rodents

	Odour donor	
Odour receiver	Adult female	Adult male
Juvenile male	Accelerate puberty (suspected only)	Delay puberty
Juvenile female	Delay puberty	Accelerate puberty (Vandenbergh effect)
Adult female	Cause prolonged anoestrus (Whitten effect) or pseudopregnancy (Lee–Boot effect) Synchronize oestrous cycles	Induce oestrus and ovulation (Whitten effect)
Adult male	Induce changes in LH and testosterone levels and sexual behaviour	Inhibit LH and testosterone levels and sexual behaviour (suspected only); increase adrenal weights (Ropartz effect)
	Stud male	Alien male
Pregnant female	Increase chance of impregnation and number of embryos implanted	Terminate pregnancy and induce oestrus (Bruce effect)
	Infants	
Lactating female	Induce prolactin and milk secretion (suspected only)	

Whitten and Bronson (1970); Whitten and Champlin (1973); Bronson and Desjardins (1974*b*); Vandenbergh (1975, 1980); Macrides (1976); Rogers and Beauchamp (1976*b*); Richmond and Stehn (1976); Macrides, Bartke, and Svare (1977); Aron (1979). K. Brown (1979); R. Brown (1979); Bronson and Coquelin (1980); Novotny, Jorgenson, Carmack, Wilson, Boyse, Yamazaki, Wilson, Beamer, and Whitten (1980); Milligan (1980); and Cowley (1980).

The function of the present chapter is to present critically the evidence for olfactory control of primer effects; to examine the similarities and differences in the odours, the receptors, and the neuroendocrine responses involved in producing these effects; and to explore the social functions of these primer effects. Because the primer effects involve reproductive success and survival of offspring, they may benefit the animal releasing the odour, the receiver, or both. Since animals are selected to respond to social signals in ways that benefit themselves, one would expect that primer effects benefit the receiver but according to the reasoning of Dawkins and Krebs (1978) the animal releasing a primer odour may be able to manipulate the responses of the receiver for its own benefit.

Who benefits may depend on the relatedness of the animals; so that primer effects between related animals may benefit both while primer effects between unrelated animals may benefit one to the disadvantage of the other. Some primer effects appear to be reciprocal, benefiting both sender and receiver, whether or not they are related, others are nepotistic, benefiting offspring and mates, while others may involve parent–offspring manipulation (see Alexander 1974; Otte 1974).

Social odours affecting puberty in rodents

Defining and measuring sexual maturation

Puberty is not an instantaneous event, but a gradual change in morphology, physiology, neuroendocrinology, and behaviour leading to sexual arousal and reproduction. In order to determine whether or not sexual maturity has occurred, or to determine the stage of puberty that a particular animal has reached, some aspects of an animal's physiology, morphology, or behaviour must be measured. Some of the most common measures of puberty are shown in Table 8.2.

Unfortunately, different measures of sexual maturation change at different rates as puberty occurs, and it is difficult to compare studies using different indices of puberty. Juvenile male Rockland–Swiss albino mice, for example, have their greatest increase in body weight between 22 and 32 days of age, their greatest increase in testes weight between 32 and 42 days of age, and develop adult plasma androgen levels between 32 and 52 days of age (Svare, Bartke, and Macrides 1978). Male hamsters reach sexual maturity in two stages. Between four and six weeks of age there are rapid increases in body weight, testes weight, plasma testosterone levels, and flank-gland size and then between six and eight

Table 8.2. Commonly used measures of puberty

I. Males

    A. *Morphological measures*

        Testes weight or length
        Prostate gland weight
        Seminal vesicle weight or length
        Preputial gland weight
        Epididymes weight
        Seminiferous tubule diameter
        Body weight
        Size of sebaceous glands

    B. *Physiological measures*

        Plasma levels of LH, FSH, and testosterone
        Sperm development
        Presence of sperm in penile smears

    C. *Behavioural measures*

        Age at first mounting, intromission, and ejaculation
        Scent-marking frequency
        Successful impregnation of female

II. Females

    A. *Morphological measures*

        Ovarian weight
        Uterine weight

    B. *Physiological measures*

        Plasma levels of LH, FSH, prolactin, and oestrogen
        Age at vaginal opening
        Age at first oestrus (by vaginal smear)
        Age at first pregnancy

    C. *Behavioural measures*

        Age at first mating (lordosis)

---

weeks of age there is an increase in sperm concentration and sexual behaviour including mounting, intromission, and ejaculation (Miller, Whitsett, Vandenbergh, and Colby 1977).

As a result of their study, Miller *et al.* (1977, p. 254) concluded that:

The great temporal variation in the development of the different characteristics associated with puberty clearly indicates that studies of a species cannot be considered comparable unless the same indices of puberty are employed. Estimates of the age at puberty based on seminal vesicle weight can neither be substantiated nor refuted by a second study in which plasma testosterone concentration or penile sperm is the only index used.

Conflicting results of the effects of olfactory signals on the timing of puberty must, therefore, be considered in light of the differences in the measures of puberty used.

Acceleration of puberty in males

Housing single 21-day-old male mice or groups of 4–5 juvenile males with an adult female results in accelerated growth of the testes, epididymes, seminal vesicles, and the preputial glands at 32–37 days of age compared to control groups of males housed without adult females (Fox 1968; Maruniak, Coquelin, and Bronson, 1978). Males housed with adult females until they reach 42–60 days of age, however, do not have larger accessory sex organs than control males (Maruniak *et al.* 1978; Bediz and Whitsett 1979). In fact, Fox (1968) found that male mice reared with adult females from 21–56 days of age had lower testes, epididymes, and seminal vesicle weights than isolated males. Vandenbergh (1971), on the other hand, found that males housed from 21 days of age with adult females had larger accessory sex organs than control males at 60 days of age. Unfortunately both Fox (1968) and Vandenbergh (1971) housed their control males in groups of four or five, a situation which may inhibit puberty (see below).

The presence of an adult female from 21–36 days of age accelerates sexual maturation to some extent, but adult females are aggressive toward juveniles of both sexes (Ayer and Whitsett 1980) and both Fox (1968) and Svare *et al.* (1978) have suggested that the aggressive behaviour of adult females may, in fact, suppress sexual maturation in juvenile males between 36 and 50 days of age.

Housing 21-day-old male mice in a partitioned cage, separated from an adult female by a mesh screen, results in accelerated growth of the seminal vesicles by 32 days of age but does not increase plasma testosterone levels or testes weights (Svare *et al.* 1978). The only studies which used an olfactory stimulus alone to accelerate puberty in male mice sprayed urine on males between 20 and 36 days of age, between 30 and 42 days of age (Maruniak *et al.* 1978) or between 21 and 56 days of age (Lawton and Whitsett 1979) and found no evidence of puberty acceleration (see Fig. 8.1).

These results led Maruniak *et al.* (1978, p. 254) to conclude that 'the adult female has the capacity to accelerate the sexual maturation of young males but we have no knowledge about the nature of her relevant cue other than it is probably not a urinary pheromone'. The acceleration of puberty in males is thus only a suspected olfactory primer effect. It is difficult to measure male sexual maturation precisely and the measures most sensitive to female odours may not yet have been found. Similarly, the 'critical period' of sensitivity to female stimuli may not have been used (see pp. 270–4).

Delay of puberty in males

At high population densities, sexual maturation, spermatogenesis, and growth of accessory sex organs are inhibited in male mice (Christian 1955*a,b*, 1971; Christian and Davis 1964). Housing 21-day-old male prairie deer mice, or albino or wild house mice with adult males of the same species results in lower testes, seminal vesicle, epididymes, and preputial gland weights and lower testosterone

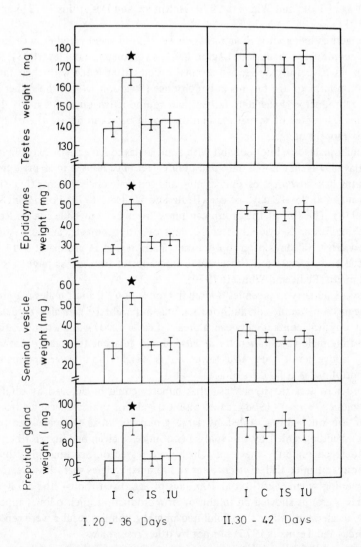

Fig. 8.1. Mean weight (± SE) of paired reproductive organs of prepubertal male house mice housed in isolation (I); cohabitation with an adult female (C); isolated and exposed to saline (IS) or isolated and exposed to female urine (IU) from 20–36 days of age (panel I) or from 30–42 days of age (panel II). Group C in panel I differs from the other three groups at the 0.01 level of significance. (Redrawn from the data of Maruniak, Coquelin, and Bronson (1978).)

levels than found in males housed without adult males when measured at 48–78 days of age (Bediz and Whitsett 1979; McKinney and Desjardins 1973; Vandenbergh 1971).

Male gerbils living with their fathers have delayed scent-gland growth and inhibited reproductive behaviour (Agren 1981; Swanson and Lockley 1978). Since these males have fully developed seminal vesicles after living with their fathers until six months of age but do not reproduce (Swanson 1980), the father may inhibit reproductive behaviour rather than reproductive physiology. To determine this, measures of sexual maturation should be taken before the offspring reach six months of age.

Juvenile males need not cohabit with adult males to have their sexual maturation delayed. Juvenile house mice and prairie deer mice housed in all male groups at weaning have lower testes, epididymes, and seminal vesicle weights than singly housed males at 35–42 days of age (Bediz and Whitsett 1979; Fox 1968; Svare *et al.* 1978). The size of the group can range from 2–11 juvenile males. Reproductive inhibition is evident after 10 days of group housing and can still be observed after 100 days of group housing. Males reared in groups of 9–11 have a reduced body weight but those reared in groups of four are as large as singly housed males (Bediz and Whitsett 1979).

Delay of puberty in juveniles housed in groups or with adult males may result from aggression, tactile, or auditory stimuli, but puberty can also be delayed without physical contact between animals. Lecyk (1967), for example, found that juvenile male voles (*Microtus arvalis*) placed adjacent to cages of crowded, sexually active adult voles had lower testes weights and fewer spermatozoa than control males at 50 days of age.

In order to test the hypothesis that puberty could be delayed by olfactory stimuli alone, Terman (1968) reared bisexual pairs of prairie deer mice from 21 days of age on bedding soiled by large groups of mice (average size of 47 animals), bedding soiled by bisexual pairs of mice or clean bedding. Rather than having delayed puberty, the mice which were reared on bedding soiled by the large groups of mice had *heavier* testes and seminal vesicles than those reared on clean bedding when sacrificed at 100 days of age. Even though the adult mice in the large groups showed an inhibition of reproduction, their odours appeared to have accelerated rather than inhibited puberty. Similar results were reported by Thomas and Terman (1975) but not by other researchers.

Rogers and Beauchamp (1976a) exposed male white-footed mice (*Peromyscus leucopus*) to the effluent from groups of 6–15 adult *Peromyscus* from birth to 100 days of age and found that these males had significantly lower testes weights than control males exposed to *Mus* odours. Similarly, Lawton and Whitsett (1979) found that prairie deer mice raised from 22–56 days of age on bedding soiled by adult males had lower testes and seminal vesicle weights than juveniles reared on clean bedding.

To further confirm that the inhibition of puberty in male mice was an olfactory primer effect, Lawton and Whitsett (1979) applied urine from adult male prairie deer mice to the nares of juveniles from 22–56 days of age. This resulted in a decrease in seminal vesicle weight, but not in testes weight (see Fig. 8.2). Juvenile males which had an olfactory bulbectomy at 22 days of age showed no inhibition of seminal vesicle development when treated with adult male urine (Lawton and Whitsett 1979).

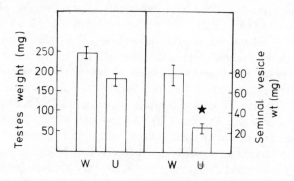

Fig. 8.2. Mean testes and seminal vesicle weights (± SE) of prepubertal male prairie deer mice exposed to water (W) or adult male urine (U). Seminal vesicle weights of group U differ from those of group W at the 0.001 level. (Redrawn from the data of Lawton and Whitsett (1979.)

Delay of puberty in females

As with males, juvenile females in large populations have delayed sexual maturity (Christian 1971; Christian and Davis 1964). The ovaries and uteri of females reared in crowded populations are smaller than those of females reared in bisexual pairs and they have fewer corpora lutea, indicating low rates of ovulation (Terman (1969a, 1973). Terman (1965) found that in asymptotic populations only 19 per cent of the females mated and the majority of these were the founding members. Only 20 per cent of the mating females were juveniles who matured and produced offspring.

Females do not need to be reared under crowded conditions to have puberty delayed. Rearing females with their parents, their male or female siblings, in all female groups or with unrelated adult females also delays puberty. Juvenile female gerbils kept with their parents have poorly developed ventral scent glands and fail to mate (Payman and Swanson 1980; Swanson and Lockley 1978). Female cactus mice (*Peromyscus eremicus*) reared with their mothers or with both parents begin reproducing at a later age than females housed alone and reproduce at a lower rate (Skryja 1975, 1978). These reproductively inhibited females showed oestrous cycles which did not differ from controls, came into oestrus and often copulated (Skryja 1978), thus the maternal inhibition of reproduction in this case does not appear to be due to delayed puberty, but to blocking of pregnancy (see pp. 308).

Female gerbils and voles (*Microtus ochrogaster*) which are reared with their brothers from weaning are less likely to mate and become pregnant than females housed with unrelated juvenile males (Agren 1981; Hasler and Nalbandov 1974). Similarly, Hill (1974) found that female *Peromyscus maniculatus* which live with their brothers from 21 days of age are slow to mate and produce fewer offspring than females mated with non-sibling males. Females who live with non-sibling males from 21 days of age have a better mating performance but not as good as females mated after puberty, suggesting that familiarity with a male before puberty inhibits reproduction, even if that male is not a relative.

Rearing female albino mice in groups of six from weaning results in an inhibition of sexual maturity. Grouped females have first oestrus at 55.1 days of age while individually housed controls mature at 33.2 days of age (Vandenbergh, Drickamer, and Colby 1972). As the size of the all female groups increase from one to 45 per group the age of vaginal opening and the age of first oestrus increase linearly (Castro 1967; Drickamer 1974*a*). Again, the measure of puberty used may be important as Bediz and Whitsett (1979) found no decrease in body weight or uterine weight in female prairie deer mice reared in groups of five from 21 days until four months of age.

When female albino mice were reared in groups of six from 21 days of age in the presence of an adult female, they showed vaginal opening and first oestrus earlier than groups of juvenile females housed without an adult female (Vandenbergh 1967). Unfortunately, it is difficult to evaluate this study because individually housed control females were not used. The results, however, seem to suggest that the presence of the adult female *advanced* puberty in these females. When Bronson and Stetson (1973) reared isolated female mice from 24–30 days of age with an adult female, however, they found lower uterine weights than in singly housed females. Drickamer (1979) found that wild house mice had their first oestrus delayed if they were housed with adult females.

Physical contact with other females is not necessary to cause a delay of puberty in juvenile females. Odours from populations of mice and odours from grouped juveniles are sufficient to delay puberty. Laboratory mice exposed to effluent from groups of adult females from birth to 21 days of age have delayed vaginal opening but the age of first oestrus is not altered (Fullerton and Cowley 1971). Similarly, very few female white-footed mice exposed to effluent from adult populations from birth to 100 days of age reached sexual maturity. These odour-exposed females had smaller ovaries and reproductive tracts than control females exposed to *Mus* odours (Rogers and Beauchamp 1976*a*).

Bedding soiled by juvenile or adult female house mice housed in groups of 4–6 is sufficient to delay vaginal opening and first oestrus in females exposed from 20 days of age, but bedding soiled by isolated females is not sufficient to delay puberty, nor is bedding soiled by females housed four per cage but separated by mesh partitions (Drickamer 1974a, 1979). These results suggest that female mice must have physical contact before they produce a puberty-inhibiting odour.

These results are in conflict with those of Terman (1968) who reared male-female pairs of prairie deer mice from 23-100 days of age on bedding soiled by populations of mice or on clean bedding. He found that females reared on soiled bedding had *larger* ovaries and a higher pregnancy rate than those reared on clean bedding. Similar results were found by Thomas and Terman (1975). One major factor in Terman's studies is that his mice were reared in male-female pairs which may respond differently to odours from large groups than animals housed in unisexual groups.

The odour which inhibits puberty in females appears to be in the urine of adult females or grouped juvenile females. Lombardi and Whitsett (1980) applied urine from adult virgin females to the nares of female prairie deer mice from 25-45 days of age. These juveniles had delayed vaginal opening, lower uterine weights, and delayed first oestrus when compared with control females who had water applied to their nares. Lombardi and Whitsett (1980) also found that urine from individually housed adult females would inhibit puberty in prairie deer mice.

In a series of experiments, McIntosh and Drickamer have attempted to identify the source of this urinary odour. McIntosh and Drickamer (1977) painted urine from adult female mice on the nares of juveniles beginning at 21 days of age. Both bladder and excreted urine from grouped adult females, but only bladder urine from individually housed adult females delayed first vaginal oestrus in juveniles. Excreted urine from individually housed females did not delay first oestrus in singly housed juveniles and when urine from grouped females was homogenized with the urethra from an individually housed female, the urine lost its ability to delay first oestrus.

Thus, the bladder urine of all adult females is able to delay first oestrus in juveniles, but in singly housed females this urine appears to be altered as it passes through the urethral tract. McIntosh and Drickamer (1977) suggest that the production of a urinary odour which delays puberty in juvenile females occurs in two stages:

First there is a maturation-delaying pheromone present in the bladder urine of all females. Second, some chemical substance is secreted along the urethra in females housed alone, which deactivates the maturation-delaying pheromone before excretion. No deactivation substance is secreted into the urethra in females housed in groups (McIntosh and Drickamer 1977, p. 1002).

Two sources have been suggested for the urinary odour which delays puberty: the ovaries and the adrenal glands. The ovaries do not appear to be involved because bladder urine from singly housed ovariectomized females delays puberty in juveniles as does urine excreted by grouped ovariectomized females (Drickamer, McIntosh, and Rose 1978). Nor does ovariectomy affect the alteration of bladder urine by the urethral tract in singly housed females. Excreted urine of singly housed ovariectomized females, like that from singly housed intact females, does not delay puberty in juveniles (Drickamer *et al.* 1978).

After adrenalectomy, however, females do not produce a puberty delaying odour in either their bladder or their excreted urine (Drickamer and McIntosh 1980; see Fig. 8.3). Adrenalectomy does not, however, affect the urethral inhibition of the puberty-delaying odour in single housed females. The urethra of a singly housed adrenalectomized female when homogenized with urine of group-housed females still prevents the delay of first oestrus (Drickamer and McIntosh 1980). The source of the urethral modification of the urine of singly housed females has not been identified, but may be the accessory glands of the urethra (e.g. clitoral glands or paraurethral glands in the urethral mucosa).

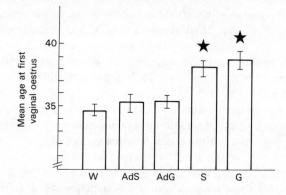

Fig. 8.3. Mean age (± SE) at first vaginal oestrus for female albino mice painted on the nares from 21 days of age with water (W); bladder urine from adrenalectomized singly housed females (AdS); adrenalectomized group-housed females (AdG); intact singly housed females (S); or intact group-housed females (G). Groups S and G differ from the other three groups at the 0.05 level of significance. (Redrawn from the data of Drickamer and McIntosh (1980), experiment 2.)

Acceleration of puberty in females: the Vandenbergh effect

The finding that male stimuli accelerate puberty in juvenile females has been termed the 'Vandenbergh effect' (Novotny *et al.* 1980). It was Vandenbergh (1967) who first found that groups of female albino mice reared from 21 days of age with an adult male showed earlier vaginal opening, first oestrus, and first matings than females housed in groups without males. Castro (1967) found that female mice housed in mixed-sex groups of seven from 21 days of age had earlier vaginal opening than females reared in unisexual groups of seven. Unfortunately, the control females in these two experiments were not individually housed, but group-housed, a condition which has been shown to delay puberty (see pp. 251-4).

The presence of an adult male has been found to accelerate puberty in rats, voles, and lemmings as well as mice. Female rats housed either in pairs with an adult male or in groups of four with an adult male have first oestrus earlier than females housed without males, and those housed in pairs reach first oestrus earlier than those housed in groups (Vandenbergh 1976). Female collared lemmings housed with adult males have earlier vaginal opening and heavier uteri than littermates housed in isolation (Hasler and Banks 1975) and female voles reared with adult males have earlier vaginal opening than females reared with juvenile males (Hasler and Nalbandov 1974). Unfortunately, this latter study did not have an isolated female control group with which the male-reared females could be compared.

As is the case with puberty delay, physical contact with adults is not necessary for puberty acceleration. Female mice housed from 21 days of age in a cage which separated them from adult males by a wire-mesh screen had their vaginal opening and first oestrus accelerated almost as much as females housed with adult males (Vandenbergh 1969). Rearing females on bedding soiled by males from 21 days of age or from birth to 21 days of age also accelerates the time of first oestrus (Fullerton and Cowley 1971; Vandenbergh 1969). Fullerton and Cowley (1971) also found that females exposed to male effluent from birth to 21 days gained weight more rapidly than controls.

The presence of an adult male is a more important determinant of the timing of puberty in female mice than nutrition, light–dark cycles, or the size of all female groups. In a study designed to examine the relative importance of male stimuli and nutrition on the timing of puberty in albino mice, Vandenbergh *et al.* (1972) recorded the age at vaginal opening, first oestrus, and first mating in a factorial experiment. While nutritional factors (8, 16, or 24 per cent protein diet) were significant, accounting for 4.8 per cent of the variance in age at puberty, male stimuli accounted for 47 per cent of the variance. Under all diet conditions, females reared with males or in the presence of bedding soiled by males reached first oestrus sooner than females housed in isolation. Both group-housed and singly housed female mice have their first oestrus accelerated by the presence of adult males or their soiled bedding, but singly housed females reach first oestrus before group-housed females in all conditions (Vandenbergh 1969; see Fig. 8.4). Despite this difference, Drickamer (1975*b*) found that group size accounted for only 9 per cent of the variance in age at first oestrus. Hours of light per day (0, 12, or 24 hours) accounted for 6 per cent of the variance and the presence or absence of an adult male for 31 per cent of the variance. Females housed with adult males had their first oestrus at an earlier age than those housed without males under all lighting conditions and in all three sizes of female groups (one, four, or seven females per group).

The phenomenon of male acceleration of puberty in juvenile females through olfactory stimuli has been extensively investigated and the majority of the research has been directed at answering two questions: what is the source of

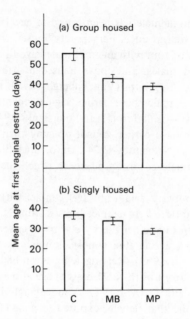

Fig. 8.4. Mean age (± SE) at first oestrus of group and singly housed female house mice which were exposed from 21 days of age to clean bedding (C); bedding soiled by adult males (MB); or the presence of an adult male (MP). (Redrawn from (A) the data of Vandenbergh (1969) and (B) the data of Vandenbergh, Drickamer, and Colby (1972).)

the odour?; and, what is the pattern of neuroendocrine responses elicited in the female by this odour?

The source of the male's puberty accelerating odour   Urine from adult males has been shown to accelerate puberty in juvenile female house mice, prairie deer mice, and prairie voles. Female deer mice exposed to the urine of intact males from 21–34 days of age have higher uterine and oviduct weights than water-treated controls, but those treated with urine from castrated males showed no acceleration (Teague and Bradley 1978). Exposure of female deer mice to male urine from 25–45 days of age, however, produces only a slight acceleration of vaginal opening and no increase in uterine weights (Lombardi and Whitsett 1980). As little as six daily applications of one drop of urine from an adult male will increase the uterine weights of 21-day-old female prairie voles (Carter, Getz, Gavish, McDermott, and Arnold 1980) and a single drop of urine from an adult male will increase the uterine weights of 25–27-day-old house mice (Wilson, Beamer, and Whitten 1980).

The urinary component which accelerates puberty in females is absent in castrated males (Colby and Vandenbergh 1974) and is present in testosterone-injected castrates in direct proportion to the amount of testosterone injected (Lombardi, Vandenbergh, and Whitsett 1976). Urine from males injected with oil or 5 μg testosterone does not increase uterine weights but doses of 50, 250, and 500 μg testosterone produce increasingly larger uterine weights (see Fig. 8.5). Similarly, urine from ovariectomized females given seven 500 μg testosterone injections contains a component which accelerates puberty in juvenile females (Lombardi *et al.* 1976). Removing the male's preputial gland does not inhibit his ability to produce oestrus acceleration, suggesting that the preputial gland is not the source of the odour (Colby and Vandenbergh 1974).

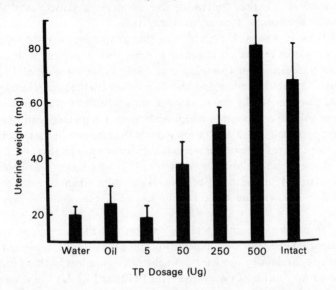

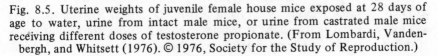

Fig. 8.5. Uterine weights of juvenile female house mice exposed at 28 days of age to water, urine from intact male mice, or urine from castrated male mice receiving different doses of testosterone propionate. (From Lombardi, Vandenbergh, and Whitsett (1976). © 1976, Society for the Study of Reproduction.)

Subordinate males, which have lighter testes and heavier adrenal glands than dominant males, do not produce a puberty-accelerating odour in their urine, nor do juvenile males reared with adult males (Lombardi and Vandenbergh 1977). Application of the urine from subordinate male mice to the nares of 28-day-old females for seven days produces no more increase in uterine weights than the application of water, while urine from dominant males produces a significant increase in uterine weights (Labov 1981a). Thus, social factors which reduce testosterone levels in males also inhibit the production of the oestrus-accelerating component in the urine.

A number of studies have attempted to find the chemical basis of the urinary odour which accelerates puberty in females. Vandenbergh, Whitsett, and Lombardi (1975) fractionated male urine by filtration, dialysis, and precipitation methods and applied each fraction to the nares of 28-day-old female mice for nine days. The fraction which accelerated uterine growth was found to be a high-molecular-weight protein. Using dialysis, ultrafiltration, and paper chromatography, Vandenbergh, Finlayson, Dobrogosz, Dills, and Kost (1976) were able to find an active fraction with a molecular weight estimated to be about 860. This fraction contained at least six compounds. Novotny *et al.* (1980) replicated and extended these chemical assays using chromatography to fractionate male urine. After applying these fractions to 23–28-day-old females, a specific fraction of urine containing five or more substances was identified as the 'active component' for oestrus acceleration.

From these bioassays, it appears that the component of male urine which accelerates oestrus in females might be a small, volatile peptide which is bound to a larger 'carrier protein' (Novotny *et al.* 1980; Vandenbergh *et al.* 1976). This peptide is androgen dependent and may be produced in the liver or kidneys. Using gas–liquid chromatography and mass spectrometry Novotny *et al.* (1980) found at least seven volatile components which were present in greater quantities in male urine than in female urine. Castration reduces the amounts of these components in the urine and testosterone injections increase them (see Fig. 8.6). Liebich, Zlatkis, Bertsch, van Dahm, and Whitten (1977) have identified two chemicals (dihydrothiazoles) in the urine of male house mice which are not present in female urine, but whether these are the source of the primer effects remains unknown.

Urine from pregnant and lactating females will also accelerate puberty in juvenile females. Female mice exposed to the urine of pseudopregnant females from birth to 20 days of age have a weight gain equivalent to that produced by exposure to male urine (Cowley and Wise 1972) and 21-day-old mice exposed to the urine of pregnant or lactating females reach first oestrus before controls exposed to water or urine from cycling females (Drickamer and Hoover 1979; see Fig. 8.7), but not as early as females exposed to urine from adult males (Drickamer 1982). The hormonal changes associated with pregnancy and lactation may result in the secretion of peptides or proteins from the liver or kidneys of these females which are similar to those manufactured by adult males.

Female hormone responses to male odours   Sexual maturation occurs as the result of neuroendocrine changes in the hypothalamic-pituitary-gonadal feedback system. The 'gonadostat theory' of puberty states that steroid-sensitive hypothalamic neurons become less sensitive to negative feedback at puberty (Davidson 1969; Raum, Glass, and Swerdloff 1980). Before puberty, small amounts of gonadal steroids inhibit the release of hypothalamic gonadotrophin-releasing hormones (luteinizing hormone releasing hormone (LH-RH), and follicle-stimulating hormone releasing hormone (FSH-RH)). Thus, in prepubertal

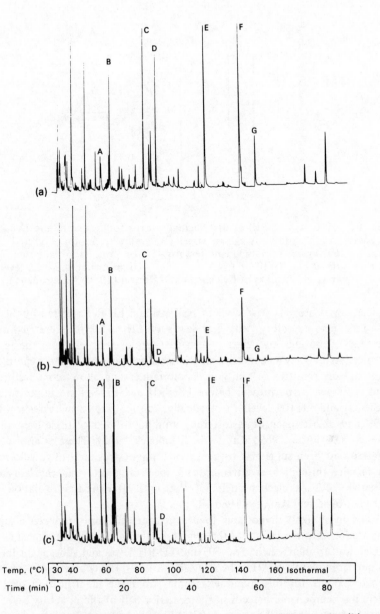

Fig. 8.6. Gas chromatograph of urine volatiles from (a) intact male; (b) castrate male, and (c) testosterone-injected castrate male *Balb/cWt* mice. (From Novotny *et al.* (1980). © Plenum Press.)

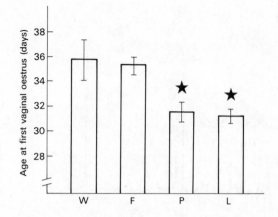

Fig. 8.7. Mean age (± SE) at first vaginal oestrus for female albino house mice exposed from 21 days of age to water (W); urine from singly housed cycling females (F); urine from pregnant females (P); or urine from lactating females (L). Groups P and L differ from the other two groups at the 0.05 level. (Redrawn from the data of Drickamer and Hoover (1979), experiment I.)

animals there are very low levels of FSH-RH and LH-RH secreted by the hypothalamic neurosecretory cells; low levels of LH and FSH secreted from the pituitary gland and low levels of gonadal steroids secreted from the testes or ovaries. As puberty occurs, the steroid-sensitive neurons in the hypothalamus become less sensitive to the gonadal hormones and allow more testosterone and oestrogen to be present before FSH-RH and LH-RH are inhibited. Thus, FSH-RH and LH-RH levels rise gradually, causing gradual increases in LH and FSH and in oestrogen or testosterone until adult levels of these hormones are present (Davidson 1969; Ruf 1973; Grumback 1975). There is also evidence that an oestrogen surge may trigger the first pro-oestrus surge of gonadotrophins in females through a positive feedback mechanism and that the 'gonadostat' negative feedback mechanism is not reset until after the first ovulation occurs (Ojeda, Advis, and Andrews 1980).

The most likely hypothesis to account for olfactorily induced changes in the timing of puberty is that the olfactory stimuli modify noradrenaline levels which lead to increases in FSH-RH and LH-RH levels and subsequent increases in circulating levels of LH and FSH (Zarrow, Estes, Denenberg, and Clark 1970). Some support for this hypothesis comes from the identification of LH-RH secreting neuroendocrine cells in the posterior half of the olfactory bulbs (Phillips, Hostetter, Kerdelhue, and Kozlowski 1980; Dluzen, Ramirez, Carter, and Getz 1981).

Using radioimmunoassays to measure plasma LH, FSH, oestradiol and progesterone levels in 25–28-day-old female mice after 72 hours of male exposure, Bronson and Stetson (1973) and Bronson and Desjardins (1974a) were able to describe the pattern of hormone changes underlying puberty acceleration. Within an hour of exposure to adult males, the females have a small burst of LH which is followed 12 hours later by a surge in oestradiol. This oestradiol surge stimulates uterine growth and weight gain (Fig. 8.8). FSH secretion is inhibited from 24-36 hours after male exposure and begins to increase after 48 hours of exposure, reaching a peak after 72 hours. A second oestradiol surge after 36 hours of male exposure stimulates an ovulatory surge of LH which produces an increase in progesterone secretion.

The second surge of LH and oestradiol 'trigger' pro-oestrus and a regular adult oestrous cycle; something which does not occur in isolated females (Stiff, Bronson, and Stetson 1974). In isolated females, vaginal and uterine changes at puberty are not well co-ordinated with FSH and LH surges, so that ovulation frequently does not occur in synchrony with first oestrus (Stiff *et al.* 1974).

The male-induced LH and oestrogen surges in juvenile female mice can be mimicked by two oestradiol injections with or without the presence of the male for a third day (Bronson 1975*a*). Thus, the presence of the male is only necessary for two days in order to produce ovulation on the third day. Bronson (1975*b*) dissociated two factors in the hormonal mediation of male-induced puberty: the release of subovulatory levels of LH and oestrogen; and positive feedback of oestrogen on the hypothalamus and pituitary resulting in a surge of LH which causes ovulation. Cohabitation with an adult male will induce a burst of LH which, although not enough to cause ovulation, produces enough oestrogen to initiate an adult oestrous cycle and increase uterine weight. This may occur in very young females (12-14 g body weight) but ovulation does not occur until a later time, when the females reach 18-19 g in weight. Thus, male-induced ovulation is not possible until the female develops a complete steroid-sensitive hypothalamic feedback system (Bronson 1975*b*), and the development of this system may be initiated by an oestrogen positive feedback system which is independent of male stimulation (Bronson 1981).

After 30 min to two hours of exposure to male urine, female mice have elevated LH levels (Bronson and Maruniak 1976; Ho and Wilson 1980). Six hours of exposure to male urine will produce an increase in prolactin levels and 24 hours of urine exposure causes a decline in FSH levels (Bronson and Maruniak 1976). Hormonal responses induced by male urine are not as pronounced as those induced by the presence of an adult male.

Discussion

There are three definite olfactory primer effects on puberty: male odours will accelerate puberty in females and delay puberty in males, and female odours will delay puberty in females (Vandenbergh 1980). There is, so far, no evidence that female odours will accelerate puberty in males. The majority of the research has focused on puberty acceleration in females, thus any models of olfactory effects on puberty must focus more on data from females than

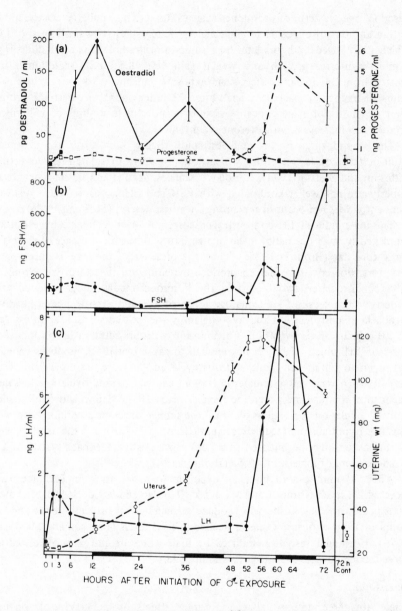

Fig. 8.8. Changes in (a) mean plasma levels (± SE) of oestradiol and progesterone; (b) serum levels of FSH, and (c) LH levels and uterine weights of juvenile albino house mice at various times after exposure to an adult male. (From Bronson and Desjardins (1974a). © The Endocrine Society.)

males. When examining the factors which mediate the primer effects of social odours on puberty, one must consider the relationship between body weight and age in the timing of puberty; the importance of species and strain differences; whether the stimuli are presented in a way that both volatile and non-volatile components are accessible to the subject; the relative importance of physical contact versus olfactory stimuli; whether or not there are critical periods for odour exposure; the relative ease of obtaining puberty delay versus acceleration; and, the degree of relatedness or familiarity of odour donors and odour receivers.

Body weight versus age as factors controlling the timing of puberty Body weight is correlated with puberty onset, but the relationship between them is complex. While body weight can be used to predict the timing of puberty, primer stimuli can alter the timing of physiological changes at puberty independent of body weight. In male hamsters, Miller *et al.* (1977) found that body weight correlated 0.85 with age, 0.76 with plasma testosterone level, 0.91 with testes weight, and 0.87 with seminal vesicle weight as an index of sexual maturation. Even though most of the work with primer odours has been done on female mice, no such study of intercorrelations among the different measures of puberty seems to have been done on females.

Juvenile male mice cohabiting with adult females have often been found to have lower body weights than controls, but larger testes (Fox 1968; Maruniak *et al.* 1978) while juveniles housed with adult males have both lower body weights and lower testes weights than controls (Vandenbergh 1971). Effluent from male populations resulted in lower testes and body weights in male white-footed mice (Rogers and Beauchamp 1976*a*). Prairie deer mice exposed to the urine of adult males or to bedding soiled by adult males had their seminal vesicle weights decreased, but did not always show a decrease in body weight. Summarizing these results on male mice, Lawton and Whitsett (1979, p. 136) state that 'the pheromonal inhibition of sexual development cannot be attributed solely to a retardation of general body growth'.

Sexual maturation in female mice can also be advanced or delayed independently of body weight. Although females exposed to males reach first oestrus at an earlier age than those not exposed to males, the two groups do not differ in weight when the same age (Vandenbergh *et al.* 1972). Colby and Vandenbergh (1974) found a correlation between body weight and first oestrus, but suggest that there is no causal relationship between body weight and puberty. Results from a number of other studies have also shown that male odours can accelerate sexual maturation in females independently of body weight (Lombardi *et al.* 1976; Lombardi and Whitsett 1980; Teague and Bradley 1978). Drickamer (1976*b*, 1981*a*) has gone on to suggest that there are two independent genetic programmes for maturation; one controlling physical growth and one controlling sexual maturation. This suggestion is supported by data showing that mice can be selectively bred for early or late sexual maturation without affecting body weight or growth rate (Drickamer 1981*a*).

When using uterine weight as a measure of puberty, however, Bronson and Desjardins (1974*a*) found body weight to be a better predictor of male-induced puberty than age. Of all 26–30-day-old females, those weighing between 17.5 and 18.5 g showed the greatest acceleration of puberty in response to males while those weighing 15.5–16.5 g showed the poorest response. Uterine growth in response to male stimuli occurs in females weighing as little as 12 g, but is not evident in all animals unless they weigh at least 16 g (Bronson 1975*b*).

Bronson and Stetson (1973) found that using age to measure the timing of puberty resulted in greater variability in male-induced hormone changes than using weight because the smaller females took longer to show male-induced puberty acceleration than the larger females (Bronson and Desjardins 1974*a*; Bronson and Maruniak 1976). While vaginal opening, first oestrus, and uterine weight may change independently of body weight, Bronson and Maruniak (1976) argue that body weight is a better predictor than age for the 'critical period' in which a male will induce an LH surge in a female and they concluded that 'the importance of body weight as an indicator of pubertal readiness in female mice cannot be overemphasized . . . ' (Bronson and Maruniak 1976, p. 1108).

Thus, while some of the events defining puberty (vaginal opening and first oestrus) may be stimulated by males independently of the female's weight (Drickamer 1981*a*), the *rate* of male-induced sexual maturation and the full synchrony of ovulation and oestrus may depend on the animal's weight as well as its age.

It is also important to know the litter size and sex ratio when considering the effects of odours on puberty. Females in litters of four pups reach first oestrus seven days before females in litters of 16 pups and females in all female litters reach puberty before those in litters which are 75 per cent males (Drickamer 1976). Females from different litter sizes all reach first oestrus at about the same body weight, suggesting that in the absence of primer stimuli, body weight is an important variable in the development of sexual maturity.

Species and strain differences   There can be three types of species and strain differences in the ability of odours to accelerate or delay puberty: adults may differ in their ability to produce odours which affect puberty; juveniles may differ in their neuroendocrine responses to odours and juveniles may differ in their sensitivity to the odours of other strains and species.

Male facilitation of PMSG- and HCG-induced ovulation occurs in females of *SJL/J, SWR/J*, and *C3H/H3J* strains of mice, but not in females of *DBA/2J* nor *C57BL/6J* strains (Zarrow, Christenson, and Eleftheriou 1971; Zarrow, Eleftheriou, and Denenberg 1972). Ho and Wilson (1980) replicated these results, finding that male odours facilitated PMSG-induced ovulation in *SWR/J* and *C3H/2Ibg* females, but not in *C57BL/6Ibg* females. It has been suggested that male *C57BL* mice do not produce a puberty-accelerating odour because they have low testosterone levels (Bartke 1974; Ho and Wilson 1980), but it is also possible that *C57BL* females are less sensitive to male odours than females of other strains.

Puberty can be delayed in both male and female laboratory and wild mice, prairie deer mice, white-footed mice, common voles, and gerbils (Table 8.3(a) and 8.3(b)) and in female cactus mice (Table 8.3(b)). Acceleration of puberty has been found in females of five strains of laboratory mice and two species of wild mice as well as rats, voles, and lemmings (Table 8.3(c)).

The ability of rodents of one species or strain to alter the timing of puberty in other species or strains has not been extensively studied. Urine from male rats will accelerate first oestrus and increase uterine weights in female mice (Colby and Vandenbergh 1974), but human male urine has no effect (Vandenbergh *et al.* 1975). On the other hand, puberty is delayed in male and female white-footed mice exposed to effluent from adults of their own species (*Peromyscus leucopus*) but is not delayed by effluent from adult house mice (Rogers and Beauchamp 1976a). Similarly, female house mice reared with deer mice have their uterine weights increased if exposed to adult male house mice but not if exposed to adult male deer mice (Kirchhof-Glazier 1979).

Odours versus physical contact Odours appear to be almost as effective as physical contact between animals in delaying puberty, but not as effective as physical contact in accelerating puberty. Exposure to the urine of adult females appears to delay puberty in females to the same extent as housing juvenile females together in groups for an equal length of time (Drickamer 1977, 1981b; McIntosh and Drickamer 1977), but only the study of Drickamer (1981b) makes a direct comparison between odour stimulation and physical contact.

Although no studies have directly compared the effect of adult male odours and physical contact on the delay of puberty in juvenile males, the results of Bediz and Whitsett (1979) and Lawton and Whitsett (1979) indicate that rearing juvenile male prairie deer mice with adult males for four weeks produces about the same decrease in testes and seminal vesicle weights as five weeks of exposure to the urine of adult males.

While male odours alone will accelerate puberty in females, the effect is not as significant as when the male is actually present. The results of Vandenbergh's (1967, 1969, 1973a) studies (see Fig. 8.4) indicate that first oestrus occurs at the earliest age when the male is present, occurs at the latest age when no male stimuli are present and occurs at an intermediate age when male bedding is present.

Drickamer (1974b) attempted to dissociate the effects of male odours and physical stimulation on puberty acceleration of female house mice. He found that females reared with adult males reached first oestrus before those housed

Table 8.3. Species and strains of rodents in which acceleration and delay of puberty based on the presence of a conspecific or their odours has been found

A. *Delay of puberty in males*

Species	Male stimulus			Reference
	Adult male	Juvenile group	Male odour	
Laboratory mice (*Mus musculus*) albino	X	X		Vandenbergh (1971) Fox (1968) Svare *et al.* (1978)
Wild *Mus musculus*	X	X		McKinney and Desjardins (1973) Svare *et al.* (1978)
Prairie deer mice (*Peromyscus maniculatus bairdii*)	X	X	X	Bediz and Whitsett (1979) Lawton and Whit- sett (1979)
White-footed mice (*Peromyscus leucopus*)			X	Rogers and Beau- champ (1976a)
Voles (*Microtus arvalis*)			X	Lecyk (1967)
Gerbils (*Meriones unguiculatus*)	X			Ågren (1981)

B. *Delay of puberty in females*

Species	Female stimulus			Reference
	Adult female	Juvenile group	Female odour	
Laboratory mice (*Mus musculus*) albino		X		Vandenbergh *et al.* (1972) Drickamer (1974a)
Wild *Mus musculus*	X		X	Drickamer (1979)
Prairie deer mice (*Peromyscus maniculatus bairdii*)		X	X	Bediz and Whitsett (1979) Lombardi and Whit- sett (1980)
White-footed mice (*Peromyscus leucopus*)			X	Rogers and Beau- champ (1976a)
Cactus mice (*Peromyscus eremicus*)	X			Skryja (1975, 1978)
Vole (*Microtus ochrogaster*)		X		Hasler and Nalbandov (1974)
Gerbils (*Meriones unguiculatus*)	X			Swanson and Lockley (1978) Ågren (1981)

C. *Acceleration of puberty in females*

Species	Male stimulus		Reference
	Presence	Odour	
Laboratory mice (*Mus musculus*)			
Swiss–Webster (SW)		X	Colby and Vanden-bergh (1974) Lombardi and Van-denbergh (1977)
Charlesworth Farms (CF-1)	X	X	Bronson (1975*a,b*) Bronson and Des-jardins (1974*a*) Bronson and Maruniak (1975, 1976)
International Cancer Research Strain (*ICR/alb*)	X	X	Drickamer (1974*b*)
Ash/CS1		X	Kennedy and Brown (1970) Cowley and Wise (1972)
Theiler Original Strain (T.O.)	X		Fullerton and Cowley (1971)
(*SJL/J* × *SWR/J*) F$_1$ hybrids	X	X	Wilson *et al.* (1980)
Wild *Mus musculus*	X	X	Drickamer (1979)
Prairie deer mice (*Peromyscus maniculatus bairdii*)		X	Teague and Bradley (1978) Lombardi and Whit-sett (1980)
Rats (*Rattus norvegicus*) Holtzman strain	X		Vandenbergh (1976)
Vole (*Microtus ochrogaster*)	X		Hasler and Nalbandov (1974)
Collared lemming (*Dicrostonyx groenlandicus*)	X		Hasler and Banks (1975)

on bedding soiled by males. Housing juvenile females with neonatally androgen-ized adult females advanced oestrus the same amount as male soiled bedding, but bedding soiled by androgenized females did not accelerate puberty. When androgenized females and male-soiled bedding were both present, females reached puberty at the same age as when a male was present. Both androgenized females and males mount juvenile females and Drickamer (1974*b*, 1975*a*) suggested that this mounting or some other contact behaviour may act in con-junction with the male odour to maximize the advance of puberty in juvenile females. Juvenile females housed with seven prepubertal males or a castrated adult male show a slight acceleration of first oestrus, but not as much as those housed with intact adult males (Drickamer and Murphy 1978). Like laboratory mice, wild house mice exposed to bedding soiled by adult males show first oestrus before females housed on clean bedding, but not as early as females housed with adult males (Drickamer 1979).

By combining adult male urine and various types of physical stimulation, Bronson and Maruniak (1975) were able to demonstrate the additive effects of male odours and tactile stimulation in the acceleration of puberty in juvenile females. Females housed with adult males had higher uterine weights than those housed in isolation, and females sprayed with male urine had intermediate uterine weights. Castrated males had no effect on uterine weights, but females sprayed with male urine and housed with castrate males had uterine weights almost as large as those of females housed with males. Females sprayed with male urine and chased around the cage by the experimenter showed a slightly greater increase in uterine weights than females sprayed with urine alone, indicating that chasing stimuli add to the olfactory effect, but not as much as mounting (see Fig. 8.9).

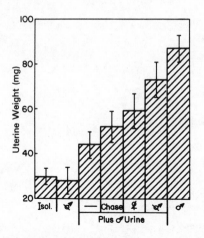

Fig. 8.9. Mean uterine weights (± SE) of 15–15.9-g female house mice after 54 hours of isolation (Isol.); exposure to castrated males (♂); exposure to intact males (♂); or exposure to the urine from intact males while isolated (−); isolated and chased by the gloved hand of the experimentor (Chase); cohabiting with an ovariectomized female (♀); or cohabiting with a castrated male (♂). (From Bronson and Maruniak (1975). © The Society for Reproduction.)

Rearing juvenile females in cages with males behind a partition and spraying male urine over the female increases uterine weight more than keeping the female in isolation, but not as much as allowing the male to contact the female (Bronson and Maruniak 1975). Finally, inhibiting the female's vision or hearing does not interfere with the increases in uterine weight produced by male contact.

Odour presentation The way in which olfactory stimuli are presented to juveniles may determine whether or not the timing of puberty is altered. If the puberty altering components act through the main olfactory system of the recipient they must be volatile, whereas if they act through the vomeronasal system they may be non-volatile and the recipient may have to contact the substance before puberty can be altered (see 318–23).

Exposing juvenile females to male-soiled bedding or to male urine seems to have an equal effect on accelerating the time of first oestrus (Colby and Vandenbergh 1974; Drickamer 1974*b*; Vandenbergh *et al.* 1972). In both of these cases, however, the female can contact the stimulus directly. Drickamer and Assmann (1981) presented male urine to 21-day-old female house mice on glass plates, on a cotton plug within a plastic tube, on soiled bedding or directly on the nares. Only those presentations in which the female could come into direct contact with the urine produced an acceleration of puberty (i.e. soiled bedding and urine applied to the nares), but this acceleration was still less than when the male himself was housed with the female. Similarly, Wilson *et al.* (1980) found that when female mice could make physical contact with male urine their uterine weights increased far more than when the odours could not be contacted. These results suggest that airborne olfactory stimuli are not sufficient to produce maximal acceleration of uterine weight and that the urine may have to contact the vomeronasal organ directly in order to accelerate puberty. It seems, therefore, that the non-volatile components of the male's urine may be more important for puberty acceleration in females than the volatile components.

Odours from adult female mice cause delay of puberty in juveniles whether or not the juvenile can make direct contact with the odour source. Exposing females to bedding soiled by adult females, to urine of adult females presented on a glass plate, in a plastic capsule, or painted on the nares results in a delay of first oestrus (Drickamer 1974*a*, 1977; Drickamer and Assmann 1981; Drickamer and McIntosh 1980). Puberty delay therefore appears to be mediated via volatile components in the urine of adult females.

Experiments showing that the urine of adult male deer mice delays puberty in juvenile males have used only bedding soiled by adult males or direct application of male urine to the nares of juvenile males (Bediz and Whitsett 1979) and the question of whether the volatile components of male urine are sufficient to delay puberty has not been examined.

Acceleration versus delay Which olfactory primer effect is easier to facilitate: acceleration or delay of puberty? In males the answer is clear cut as it does not seem possible to accelerate puberty using female odours alone while puberty is delayed after exposure to urine from adult males. Delay of puberty in females appears to be easier to elicit than acceleration, as puberty delay is produced by volatile chemicals which the female need not contact, while puberty acceleration requires physical contact with the urine, plus contact with the male for the maximal effect.

When identical stimulus presentation is used, puberty in females can be delayed to a greater extent than it can be accelerated. Bedding soiled by adult male wild house mice will advance first oestrus by 8.3 days while the same quantity of bedding soiled by adult females delays first oestrus by 9.5 days (Drickamer 1979). Similarly, placing urine from adult males directly on the nares of laboratory mice accelerates their first oestrus by an average of 3.9 days while the same quantity of urine from adult females delays first oestrus by 4.5 days.

In a more direct comparison of the power of urinary odours to accelerate or delay puberty, Drickamer (1982) mixed urine from adult males, pregnant and lactating females, and grouped adult females. Whereas mixtures of urine from adult males and pregnant and lactating females all advanced puberty, equal parts of urine from grouped females mixed with any or all of these resulted in a delay of puberty. The urinary odour which delays puberty is therefore able to override any odours which accelerate puberty in female mice. Thus, in crowded populations, the odours of adult females will over-ride those of adult males and pregnant or lactating females and prevent juveniles from achieving early puberty.

The potency of urine from adult females to delay puberty may also change as population density increases or decreases. Urine collected from wild female mice during population surges will delay first oestrus in laboratory mice, whereas urine collected from the same mice during periods of low population density does not delay first oestrus (Massey and Vandenbergh 1980). Male urine, however, will accelerate first oestrus whether the urine is taken from males in low or high-density populations (Massey and Vandenbergh 1981).

Increases in population density cause increases in adrenal weights (Christian and Davis 1964), and since the adrenal gland appears to regulate the female puberty-inhibiting odour (Drickamer and McIntosh 1980) changes in population density may affect the timing of puberty in juvenile females by modulating odour production in adult females through changes in the adrenal gland.

In both high- and low-population densities adult males produce odours which will accelerate female puberty. In low-density populations, the adult female does not produce a puberty-inhibiting odour, and the male odour will predominate, inhibiting puberty in males and accelerating puberty in females. When population density increases, however, adult females produce a puberty-inhibiting odour which is more potent than the male's odour so that puberty is inhibited in females. Thus, the potency of adult odours may vary under different population conditions, and the relative acceleration or delay of puberty in juvenile females may be modulated by the relative potency of adult male and female odours.

Critical periods and the duration of odour exposure   A critical period is a time during which an organizational process proceeds at the most rapid rate and when the process can most easily be modified or altered (Scott, Stewart, and DeGhett 1974). Three questions might be asked about the critical period for the effects of odours on puberty: when should the animal first be exposed to the odour? How long should the exposure last? How much olfactory stimulation is needed?

The different physiological and morphological changes at puberty mature at different rates (Miller *et al.* 1977; Svare *et al.* 1978) and may have different critical periods which overlap in time, or require different periods of time to mature. During puberty the body growth spurt, testes or ovarian growth, accessory sex gland growth, and sperm or ova maturation may begin at different times, proceed at different rates and end at different times. Underlying these processes are the changes in the hypothalamic–pituitary–gonadal hormone system which 'trigger' ovulation and sperm production and the critical period for altering the time of puberty may depend on the readiness of this hormonal system to respond to external stimulation.

Modification of the neuroendocrine system controlling puberty may be attempted before, during, or after the 'critical period' (see Fig. 8.10(a)). For measures such as first oestrus, testes, ovary, uterine, or seminal vesicle weight, there is an upper threshold at which a maximum is reached and in order to show an acceleration effect the odour must be presented as the rate of development begins to accelerate (A or B in Figure 8.10(a)). At time C or D the critical period is over and increased development is no longer possible.

A possible reason for the lack of evidence for olfactory acceleration of puberty in males is that odour presentation or physiological measurements have been made during inappropriate developmental periods. From the studies reported on pages 248–56, period B is about 20–21 days of age, period C, 32–36 days of age, and period D, 50–60 days of age. Studies which report acceleration due to contact or olfactory stimuli administered the treatments when males were between 21 and 36 days of age and took physiological measurements between 32 and 36 days of age. Studies finding no puberty acceleration in males took measurements after 50 days of age. The neuroendocrine response to female odours is fully developed in 36-day-old males, which show a rapid elevation of LH after 30 min of exposure to female urine, but this response does not occur in 12- and 24-day-old males (Maruniak *et al.* 1978). Whether exposure to female odours during periods A or B (see Fig. 8.10(a)) will accelerate the development of the male's neuroendocrine system does not seem to have been investigated.

Exposure of females to male stimuli between birth and 21 days of age (Fullerton and Cowley 1971; Cowley and Wise 1972); four and 18 days of age (Kennedy and Brown 1970); and 21 and 34 days of age (Teague and Bradley 1978) has been reported to accelerate puberty, while exposure between 25 and 45 days of age did not accelerate puberty (Lombardi and Whitsett 1980). Male exposure at different periods and for different lengths of time, however, may accelerate puberty at different rates. Vandenbergh (1967), for example, found that females reared with adult males from 21 days of age had a greater acceleration of puberty than those reared with adult males from 2-21 days of age or 30-38 days of age.

In female rats there is a burst of FSH secreted at 12-15 days of age which initiates follicular development. In the mouse this FSH burst appears at 10 days of age (Ojeda and Ramirez 1972; Cowley 1980). This FSH surge may define the beginning of the critical period for acceleration of puberty in females. K. Brown (1979, p. 616) for example, found that daily one hour exposures of female mice to odours from an intact male from 7-13 days of age accelerated vaginal opening whereas exposure from day 15-21 had no effect. Wilson *et al.* (1980), on the other hand, found that exposure of female mice to male urine for 48 h at 25-29 days of age produced a greater increase in uterine weights, than urine-exposure at 23 days of age.

Bronson and Maruniak (1976; see also Bronson 1981) argue that body weight is a more important factor in defining the 'critical period' for puberty accelera-tion than age because body weight determines the timing of first ovulation (see pp. 246-7) and male stimuli cannot advance puberty until females reach a certain weight. Females weighing 12 g require 10 or more days of male exposure before puberty occurs; 15-g females require 5-8 days of exposure, 18-g females require 2-4 days, and 20-g females only 1-2 days of male exposure. Thus, males may not accelerate any of the mechanisms controlling puberty; male stimuli may facilitate the occurrence of puberty as soon as the hypothalamic-pituitary-gonadal system matures and this maturation appears to be independent of male stimulation (Bronson 1981).

Some support for the critical weight theory comes from an experiment in which female *ICR/Albino* mice were selectively bred for early or late puberty. After six generations, those females bred for accelerated puberty reached first oestrus at 28 days of age, 7.5 days before the random-bred controls and weighed about 17.5 g. These females failed to show further puberty acceleration when housed with adult males or painted with male urine (Drickamer 1981*a,b*).

Those females bred for late puberty reached first oestrus at 47.9 days of age, 12.4 days later than the random-bred controls and weighed about 23 g when first oestrus occurred (Drickamer 1981*a*). These females did show puberty acceleration when exposed to adult males or their urine, reaching first oestrus 6-10 days before control females housed alone (Drickamer 1981*b*).

From this evidence it appears that olfactory stimulation from 21 days (period B in Fig. 8.10(a)) is the critical period for accelerating puberty in female mice because the female's neuroendocrine system has matured sufficiently for the male stimuli to induce an early ovulatory cycle. This period corresponds with a minimum weight which must be reached before ovulation will occur (Bronson 1981). The earliest time for first oestrus appears to be about 27 days of age or 17.5 g in weight.

Delay or inhibition of puberty may occur in a number of ways, so may be easier to detect than acceleration (see Fig. 8.10(b)). Stimuli presented during

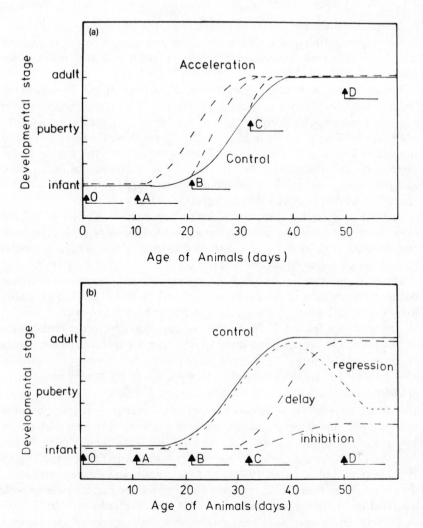

Fig. 8.10. Hypothetical critical periods for (a) accelerating and (b) delaying puberty. See text for explanation.

period B could delay sexual maturation or inhibit it completely. Stimuli presented during period C, when sexual maturation is almost complete, may either inhibit maturation (dashed line) or cause regression (dotted line) which could be mistaken for inhibition if measurements were only taken during period D.

Rearing males in large groups or with adult males during periods B or C inhibits testes growth while male odours presented from period B to D also cause inhibition of puberty (Lawton and Whitsett 1979). This is a long period of stimulation and no attempt has been made to present juvenile males with male odours for shorter periods at different ages (for example, periods 0, A, B, or C in Figure 8.10(b)). The critical period for presenting male odours and the duration of odour exposure required to delay or inhibit puberty in males do not seem to have been investigated.

Female stimuli presented to juvenile females between 21 and 24 days of age delay first oestrus if exposure lasts for seven days, but not if there are only three days of exposure. This delay of first oestrus occurs in juveniles housed with adult females, housed in all female groups or painted on the nares with urine from adult females (Drickamer 1977). Painting urine from adult females on the nares of female mice from birth to 21 days of age delays vaginal opening and first oestrus (Cowley and Wise 1972) but exposing 30-day-old females to adult female stimuli does not delay first oestrus (Drickamer 1977).

Females bred for late puberty reach first oestrus at 44–47 days of age and show no further delay if exposed to other juvenile females or urine from adult females whereas these treatments delay first oestrus by 5 to 6 days in females bred for early puberty (Drickamer 1981*a,b*).

Delay of puberty can be produced by exposing females to adult female odours for seven days or more between birth and 24 days of age, but no studies have investigated whether, for example, exposure to urine from adult females at 21 days of age (period B) delays puberty more than exposure at birth, seven, or 14 days of age (periods 0 and A in Figure. 8.10(b)). Exposure to adult odours at 30 days of age (period C) does not delay puberty but exposure at period D, after puberty has occurred, may produce regression of the gonads and thus inhibition of sexual behaviour and reproduction (see 276–80).

The degree of relatedness of odour donors and receivers   Rearing juveniles with relatives inhibits reproduction more than rearing with non-relatives. In asymptotic populations of mice the parents breed while the offspring do not (Terman 1965). Female gerbils do not mature if kept with their parents (Agren 1981; Payman and Swanson 1980) and female voles have delayed maturation if they are reared from 21 days of age with their brothers, but not if reared with unrelated males (Hasler and Nalbandov 1974; McGuire and Getz 1981). California voles and prairie voles kept continuously with siblings of the opposite sex from weaning have delayed puberty and do not mate (Batzli, Getz, and Hurley 1977), but siblings which are kept apart for eight days will mate when reintroduced (McGuire and Getz 1981). In prairie voles, reproductive inhibition occurs when siblings are kept in separate cages and have only olfactory contact.

The inhibition of mating which occurs between related juveniles and juveniles and their parents may occur because of familiarity, not because of relatedness, and continuous contact is required to maintain familiarity with relatives (see Bekoff 1981). Fathers and brothers may provide individual or family odours which serve as 'reference odours' so that juvenile females can recognize novel (unrelated) males. After a period of separation or when housed in a novel environment the odours of relatives will accelerate puberty (Carter *et al.* 1980).

Who benefits from puberty acceleration or delay? Puberty acceleration appears to ·benefit both the adult who produces the odour and the juvenile whose puberty is being accelerated as both will mate and produce offspring. The presence of a male is important for organizing the first oestrus and ovulation of females since those reared without being able to contact males or their odours show prolonged and disorganized first oestrous cycles in which ovulation often fails to occur due to a lack of gonadotrophin bursts (Stiff *et al.* 1974; Bronson and Desjardins 1974*a*).

Male acceleration of puberty in females is the primer effect which seems to have the highest probability of occurrence in wild populations. Acceleration of puberty in females may be most important (i) when a population is vigorously expanding into a new habitat, or (ii) when a dominant stud male has a stable deme in a population of low to intermediate density in which mother–young groups stay together (Rogers and Beauchamp 1976*b*, p. 189).

Female induction of early puberty in juvenile males may stimulate male reproductive behaviour which will increase the female's chances of being impregnated, thus producing offspring at a maximum rate. In this case the female-male external feedback loop postulated by Bronson (1976*b*, 1979) may be operating: females may induce early puberty in males who then produce primer odours which stimulate oestrus in the female and mating will then occur. The acceleration of puberty in males, however, requires that the female be present, not just her odour (see p. 248).

Suppression of puberty may be advantageous in (i) dispersing populations or (ii) social conditions in which fertile males are scarce (Bronson 1979; McClintock 1981). Who benefits from this delay of puberty may depend on the relationship of the adults and juveniles. When they are unrelated puberty suppression may be a mechanism of male–male or female–female competition so that only the dominant adult has the benefit of mating. Strange juveniles are attacked by both adult females and adult males (Ayer and Whitsett 1980; Whitsett, Gray, and Bediz 1979), thus aggressive behaviour of adults as well as their odours may inhibit puberty in strange juveniles of the same sex and benefit the adult's relative fitness by reducing the juvenile's ability to mate.

If the adults and juveniles are relatives, both may benefit from delay of puberty. Adults may be able to rear more litters if their offspring do not mate and thus compete for resources, while the offspring might benefit by waiting for suitable environmental conditions before mating. It is possible that infants may not reach puberty as early as possible because they benefit by remaining a juvenile. These benefits may include greater adult size, increased time for learning and conservation of reproductive energy (Alexander 1974). On the other hand, the ability of parents to manipulate the time at which their offspring mate may benefit the parent more than the juvenile. By delaying puberty in their offspring, parents reduce competition for mates, gain some help in rearing future offspring and thus may improve their reproductive success (see Alexander 1974).

Social odours affecting reproduction in adult rodents

Spontaneous versus induced ovulation

Many species of laboratory rodents such as the rat, mouse, and hamster are spontaneous ovulators which have 4-6-day oestrous cycles consisting of pro-oestrus, oestrus, metoestrus, and dioestrus phases. During the oestrous cycle the vaginal epithelium shows characteristic changes in cell structure (Young, Boling, and Blandau 1941). In reality, the different stages of the oestrous cycle contain a mixture of cell types and the phases of oestrus are determined by the predominant type of cell (McClintock 1978). Behavioural oestrus (lordosis) and mating are highly correlated with the oestrus phase of the cycle. Ovulation occurs spontaneously early in the oestrus phase. During metoestrus and dioestrus lordosis occurs only rarely and the male is usually rejected (Hardy 1970).

In reflex or induced ovulators, the female is often a seasonal breeder who shows behavioural oestrus but does not ovulate without some form of external stimulation. The most potent stimulation is copulation including intromission and ejaculation, but female–female mounting, handling, and non-specific stressors may induce an LH surge and ovulation. According to Conaway (1971, p. 242), '. . . both induced and spontaneous ovulation are the extremes of a single continuum'. A number of factors such as light cycles and nutrition affect oestrus and ovulation in both spontaneous and induced ovulators, but the most potent factors are social stimuli.

Inhibition of oestrus in adult females

In crowded populations many adult females cease reproducing. Terman (1973) found that only 38 per cent of the females in dense populations became pregnant compared with 83 per cent of the females kept in bisexual pairs. Females in dense populations have lower body weights and ovarian weights than those housed in bisexual pairs, but show no differences in plasma progesterone levels (Albertson, Bradley, and Terman 1975). Reproductive cessation in grouped females may be due to prolonged dioestrus, pseudopregnancy, or anoestrus and may be caused by physical contact with other females or by female odours.

Pseudopregnancy in grouped females: the Lee–Boot effect   While researching mammary cancer in *C3H* mice, Andervont (1944) noticed that those housed alone came into oestrus more frequently than those housed in groups of eight. Lee and Boot (1955) found that between 14 and 26 per cent of $(C57 \times DBA)F_1$ hybrid female mice housed in groups of four had oestrous cycles longer than eight days. The increased length was due to prolonged dioestrus and, since deciduoma formed after traumatization of the uterus, it appeared that these females were pseudopregnant. That these females were truly pseudopregnant was shown by transplanting embryos and finding that they were carried to birth (Lee and Boot 1956). Such pseudopregnancies occurred more frequently in 6–10-month-old females than younger females and occurred at a very low rate in isolated females (Lee and Boot 1956). Ablation of the olfactory bulbs of grouped females reduced the number of pseudopregnancies from 27 to 2 per cent, suggesting that an olfactory stimulus was involved (Lee and Boot 1956).

Housing female mice in groups ranging from 3-20 in size results in 20-47 per cent of the mice becoming pseudopregnant with most pseudopregnancies occurring in 7-10-month-old females (Dewar 1959; Mody 1963). Since the length of pseudopregnancy ranged from 9-30 days, it is possible that some of these females were anoestrus, rather than pseudopregnant (see below). In contrast, only about 2 per cent of singly housed females which were handled each day became pseudopregnant. While the Lee-Boot effect is caused, like delayed puberty, by housing females in groups, there is no evidence that female odours alone can induce pseudopregnancy.

Anoestrus in grouped females   Female mice housed in groups of 30 have lower ovarian weights than controls, reduced frequency of ovulation, and may become anoestrus for periods up to 40 days (Whitten 1957, 1959). These mice did not form deciduoma when the uterus was traumatized, thus Whitten (1959) concluded that they were not pseudopregnant, but anoestrus. In groups of 4-8, and even when housed in pairs, female mice have prolonged dioestrus, resulting in oestrous cycles of 11-13 days rather than the 5-6 day cycles of isolated females (Champlin 1971; Lamond 1959). When these grouped females are singly housed they resume a normal oestrous cycle.

In order to discover whether females in groups of 20 show anoestrus or pseudopregnancy, Ryan and Schwartz (1977) examined vaginal smears, ovarian and uterine weights, deciduoma formation, and progesterone secretion. Since most of the grouped females had oestrous cycles of 10-12 days in length, showed evidence of corpora lutea in the ovaries, produced a deciduoma response in the uterus, and had serum progesterone levels similar to those of pseudopregnant females, Ryan and Schwartz concluded that grouping caused successive pseudopregnancies rather than anoestrus. The stage of the oestrous cycle in which grouping occured determined the number of females becoming pseudopregnant with a maximum of 83 per cent pseudopregnancy in females grouped when in oestrus.

Two factors have been suggested to account for anoestrus in grouped females: stress resulting from crowding and an olfactory stimulus which inhibits ovulation. Since stress levels increase in crowded groups of mice, resulting in increased adrenal weights and increased plasma corticosteroid levels (Christian and Davis 1964), it is possible that anoestrus in grouped females is caused by a stress response. It appears, however, that isolated females may be under more stress than grouped females. Isolated female *C57BL/6J* mice have higher adrenal weights and higher plasma corticosteroid levels than females in groups of four (Bronson and Chapman 1968). On the other hand, Sung, Bradley, and Terman

(1977) found that females in dense populations had lower adrenal weights than females in bisexual pairs, but had much higher serum corticosterone levels.

The increases in corticosteroid levels in grouped females may result in a urinary odour that suppresses the oestrous cycle of adult females in the same way that adrenal mediated urinary odours of grouped females inhibit puberty in juveniles (Drickamer and McIntosh 1980; see pp. 251-4). Blind female mice housed in groups show oestrus suppression, indicating vision is not necessary and females housed in groups of 30 but isolated in individual compartments to prevent tactile contact show some degree of oestrus suppression (Whitten 1959; see Fig. 8.11). Isolated females exposed to bedding soiled by a group of four females also showed a significant suppression of oestrus (Champlin 1971). The odours of grouped females, however, are not as effective as group housing for suppressing oestrous cycles (see Fig. 8.11). Singly housed females come into oestrus every 6-7 days; females in groups of four, every 11-14 days, and females exposed to soiled bedding from females housed in groups of four come into oestrus every 8-10 days (Champlin 1971).

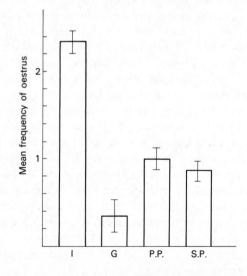

Fig. 8.11. Mean frequency of oestrus ($\pm$ SE) in adult female house mice housed for 16 days in isolation (I); in groups of 30 (G), or in groups of 30 but separated at perforated partitions (P.P.) or solid partitions (S.P.). (Redrawn from the data of Whitten (1959), table 6.)

When female CFLP mice have been kept in groups of eight for three weeks they show anoestrus, but after removal of the vomeronasal organ there is a resumption of oestrous cycles in the majority of these females (Reynolds and Keverne 1979). This suggests that the accessory olfactory system may mediate oestrus suppression in grouped female mice.

Female hamsters housed in groups of 4-12 have regular 4-5 day oestrous cycles and ovulate but do not show lordosis and are not sexually receptive to males (Lisk, Reuter, and Raub 1974). Since group-housed female hamsters which are prevented from physical contact by wire mesh barriers have no decrease in sexual receptivity, this blocked receptivity appears to result from physical contact and not from an olfactory primer effect (Brown and Lisk 1978).

Olfactory bulbectomy   Olfactory bulbectomy results in atrophy of the ovaries and uteri, suppression of oestrous cycling, and reduced likihood of pregnancy (Gandelman, Zarrow, and Denenberg 1972; Lamond 1959; Mody 1963; Whitten 1956b). Peripheral anosmia induced by infusion of $ZnSO_4$ into the nares also results in prolonged oestrous cycles in female mice, but does not completely inhibit ovulation or reproductive behaviour (Vandenbergh 1973b).

There are considerable strain differences in the effects of olfactory bulbectomy on the oestrous cycle: *SJL/J* females show a severe suppression of oestrous; $(C \times SJL)$ F_1 females a slight suppression; and *balb/cWt* females show no suppression after olfactory bulbectomy (Champlin 1977). Olfactory bulbectomy also interferes with oestrous cycling in hamsters (Carter 1973), but does not effect oestrous cycling nor ovarian weight in rats (Aron, Roos, and Asch 1970; Moss 1971). Courtship behaviour in female rats, mice, and hamsters is reduced by olfactory bulbectomy, but mating is only slightly affected (see Murphy 1976).

Hormonal basis of oestrus suppression in grouped females   Various suggestions have been made as to the hormonal basis of oestrus suppression in grouped females. These include inhibition of LH and FSH (Whitten 1959); inhibition of FSH and prolactin (Avery 1959); elevation of FSH or prolactin (Ryan and Schwartz 1977); and elevation of progesterone (Bloch 1976).

Using radioimmunoassay to examine plasma hormone levels, Bronson (1976a) found that adult females in groups of six had lower LH levels and higher prolactin levels than isolated females, while FSH levels did not differ. Since anoestrus grouped females come into oestrus 3-4 days after an injection of the prolactin-inhibiting drug bromocriptine (Milligan 1980) and injections of the drug haloperidol, which elevates prolactin levels, causes anoestrus in cycling females (Reynolds and Keverne 1979), it appears that oestrus suppression in grouped females may be caused by elevated prolactin levels (Milligan 1980). Since prolactin acts as a luteotrophic hormone, it maintains progesterone secretion, thus producing long periods of dioestrus, anoestrus, or pseudopregnancy.

Who benefits from oestrus suppression?   Some authors suggest that oestrus suppression is a laboratory artefact (Bronson 1979) or a hold-over from the female-female inhibition of puberty (Bronson and Coquelin 1980). Rogers and Beauchamp (1976b, p. 186) state that oestrus suppression 'is of questionable significance and occurrence in wild populations' because of the 'probable rarity of all-female groups in nature'. Since female-only groups are rarely encountered, the inhibition of oestrous cycling in adult female *Mus* may reflect a dependence on the presence of a male for normal ovarian cyclicity (McClintock 1981).

In crowded populations, only the older dominant females produce offspring; young and subordinate females fail to reproduce. High-ranking females are more active, move about more frequently and over a wider range than subordinate females, and are seen more frequently to associate with the highest ranking males (Lloyd and Christian 1969).

Oestrus suppression does not occur in all grouped females and those that remain cyclic may be dominant females which are able to suppress the cycles of the rest of the group. Thus, 'the dominant female would be the first to mate with the available males, while the remainder would benefit from the facilitation of group mating' (McClintock 1981, p. 253). The subordinate females, however, may not benefit but have their cycles continuously inhibited by the dominant females.

If the suppressed females are unrelated to the dominant female, oestrus suppression could be a mechanism of female–female competition, with the dominant female benefitting by being able to breed, while breeding is suppressed in submissive females. If, on the other hand, a dominant female is likely to kill strange pups or pups of subordinate females, which occurs in collared lemmings (Mallory and Brooks 1980), it may benefit the subordinate female not to mate in the presence of a dominant female.

If the suppressed females are related to the dominant female (as sisters or daughters), all could benefit from oestrus suppression. Unmated sisters could act as helpers at the nest, promoting the development of their nieces and nephews (and thus increasing their own fitness). If daughters have oestrus suppressed by the mother it could be a case of parental manipulation, a prevention of incest (father–daughter matings) or a signal that the daughters are too young to mate. Young females have smaller litters and are poorer mothers than older females and thus delayed mating may increase the pups' chances of survival. A mother who prevents her daughter from mating too early may improve the chances of her grandchildren's survival. Thus, both the mother and the daughter would benefit in terms of survival of offspring and increased fitness. Such parental manipulation has been described by Alexander (1974).

Male-induced oestrus in grouped females: the Whitten effect

It was Whitten (1956*a*) who first reported that pairing a female from a large group of anoestrus female mice (28–40 females per group) with a male led to a peak of mating on the third day after pairing. Most females housed in groups of 30 come into oestrus on the third day after a male is introduced while singly housed females show equal frequency of oestrus on each of the first four days of cohabitation (Whitten 1959; see Fig. 8.12). Similar results occur in wild house mice (Chipman and Fox 1966*a*).

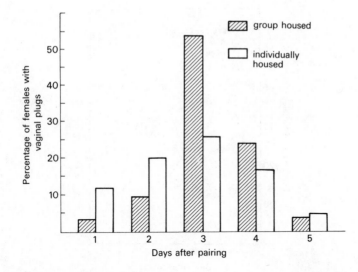

Fig. 8.12. Percentage of group-housed and individually housed female house mice mating on each of the first five days after being paired with a male. (Redrawn from the data of Whitten (1959), table 4.)

The majority of female mice with five-day oestrous cycles come into oestrus and mate on the third and fourth days after the introduction of a male (Lamond 1959). The peak in mating on the third day after pairing male and group-housed female prairie deer mice is most pronounced if the females are housed in groups of 10 before pairing; groups of four show more matings on days 1 and 2, but not as many as females housed alone (Bronson and Marsden 1964).

As well as inducing oestrus in grouped females, the presence of the male shortens the female oestrous cycle. In the presence of a male, female mice have oestrous cycles averaging 4.6 days in length, while individually housed females have cycles averaging six days in length. Many isolated females have prolonged oestrous cycles (10–16 days) while no females living with males have cycles longer than seven days (Whitten 1958; Whitten and Bronson 1970). Oestrous cycle length is reduced most rapidly if males are present for 48 hours beginning immediately after metoestrus. The presence of males for 12 or 24 hours does not shorten the cycles (Whitten 1958).

Although physical contact with a male is the most potent stimulus for inducing oestrus in group-housed females, male odours alone have some effect. Group-housed females exposed to males for two days, but prevented from contacting them by a wire-mesh barrier show a peak in mating as soon as the barrier is removed on the third day. Females housed for two days on bedding soiled by adult males have two peaks in mating activity when males are introduced: one on the first night of contact and one on the third night (Whitten 1956*a*;

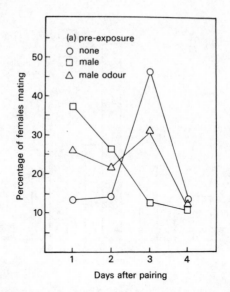

Fig. 8.13(a). Percentage of adult female house mice mating on each of the four days after being paired with a male following two days pre-exposure to: a male behind a wire-mesh screen (□), male excreta (△), or no pre-exposure (○). (Redrawn from the data of Whitten (1956*a*).)

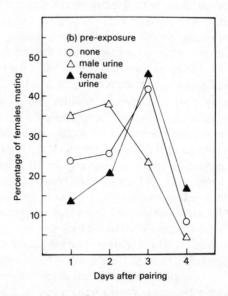

Fig. 8.13(b). Percentage of adult female house mice mating on each of the four days after being paired with a male following: two days pre-exposure to male urine (△), female urine (▲), or no pre-exposure (○). (Redrawn from the data of Marsden and Bronson (1964).)

see Fig. 8.13(a)). The presence of the male or his odour for 48 hours will, therefore, induce oestrus in group-housed females but olfactory stimuli are less effective than the male's presence.

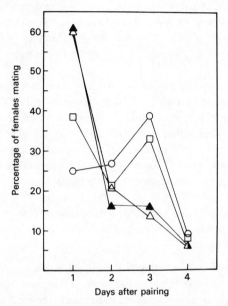

Fig. 8.14. Percentage of group-housed female *C57BL/6J* mice mating on each of the four days after being paired with a male following: 48 hours of pre-exposure to urine from male *CBA/J* mice (□), no urine (○); 48 hours of isolation (△) or 48 hours of isolation plus *C57BL/6J* male urine (▲). (Redrawn from the data of Marsden and Bronson (1965*a*).)

The source of the male's oestrus-inducing odour   Grouped female mice which have male urine painted on their nares for two days show the highest frequency of mating on the first two nights after being paired with an adult male. Females painted with urine from adult females or no urine have mating peaks on the third day after pairing (Marsden and Bronson 1964; see Fig. 8.13). Similar results were found in grouped female deer mice exposed to male urine for two days (Bronson and Marsden 1964). The dose of urine used to induce oestrus in grouped female deer mice does not seem to be important: doses of 0.17, 0.50, and 1.50 ml urine/day result in 48–50 per cent mating on the first day of pairing while the presence of an adult male results in 62 per cent mating (Bronson and Dezell 1968).

Although Parkes and Bruce (1961) reported that castrated males could induce oestrus in grouped females; this study was confounded by changing the size of the groups of females in the no-male condition. Bruce (1965) has shown that when female group size is controlled, castrated males behind wire barriers do not induce oestrus while intact males stimulate an increase in number of periods of oestrus. Urine from castrated males does not induce oestrus in grouped females, but urine from androgenized females has the same effect as urine from intact males (Bronson and Whitten 1968). Thus the male odours which induce oestrus in grouped females are androgen dependent.

The active component of male urine which induces oestrus in grouped females appears in the lipid portion of urine extract (Monder, Lee, Donovick, and Burright 1978) but the identity of this active component has yet to be discovered.

The importance of the preputial glands in the production of the male's oestrus-inducing odour is unclear. Whereas Bronson and Whitten (1968) found that bladder urine from intact males induced oestrus in adult females and McKinney (1972) found that preputial gland odours alone had little effect, Chipman and Albrecht (1974) found that preputial gland secretions induced oestrus in the same number of group-housed females as urine from intact males. Urine from preputialectomized males was less effective for inducing oestrus and Chipman and Albrecht (1974, p. 95) state that 'the oestrus-accelerating pheromone is closely related to and/or augmented by the preputial gland secretion'. Why there should be a discrepancy between this study and that of McKinney (1972) is unclear.

Urine from pregnant and lactating females   As has been found with acceleration of puberty in juvenile females, the urine of pregnant and lactating females will induce oestrus in adult females. Individually housed adult female mice which have urine from pregnant or lactating females painted on their nares for 21 days are in oestrus more often than females painted with water or urine from singly housed females (Hoover and Drickamer 1979). Similar results occurred if the urine was placed on cotton balls within perforated plastic capsules, indicating that the increased periods of oestrus are caused by a volatile urinary odour. Females exposed to urine from pregnant or lactating females do not have an increased number of oestrous cycles, but have longer periods of oestrus within each cycle. This provides further evidence that the urine of pregnant and lactating females contains a chemical similar to that produced by intact males.

Isolation versus male odours   In Whitten's (1956a,b, 1958, 1959) experiments grouped females were both isolated and presented with males or their odours in order to induce oestrus. There may, therefore, be two components involved in oestrus acceleration in grouped females: release from the suppressive effect of female stimuli and stimulation by male stimuli.

Marsden and Bronson (1965a) found that females which were isolated for 48 hours before being paired with a male showed the majority of matings on the first day of male exposure, rather than the third. Furthermore, these isolated females showed a higher response than females exposed to male urine for 48 hours (compare Fig. 8.14 and Fig. 8.13(a) and 8.13(b)).

The male odour does, however, stimulate oestrus in grouped females that are not isolated before exposure. Urine from intact males dripped on to groups of 6–10 females causes a peak in oestrus on days 3 and 4, while urine from castrated males is not effective (Bronson and Whitten 1968). Using a wind tunnel, Whitten, Bronson, and Greenstein (1968) found that groups of 8–10 females living downwind of males had a higher frequency of oestrus than those housed upwind or outside the chamber. Females housed downwind from males showed the same degree of oestrus acceleration as females receiving male urine.

It therefore appears that when grouped females are paired with a male for two days there are two independent effects which cause a peak in mating on the third day: the female's recovery from oestrus suppression and stimulation by the male and his odour. Whether these effects are additive or independent has not been investigated.

Species and strain differences Male-induced oestrus has been found in *C57BL/6J*, *SJL/J*, *SWR/J*, and *129/J* strains of mice (Marsden and Bronson 1965a; Bronson and Whitten 1968). Although Bartke and Wolff (1966) reported that male mice of the inbred *Ys/ChWf* strain which are heterozygous for the lethal yellow allele (A^ya) do not induce oestrus in grouped females, Whitten (1969) found that these males were able to induce oestrus.

Balb/cDg females do not respond to the male's oestrus-inducing pheromone. These females have short oestrous cycles and do not become acyclic after olfactory bulbectomy (Whitten and Bronson 1970). Oestrous cycling of *Balb* female mice may not be sensitive to olfactory stimuli (Champlin 1977) or the cycles may be so short that they cannot be reduced in length by any stimuli. Group-housed *Balb* mice show a peak in mating on the first night after being exposed to a male, much like induced ovulators (see pp. 286–7). Seven other strains of mice showed third night peaks in mating as did the F_1 hybrids of *Balb/CWt* and these other strains, indicating that 'the tendency for the third night mating is inherited in a dominant manner and that the tendency for the first night mating is inherited in a recessive manner' (Champlin, Beamer, Carter, Shire, and Whitten 1980, p. 167).

There are only a few studies examining species and strain specificity of the male's oestrus-stimulating odour and these indicate that the primer effect is species specific, but not strain specific. Oestrus is induced in *C57BL/6J* females which are exposed to urine from *CBA/J* males, but does not occur if urine from *Peromyscus* males is used (Marsden and Bronson 1965a). Further evidence for species specificity was found by Bronson and Dezell (1968) who found that urine from male deer mice would induce oestrus in grouped female deer mice, but urine from *C57BL/6J* males had no effect. Similarly, *SJL/J* females do not show oestrus synchrony if presented with urine from human males (Bronson and Whitten 1968).

Female *Mus* reared from birth until 18 days of age with deer mice (*Peromys-*

cus) and then housed in all-female groups of *Mus* show a peak in mating on the third day after being paired with an adult male *Mus* (Kirchhof-Glazier 1979). While these results suggest that early rearing experience with another species does not alter the Whitten effect, the fact that these females lived with other *Mus* females rather than *Peromyscus* females after they reached 18 days of age may have nullified any effects of rearing with *Peromyscus*.

Female hormonal responses to male odours   The hormonal changes induced in grouped females by male stimuli are not precisely known. Based on the indirect evidence that females exposed to males behind wire mesh barriers have a reduced number of pituitary acidophils, Avery (1969) suggested that the male odour triggers the secretion of both FSH and prolactin. After 24 hours of exposure to male urine, isolated *CF-1* female mice have an increase in plasma levels of LH but no change in either FSH or prolactin levels (Bronson 1976*a*). Since grouped females have higher prolactin levels than isolates (Bronson 1976*a*), merely isolating the females may cause a decrease in prolactin, while male stimulation increases LH secretion. Injections of bromocriptine, a prolactin inhibiting drug, produce oestrus in 3-4 days in grouped females, again suggesting that before oestrus occurs prolactin levels must be reduced (Milligan 1980).

Male-induced ovulation in voles   Most voles are induced ovulators and require external stimulation in order for vaginal opening, oestrus, and ovulation to occur (Richmond and Stehn 1976). Physical contact with a male is the most potent stimulus for inducing oestrus and ovulation in many vole species including prairie voles (*Microtus ochrogaster*); montane voles (*Microtus montanus*) and short-tailed field voles (*Microtus agrestis*) (Gray, Davis, Zerylnick, and Dewsbury 1974; Milligan 1974, 1975*a*; Richmond and Conaway 1969).

One hour of exposure to an adult male will cause an increase in the uterine weights of female prairie voles (Carter *et al.* 1980), but 48 hours of male exposure are required for maximal increases in uterine growth (Hasler and Conaway 1973). The presence of a male for 48 hours induces pro-oestrus and the maximum number of females come into oestrus after 72 hours of male exposure (Hasler and Conaway 1973).

Oestrus and ovulation can be induced in some species of voles without the female having physical contact with the male. Housing female field voles next to novel males (Chitty and Austin 1957) and housing female prairie voles and field voles in cages where they are separated from males by wire mesh screens results in ovulation in over 70 per cent of these females (Milligan 1975*a*; Richmond and Conaway 1969). Anosmic (bulbectomized) prairie voles remain anoestrus when housed next to males, suggesting that the olfactory system must be intact for male stimuli to induce oestrus (Horton and Shepherd 1977; Richmond and Stehn 1976). Female montane voles, on the other hand, do not ovulate when merely housed next to males (Gray *et al.* 1974) and Carter *et al.* (1980)

found no increases in uterine weight in female prairie voles housed across a wire mesh barrier from males.

Exposing female prairie voles or short-tailed field voles to bedding soiled by adult males induces ovulation in some females, but exposure to males behind wire mesh screens induces ovulation in many more females (Milligan 1975b; Richmond and Conaway 1969).

Painting urine from adult males on to the nares of juvenile female prairie voles for six days results in a significant increase in uterine weights when compared to controls, but not as much of an increase as six days of exposure to adult males (Carter *et al.* 1980). Urine from castrated males produces only a slight increase in uterine weights (Carter *et al.* 1980). A single drop of male urine placed on the upper lip of an anoestrus female prairie vole increases noradrenalin and LH-RH levels in the olfactory bulb within 60 minutes (Dluzen *et al.* 1981).

These results indicate that, whereas physical contact with an adult male induces ovulation in the majority of females, male odours which contact the female directly, and may therefore stimulate the vomeronasal organ, are capable of elevating hormone levels and inducing physiological changes in some females. Although urine has been considered as the source of these male odours in voles, Netto and Pederson (1976) suggested that the oestrus-inducing odour of the male prairie vole is produced in the preputial glands. There seem to have been no experiments to examine the role of the preputial glands versus urine in inducing ovulation in voles.

While male stimuli appear to be the most important factors for inducing ovulation in female voles, simply moving females to new cages or moving them from isolation to all female groups or vice versa will induce oestrus in 17-38 per cent (Richmond and Conaway 1969). That disturbance is responsible for some proportion of oestrus induction in females was suggested by Petrusewicz (1958) who found that transferring large groups of mice to large cages, different cages of the same size, or even smaller cages resulted in an increase in reproduction. This disturbance effect may be due to the novelty of the new environment or to a change in the olfactory environment rather than to disturbance per se, but this has not been investigated.

Male-induced ovulation in rats Although the rat is a spontaneous ovulator, with a 4-6 day oestrous cycle, ovulation can be induced in females under some conditions. Female albino rats kept in constant light show an increase in periods of vaginal and behavioural oestrus, but these measures are not highly correlated as occurs in normally cycling females on a 12:12 L:D cycle (Hardy 1970). Constant light appears to produce a low but constant release of oestrogen which is sufficient to produce vaginal cornification and lordosis, but not sufficient to induce the LH surge and ovulation. Light-induced constant oestrus has not been found in pigmented (hooded) rats (Brown-Grant, Davidson, and Grieg 1973).

When constant light-induced oestrus female rats are exposed to males, ovulation is induced and elevated levels of LH and FSH can be detected within 40 minutes of mating (Brown-Grant *et al.* 1973). Mating is not necessary to induce ovulation in these females. Bedding soiled by male rats or male urine sprayed on bedding will induce ovulation in 38–54 per cent of light-induced oestrous female rats (Johns *et al.* 1978, 1980). Ovulation occurs only when the females can contact the urine directly and does not occur if the females are separated from the urine by a wire mesh screen. Lesions of the vomeronasal organ prevent urine-induced ovulation, indicating that this effect is mediated by the accessory olfactory system rather than the main olfactory system (Johns, Feder, Komisaruk, and Mayer 1978).

Female albino rats which are put on restricted diets and which lose 10–15 per cent of their body weight show prolonged periods of dioestrus and may become anoestrus. The introduction of a male shortens the oestrous cycles of these underfed females and induces oestrus (Cooper and Haynes 1967). Oestrous cycles are also shortened when males are caged with underfed females but separated by a wire mesh barrier (McNeilly, Cooper, and Crighton 1970). If new males are presented behind the barrier every day, the oestrous cycle remains shortened, but if a single male is kept behind the barrier, the female's cycle gradually lengthens and by the third cycle many females become anoestrus (Purvis, Cooper, and Haynes 1971). Simply removing and replacing the same male does not maintain the oestrous cycle in these underfed females (Cooper, Purvis, and Haynes 1972).

While male stimuli induce ovulation in these rats with altered oestrous cycles, males can also effect the oestrous cycle of normally cycling females. When group-housed female rats are paired with males the length of their oestrous cycle is reduced and the majority of the matings occur on the third night after pairing (Hughes 1964).

Five-day cycling female rats have their cycle reduced to four days, accelerating receptivity and ovulation by 24 hours, as a result of early mating (Aron, Asch, and Roos 1966). Olfactory bulbectomy inhibits this male-induced oestrus acceleration but does not affect vaginal cyclicity in five-day cyclic rats, nor is normal oestrus receptivity affected (Aron *et al.* 1970). Because a single injection of oestradiol will also advance oestrus by 24 hours, Aron *et al.* (1970) suggested that the effect of male stimulation was to cause an increase in oestrogen secretion in females.

Physical contact with a male is not necessary to reduce the female's oestrous cycle from five to four days. Male urine presented for 8–10 days on the female's bedding or behind a wire mesh screen, female urine and urine from gonadectomized males and females will all reduce the female's oestrous cycle to four days (Aron 1975; Aron and Chateau 1971; Chateau, Roos, and Aron 1972). Both male and female urine act to shorten the dioestrous phase of the female's cycle by reducing progesterone secretion and inducing follicular development (Chateau, Roos, Plas-Roser, Roos, and Aron 1976).

Olfactory bulbectomized female rats have their oestrous cycles shortened to four days only if they can make direct contact with male urine (Chateau, Roos and Aron 1977). It is odd that bulbectomized females respond to urine with a shortened oestrous cycle, and suggests that some receptors which are sensitive to urine send afferents to olfactory areas which are not destroyed by olfactory bulbectomy (see pp. 318-22).

The ventromedial nucleus (VMN) of the hypothalamus is capable of regulating the five-day oestrous rhythm of female rats after bulbectomy (Chateau, Carrer, Roos, and Aron 1973) and lesions of the VMN inhibit the shortening of the oestrous cycle from five to four days when females are exposed to urine (Chateau and Aron 1977; Chateau *et al.* 1977). Large lesions of the VMN increase oestrous cycles from four to five days in Wistar female rats (Carrer, Asch, and Aron 1973). Progesterone injections produce cycle lengthening more in VMN-lesioned rats than controls (Chateau, Roos, and Aron 1976), indicating a role for the VMN in controlling the length of the oestrous cycle, possibly through its sensitivity to progesterone.

Male-induced oestrus and ovulation in hysticomorph rodents   Isolated and unisexually housed cuis (*Galea musteloides*) seldom show spontaneous vaginal opening while exposure to an adult male for 48 hours induces oestrus in most of these females (Weir 1971). The importance of olfactory stimuli for induced oestrus has yet to be determined. Females kept from contacting males by a wire mesh partition fail to show vaginal opening as do those exposed to male urine and faeces. The secretion of the male's chin gland, however, may play a small role in inducing ovulation as vasectomized males with their chin gland removed induce ovulation in fewer females than those with an intact chin gland (Weir 1973). There is, as yet, no evidence that the chin gland secretion from the male cuis alone will induce ovulation in females.

Oestrus synchrony in grouped female rats, hamsters, and guinea pigs   Female rats do not become anoestrus nor pseudopregnant when housed in all-female groups. Female Wistar rats housed in groups of three or four or in groups as large as 40-50 continue to show regular oestrous cycles (Aron, Roos and Roos 1971). In fact, grouped females tend to show a higher proportion of four-day cycles than isolated females (Aron *et al.* 1971) suggesting that grouping shortens the oestrous cycle from five to four days. That this effect may be caused by olfactory stimuli is suggested by two findings: olfactory bulbectomized females do not show this reduction of oestrous cycle length when grouped (Aron *et al.* 1971) and urine from female rats acts to reduce the length of the oestrous cycle in other females from five to four days (Chateau *et al.* 1972).

Grouped female Sprague-Dawley rats not only have shortened cycles but tend to show oestrus synchrony, i.e. they tend to come into oestrus at the same time (McClintock 1978). Oestrus synchrony also occurs in isolated female rats which are in olfactory contact with four other rats through cages connected with plastic tubing (McClintock 1978). Complete synchrony of oestrus is rarely achieved and it takes 3–4 cycles (12–16 days) for synchrony to occur. The synchrony appears to be a change in the length of the follicular phase of the cycle which brings the timing of ovulation in closely associated females closer together. Females which live in the same environment, but which are not socially or olfactorily associated, do not show oestrus synchrony.

Further evidence that olfactory stimuli from grouped female rats will induce oestrus synchrony is given by McClintock and Adler (1978) who placed groups of light-induced constant-oestrus females in a wind-tunnel. Females in groups of six living downwind from females in oestrus showed induced oestrus, being in oestrus 56 per cent of the test days as opposed to 25 per cent for controls which lived upwind of the oestrous females. When oestrous females were re-placed by random cycling females, oestrus synchrony did not occur in the upwind, nor the downwind females. Odours from acyclic females, however, desynchronized the cycles of females showing oestrus synchrony (McClintock 1981).

In isolated female rats, vaginal cornification (vaginal oestrus) and lordosis (behavioural oestrus) become dissociated from each other and become acyclic (McClintock 1981). While olfactory signals are able to regulate synchrony of vaginal oestrus, they do not regulate the lordosis reflex, which may depend on physical contact with other females or on male stimuli. In grouped females there is synchrony of both the vaginal and behavioural components of oestrus (McClintock 1981; see Fig. 8.15).

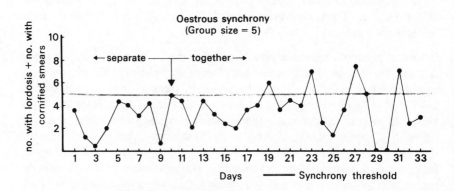

Fig. 8.15. The development of oestrus synchrony within a group of five female albino rats, after regrouping from the colony. Oestrous score: lordosis reflex = 1 point. Cornified vaginal smear = 1 point. Highest possible group total = 10 points. Synchrony threshold = 5 points. Scores above this value indicate that the majority of the group is in oestrus. (From McClintock (1981). © The American Society of Zoologists.)

Oestrus synchrony also occurs when female hamsters are housed in groups by four but separated by wire mesh, or housed in pairs (Handelmann, Ravizza, and Ray 1980). When female hamsters are housed in pairs, the oestrous cycle of the subordinate female is accelerated or delayed until it is entrained to that of the dominant female (Handelmann *et al.* 1980). Whether this oestrus synchrony in female hamsters is due solely to olfactory cues is unknown, but given the heavy reliance on olfaction in hamster social behaviour, it is likely that olfaction plays a large role.

Housing female hamsters (which show a spontaneous four-day oestrous cycle) with female rats alters the oestrous cycle of the rats by decreasing the number of acyclic (anoestrus or pseudopregnant) females (Weizenbaum, Mc-Clintock, and Adler 1977). Whether the female hamster provides olfactory information or some other stimulus such as ultrasound or tactile cues which induce oestrous cycling in female rats is unknown.

Although Donovan and Kopriva (1965) reported that female guinea pigs caged together showed oestrus synchrony, Harned and Casida (1972) found no evidence of oestrus synchrony when randomly cycling adult females were housed together for two oestrous cycles. While Harned and Casida (1972) suggest that the observation of oestrus synchrony occurred in littermates which reached puberty at the same time, it is possible that oestrus synchrony in the guinea pig takes longer than two cycles to occur (see Fig. 8.15, for example).

Who benefits from induced oestrus and oestrus synchrony?   According to Rogers and Beauchamp (1976b) oestrus induction may be important in two specific situations: the timing of the first oestrus in juvenile females and stimulation of the first oestrus of the breeding season in seasonal breeders. Oestrus induction by male stimuli may also be important when nomadic females enter a new deme or colony. For induced ovulators, such as the Microtine rodents, the necessity for male stimulation ensures that ovulation does not occur in the absence of male insemination.

Since induced oestrus results in both the male who released the odour and the female receiver mating, both appear to benefit. Since males are usually able to mate throughout the breeding season, the use of odours to induce oestrus may shorten the dioestrus phase of the female's cycle so that she will mate as soon as possible after encountering the male. Oestrus induction may, therefore, be a male strategy for inseminating as many females as possible.

The ability to accelerate the oestrous cycle will also benefit females who live away from males or do not get the chance to mate because of female–female conflict. In seasonally breeding females, it may be advantageous to mate as early as possible to ensure a plentiful food supply during lactation. Breeding early may also permit the female to have a second litter in one season from post-partum mating. Early breeding may therefore benefit a female in three ways: (i) if she does not breed early, she may not live to breed at all; (ii) if she breeds early, she may be able to have two litters in one breeding season; and, (iii) the offspring from the first litter may become sexually mature and breed in the season of birth (Fairbairn 1977). On the other hand, males which induce oestrus in females which are very young or in adult females very early in the breeding season may be manipulating them into mating when it is not optimal for them to do so. Early breeding results in a lower rate of litter survival and variable reproductive success (see Fairbairn 1977).

Oestrus induction appears to depend on the presence of a novel male. Female voles living continuously with their father do not become oestrus, but if he is removed for eight days and then replaced, females will show induced oestrus (Richmond and Stehn 1976). As with induced puberty in juvenile females, a relative may provide a 'reference odour' by which the female can identify novel males.

Oestrus induction in grouped females by male odours and oestrus synchrony in grouped females both result in synchronous breeding. McClintock (1981) has suggested a number of adaptive functions for synchronous breeding, most of which, as yet, have little or no experimental confirmation.

Synchronization of oestrus may produce more oestrus females than can be mated by the dominant male thus allowing subordinate males to mate, producing some heterogeneity in the deme. The presence of several oestrous females may also facilitate male sexual performance, and oestrus synchrony may increase male parental investment (Knowlton 1979). Synchronized breeding may signal that environmental stimuli are suitable for litter survival so that oestrus synchrony may provide a social signal which acts with environmental stimuli (light cycles, temperature, and rainfall) to signal the breeding season. Finally, breeding synchrony will result in synchronous births which may, in turn, reduce infanticide; reduce the proportion of infants lost to predators as more infants would be born than could be taken by predators; and increase the level of social stimulation available for developing juveniles.

Inhibition of reproduction in adult males

As population density increases, the size of the male mouse's accessory sex organs (preputial glands and seminal vesicles) decrease and the size of the adrenal gland increases (Christian 1955a,b; Christian and Davis 1964). Deer mice in crowded groups show decreases in testes weight as well as in weights of the accessory sex organs (Terman 1968, 1973) whereas the albino mice in Christian's studies showed no reduction of testes weights.

Steinach (1936) reported that male rats housed for several months, either in isolation or in groups of four males, in rooms without females, showed testicular atrophy, decreased weight of the accessory reproductive glands, and a loss of sex drive. Attempts to replicate these findings have given rise to contradictory results. While Folman and Drori (1966) found that grouped male rats had no decrease in weights of the accessory sex glands, Purvis and Haynes (1972) reported that males living in groups have lower epididymes, prostate, and seminal vesicle weights than isolated males or those cohabiting with females. These Wistar rats spent 98 days (from 21-120 days of age) in segregation while Folman and Drori's albino rats spent 101 days, so the reason for the difference in results is unclear.

External stressors such as electric shocks and forced swimming will reduce plasma testosterone levels in rats (Bliss, Frishat, and Samuels 1972) and the stress of defeat in agonistic encounters will inhibit gonadal hormone secretion and increase adrenocortical hormone levels in mice. Testosterone, LH, and FSH levels begin to decline 30-60 minutes after males are paired (Bronson 1973, 1979; Bronson, Stetson, and Stiff 1973; Maruniak *et al.* 1978) and remain low in the subordinate males (McKinney and Desjardins 1973; Bartke and Dalterio 1975; Bronson and Desjardins 1971). Subordinate males also have lower preputial gland, prostrate gland and seminal vesicle weights than dominant males (Bronson 1973; Brain 1972; Brain and Nowell 1971). Dominant males have increased preputial gland weights, (Bronson 1973) but show no increase in testes weights or testosterone levels (Brain 1972; Bronson and Desjardins 1971).

Defeat in an aggressive encounter results in increased adrenal weights and adrenal corticosteroid levels in male mice (Archer 1970; Bronson and Eleftheriou 1964, 1965a; Louch and Higginbotham 1967; Brain 1972; Brain and Nowell 1971). Males which are victorious in aggressive encounters show no increase in adrenal weight (Archer 1970), but do show a slight increase in corticosteroid levels which occurs within 60 minutes of a fight (Louch and Higginbotham 1967; Bronson 1973, 1979). Male mice can be conditioned to release corticosteroids by being paired repeatedly with a 'trained fighter' male (Bronson and Eleftheriou 1965b). As well as corticosteroids, submissive responses are also mediated by increased levels of ACTH and vasopressin (Nock and Leshner 1976; Roche and Leshner 1979).

Dominant male collared lemmings (*Dicrostonyx groenlandicus*) have higher testes, seminal vesicle, and preputial gland weights and higher plasma testosterone levels than subordinates (Buhl, Hasler, Tyler, Goldberg, and Banks 1978). In one-third of the 'active subordinates' in this study spermatogenesis was inhibited, even though there were low levels of aggressive behaviour and little severe wounding of these animals.

Male odours alone will produce some of the effects caused by aggressive behaviour in mice (termed 'the Ropartz effect' by Wilson (1975)). Male mice exposed to air currents containing the odours of grouped males or to bedding soiled by grouped males have increased adrenal weights (Ropartz 1966; Archer 1969) and the increase in adrenal weights of isolated males presented with bedding soiled by grouped males is as large as that of males actually living in groups (Wuensch 1979). While these results show that male odours can produce increased adrenal weights, there is as yet no evidence that male odours alone will cause decreases in testes or accessory sex organ weights or testosterone levels in rodents.

Who benefits from inhibition of reproduction in males?   The most likely hypothesis to account for inhibition of male hormones by other males (or their odours) is male-male competition. Crowding does not affect all members of a population equally: dominant males or territory holders are unaffected (Bronson 1979; Lloyd and Christian 1969). Dominant males may therefore increase their reproductive success at the expense of subordinates by behaviourally preventing the subordinates from gaining access to oestrous females; or by inhibiting gonadal hormone secretion in the subordinates which inhibits their sex odour production and their reproductive behaviour. Finally, if the subordinates do manage to mate and inseminate a female, this pregnancy may be blocked by the presence of the dominant male or his odour (see pp. 299–303). The odour of the dominant male may, therefore, be used to manipulate the reproductive condition of other males, resulting in a decrease in competition for mates and thus an increase in reproductive success for the male producing the odour.

Submissive males may obtain some benefits from having their reproductive behaviour inhibited. These may include reduced attack from dominant males and the potential to mate at a later time (see Alexander 1974).

Induction of reproductive behaviour in adult males

Females and their odours stimulate the reproductive system of male rats, hamsters, and mice. Steinach (1936) claimed that testicular atrophy of group-housed male rats was reversed by contact with females or their odours, but he did not distinguish between female contact and odours (Purvis and Haynes 1972). Male albino rats which cohabit with females from weaning to 100 or 286 days of age have heavier seminal vesicles, coagulating glands, bulbourethral glands, and penises than males living together in groups of four or five, but show no differences in testes size (Drori and Folman 1964).

When isolated male rats, grouped males (3–8 per group) and cohabiting males are compared, the reproductive organs of the grouped and isolated males do not differ and both groups of unmated males have lower penis, seminal vesicle, and bulbourethral gland weights than males cohabiting with females (Folman and Drori 1966). Housing grouped and isolated male rats in the same animal room as females, thus subjecting them to female odours, does not produce an increase in the weights of any of the accessory reproductive organs (Folman and Drori 1966).

Drori and Folman (1964, p. 358) concluded that '. . . cohabitation is necessary to maintain testicular androgen secretion at a normal level' in male rats. Thomas and Neiman (1968) supported this statement by finding that intromission and ejaculation were necessary to maintain the weights of the male rat reproductive system; neither the odours of females nor mounting without intromission were sufficient to stimulate the reproductive organs of these males which were smaller than those of males having sexual activity. Males which mated had higher testosterone levels (in the testes) than unmated males (Herz,

Folman, and Drori 1969). Not all results are in agreement with these, however. Purvis and Haynes (1972) found that male rats living for 98 days with females had lower LH and testosterone levels than isolated males, but prostrate gland, epididymes, and seminal vesicle weights did not differ.

Plasma testosterone levels of male rats housed with females from weaning until 60 or 180 days of age are higher than those of males reared in isolation or in unisexual groups. Heterosexually reared males also have lower rates of testosterone aromatization in the brain than isolates or unisexually reared males (Dessi-Fulgheri, Lupo di Prisco, and Verdarelli 1976; Lupo di Prisco, Lucarini, and Dessi-Fulgheri 1978). These results may be due to an inhibition of sexual development in the isolated and unisexually reared males, but it is more likely that heterosexual rearing results in advanced sexual development and increased stimulation of the neuroendocrine system of adult males. Olfactory bulbectomy, for example, reduces the testosterone levels of sexually experienced males housed with females but has no effect on testosterone levels of sexually inexperienced males (Dessi-Fulgheri, Dahlof, Larsson, Lupo di Prisco, and Tozzi 1980).

Contact with adult females is not necessary to produce increases in testosterone levels in male rats. Exposing unisexually reared males to bedding soiled by adult females from 25-65 days of age increases their testosterone levels and reduces the rate of testosterone aromatization in the brain. Exposure to bedding soiled by adult male rats is not as effective at changing hormone levels in these males as exposing them to female-soiled bedding (Dessi-Fulgheri and Lupo 1982).

Long-term exposure to females or their odours is not necessary to produce neuroendocrine responses in males. Four days of exposure to a novel female presented behind a wire-mesh barrier will result in an increase in LH and testosterone levels and an increase in prostate gland, epididymes, and seminal vesicle weights in males housed in unisexual groups or cohabiting with females. Socially isolated males have an increase in LH levels but not in the weights of their reproductive organs (Purvis and Haynes 1972).

When male Wistar rats are exposed to females, but separated from them by a wire-mesh barrier, they show an increase in testosterone levels within five minutes. If they are then allowed to mate, they show a peak in plasma testosterone within 10 minutes of exposure to the female whereas if they are not allowed to mate, but kept separated from the female, testosterone levels peak after 30-60 minutes of exposure to the female (Bliss *et al.* 1972; Purvis and Haynes 1974).

Sexually experienced male rats show an increase in testosterone, LH, and prolactin levels when exposed to females whether or not they are receptive, but sexually naive males have much less pronounced hormonal responses to female stimuli (Kamel, Mock, Wright, and Frankel 1975).

The presence of the female, however, is not necessary to stimulate hormonal secretion. Twenty minutes exposure to the odour of female rats produces a significant increase in LH levels in sexually experienced male rats and a large but not statistically significant increase in testosterone levels, but no increase in prolactin or FSH levels (Kamel, Wright, Mock, and Frankel 1977). Four days of exposure to female urine, however, results in an elevation of testosterone levels in both sexually experienced and naive male rats (Purvis and Haynes 1978).

Immediately after mating, male rats have elevated levels of testosterone, LH, and prolactin (Kamel *et al.* 1975), but these then decline after an hour of mating (Bliss *et al.* 1972; Kamel *et al.* 1977). Thus, male rats have increased hormone levels when exposed to novel females, a second elevation during copulation and then a decline in hormone levels after continued copulation. Sexually experienced male rats have increased hormone levels when they are placed in arenas in which they have previously mated (Kamel *et al.* 1975) and this hormonal response may have been conditioned to olfactory or visual cues of the test arena associated with mating. Male rats can be conditioned to respond to the odour of methyl salicy-late with increases in LH and testosterone levels by pairing the methyl salicylate with a receptive female for 14 daily trials in a Pavlovian conditioning paradigm (Graham and Desjardins 1979, 1980).

Like rats, male hamsters show hormonal responses to female odours. Introducing an oestrous female hamster or a bottle containing female vaginal discharge into a male hamster's home-cage for 30 min significantly increases his plasma testosterone level (Macrides, Bartke, Fernandez, and D'Angelo 1974), but female stimulation of the reproductive organs of male hamsters depends on the day-night cycle. The presence of an oestrous female for 18 h has no effect on male testes or seminal vesicle weight during long days or short days, but twice weekly presentations of a female or bedding soiled by a female during an eight-week period of short days causes an accelerated growth of testes, seminal vesicles, and flank glands when the males are switched to long days (Vandenbergh 1977). Thus, in hamsters, it appears that female odours must be present during the transition from short to long days in order to stimulate the male reproductive system.

Male house mice which are housed in isolation or in cohabitation with a female show increased testosterone levels within 60 minutes of the presentation of a novel female (Bartke and Dalterio 1975; Macrides, Bartke, and Dalterio 1975). During mating, male mice show rapid elevations in LH within five minutes of female presentation and males which have long ejaculation latencies shown a second LH peak after ejaculation (Coquelin and Bronson 1980).

Placing oestrus females behind a wire mesh barrier causes an increase in testosterone level in males of some strains of mice, but not in others. Males of strains which have high or medium rates of sexual activity (*DBA/2J, C57BL/6Fa*, and the (*C57 X DFA*) *BD* F_1 hybrid) show significant increases in testosterone levels in the presence of a female, but males from strains which have low rates of sexual activity (*Balb/c* and *CBA/H*) fail to show increased testosterone levels in the presence of females (Batty 1978). Testosterone levels are higher in males which copulate than in non-copulators and are highest at the initiation of mounting then decline during copulation (Batty 1978).

When male *CF-1* mice are exposed to female urine they show a dramatic elevation of serum LH within 15 minutes and a lesser elevation of FSH levels (Maruniak and Bronson 1976). The elevated LH levels persist for over two hours and are slightly, though not significantly, higher in sexually experienced than in sexually naive males. The magnitude of LH release increases with the duration of urine-exposure; LH levels are higher after 30 min of exposure than after 2 or 10 min although a significant rise in LH occurs after 10 min (Maruniak and Bronson 1976). Neither urine from female hamsters, nor urine from male mice produces an increase in LH but urine from intact female mice, whether in oestrus or dioestrus, from ovariectomized females and from pregnant females will all stimulate LH surges in male mice (Maruniak and Bronson 1976). Whether it does so through the primary olfactory receptors, the vomeronasal system or through other receptors, there is evidence that female urine contains an odour which activates the male's neuroendocrine system. While the female odour does not appear to depend on ovarian hormones, the chemical nature of this odour appears to be unknown.

Some aspects of the male's LH response appear to involve identification of the individual female. If a male mouse is exposed to a dioestrous female on three occasions, the LH response decreases on each presentation. If a novel female is then introduced, the LH response is greatly enhanced, but if the original female is re-introduced the LH response is very low (Coquelin and Bronson 1979). This habituation effect may explain why male rats cohabiting with females show no higher levels of LH than isolated males (Purvis and Haynes 1972) but both show acute increases in response to novel females.

The LH response in males can therefore be seen to be closely associated with social factors: the introduction of a novel female stimulates LH, continued cohabitation produces a habituation of the LH response, and presentation of a novel female produces another LH surge. Copulation produces a rise in LH after ejaculation and stimuli associated with mating acquire a stimulatory value through conditioning and can then cause an elevation of LH. It appears that the most effective female stimulus is her odour, but conclusive studies on conditioning hormonal responses to female odours have yet to be completed.

Who benefits from induced reproductive behaviour in males?　According to Bronson and Coquelin (1980, p. 251), 'one would expect little if any effect of a transiently elevated LH titer on latency to mate or on ejaculatory potency in a normally active male', but Harding (1981) has suggested a number of functions for acute changes in male hormone levels. Increased gonadotrophin levels, for example, may be responsible for increasing sexual behaviour. While males which fail to mate also fail to show increases in gonadotrophin and testosterone levels *after* testing, non-mating males do show increases in these hormones when first exposed to females, suggesting that an individual animal's previous sexual experience may determine his hormonal response to a female (Harding 1981).

Examples of correlations between rapid changes in aggressive behaviour in mice and changes in ACTH, vasopressin, and testosterone (Nock and Leshner 1976; Roche and Leshner 1979) indicate that changing an animal's hormone levels during social interactions can affect ongoing behaviour as well as influence 'social learning', i.e., the animal's response to similar situations in the future (Harding 1981). Thus, hormone changes may be more important in the modulation of an individual's ongoing social behaviour than was previously thought.

Bronson (1976*b*, 1979; Bronson and Coquelin 1980) has attempted to develop a coherent theory of the adaptive function of the primer effects of rodent social odours. This theory, as depicted in Fig. 8.16, is that 'the mouse's

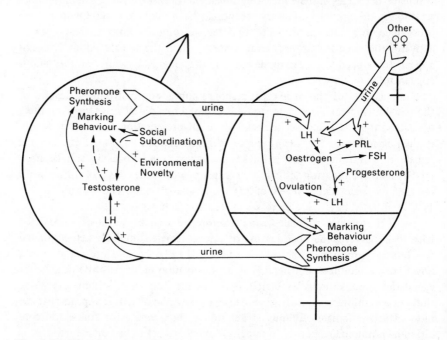

Fig. 8.16. The pheromonal cueing system of the house mice. + indicates stimulation, − indicates inhibition. (From Bronson (1979). © State University of New York.)

system of pheromonal cueing is designed strictly for the temporal regulation of ovulation, pubertal or otherwise' (Bronson 1979, p. 284). This viewpoint suggests that the male's odour is essential for the proper organization of puberty in females and for the induction of ovulation in adult females. The function of female odours which induce LH secretion in males is to 'increase pheromone synthesis in the male', and thus 'allow a mutual feedback loop, the objective of which is still the stimulation of ovulation in the female' (Bronson 1979, p. 284). In this view, the sole function of male odours is to induce oestrus and the function of female odours is to induce the male to produce and disseminate his odours, which will then induce oestrus in the female.

Activation of male hormone levels may be a female strategy for ensuring the presence of males during her oestrus period, as oestrous females scent-mark more than dioestrous females (Birke 1978). But since both animals will mate, it appears that both will benefit by increased reproductive success when the female stimulates male hormone levels.

Social odours affecting pregnancy

Pregnancy block: the Bruce effect

In 1959, Hilda Bruce reported that female albino mice which were placed with a strange male within 24 hours of mating and left for 7–10 days had their first pregnancy blocked, came into oestrus, mated with the new male and produced a new litter. If the second male was an albino (the same strain as the original stud male), 28 per cent of the females had blocked pregnancies. If a pigmented male mouse of the *CBA* strain was presented, however, pregnancy was blocked in 71 per cent of albino females. Removing and then replacing the original stud male after 24 hours or pairing with another female failed to cause pregnancy block. Only about 8 per cent of females not introduced to a second male return to oestrus 4–5 days after the first mating (Bruce 1959, 1960*a*). Using eye colour as a genetic marker, it was shown that all females who mated with albino males and then had their pregnancies blocked by and mated with the pigmented male had black eyed offspring while all females whose pregnancies were not blocked had albino offspring (Bruce 1959, 1960*b*). When pregnancy block occurs, vaginal cornification and the return to oestrus happens 3–4 days after the introduction of male stimuli (Bruce 1961, 1963*a*). Females can be mated and have their pregnancy blocked from 2–12 times in succession, and then give birth to a litter when the final pregnancy is left undisturbed (Bruce 1962).

Both the latency and duration of exposure to this male influence his effectiveness in blocking pregnancy. For the maximum pregnancy block (85 per cent) exposure to the strange male must occur within 24 hours of mating; exposure to strange males two, three, or four days after mating produces pregnancy block in 65 per cent of the females, exposure on day 5 results in a 40 per cent pregnancy block and exposure after seven days produces no block to pregnancy (Bruce 1961). Exposure to the strange male must occur for at least 48 hours in order to block the maximum number of pregnancies (76–81 per cent). Exposure for 12 or 24 hours results in 46 and 52 per cent pregnancy block, respectively (Bruce 1961). However, if females are exposed to strange males for three 15-min periods per day for four days, significant pregnancy block (50 per cent) occurs (Chipman,

Holt, and Fox 1966). Female deer mice exposed to strange males for two hours after mating also have their pregnancies blocked (Dewsbury 1982; Dewsbury and Baumgardner 1981).

Contact between the female and the strange male is not necessary as females housed in small cages within the cage of a novel male will show the same degree of pregnancy block as those able to contact males (Bruce 1960b). Exposure to bedding soiled by strange males will induce the same extent of pregnancy block as the presence of an adult male (Parkes and Bruce 1962; see Fig. 8.17). Exposing female deermice to bedding soiled by strange males for two hours after they have mated will also cause pregnancy block (Dewsbury 1982).

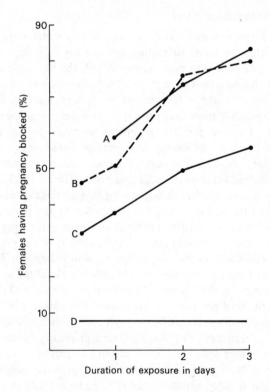

Fig. 8.17. Percentage of house mice with pregnancies blocked following: (a) placement for 1–3 days in boxes with male soiled bedding renewed twice a day; (b) exposure to alien males for 12 hours to three days; (c) placement for 12 hours to three days in boxes with male soiled bedding replaced once a day; (d) disturbance by transfer to clean boxes. (From Parkes and Bruce (1962), © The Biochemical Society.)

Increasing the number of strange males to which a pregnant female is exposed from 2–8 does not cause a greater degree of pregnancy block than exposure to one male (Bruce 1963b; Bronson and Eleftheriou 1963) but when females are exposed to male excreta, the more intense the male odour, the more pregnancy block occurs. Excreta from six males blocks pregnancy in 85 per cent of exposed females while only 58 per cent of females exposed to the excreta of one male have their pregnancy blocked (Chipman and Fox 1966b).

Increasing the duration of exposure to male odours also increases the number of females which have their pregnancies blocked. Twenty-one to 24 hours of exposure results in 46 per cent pregnancy block while 72 hours of exposure results in 63 per cent pregnancy block. If male odours are renewed every 24 hours for three days, 50 per cent of females show a pregnancy block but if the male odours are renewed every 12 hours for three days, 80 per cent of pregnancies are blocked (Parkes and Bruce 1962; see Fig. 8.17).

Research on the olfactory block to pregnancy has been concerned with a number of problems. These problems include the ability of females to discriminate between stud and alien males; determination of the source of the male's pregnancy blocking odour; the nature of the female's neuroendocrine response to male odours; and the variables which modify the female's susceptibility to pregnancy block.

Female discrimination between stud and novel males The Bruce effect depends on the female identifying the odour of novel males. In laboratory mice, males of the same strain as the stud male block 30 per cent of pregnancies; males of other strains block up to 80 per cent of pregnancies and the stud male does not block any pregnancies (Bruce 1960a,b; Parkes and Bruce 1961). In female deer mice, however, strange male deer mice block up to 56 per cent of pregnancies while male *Mus* (*C57BL/10J* strain) block only 32 per cent (Bronson and Eleftheriou 1963).

These results indicate that there is a difference in the female's response to males of different species versus those of different strains of the same species. In order to explain the female's ability to make such a discrimination Parkes and Bruce (1961, p. 1051) suggested that:

It is most unlikely that every male mouse produces a different odorous substance. Probably a 'spectra' of odours are involved, which differ slightly between individuals of the same strain and markedly between certain strains.

It is possible, therefore, that individual males of inbred strains are very difficult to differentiate because their odours are so very similar, whereas individual males of wild strains, which are more heterogenous, may be much easier to identify. Thus, females mated with a male of inbred strain A may have more pregnancies blocked if the alien male is from a second strain B because the strain difference may provide the female with a better basis to discriminate between the two males. Some strains may smell more alike, where others may smell much different (see Bruce 1963*b*). Males of a different species from the female, however, may provide an odour which is easy to discriminate and thus not effective in blocking pregnancies.

Males of some inbred strains do not appear to produce a pregnancy-blocking odour. In particular, males of the *C57BL* strain are low producers of odours (Chapman and Whitten 1969; Marchlewska-Koj 1977). Although Bartke and Wolff (1966) suggested that male yellow-lethal heterozygous mice (A^ya) do not produce odours which alter female hormone levels, Kakihana, Ellis, Gerling, Blum, and Kessler (1974) found that male A^ya mice produced pregnancy block in up to 80 per cent of the females exposed to these males.

Marsden and Bronson (1965*b*) reported failure to achieve pregnancy block in five strains of inbred mice (*C57BL/6J, CBA/J, CBA/Ca, SWR/J*, and *129/J*), but many experiments have used *CBA* males to produce pregnancy block (Chapman and Whitten 1969; Chapman, Desjardins, and Whitten 1970). Males of the *SJL/J, Balb/CWT*, and *Balb/cJ* strains have also been shown to produce pregnancy block (Chapman and Whitten 1969; Chapman *et al.* 1970). Using cross-breeding methods and testing backcross males for pregnancy-blocking ability, Chapman and Whitten (1969) concluded that the difference between *SJL/J* males and *C57BL/10Dg* males (which do not block pregnancies) is a genetic trait controlled by a single recessive gene.

As the difference between the stud and alien males increases, the number of females whose pregnancies are blocked increases. Since the stud male may be of a different strain than the female, and a male of the same strain as the female will block her pregnancy, it appears that some characteristic of the stud male affects the female's response to other males. Whether the original stud male is an albino, *CBA* or Dutch, the albino female is able to 'remember' his characteristics for at least 24 hours and identify a male of a different strain when he is introduced (Parkes and Bruce 1961).

Lott and Hopwood (1972) examined two hypothesis to account for this 'memory'. The *habituation hypothesis* states that a pregnant female will have her pregnancy blocked by any male, but because she is familiar with the stud male she is habituated to his odour and thus not responsive to him. The *sensitization hypothesis* states that a pregnant female will not have her pregnancy blocked by any male, but because she lives with the stud male, she becomes sensitive to other males. Using short (3-h) and long (24-h or more) exposure to stud males, Lott and Hopwood (1972) found that the longer the female was exposed to the stud male, the more pregnancy blocking occurred, suggesting that the female is sensitized by the stud male to respond to other males.

Another interpretation of these results, however, is that the female requires a certain amount of time to 'learn' the odour of the stud male. The more time the stud male is present, the better she learns his odour, and the better she can identify strange males. Using this *discrimination hypothesis*, one would predict that the longer the stud male was present, the easier it would be for the female to discriminate between closely related stud and alien males. Two other predictions from this hypothesis are that the less exposure the female has to the stud male, the more likely the stud male will be to cause pregnancy block and the more familiar the 'novel' male, the less likely it is that he will cause pregnancy block.

Parkes and Bruce (1961) found that when the female was pre-exposed to the alien male the frequency of pregnancy block dropped from 60 per cent to 40 per cent of females. Bloch (1974) has also examined the effect of familiarity of the 'novel' male. In her experiments females lived with the stud males (male A) for three weeks and were then mated at post-partum oestrus with a novel male (male B) for three days, after which the original stud male (male A), was used to test for pregnancy block. As would be expected by the discrimination hypothesis, the rate of pregnancy block in these females exposed to familiar males did not differ from that of females not exposed to an alien male (33 per cent versus 27 per cent of controls), but no control group was run to ensure that an alien male (male C), would block more pregnancies.

Bloch (1974) also found that if the stud male was with the female for only three days, and then removed for 24 hours and replaced, he produced 51 per cent pregnancy block. Thus, the shorter time that the stud male spends with the female, the more likely he is to produce pregnancy block. Lott and Hopwood (1972), on the other hand, found that stud males which were allowed to mate for only three hours and were then removed for 24 hours and replaced did not produce pregnancy block.

Research on pregnancy block in voles (see below) suggests that the female must learn and remember the characteristics of the stud male in order to recognize a novel male.

Pregnancy block in voles   Exposure to strange males after mating causes pregnancy block in field voles (*Microtus agrestis*), meadow voles (*Microtus pennsylvanicus*), prairie voles (*Microtus ochrogaster*), montane voles (*Microtus montanus*), and pine voles (*Pitymys pinetorum*) (Clulow and Clarke 1968; Clulow and Langford 1971; Stehn and Jannett 1981). Exposure to strange males for 24 hours is sufficient to block 75 per cent of pregnancies in field voles (Clulow and Clarke 1968). These females abort their first litter within 36 hours of exposure to a strange male and come into oestrus and mate within two days of aborting (Stehn and Richmond 1975).

The period of sensitivity to novel males varies among different species of voles. Some, like mice, are sensitive only before implantation occurs while others are sensitive throughout most of their pregnancy. Female prairie voles will respond to novel males for up to 17 days after mating (Stehn and Richmond 1975) but meadow voles fail to show pregnancy block if exposed to males 14 days after mating (Kenney, Evans, and Dewsbury 1977). Wild-trapped meadow voles do not appear to have their pregnancies blocked when exposed to strange males (Mallory and Clulow 1977). Sagebrush voles (*Lagurus curtatus*) have their pregnancies blocked if exposed to strange males before implantation occurs (Jannett 1979), but if exposed to strange males 8–14 days after mating, pregnancy block does not occur (Stehn and Jannett 1981).

When a female vole mates, the vaginal stimulation has two hormonal effects: ovulation is stimulated and a 'rhythm' of prolactin secretion is induced. This luteotrophic rhythm is maintained for up to 10 days after copulation in the vole (Milligan, Charlton, and Versi 1979) and also occurs in the rat, a spontaneous ovulator (Freeman, Smith, Nazian and Neill 1974). Within 48 hours of exposure to a strange male this luteotrophic rhythm is inhibited, the corpus-luteum of the female vole degenerates, Graafian follicles develop and the vagina shows cornification indicative of oestrus (Milligan 1976b; Mallory and Clulow 1977). Maintenance of the luteotrophic rhythm may act as a 'luteal mnemonic system' which enables the female to 'remember' the stud male (Freeman *et al.* 1974; Milligan *et al.* 1979). Since the corpus luteum of the vole is maintained by placental lactogen after implanation, rather than by prolactin, stimuli which produce pregnancy block after implantation occurs may have a different neuro-endocrine basis than those which produce preimplantation pregnancy block (see Milligan 1980).

Female field voles exposed to stud males for one hour or 48 hours both have their pregnancies blocked to the same extent when exposed to a new male and females made pseudopregnant by mechanical stimulation return to oestrus when exposed to males (Milligan 1979). These results suggest that the female is not sensitized by the stud male to respond to novel males, but must learn the characteristics of the stud male and discriminate between the stud and alien males. Female voles appear to be able to learn the characteristics of the stud male after one hour of exposure.

The particular characteristics which the female vole uses to discriminate between stud and alien males are testosterone dependent as castrated males fail to cause pregnancy block while testosterone-injected castrates are as effective as intact males (Milligan 1976a). An olfactory stimulus may be involved as females housed under cages of males have their pregnancies blocked, but exposure to bedding soiled by males, or to male urine and faeces is not sufficient to block pregnancies (Milligan 1976a). It is possible that secretions from the posteriolateral glands of male voles are necessary for olfactory stimulation of pregnancy block, but this has not been tested.

Pregnancy block in rats   Both Hughes (1964) and Davis and de Groot (1964) failed to demonstrate pregnancy block in rats. No other studies on rats seem to have been done.

The source of the male's pregnancy-blocking odour   Most of the evidence indicates that the pregnancy-blocking odour of the male mouse is an androgen-dependent component of the urine. Bruce (1960a; see also Parkes and Bruce

(1961)) indicated that castrated male albino mice produced a pregnancy block in as many females as intact males, but subsequent experiments (Bruce 1965) have shown that castrated males produce rates of pregnancy block which are significantly lower than produced by intact males.

Although submissive male mice have lower testes weights and testosterone levels than dominant males, both submissive and dominant males are able to cause pregnancy block at about the same rate (Labov 1981*a*). Thus, while the odours of dominant and subordinate males may differ (Carr, Martorano, and Krames 1970), their behaviour when paired with a female may not differ. A recurring problem with studies of the Bruce effect (Labov 1981*a*; Bloch 1971, 1973) is that, while an olfactory effect is postulated, the presence of the whole animal is used as a stimulus rather than urine or some other olfactory stimulus, and females may be responding to ultrasonic vocalizations or tactile and visual signals as well as olfactory cues.

Females exposed to bedding soiled by males or to male excreta have their pregnancies blocked (Bruce 1960*b*; Dominic 1966*b*; Parkes and Bruce 1961, 1962) whereas females with olfactory bulbectomy do not have their pregnancies blocked when exposed to male odours (Bruce and Parrott 1960). The urine of male *CBA* mice blocks up to 84 per cent of pregnancies in P-strain females while the presence of the male blocks up to 83 per cent. The urine of spayed-androgenized *CBA* females (testosterone implants) produces 81 per cent pregnancy block (Dominic 1965).

Exposure of pregnant females to male urine from a drip cage or to male urine applied to the nares results in 74-84 per cent pregnancy blockage. Urine stored for a week (using preservatives) blocks pregnancy in 54-64 per cent of females but after two weeks of storage, urine loses its ability to block pregnancy (Dominic 1966*b*). Exposure of females to male urine produces the same time course of pregnancy block as exposure to the presence of males: the majority of females return to oestrus four days after exposure to the male urine with fewer females returning to oestrus on days 3 and 5 (Dominic 1966*b*).

Injections of testosterone into male and female mice of the *SJL* strain, which produce normal amounts of the pregnancy blocking odour and into *C57BL/10Wt* males, which produce little odour, significantly increases the number of exposed females whose pregnancies are blocked (Hoppe 1975). A wide range of other androgens, including androstenedione, androsterone, and epiandrosterone increase the male's ability to block pregnancies (Hoppe 1975).

Attempts to determine the component of the male's urine which is active in blocking pregnancy suggest that it is a protein or a peptide of low molecular weight which may be bound to a protein (Marchlewska-Koj 1980*a*, 1981). Adult male mice have significantly higher levels of urinary proteins than females or castrated males and testosterone injections raise the level of urinary proteins in castrates to that found in intact males, indicating that the quantity, if not the quality of these urinary proteins is androgen dependent (Marchlewska-Koj

1977). Urinary proteins from both the excreted urine and bladder urine of intact males induce pregnancy block when dropped on the nares of females for three days, but proteins from castrated males or females have no effect (Marchlewska-Koj 1977). Male *C57BL/kw* mice produce less urinary protein than outbred males (Marchlewska-Koj 1977) and this may explain why these mice are not as effective as males of other strains in blocking pregnancies.

Whether or not the preputial gland is a source of a pregnancy blocking odour is unclear. Parkes and Bruce (1961) reported that preputialectomized males produced pregnancy block at the same rate as intact males, and *Tabby-J* male mice, which have no preputial glands, still secrete a pregnancy blocking odour (Hoppe 1975) indicating that the preputial gland is unnecessary for pregnancy block to occur. Marchlewska-Koj (1977), however, found that preputial gland homogenate from outbred male mice did produce pregnancy block.

The female's neuroendocrine response to the alien male   Ablation of the olfactory bulbs abolishes the pregnancy block effect in female mice (Bruce and Parrott 1960) but this procedure abolishes both the main and accessory olfactory bulbs as well as non-sensory tissue. Excision of the vomeronasal organ alone blocks the Bruce effect, indicating that the neuroendocrine response to the male's odour involves the vomeronasal–accessory olfactory system (Bellringer, Pratt, and Keverne 1980).

When pregnancy is blocked, implantation of the blastocyst fails to occur, the corpora lutea do not develop and no progesterone is secreted (Bruce 1960*a*; Dominic 1970; Parkes and Bruce 1961). This appears to be caused by reduced secretion of luteotropic hormones. If pregnant female mice are given prolactin injections after mating and exposed to strange males at the same time, pregnancy block occurs in very few females (Bronson, Eleftheriou, and Dezell 1969; Bruce and Parkes 1960; Dominic 1966*c*; Marchlewska-Koj and Jemiolo 1978). Lactating females, which have high prolactin levels, do not have their pregnancies blocked by strange males (Bruce and Parkes 1961).

Other evidence further supports this luteotropic hormone theory. Pregnant females having ectopic pituitary grafts fail to have their pregnancies blocked when exposed to male urine (Dominic 1967) and injections of reserpine, a drug which stimulates prolactin secretion, prevents pregnancy block (Dominic 1966*a,c*). Pimozide, another drug which stimulates prolactin release, also prevents pregnancy block by strange males (Marchlewska-Koj and Jemiolo 1978) and injection of bromocriptine, which inhibits prolactin secretion, produces a block to pregnancy (Bellringer *et al.* 1980; Marchlewska-Koj and Jemiolo 1978).

Drugs which decrease prolactin levels (e.g. bromocriptine) act by stimulating the neurotransmitter dopamine in the hypothalamus while drugs which increase prolactin (e.g. pimozide and reserpine) act by inhibiting the action of dopamine in the hypothalamus. Thus, one way in which the male can cause pregnancy block is by stimulating an increase in dopamine secretion which stimulates the release of hypothalamic prolactin release inhibiting factor (PIF) and this inhibits the release of prolactin from the adenohypophysis (Fig. 8.18).

Lack of prolactin fails to maintain progesterone secretion from the corpus luteum and without progesterone the uterus cannot accept the embryo or allow implantation to occur. Females given progesterone injections before being exposed to alien males are less likely to have their pregnancies blocked than control females (Dominic 1966c; Bruce and Parkes 1960) indicating that both prolactin and progesterone secretion is inhibited in females whose pregnancies are blocked.

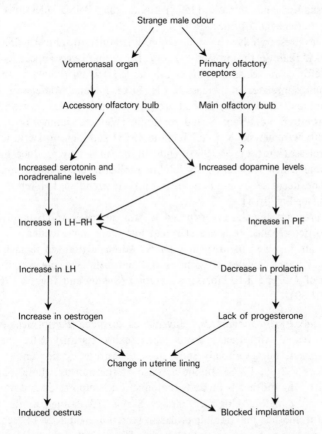

Fig. 8.18. Possible neuroendocrine mechanisms underlying the Bruce effect. See text for explanation.

There is, also, a second possible neuroendocrine mechanism underlying pregnancy block based on the fact that females come into oestrus immediately after their pregnancy is blocked. Chapman *et al.* (1970) found that pregnant females had a depletion of pituitary LH within 24 hours of exposure to an alien male and an increase in pituitary prolactin levels after 48 hours of male exposure. This suggests that the alien male produces both a release of LH and an inhibition of prolactin secretion. Based on the hypothesis that pregnancy block results from a male-induced surge of LH which stimulates ovulation, rather than a male-induced inhibition of prolactin which causes failure of the corpus luteum, Hoppe and Whitten (1972) found that injections of pregnant mares serum gonadotrophin (PMSG) blocked pregnancy when given 48 hours after mating. Injections of other gonadotrophins (LH, FSH, HCG), however, failed to block pregnancy (Hoppe and Whitten 1972). In a partial replication of this study, Marchlewska-Koj and Jemiolo (1978) found that PMSG injections blocked pregnancy in female *CBA* mice.

Much less research has been done on this gonadotrophin theory of pregnancy block than on the prolactin theory. Dopamine levels appear to have the opposite effects on gonadotrophic hormones (LH) as on prolactin: an increase in dopamine stimulates an increase in LH (Fuxe, Ferland, Andersson, Eneroth, Gustafsson, and Skett 1978). Gonadotrophin secretion is also controlled by the neurotransmitters noradrenalin and serotonin (Wuttke, Honma, Lamberts, and Höhn 1980; Marchlewska-Koj and Jemiolo 1978) and high prolactin levels suppress LH release (Wuttke *et al.* 1980). Thus, an alien male may produce an increase in dopamine which simultaneously inhibits prolactin release and stimulates LH release. The decrease in prolactin may further stimulate LH release by disinhibition (see Fig. 8.18).

When pregnant females are exposed to strange males ACTH is released and plasma corticosterone levels are elevated, but there is no evidence that these hormones are involved in pregnancy block. Adrenalectomized females respond the same as intact females in both control and male-exposure conditions, and injections of ACTH fail to block implantation (Snyder and Taggart 1967; Bronson *et al.* 1969).

Female-induced pregnancy block   Juvenile cactus mice (*Peromyscus eremicus*) which are reared with their parents reach sexual maturity later than singly housed females, but eventually show normal oestrous cycling and mate with their fathers (Skryja 1978). In the absence of the mother, these matings are fertile, but if the mother is present, reproduction is suppressed and no juveniles give birth after mating with their fathers. Skryja (1978) suggests that the presence of the mother may inhibit prolactin secretion or induce LH secretion in the same manner as a strange male. Whether females other than the mother can produce pregnancy block and whether this is an olfactory effect have yet to be investigated.

Variables affecting the females' susceptibility to pregnancy block   All females do not seem equally susceptible to pregnancy block. There are differences due to species, age, previous sociosexual experience, exposure to other females, and non-specific disturbance. In an examination of the species differences,

Bronson, Eleftheriou, and Garick (1964) found that 62 per cent of pregnant deer mice had their pregnancies blocked by strange male deer mice, 55 per cent were blocked by male *Peromyscus gracilis* (an inter-fertile subspecies), and 44 per cent were blocked by male *Mus* (*C57* strain). Unfortunately, the control females had a 25 per cent pregnancy block and exposure to other female mice produced a 31-48 per cent pregnancy block. This is far higher than the 8-16 per cent found in controls by Bronson and Eleftheriou (1963) and the 8 per cent found by Bruce (1960a; see Parkes and Bruce 1961) when females were presented rather than males.

Changing the cages of pregnant female deer mice or moving these mice to new quarters during the first 24 hours after mating produces up to 70 per cent pregnancy failure while introduction of a strange male produces 70-85 per cent pregnancy block (Eleftheriou, Bronson, and Zarrow 1962). House mice (*Mus*) have pregnancies blocked in up to 84 per cent of females exposed to strange males, but 60-88 per cent of pregnancies are blocked if the females are moved and handled often (Chipman and Fox 1966a). Disturbing (handling) pregnant collared lemmings reduces the success of pregnancy to 10 per cent (Mallory and Brooks 1980). Wild females may, therefore, be more sensitive to changes in the physical environment during the pre-implantation period than laboratory mice.

The influence of parity and age of females on pregnancy block have often been confused. Bruce (1960a, p. 102), for example, stated that 'when parous females were used, pregnancy appeared to be less readily blocked by the presence of a strange male'. Chipman and Fox (1966b) found that virgin and parous females of the same age were equally susceptible to pregnancy blocking, but young females (2-3 months of age) were more susceptible to pregnancy block than older females (six months of age). Terman (1969b), found that pregnancy block was significantly more common in nulliparous female deer mice than in parous females. Unfortunately, the ages of these females differed, the nulliparous being 2-3 months old and the parous being over five months of age. Multiparous voles (which are older) are less likely to have their pregnancies blocked than primiparous voles (which are younger) (Stehn and Jannett 1981). Until this confounding age variable has been examined it can not be concluded that nulliparous females are more susceptible to pregnancy block than parous females.

When pregnant females are grouped before being exposed to male stimuli, there is a decrease in the number of females showing pregnancy block (Bruce 1963b). Since grouped, unmated females have increased prolactin levels (see p. 279), Bruce (1963b) suggested that stimuli from other pregnant females prevent the decline in prolactin levels caused by strange males. No further work seems to have been directed toward this finding.

Who benefits from pregnancy block? Bronson and Coquelin (1980) argue that because the neuroendocrine mechanism underlying pregnancy block appears to be the same as that required to induce ovulation; because it is difficult to visualize the occurrence of the conditions necessary to produce pregnancy block in a natural population; and because the adaptive functions suggested so far are 'not convincing', pregnancy block is merely a laboratory artefact. Stehn and Jannett (1981), who found pregnancy block in four species of voles, both laboratory reared and wild-caught, suggest that rather than being a laboratory artefact, pregnancy block may be a side-effect of an adaptive mechanism which selects for the rapid initiation of oestrus and acceleration of puberty in rodents.

Schwagmeyer (1979) examined three theories regarding the function of the Bruce effect and suggested that pregnancy block is a mechanism whereby a female can select one mate in preference to another. This hypothesis, however, implies that females may choose to prevent pregnancy block from occurring. One mechanism for such prevention is to maintain contact with the stud male (see pp. 312–13). According to Dawkins (1976) the Bruce effect may have evolved to the advantage of females deserted by their mates, thus enabling them to remate with a male who will invest in their offspring. Labov (1981b) has also postulated that the Bruce effect benefits females and suggests that 'minimizing the consequences of encountering infanticidal males may have served as a primary selective pressure for the evolution of the Bruce effect as a female strategy' (Labov 1981b, p. 363). Pregnancy block should thus occur in species in which males show paternal care for their own offspring, but kill or ignore the offspring of other males and in which a female alone is not as successful in rearing the pups as the male and female together (Schwagmeyer 1979). The Bruce effect may thus confer two advantages on the female: it reduces investment in young that would be killed by an alien male, and it reduces the time required to produce a new litter which the male would accept (Labov 1981b). To support these hypotheses, Labov (1980) showed that wild male house mice killed more unrelated than related pups, and that males that had mated killed fewer pups than males which did not mate. In many of the rodents in which the Bruce effect has been demonstrated the males also show paternal behaviour (see Hartung and Dewsbury 1979).

Pregnancy block in the wild may occur under high population densities where normal social structure breaks down or at the start of the breeding season when females may meet strange males. Under high population density, it may be advantageous for the female to have her pregnancy blocked because if the litter is born it is likely to suffer a high rate of mortality (Fairbairn 1977) so the female might conserve her reproductive effort until the population structure is more favourable to pup survival (Milligan 1980). Newly pregnant females would be most likely to encounter strange males if they left the original deme or if a male immigrant displaced the stud male (Rogers and Beauchamp 1976b).

Older females, which have lower reproductive value (potential to produce further litters; Trivers 1974) are less likely to have their pregnancies blocked than young females (Chipman and Fox 1966b; Stehn and Jannett

1981), so the advantage of pregnancy block to the female may decrease as she gets older.

The Bruce effect may also be a mechanism for post-copulatory male–male competition (Trivers 1972). Male–male competition for reproductive success may continue after mating in a number of ways. A male may kill the newborn pups of another male (infanticide); block the pregnancy caused by another male (Bruce effect); or protect a female he has inseminated and their offspring (paternal behaviour). Using collared lemmings, Mallory and Brooks (1978) found that if the stud male was introduced to the female 24 hours after the birth of their pups, he killed none of the pups but showed paternal behaviour so that 95 per cent of the pups survived to weaning. If strange males were introduced, however, the females were extremely aggressive. If they became less aggressive, the strange males killed up to 50 per cent of the pups. Pup killing may be advantageous for alien males in two ways: it eliminates the progeny of another male, and shortens the gestation period for pups conceived in the post-partum oestrus. The gestation period of females who mate with a male who does not commit infanticide is 28 days, while the gestation period of females whose first litter is killed by an alien male is 21 days. Collared lemmings have their pregnancies blocked if exposed to strange males and thus could increase their own reproductive success by avoiding infanticide and mating with the novel male.

Another way in which pregnancy block may function in male–male competition is to reduce the fertility rate of 'sneaky maters' (Dawkins and Krebs 1978). These are subordinate males who manage to mate when the dominant male cannot service all of the oestrous females because many females come into oestrus synchronously, or because of male–male conflicts during mating.

According to Schwagmeyer (1979, p. 933), however, pregnancy block 'could evolve as a trait exclusively advantageous to males only if females were somehow incapable of preventing its occurrence'. Females may be able to influence the occurrence of pregnancy block by avoiding novel males after mating or by approaching novel males. In this way, pregnancy block may be a female strategy for keeping stud males nearby, as the presence of stud males increases the number of successful pregnancies (Mallory and Brooks 1980).

A final hypothesis is that pregnancy block promotes exogamy (Bruce and Parrott 1960). The advantages gained by the female through pregnancy block include breeding with rare males thus producing outbred offspring which have genetic heterogeneity and which might be more viable in an unstable environment in which dispersal and colonization of new environments were important for survival. Because the female's identification of the novel male is dependent on the characteristics of the stud male, however, she may end up mating with more closely related males, so this exogamy hypothesis seems unlikely (see Schwagmeyer 1979, pp. 934-5). Females have their pregnancies blocked when exposed to the brothers of the stud males and even when exposed to their own brothers (Bloch and Wyss 1972).

Male facilitation of pregnancy

When the stud male remains with the female after insemination, there is a higher proportion of pregnancies than if the stud male is removed immediately after impregnation (Terman 1969*b*). Parkes and Bruce (1961, p. 1050) reported that the presence of the stud male during exposure to an alien male reduced the number of females showing pregnancy block. The presence of the stud male may prevent the female from contacting the new male or his odour and may induce hormonal changes which enhance insemination. Male facilitation of pregnancy has been found in mice, rats, and voles.

If PMSG-injected female mice of some strains are exposed to males which are prevented from physical contact by a wire-mesh screen, the number of females ovulating and the number of ova released are increased (Eleftheriou, Christenson, and Zarrow 1973; Marchlewska-Koj 1980*b*). Male odours increase the levels of uterine enzymes in these females and this may facilitate implantation so that more pups are born in each litter (Marchlewska-Koj 1980*b*; Marchlewska-Koj, Jemiolo, Wozniacka, and Kozlowski 1980).

Pregnant female deer mice which are exposed to the stud male or his soiled bedding for two hours are more likely to become pregnant than females left alone (Dewsbury 1982). The stud male may facilitate implantation by stimulating secretion of luteotropic hormones, but this has not been investigated.

Exposure of lactating females which become pregnant during post-partum oestrus to novel males facilitates implantation rather than causing pregnancy block (Bloch 1971). Implantation in lactating pregnant females is normally delayed until the first litter is weaned but oestrogen injection will cause immediate implantation (Bloch 1971). This result is the opposite of that found in non-lactating pregnant mice in which oestrogen injections block implantation. The presence of a strange male (and possibly his odour) after post-partum copulation therefore appears to stimulate FSH and oestrogen secretion. In non-lactating females this would result in pregnancy block, but in lactating females it accelerates implantation. This elevation of FSH and oestrogen levels may occur simultaneously with a decrease in prolactin levels (Bloch 1971, 1973).

The copulatory behaviour of the male rat can advance ovulation and increase the number of ova released by the female rat (Rogers 1971). The more intromissions achieved by male rats during copulation, the more likely the female will become pregnant as copulation induces a neuroendocrine reflex in the female rat which results in prolactin surges and increased progesterone levels required to maintain pregnancy. If there are few intromissions before ejaculation this neuroendocrine reflex may not be activated and the female is less likely to become pregnant (see review by Adler 1979).

The role of the odour of the stud male in maintaining pregnancy in the rat has not been investigated. Since male rats produce ultrasounds and urinate after ejaculation (Anisko, Adler, and Suer 1979) either or both of these may stimulate pregnancy.

In voles, the presence of the stud male after implantation also facilitates pregnancy. If male prairie voles remain with the female for four or more days after insemination, 95–100 per cent of the females produce litters, whereas removal of the male after 24 hours results in a conception rate of only 35 per cent (Richmond and Stehn 1976). In the montane vole, removal of the male two hours after copulation results in 33 per cent pregnancy, while placing the stud male in a wind tunnel and blowing his odours onto the inseminated female produces a significant increase in pregnancy (Kranz and Berger 1975; cited in Richmond and Stehn 1976).

The presence of the stud male may be necessary for the maintenance of the corpus luteum so that implantation can occur. In order for functional corpora lutea to develop in the vole, a prolactin reflex must be initiated by mating and this may be facilitated by male odours. Limited mating (one intromission) may not be sufficient to initiate prolactin secretion and the corpora lutea degenerate. Longer bouts of mating (many intromissions with ejaculation) initiate the prolactin reflex and a functional corpus luteum develops (Milligan 1975a; Milligan and MacKinnon 1976).

The benefits of enhanced pregnancy If the female can keep the stud male nearby after copulation she will enhance her probability of becoming pregnant and increase her litter size. Oestrous females may therefore produce odours that are attractive to males and allow them to copulate longer than is necessary for impregnation.

The amount of parental investment of males which do not show paternal behaviour is limited to sperm production and insemination. But even prolonged contact with the inseminated female may be considered as male parential investment because the longer the stud male (or his odour) is present, the more implantations will occur and the more likely the embryos will survive. Such prolonged contact with a female which he has inseminated may also guard against other males approaching the female and blocking the pregnancy.

The ability of the stud male to increase the fertility and fecundity of his mate will increase the fitness of both male and female, especially if the male shares in pup care. The optimal litter size for situations in which the male and female co-operate in pup care may be larger than the litter size which can be cared for by a female alone.

Social odours affecting lactation

Female mice 'primed' by the odours of 4–6-day-old pups show increased pup-licking in a standard test for maternal behaviour (Noirot 1969). It is possible that the odours from these pups alter hormone levels in the female and thus stimulate lactation. When pups suckle from a lactating female, prolactin is released and levels of pituitary prolactin decline. Grosvenor (1965) showed that placing 14-day-old Sprague-Dawley rat pups beneath a wire-mesh screen under their lactating mother caused the same drop in pituitary prolactin levels as suckling. Moltz, Levin, and Leon (1969) showed that post-partum females whose nipples were excised, so that nursing pups could not maintain prolactin secretion, had an elevated prolactin level for up to 16 days in response to the sight, sound, odour, or tactile stimuli from the pups. Mena and Grosvenor (1971) suggested that olfactory stimuli from the pups were more important than visual or auditory stimuli in eliciting prolactin release. Possible sources of pup odours include skin secretions, byproducts of digestion, and excretions of the infant (Cowley 1980).

There appears to be a range in the sensitivity of lactating females to sensory stimuli from their pups. Seven-day lactating primiparous females do not release prolactin to stimuli from seven-day-old pups but do release prolactin in response to 14-day-old pups. Day-14 lactating females release prolactin to their own or alien 14-day-old pups, but not to seven- or 21-day-old pups (Grosvenor, Maiweg, and Mena 1970). Twenty-one-day lactating females release prolactin in response to other lactating females and their pups if their own pups have been removed (Mena and Grosvenor 1972).

Grosvenor *et al.* (1970) suggested that between seven and 14 days post-partum the lactating female becomes conditioned to associate pup stimuli with suckling and thus prolactin release occurs to pup stimuli alone. Between 14 and 21 days post-partum the lactating female may become hyperresponsive, secreting prolactin not only to stimuli from pups but also to stimuli from other lactating females and their pups.

In order to investigate this phenomenon, Grosvenor and Mena (1973) exposed 14- or 21-day post-partum females to either their own pups or to other lactating females and their pups. Day-14 females released prolactin in response to their own pups placed below a wire-mesh screen under their cage, but did not respond to other lactating females. Day-21 lactating females, however, released prolactin to both their own pups and to other lactating females.

Although prolactin is stimulated in both 14- and 21-day post-partum females exposed to pup stimuli, milk secretion is stimulated only in 14-day mothers (Grosvenor and Mena 1973). On the other hand, the presence of other lactating females induced milk secretion in 21-day but not in 14-day lactating females (Grosvenor and Mena 1973). Thus, it appears that the mother–pup relationship has altered by day 21 so that the pup stimuli continue to elicit prolactin release, but no longer stimulate milk secretion. One suggestion is that 21-day old pups activate catecholamine secretion which blocks the stimulatory effects of prolactin so that milk is not synthesized (Grosvenor and Mena 1973).

Exposure of 21-day lactating females to their own pups for three or four minutes inhibits the milk secretion response to other lactating females (Grosvenor and Mena 1973). Using this paradigm, Grosvenor, Mena, and Whitworth (1977) found that olfactory and auditory stimuli from 21-day-old pups inhibited milk secretion while visual stimuli had no effect. While Grosvenor *et al.* (1977)

tend to believe that pup odours are the primary stimulus for activating prolactin release in 14-day post-partum females and inhibiting milk secretion in 21-day lactating females, Terkel, Damassa, and Sawyer (1979) demonstrated that ultrasonic vocalizations from seven-day-old pups would stimulate prolactin release in 5-19-day post-partum females. The relative effectiveness of pup odours and ultrasounds in stimulating prolactin release from lactating females has not been investigated.

Apart from the experiments of Deis (1968), there are few studies on the role of pup stimuli on the release of oxytocin in the lactating female. Deis found that stimuli from 9-14-day-old pups caused increased milk secretion in lactating females and he suggested that this was due to increased oxytocin levels, but hormone levels were not measured. Because deafened females did not increase milk secretion in response to pups, Deis (1968) concluded that auditory stimuli were most important. This study should be repeated in order to examine oxytocin levels directly and to compare the effectiveness of pup odours and ultrasounds.

In addition to increases in prolactin and oxytocin levels, lactating female rats show elevated plasma corticosterone levels when their pups are returned after a three-hour separation. Placing pups behind a wire-mesh barrier also elevates corticosterone levels in the mother. While blinded and deafened females respond to pups with corticosterone increases, those with olfactory bulbectomy show no corticosterone response (Zarrow, Schlein, Denenberg, and Cohen, 1972). The primary sensory modality for female corticosterone responses to pups thus appears to be olfactory. Lactating female rats and mice have a greater elevation in corticosterone levels when their pups have been shocked or handled than when the pups are simply separated from the mother (Hennessy, Hollister, and Levine 1981; Smotherman, Brown, and Levine 1977; Smotherman, Wiener, Mendoza, and Levine 1977). Whether or not this increased corticosterone response is due to changes in the odour of the pups has not been investigated.

The benefits of inducing lactation   Although an infant's rate of growth depends on genetic factors, the infant's ability to find, grasp, and suckle the nipple and the maternal behaviour of the mother, the most important factor determining growth rate may be the mother's efficiency in lactation (Cowley 1980). The infant's ability to stimulate hormone secretion in the mother may therefore be a way in which the infant can manipulate its mother for its own benefit, increasing the amount of milk available and thus increasing its growth rate and thus its likelihood of survival and reproduction.

The ability of the pups to influence the female's lactation period may result in a parent–offspring conflict. The cost of parental investment is measured only in terms of decreased ability to produce future offspring, thus an offspring, to compete with its parent, 'should attempt to *induce* more investment than the parent wishes to give' (Trivers 1974, p. 257). Since lactating females increase prolactin and corticosteroid secretions when pups are removed and then replaced, pups might manipulate the mother by leaving the nest or producing ultrasounds or odours of distress which stimulate hormones and thus milk secretion. The ability of infants to manipulate their mother in this way may help to balance the asymmetry inherent in parent–offspring conflicts (see Alexander 1974).

Offspring who induce more prolactin secretion from their mother compete in two ways: gaining more nutrition and delaying the production of a new litter, thus reducing the mother's fitness. The longer the female lactates, the longer the gestation period for her next litter if she became pregnant during post-partum oestrus (Mallory and Brooks 1978), and the longer before she comes into oestrus if she was not impregnated post-partum. Thus a pup who causes its mother to lactate longer may increase its own fitness at the cost of its mother's fitness. Post-partum oestrus may have evolved to overcome this mother–offspring conflict as a lactating pregnant female can invest in present and future offspring at the same time.

General discussion

Some of the olfactory primer effects discussed in this chapter have been studied extensively (puberty acceleration in females, inhibition of oestrus in grouped females, induction of oestrus in grouped females, and pregnancy block) while other primer effects have received less attention (puberty delay in males, puberty delay in females, stimulation of hormone release in adult males, and enhancement of pregnancy) and still others are merely suspected olfactory primer effects (acceleration of puberty in males; inhibition of reproduction in adult males, and induction of milk secretion in lactating females).

The study of these olfactory primer effects involves a number of common problems. These include the relative importance of olfaction versus other stimuli; the source and chemical structure of the odours which produce the primer effects; the nature of the olfactory receptors receiving these odours; the neuro-endocrine responses to the odours; developmental and experiential factors influencing the primer effects and, the role of primer effects in the animal's natural environment.

The importance of olfactory stimuli

In many of the primer effects contact with another animal is a more potent stimulus than an odour alone. This suggests that odours may act in conjunction with tactile, auditory, or visual cues to stimulate physiological changes. Advancing puberty in juvenile males occurs when an adult female is present, but not when the female's odour is the only stimulus (see Fig. 8.1). Likewise, induced pseudopregnancy has only been shown to occur when females can contact each other.

While the delay of puberty in both male and female rodents appears to be produced equally by contact or olfactory stimuli, puberty acceleration in females is faster if both contact and olfactory stimuli are present (see Fig. 8.4). Oestrus suppression in grouped females is maximal when females can contact each other and is not as great when grouped females are separated by partitions (Fig. 8.11). Similarly, the presence of the male is more potent than his odour at inducing oestrus in grouped females (Whitten 1956*a*). While female odours will induce acute increases of LH in males, there is no evidence that male odours alone will inhibit gonadotrophin levels in males; this appears to require contact stimulation. The presence of a male is more potent than the odour of a single male at inducing pregnancy block, but if the odours are collected from a number of males, and are thus more intense, they will cause pregnancy block as often as the presence of the male (Parkes and Bruce 1962; Bruce 1963*a*; Chipman and Fox 1966*b*; see Fig. 8.17). Odours alone do not appear to be as potent as pup contact at inducing prolactin secretion in lactating rats (Mena and Grosvenor 1971).

Very few studies have compared the roles of odours and ultrasonic vocalizations in the causation of 'pheromonal' effects. If this were done there might be a number of additive effects as were shown by Bronson and Maruniak (1975) for the interaction of male urine and tactile stimuli in accelerating puberty in female mice (see Fig. 8.9).

Source and chemical structure of odours producing primer effects

The search for the source and chemical structure of 'primer pheromones' has focused almost entirely on the components of the urine of male mice which are able to accelerate puberty, induce oestrus, and block pregnancies in females. Little attention has been paid to the chemicals in female urine or vaginal secretions which alter male hormone levels. Chemical studies of female hamster vaginal secretions for example, have stressed male attraction rather than primer effects (Singer, Macrides, and Agosta 1980). Likewise, the urine of pregnant and lactating females, which will accelerate puberty and oestrus in other females, has not been analysed for its active components. Little work seems to have been done on the source of the female odours which produce anoestrus in grouped females, nor has there been an examination of the source of the female odours which induce oestrus synchrony in female rats.

There have been no attempts to search for components of skin glands in voles and other rodents which produce primer effects. Innumerable rodent skin glands have been identified (see Chapter 9), and there is evidence that the chemical components of many of these glands show differences due to the age, sex, and reproductive state of the animals (see, for example, Stoddart 1973). Weir (1973) suggested that the secretion of the chin gland of the male cuis might be involved in oestrus induction, but no research has been directed at the study of primer effects caused by the secretion of this or any other skin gland.

Given that male urine will accelerate puberty, induce oestrus, and block pregnancies, how many different odours or components are produced? Is each primer effect caused by a specific chemical or are all three effects caused by the same chemical(s)? Bronson (1971, p. 352) suggested that the male mouse produces 'a single primer that induces estrus, accelerates the attainment of sexual maturity, and when coupled with strangeness, will induce estrus at the expense of implantation'. Monder *et al.* (1978) found that the lipid fraction of the male's urine induced oestrus in grouped females to a greater extent than whole urine, but was not as powerful in blocking pregnancies as whole urine and thus concluded that 'pregnancy block is mediated by different pheromones than those involved in estrus acceleration' (Monder *et al.* 1978, p. 451).

Marchlewska-Koj (1977), however, found that urinary proteins from male mice would accelerate puberty in juvenile females, induce oestrus in grouped females and block pregnancy in recently impregnated females. She suggested that one male pheromone is responsible for all three effects, depending on the physiological state of the female. While these results indicate that urinary proteins from male urine produce the three types of primer effects seen in females, there is no evidence that the *same* protein produces all three effects. The active components may be peptides (Vandenbergh *et al.* 1976; Marchlewska-Koj 1980*a*) or combinations of two or more compounds (Novotny *et al.* 1980).

Olfactory receptors mediating primer effects, their neural projections, and interaction with the neuroendocrine system

Olfactory receptors   The mammalian olfactory receptors consist of five different components which are interconnected in an unknown way. These components are the main olfactory epithelium; the vomeronasal or Jacobson's organ; the terminal endings of the nervus terminalis; and the septal organ of Masera. Each of these five morphologically distinct organs acts as an odour receptor and each receptor has separate pathways to different brain centres (Fig. 8.19). A historical review of research on these receptors is given by Graziadei (1977).

Very little is known about the role of the trigeminal nerves, the nervus terminalis, or the organ of Masera in odour detection and the control of primer effects. The trigeminal nerve appears to innervate both the olfactory mucosa and the vomeronasal organ and projects to the olfactory bulb, the anterior septum, and the nasopalatine canal. The trigeminal nerve is sensitive to low concentrations of many chemicals (Cain 1974; Silver and Moulton 1982) and may be important in reception of social odours. The connections and olfactory functions of the nervus terminalis are in dispute (see Wysocki 1979). One branch of the nervus terminalis follows the vomeronasal nerve while another innervates a vestibular organ in the anterior septum of the nose (Bojsen-Møller 1975). The septal organ of Masera is a small nerve which projects to the caudal area of the olfactory bulb and the accessory olfactory bulb (Bojsen-Møller 1975).

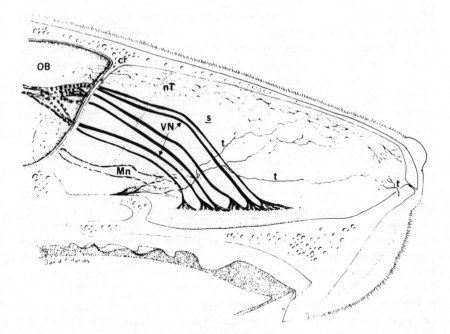

Fig. 8.19. Sagittal section through the nose of a rat at the level of the septum (s), to show the five nerve components discussed in the text. OB, olfactory bulb; nT, nervus terminalis; VN, vomeronasal nerve; t, branches of the trigeminal nerve; Mn, nerves originated from the septal organ of Masera. The region of the septum close to the olfactory bulb and lamina cribrosa (cr) which is densely shaded represents the olfactory region proper. (From Graziadei (1977). © Plenum Press.)

Adams and McFarland (1971) have described the septal olfactory organ in *Peromyscus* and suggest that it functions to determine food palatability.

The main olfactory neuroepithelium or 'primary olfactory system' and the vomeronasal organ or 'accessory olfactory system' have been more extensively studied. The olfactory epithelium contains receptor cells, supporting cells, and basal cells. Each receptor cell is a primary neuron which sends an axon directly to the olfactory bulb, without an intermediate synapse. These axons form the main olfactory nerve (Moulton and Beidler 1967). The epithelium of the vomeronasal organ resembles the olfactory epithelium and the vomeronasal receptor cells are primary neurons which project to the accessory olfactory bulb (Graziadei 1977).

To stimulate the vomeronasal organ animals must make contact with the odorous chemical, which may be a non-volatile, high-molecular-weight substance. To stimulate the main olfactory receptors, however, the animal need not contact the chemical directly, but need only inhale its volatile chemicals

(Winans and Powers 1977). Species which ingest urine, vaginal secretions, or glandular scretions of conspecifics may, therefore, be analysing these chemicals with the vomeronasal organ. Wysocki, Wellington, and Beauchamp (1980; see also Beauchamp, Wellington, Wysocki, Brand, Kubie, and Smith 1980), have shown that when male guinea pigs are presented with female urine labelled with the fluorescent dye, rhodamine hydrochloride, the dye is taken up by the vomeronasal organ. Thus the method of odour presentation may determine whether the vomeronasal organ, the main olfactory epithelium, or both are stimulated. If an animal cannot contact an odour source directly it may not be able to draw the chemical into contact with the vomeronasal organ and no primer effect will be observed.

Rather than forcing chemical stimuli into the vomeronasal organ using a flehmen reflex as occurs in ungulates and some carnivores (Estes 1972) the vomeronasal organ of the hamster acts like a pump. Efferent stimulation from the nasopalatine nerve causes both contraction (sucking in fluids) and dilation (expelling fluids) of the vascular tissue of the vomeronasal duct (Meredith 1980; Meredith and O'Connell 1979; Meredith, Marques, O'Connell, and Stern 1980).

In those studies where sensory receptors have been closely examined, there is increasing evidence that the vomeronasal system is involved in primer effects. The vomeronasal organ has been implicated in the acceleration of puberty in females (Kaneko, Debski, Wilson, and Whitten 1980), the inhibition of oestrus in grouped females (Reynolds and Keverne 1979), induced oestrus in female rats exposed to continuous light (Johns 1980; Johns *et al.* 1978), and the Bruce effect (Bellringer *et al.* 1980). There has been no study of the olfactory receptors mediating puberty delay in males; puberty delay in females; oestrus suppression in grouped females; nor the induction of prolactin secretion in lactating females by pup odours. The female's prolactin response to the odours of infants may be mediated by the vomeronasal system, as Fleming, Vaccarino, Tambosso, and Chee (1979) have shown that the vomeronasal organ is involved in maternal behaviour in the rat.

While the vomeronasal organ is important in mediating the male hamster's preference for female vaginal secretions (Powers, Fields, and Winans 1979) and in controlling male sexual arousal (Murphy and Schneider 1970; Powers and Winans 1975), there is, so far, no direct evidence that the vomeronasal organ mediates hormonal changes in males in response to oestrous female odours. Since volatile odours carried through a wind tunnel produce oestrus synchrony in female rats (McClintock 1981) and increase implantation in voles (Kranz and Berger 1975), the primary olfactory system rather than the vomeronasal receptors may mediate these effects.

Neural projections   The olfactory brain consists of the main and accessory olfactory bulbs, their afferent and efferent nerve pathways, the olfactory cortex, and much of the limbic system (Broadwell 1977). There is little information on the afferent and efferent connections of the nervus terminalis, the trigeminal nerve or the organ of Masera (see Wysocki 1979). The central connections of the main olfactory nerves and the vomeronasal nerves have been well documented and are shown in Fig. 8.20. As is evident from this figure, the olfactory system has two separate sets of neural projections: one associated with the main olfactory bulb and one with the accessory olfactory bulb (Raisman 1972; Scalia and Winans 1975, 1976; Keverne 1979; Barber and Raisman 1974).

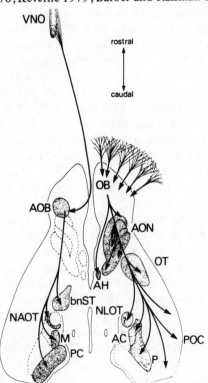

Fig. 8.20. Schematic diagram of a horizontal view of the rat brain comparing the independent primary and secondary projections (ipsilateral only) of the olfactory and vomeronasal systems (right and left halves, respectively). For clarity, secondary contralateral projections and many olfactory bulbar projections to rostral mideline structures are not shown. Abbreviations: AC = anterior cortical nucleus of the amygdala; AH = anterior hippocampus; AOB = accessory olfactory bulb; AON = anterior olfactory nucleus; bnST = bed nucleus of the stria terminalis; M = medial nucleus of the amygdala; NAOT = nucleus of the accessory olfactory tract; NLOT = nucleus of the lateral olfactory tract; OB = olfactory bulb; OT = olfactory tubercule; P = periamygdaloid region; PC = posterior cortical nucleus of the amygdala; POC = primary olfactory cortex; VNO = vomeronasal organ. (From Wysocki (1979), © ANKHO International Inc.)

Secondary and tertiary nerve pathways connect both the main and accessory olfactory bulbs to the hippocampus and the amygdala. The lateral olfactory tract (LOT) connects the main olfactory bulb to the olfactory peduncle, olfactory tubercule, lateral corticomedial amygdala, prepriform cortex, and ventrolateral entorhinal cortex. Through the medial forebrain bundle the olfactory tubercule and anterior olfactory nucleus are connected to the lateral hypothalamus and septal area. The lateral hypothalamus, hippocampus, and septal area are connected through the precommisural fornix (Broadwell 1977; Wysocki 1979). Many of the telencephalic and neocortical olfactory areas (e.g. anterior olfactory nucleus and olfactory cortex) receive their afferents from the ventral amygdala. There is very little innervation of the hypothalamus by the main olfactory pathways (Wysocki 1979).

The accessory olfactory tract connects the accessory olfactory bulb to the medial corticomedial amygdala and this is connected to the medial hypothalamus through the stria terminalis (Broadwell 1977; Wysocki 1979) (see Fig. 8.20). As well as these afferent pathways, there are efferent or centrifugal fibres from the nucleus of the lateral olfactory tract, the prepyriform cortex and the anterior septal area to the main olfactory bulb. The accessory olfactory bulb receives efferents from the corticomedial nucleus of the amygdala, the bed nucleus of the stria terminalis, and the bed nucleus of the accessory olfactory tract (Broadwell 1977; Wysocki 1979).

In summary, the main olfactory system and the vomeronasal system are relatively independent in their primary and secondary neural connections. The main olfactory system projects mainly to thalamic and telencephalic cortical areas while the accessory olfactory system is connected mainly to the amygdala, hypothalamus, and stria terminalis of the diencephalon.

Until the 1970s it was assumed that the main olfactory epithelium and the main olfactory bulb were responsible for the majority of olfactory influences on behaviour. In the last decade, however, evidence has accumulated to suggest that the vomeronasal organ and its neural projections are responsible for the neuroendocrine changes underlying the primer effects which are produced by social odours (Johns 1980).

Interest in the role of the vomeronasal organ in reproductive physiology was greatly stimulated by Estes (1972) who provided a comparative study of mammalian vomeronasal organs and suggested that:

the mammalian vomeronasal organ functions primarily as a specialized chemoreceptor of excreted sex hormones and/or their breakdown products. As such, it is involved in the detection of oestrus and in the release, control and co-ordination of sexual activity (Estes 1972, p. 316).

More recent studies have focused on the involvement of the vomeronasal system with the primer effects of mammalian social odours (see reviews by Graziadei 1977; Wysocki 1979; Johns 1980).

The relationship between the olfactory receptors and the neuroendocrine system   Many of the neural areas receiving afferents from the accessory olfactory bulb are also involved in hormone release or reception. In the guinea pig, LH-RH is contained in cells of the medial basal hypothalamus, the medial preoptic area, the septal area (Silverman 1976; Silverman, Krey, and Zimmerman 1979). Fibre tracts containing LH-RH occur in the medial preoptic area and the medial basal hypothalamus, the stria terminalis, and the medial amygdaloid nucleus (Silverman 1976; Silverman *et al.* 1979). In the hamster, cell bodies containing LH-RH are found in the olfactory peduncle, the bed nucleus of the stria terminalis and the medial preoptic area. LH-RH-containing fibres are found in the main and accessory olfactory bulbs, especially in the area of the vomeronasal nerve (Phillips *et al.* 1980). Fibres of the stria terminalis and areas of the medial nucleus of the amygdala also contain LH-RH. These areas receive direct input from the accessory olfactory bulb and project to the preoptic anterior-hypothalamic area (Scalia and Winans 1975; Phillips *et al.* 1980).

Two of the major areas for testosterone receptors are the medial nucleus of the amygdala and the bed nucleus of the stria terminalis (Sar and Stumpf 1973; Sheridan 1979). Testosterone receptors also occur in the ventromedial hypothalamus, the medial preoptic area, the lateral septum, and the hippocampus (Sar and Stumpf 1973). There are oestrogen receptors in the hypothalamus, preoptic area, amygdala, and septum (McEwen, Denef, Gerlach, and Plapinger 1974). It is, therefore, not surprising that manipulation of oestrogen and testosterone levels can alter a rodent's sensitivity to odours (Pfaff and Pfaffman 1969; Pietras and Moulton 1974).

Corticosterone receptors are found in the hippocampus, amygdala, and medial preoptic area (McEwen *et al.* 1974). There are receptors for hypothalamic peptide hormones such as LH-RH, thyrotophin releasing hormone (TRH), and somatostatin in olfactory related brain areas. LH-RH receptors occur in the hypothalamus, preoptic area, and septum; TRH receptors in the preoptic area, septum, and the medial basal and ventromedial hypothalamus. Somatostatin receptors occur in the ventromedial hypothalamus (Moss 1979). There may also be ACTH receptors associated with olfactory brain areas as ACTH levels affect olfactory sensitivity (Henkin 1975).

These intricate relationships between the olfactory system and the neuroendocrine system demonstrate the interdependencies of the hormonal and 'pheromonal' systems. The projections of the olfactory pathways are particularly closely associated with the neural control of LH-RH and neural receptors for oestradiol and testosterone are closely associated with the olfactory projections, so it is not surprising that these hormones are those that are most often associated with olfactory primer effects.

Neuroendocrine responses to social odours

In those primer effects where the neuroendocrine responses have been identified, LH and prolactin appear to be the primary pituitary hormones involved. An elevation of LH has been shown to underlie puberty acceleration in juvenile females (Bronson and Stetson 1973); oestrus induction in grouped females (Bronson 1976a); and the return to oestrus that follows pregnancy block (Chapman *et al*. 1970).

Surges of LH also underlie male-induced ovulation in female voles (Dluzen *et al.* 1981) and in light-induced constant oestrus rats (Brown-Grant *et al.* 1973). The oestrus synchrony of group-housed female rats may be due to earlier LH surges, shortening the follicular phase of the oestrous cycle (McClintock 1981). In male mice, the odour of novel females causes an increase in LH levels (Maruniak and Bronson 1976). Decreases in LH levels underlie the prolonged dioestrus of grouped female mice (Bronson 1976a) and the inhibition of reproductive behaviour of subordinate males (Bronson 1973).

Changes in FSH secretion may also be associated with primer effects. In females whose puberty is accelerated by male stimuli there is a decrease in FSH and then a later increase (Bronson and Desjardins 1974a; see Fig. 8.8). FSH levels do not appear to change in group-housed anoestrus females (Bronson 1976a) nor in females with male-induced oestrus (Bronson 1976a).

Increased prolactin levels may underly the Lee–Boot effect, but no hormone assays appear to have been done on pseudopregnant grouped females. Anoestrus grouped females have elevated levels of prolactin (Bronson 1976a) and stud males may promote prolactin secretion in recently inseminated females (Milligan 1975a). Pup stimuli also elevate prolactin secretion in lactating females (Mena and Grosvenor 1971). Inhibition of prolactin secretion may cause male-induced oestrus in grouped females (Milligan 1980) and pregnancy block (Dominic 1966a,c).

Elevated levels of ACTH and adrenal corticosteroids have been found in submissive males exposed to stimuli from dominant males (Bronson 1979). Puberty delay in males may be due to increased corticosteroid levels but this has not yet been investigated nor have the hormonal changes mediating puberty acceleration in males exposed to adult females or in juvenile females which have delayed puberty.

Novelty and familiarity as factors influencing primer responses to odours

Genetic differences between species or strains of rodents, age, parity, social status, kinship relations between odour donors and receivers, and some aspects of developmental experiences all influence on animal's primer responses to some degree.

One of the most important factors determining primer responses to odours appears to be whether the odour donor is novel to the receiver or familiar. Some primer effects depend on novelty, others on familiarity. Novelty is impor-

tant for puberty acceleration, induced oestrus, activation of male sexual behaviour, and blocking pregnancy.

Familiarity of the odour receiver with the odour donor appears to be important for delay of puberty, oestrus suppression in grouped females, inhibition of
reproduction in subordinate males, increased implantation in pregnant females
caused by the stud male, and elevation of prolactin levels in lactating females.
All of these effects require prolonged exposure of the odour receiver to stimuli
from the same animal or animals.

That familiarity and novelty may underlie the primer effects indicates that
animals must always make discriminations betwen novel and familiar animals and
their odours. This may require discrimination between individuals as must occur
in the Bruce effect, between sexes as occurs in the Whitten effect, and between
colony members (relatives) and non-colony members (non-relatives) as occurs in
puberty acceleration and delay. An important future development in the study of
the primer effects will be to understand how these discriminations are made.

Primer effects under natural conditions

Rogers and Beauchamp (1976*b*) suggested that in natural rodent populations the
primer effects function to regulate population density and to regulate gene flow
between breeding units (demes). The study of primer effects in microtine
rodents has been directed primarily at understanding the cyclic fluctuations
in population density (Richmond and Stehn 1976). According to Mallory and
Clulow (1977, p. 16) 'it seems reasonable to hypothesize that pregnancy blockage is one of the density-dependent factors that decreases reproduction and
ultimately the recruitment of young animals to the populations. This phenomenon
could reach a climax during peak years and might contribute to a marked
decline in the next seasons population owing to the normal die-off of older
animals'. According to this view, primer effects which result in accelerated
puberty, induced oestrus and mating in males, and thus increase population size
may be important at low population densities while primer effects which result
in delay of puberty or inhibition of reproduction will decrease population size
and thus be important at high population density.

Since all of the primer effects deal with reproduction and parental behaviour,
they all concern selection or fitness of the sender and receiver. Whitten and
Bronson (1970, p. 320) noted that 'the ability to induce estrus in a female even
at the expense of another male's insemination must exert considerable selection
pressure'. The proper level of analysis of the function of primer effects in
natural populations is, therefore, to understand which individuals benefit by
releasing and responding to stimuli which produce primer effects. According
to Alexander (1974, p. 330) 'social behavior evolves because of effects upon the
reproductive competition of group members' and the odours producing the
primer effects play a prominent role in determining who is successful in this
reproductive competition.

Acknowledgements

I would like to thank Diane Elrick, Patti Lynch, Mary MacConnachie, and the librarians of the Macdonald Science Library, Dalhousie University for their help in the preparation of this chapter. I am grateful to the National Science and Engineering Research Council of Canada (Grant A7441) for supporting this work.

References

Adams, D. R. and McFarland, L. Z. (1971). Septal olfactory organ in *Peromyscus. Comp. Biochem. Physiol. Comp. Physiol.* **40**, 971-4.

Adler, N. T. (1979). On the physiological organization of social behavior: Sex and aggression. In *Handbook of behavioral neurobiology*, Vol. 3 *Social behavior and communication* (ed. P. Marler and J. G. Vandenbergh) pp. 29-71. Plenum Press, New York.

Agren, G. (1981). Two laboratory experiments on inbreeding avoidance in the Mongolian gerbil. *Behav. Process.* **6**, 291-7.

Albertson, B. D., Bradley, E. L., and Terman, C. R. (1975). Plasma progesterone concentrations in prairie deermice (*Peromyscus maniculatus bairdii*) from experimental laboratory populations. *J. Reprod. Fert.* **42**, 407-14.

Alexander, R. D. (1974). The evolution of social behavior. *A. Rev. Ecol. Syst.* **5**, 325-83.

Andervont, H. B. (1944). Influence of environment on mammary cancer in mice. *J. natn. Cancer Inst.* **4**, 579-81.

Anisko, J. J., Adler, N. T., and Suer, S. (1979). Pattern of postejaculatory urination and sociosexual behavior in the rat. *Behav. neural Biol.* **26**, 169-76.

Archer, J. E. (1969). Adrenocortical responses to olfactory social stimuli in male mice. *J. Mammal.* **50**, 839-41.

— (1970). Effects of aggressive behavior on the adrenal cortex in male laboratory mice. *J. Mammal.* **51**, 327-32.

Aron, C. (1975). Olfactory stimuli and their role in the regulation of estrous cycle duration and sexual receptivity in the rat. In *Olfaction and taste V* (ed. D. A. Denton and J. P. Coghlan) pp. 397-402. Academic Press, New York.

— (1979). Mechanisms of control of the reproductive function by olfactory stimuli in female mammals. *Physiol. Review* **59**, 229-84.

— Asch, G., and Roos, J. (1966). Triggering of ovulation by coitus in the rat. *Int. Rev. Cytol.* **20**, 139-72.

— and Chateau, D. (1971). Presumed involvement of pheromones in mating behaviour in the rat. *Horm. Behav.* **2**, 315-24.

— Roos, J., and Asch, G. (1970). Effect of removal of the olfactory bulbs on mating behaviour and ovulation in the rat. *Neuroendocrinology* **6**, 109-17.

— — and Roos, M. (1971). Olfactory stimuli and their function in the regulation of the duration of the oestrous cycle in the rat. *J. interdiscip. cycle Res.* **2**, 239-46.

Avery, T. L. (1969). Pheromone-induced changes in the acidophil concentration of mouse pituitary glands. *Science, NY* **164**, 423-4.

Ayer, M. L. and Whitsett, J. M. (1980). Aggressive behaviour of female prairie deermice in laboratory populations. *Anim. Behav.* **28**, 763-71.

Barber, P. C. and Raisman, G. (1974). An autoradiographic investigation of the projection of the vomeronasal organ to the accessory olfactory bulb in the mouse. *Brain Res.* 81, 21-30.

Bartke, A. (1974). Increased sensitivity of seminal vesicles to testosterone in a mouse strain with low plasma testosterone levels. *J. Endocr.* 60, 145-8.

— and Dalterio, S. (1975). Evidence for episodic secretion of testosterone in laboratory mice. *Steroids* 26, 749-56.

— and Wolff, G. L. (1966). Influence of the lethal yellow (A^y) gene on estrous synchrony in mice. *Science, NY* 153, 79-80.

Batty, J. (1978). Acute changes in plasma testosterone levels and their relation to measures of sexual behaviour in the male mouse (*Mus musculus*). *Anim. Behav.* 26, 349-57.

Batzli, G. O., Getz, L. L., and Hurley, S. S. (1977). Suppression of growth and reproduction of microtine rodents by social factors. *J. Mammal.* 58, 583-91.

Beauchamp, G. K., Wellington, J. L., Wysocki, C. J., Brand, J. G., Kubie, J. L., and Smith, A. B. (1980). Chemical communication in the guinea pig: urinary components of low volatility and their access to the vomeronasal organ. In *Chemical signals: vertebrates and aquatic invertebrates* (ed. D. Müller-Schwarze and R. M. Silverstein) pp. 327-39. Plenum Press, New York.

Bediz, G. M. and Whitsett, J. M. (1979). Social inhibition of sexual maturation in male prairie deermice. *J. comp. physiol. Psychol.* 93, 493-500.

Bekoff, M. (1981). Vole population cycles: kin selection or familiarity? *Oecologia, Berl.* 48, 131.

Bellringer, J. F., Pratt, H. P. M., and Keverne, E. B. (1980). Involvement of the vomeronasal organ and prolactin in pheromonal induction of delayed implantation in mice. *J. Reprod. Fert.* 59, 223-8.

Birke, L. I. A. (1978). Scent-marking and the oestrous cycle of the female rat. *Anim. Behav.* 26, 1165-6.

Bliss, E. L., Frishat, A., and Samuels, L. (1972). Brain and testicular function. *Life Sci.* 11, 231-8.

Bloch, S. (1971). Enhancement of on-time nidations in suckling pregnant mice by the proximity of strange males. *J. Endocr.* 49, 431-6.

— (1973). Nidation induced in mice during the lactational delay by the presence of strange males. *J. Endocr.* 57, 185-6.

— (1974). Observations on the ability of the stud male to block pregnancy in the mouse. *J. Reprod. Fert.* 38, 469-71.

— (1976). A progesterone-dependent pheromone of the female mouse. *Experientia* 32, 937-8.

— and Wyss, H. I. (1972). Blokierung der Gravidität bei der Maus durch den Geruch männlicher Wurfgeschwister. *Experientia* 28, 703.

Bojsen-Møller, F. (1975). Demonstration of terminalis, olfactory, trigeminal and perivascular nerves in the rat nasal septum. *J. comp. Neurol.* 159, 245-56.

Brain, P. F. (1972). Endocrine and behavioral differences between dominant and subordinate male house mice housed in pairs. *Psychonom. Sci.* 28, 260-2.

— and Nowell, N. W. (1971). Isolation versus grouping effects on adrenal and gonadal function in albino mice. I. The male. *Gen. comp. Endocr.* 16, 149-54.

Broadwell, R. D. (1977). Neurotransmitter pathways in the olfactory system. In *Society for neuroscience symposia*, Vol. 3 *Aspects of behavioral neurobiology* (ed. J. A. Ferrendelli) pp. 131-66. Society for Neuroscience, Bethesda, Maryland.

Bronson, F. H. (1968). Pheromonal influences on mammalian reproduction. In *Perspectives in reproduction and sexual behavior* (ed. M. Diamond) pp. 341-61. Indiana University Press, Bloomington.

— (1971). Rodent pheromones. *Biol. Reprod.* **4**, 344–57.

— (1973). Establishment of social rank among grouped male mice: relative effects on circulating FSH, LH, and corticosterone. *Physiol. Behav.* **10**, 947–51.

— (1975a). Male-induced precocial puberty in female mice: confirmation of the role of estrogen. *Endocrinology* **96**, 511–14.

— (1975b). A developmental comparison of steroid-induced and male-induced ovulation in young mice. *Biol. Reprod.* **12**, 431–7.

— (1976a). Serum follicle stimulating hormone, luteinizing hormone, and prolactin in adult ovariectomized mice bearing silastic implants of estradiol: responses to social cues. *Biol. Reprod.* **15**, 147–52.

— (1976b). Urine marking in mice: causes and effects. In *Mammalian olfaction, reproductive processes, and behavior* (ed. R. L. Doty) pp. 119–41. Academic Press, New York.

— (1979). The reproductive ecology of the house mouse. *Q. Rev. Biol.* **54**, 265–99.

— (1981). The regulation of luteinizing hormone secretion by estrogen: relationships among negative feedback, surge potential and male stimulation in juvenile, peripubertal and adult female mice. *Endocrinology* **108**, 506–16.

— and Chapman, V. M. (1968). Adrenal-oestrus relationships in grouped or isolated female mice. *Nature, Lond.* **218**, 483–4.

— and Coquelin, A. (1980). The modulation of reproduction by priming pheromones in house mice: speculations on adaptive function. In *Chemical signals: vertebrates and aquatic invertebrates* (ed. D. Müller-Schwarze and R. M. Silverstein) pp. 243–65. Plenum Press, New York.

— and Desjardins, C. (1971). Steroid hormones and aggressive behavior in mammals. In *The physiology of aggression and defeat* (ed. B. E. Eleftheriou and J. P. Scott) pp. 43–63. Plenum Press, New York.

— — (1974a). Circulating concentrations of FSH, LH, estradiol and progesterone associated with acute male-induced puberty in female mice. *Endocrinology* **94**, 1658–68.

— — (1974b). Relationships between scent marking by male mice and the pheromone-induced secretion of the gonadotrophic and ovarian hormones that accompany puberty in female mice. In *Reproductive behavior* (ed. W. Montagna and W. A. Sadler) pp. 157–78. Plenum Press, New York.

— and Dezell, H. E. (1968). Studies on the estrus-inducing (pheromonal) action of male deermouse urine. *Gen. comp. Endocr.* **10**, 339–43.

— and Eleftheriou, B. E. (1963). Influence of strange males on implantation in the deermouse. *Gen. comp. Endocr.* **3**, 515–8.

— — (1964). Chronic physiological effects of fighting in mice. *Gen. comp. Endocr.* **4**, 9–14.

— — (1965a). Relative effects of fighting on bound and unbound corticosterone in mice. *Proc. Soc. exp. Biol. Med.* **118**, 146–9.

— — (1965b). Adrenal response to fighting in mice: separation of physical and psychological causes. *Science, NY* **147**, 627–8.

— — and Dezell, H. E. (1969). Strange male pregnancy block in deermice: prolactin and adrenocortical hormones. *Biol. Reprod.* **1**, 302–6.

— — and Garick, E. I. (1964). Effects of intra- and interspecific social stimulation on implantation in deermice. *J. Reprod. Fert.* **8**, 23–7.

— and Marsden, H. M. (1964). Male-induced synchrony of estrus in deermice. *Gen. comp. Endocr.* **4**, 634–7.

— and Maruniak, J. A. (1975). Male-induced puberty in female mice: evidence for a synergistic action of social cues. *Biol. Reprod.* **13**, 94–8.

— — (1976). Differential effects of male stimuli on follicle-stimulating hormone, luteinizing hormone, and prolactin secretion in prepubertal female mice. *Endocrinology* 98, 1101–8.

— and Stetson, M. H. (1973). Gonadotropin release in prepubertal female mice following male exposure: a comparison with the adult cycle. *Biol. Reprod.* 9, 449–59.

— — and Stiff, M. E. (1973). Serum FSH and LH in male mice following aggressive and nonaggressive interaction. *Physiol. Behav.* 10, 369–72.

— and Whitten, W. K. (1968). Oestrus-accelerating pheromone of mice: assay, androgen dependency and presence in bladder urine. *J. Reprod. Fert.* 15, 131–4.

Brown, K. (1979). Chemical communication between animals. In *Chemical influences on behaviour* (ed. K. Brown and S. J. Cooper) pp. 599–649. Academic Press, London.

Brown, R. E. (1979). Mammalian social odors: a critical review. *Advances in the study of behavior*, Vol. 10 (ed. J. S. Rosenblatt, R. A. Hinde, C. Beer, and M. C. Busnel) pp. 103–62. Academic Press, New York.

Brown, S. M. and Lisk, R. D. (1978). Blocked sexual receptivity in grouped female golden hamsters, the result of contact induced inhibition. *Biol. Reprod.* 18, 829–33.

Brown-Grant, K., Davidson, J. M., and Grieg, F. (1973). Induced ovulation in albino rats exposed to constant light. *J. Endocr.* 57, 7–22.

Bruce, H. M. (1959). An exteroceptive block to pregnancy in the mouse. *Nature, Lond.* 184, 105.

— (1960a). A block to pregnancy in the mouse caused by proximity of strange males. *J. Reprod. Fert.* 1, 96–103.

— (1960b). Further observations on pregnancy block in mice caused by the proximity of strange males. *J. Reprod. Fert.* 1, 311–12.

— (1961). Time relations in the pregnancy-block induced in mice by strange males. *J. Reprod. Fert.* 2, 138–42.

— (1962). Continued suppression of pituitary luteotrophic activity and fertility in the female mouse. *J. Reprod. Fert.* 4, 313–18.

— (1963a). A comparison of olfactory stimulation and nutritional stress as pregnancy-blocking agents in mice. *J. Reprod. Fert.* 6, 221–7.

— (1963b). Olfactory block to pregnancy among grouped mice. *J. Reprod. Fert.* 6, 451–60.

— (1965). The effect of castration on the reproductive pheromones of male mice. *J. Reprod. Fert.* 10, 141–3.

— (1969). Pheromones and behavior in mice. *Acta neurol. belg.* 69, 529–38.

— and Parkes, A. S. (1960). Hormonal factors in exteroceptive block to pregnancy in mice. *J. Endocr.* 20, xxix–xxx.

— — (1961). The effect of concurrent lactation on the olfactory block to pregnancy in the mouse. *J. Endocr.* 22, vi–vii.

— and Parrott, D. M. V. (1960). Role of olfactory sense in pregnancy block by strange males. *Science, NY* 131, 1526.

Buhl, A. E., Hasler, J. F., Tyler, M. C., Goldberg, N., and Banks, E. M. (1978). The effects of social rank on reproductive indices in groups of male collared lemmings (*Dicrostonyx groenlandicus*). *Biol. Reprod.* 18, 317–24.

Cain, W. S. (1974). Contributions of the trigeminal nerve to perceived odor magnitude. *Ann. N.Y. Acad. Sci.* 237, 28–34.

Carr, W. J., Martorano, R. D., and Krames, L. (1970). Responses of mice to odors associated with stress. *J. comp. physiol. Psychol.* 71, 223–8.

Carrer, H., Asch, G., and Aron, C. (1973). New facts concerning the role played by the ventromedial nucleus in the control of estrous cycle duration and sexual receptivity in the rat. *Neuroendocrinology* **13**, 129–38.

Carter, C. S. (1973). Olfaction and sexual receptivity in the female golden hamster. *Physiol. Behav.* **10**, 47–51.

— Getz, L. L., Gavish, L., McDermott, J. L., and Arnold, P. (1980). Male-related pheromones and the activation of female reproduction in the prairie vole *(Microtus ochrogaster)*. *Biol. Reprod.* **23**, 1038–45.

Castro, B. M. (1967). Age of puberty in female mice: relationship to population density and the presence of adult males. *Ann. Acad. bras. Cienc.* **39**, 289–91.

Champlin, A. K. (1971). Suppression of oestrus in grouped mice: the effects of various densities and the possible nature of the stimulus. *J. Reprod. Fert.* **27**, 233–41.

— (1977). Strain differences in estrous cycle and mating frequencies after centrally produced anosmia in mice. *Biol. Reprod.* **16**, 513–16.

— Beamer, W. G., Carter, S. C., Shire, J. G. M., and Whitten, W. K. (1980). Genetic and social modifications of mating patterns of mice. *Biol. Reprod.* **22**, 164–72.

Chapman, V. M., Desjardins, C., and Whitten, W. K. (1970). Pregnancy block in mice: changes in pituitary LH, LTH, and plasma progesterone levels. *J. Reprod. Fert.* **21**, 333–7.

— and Whitten, W. K. (1969). The occurrence and inheritance of pregnancy block in inbred mice. *Genetics* **61**, 59. (Abstr.)

Chateau, D. and Aron, C. (1977). Does the hypothalamic ventromedial nucleus mediate the action of olfactory stimuli on estrous rhythm in the rat? *J. indiscip. cycle Res.* **8**, 297–300.

— Carrer, H., Roos, J., and Aron, C. (1973). Modifications de la durée du cycle oestral chez des rattes privées de leurs bulbes olfactifs et porteuses de lésions du noyau ventromédian. *C. r. Séanc. Soc. Biol. Filial.* **167**, 1964–8.

— Roos, J., and Aron, C. (1972). Action de l'urine mâle ou femelle provenant de rats normaux ou castrés sur la durée du cycle oestral chez la ratte. *C. r. Séanc. Soc. Biol. Filial.* **166**, 1110–13.

— — — (1976). Progesterone action on estrous rhythm in the rat following ventromedial nucleus lesions. *Neuroendocrinology* **21**, 157–64.

— — (1977). Neuroendocrine mediation of the effects of olfactory stimuli on estrous rhythm regulation in the rat. In *Olfaction and taste VI* (ed. J. LeMagnen and P. MacLeod) pp. 149–55. Information Retrieval, London.

— — Plas-Roser, S., Roos, M., and Aron, C. (1976). Hormonal mechanisms involved in the control of oestrous cycle duration by the odour of urine in the rat. *Acta endocr. Copenh.* **82**, 426–35.

Chipman, R. K. and Albrecht, E. D. (1974). The relationship of the male preputial gland to the acceleration of oestrus in the laboratory mouse. *J. Reprod. Fert.* **38**, 91–6.

— and Fox, K. A. (1966a). Oestrus synchronization and pregnancy blocking in wild house mice *(Mus musculus)*. *J. Reprod. Fert.* **12**, 233–6.

— — (1966b). Factors in pregnancy blocking: age and reproductive background of females: numbers of strange males. *J. Reprod. Fert.* **12**, 399–403.

— Holt, J. A., and Fox, K. A. (1966). Pregnancy failure in laboratory mice after multiple short-term exposure to strange males. *Nature, Lond.* **210**, 653.

Chitty, H. and Austin, C. R. (1957). Environmental modification of oestrus in the vole. *Nature, Lond.* **179**, 592–3.

Christian, J. J. (1955a). Effect of population size on the weights of the reproductive organs of white mice. *Am. J. Physiol.* 181, 477–80.
— (1955b). Effect of population size on the adrenal glands and reproductive
organs of male mice in populations of fixed size. *Am. J. Physiol.* 182,
292–300.
— (1971). Population density and reproductive efficiency. *Biol. Reprod.* 4,
248–94.
— and Davis, D. E. (1964). Endocrines, behavior and population. *Science, NY*
146, 1550–60.
Clulow, F. V. and Clarke, J. R. (1968). Pregnancy block in *Microtus agrestis*
and induced ovulator. *Nature, Lond.* 219, 511.
— and Langford, P. E. (1971). Pregnancy-block in the meadow vole, *Microtus
pennsylvanicus. J. Reprod. Fert.* 24, 275–7.
Colby, D. R. and Vandenbergh, J. G. (1974). Regulatory effects of urinary
pheromones on puberty in the mouse. *Biol. Reprod.* 11, 268–79.
Conaway, C. H. (1971). Ecological adaptation and mammalian reproduction.
Biol. Reprod. 4, 239–47.
Cooper, K. J. and Haynes, N. B. (1967). Modification of the oestrous cycle of
the underfed rat associated with the presence of the male. *J. Reprod. Fert.*
14, 317–20.
— Purvis, K., and Haynes, N. B. (1972). Further observations on the ability
of the male to influence the oestrous cycle of the underfed rat. *J. Reprod.
Fert.* 28, 473–5.
Coquelin, A. and Bronson, F. H. (1979). Release of luteinizing hormone in male
mice during exposure to females: habituation of the response. *Science, NY*
206, 1099–101.
— — (1980). Secretion of luteinizing hormone in male mice: factors that
influence release during sexual encounters. *Endocrinology* 106, 1224–9.
Cowley, J. J. (1980). Growth and maturation in mice (*Mus musculus*). *Symp.
zool. Soc. Lond.* 45, 213–50.
— and Wise, D. R. (1972). Some effects of mouse urine on neonatal growth
and reproduction. *Anim. Behav.* 20, 499–506.
Davidson, J. M. (1969). Feedback control of gonadotropin secretion. In *Frontiers
in neuroendocrinology* (ed. W. F. Ganong and L. Martini) pp. 343–88. Oxford
University Press, New York.
Davis, D. L. and de Groot, J. (1964). Failure to demonstrate olfactory inhibition
of pregnancy ('Bruce effect') in the rat. *Anat. Rec.* 148, 336. (Abstr.)
Dawkins, R. (1976). *The selfish gene.* Oxford University Press.
— and Krebs, J. R. (1978). Animal signals: information or manipulation? In
Behavioural ecology: an evolutionary approach (ed. J. R. Krebs and N. B.
Davies) pp. 282–309. Sinauer, Sunderland, Mass.
Deis, R. P. (1968). The effect of an exteroceptive stimulus on milk ejection in
lactating rats. *J. Physiol. Lond.* 197, 37–46.
Dessi-Fulgheri, F., Dahlof, L.-G., Larsson, K., Lupo di Prisco, C., and Tozzi, S.
(1980). Anosmia differently affects the reproductive hormonal pattern in
sexually experienced and inexperienced male rats. *Physiol. Behav.* 24, 607–11.
— and Lupo, C. (1982). Odour of male and female rats changes hypothalamic
aromatase and 5α-reductase activity and plasma sex steroid levels in unisexually reared male rats. *Physiol. Behav.* 28, 231–5.
— Lupo di Prisco, C., and Verdarelli, P. (1976). Effects of two kinds of social
deprivation on testosterone and estradiol-17β plasma levels in the male rat.
Experientia 32, 114–15.

Dewar, A. D. (1959). Observations on pseudopregnancy in the mouse. *J. Endocr.* 18, 186–90.

Dewsbury, D. A. (1982). A pregnancy block resulting from multiple-male copulation or exposure at the time of mating in deermice (*Peromyscus maniculatus*). In *Chemical signals III* (ed. D. Müller-Schwarze and R. M. Silverstein). Plenum Press, New York.

— and Baumgardner, D. J. (1981). Studies of sperm competition in two species of muroid rodents. *Behav. Ecol. Sociobiol.* 9, 121–33.

Dluzen, D. E., Ramirez, V. D., Carter, C. S., and Getz, L. L. (1981). Male vole urine changes luteinizing hormone-releasing hormone and norepinephrine in female olfactory bulb. *Science, NY* 212, 573–5.

Dominic, C. J. (1965). The origin of pheromones causing pregnancy block in mice. *J. Reprod. Fert.* 10, 469–72.

— (1966a). Reserpine inhibition of olfactory blockage of pregnancy in mice. *Science, NY* 153, 1764–5.

— (1966b). Observations on the reproductive pheromones of mice. I. Source. *J. Reprod. Fert.* 11, 407–14.

— (1966c). Observations on the reproductive pheromones of mice. II. Neuroendocrine mechanisms involved in the olfactory block to pregnancy. *J. Reprod. Fert.* 11, 415–21.

— (1967). Effect of ectopic pituitary grafts on the olfactory block to pregnancy in mice. *Nature, Lond.* 213, 1242.

— (1970). Histological evidence for the failure of corpus luteum function in the olfactory block to pregnancy in mice. *J. anim. Morph. Physiol.* 17, 126–30.

Donovan, B. T. and Kopriva, P. C. (1965). Effect of removal or stimulation of the olfactory bulbs on the estrous cycle of the guinea pig. *Endocrinology* 77, 213–17.

Drickamer, L. C. (1974a). Sexual maturation of female house mice: social inhibition. *Dev. Psychobiol.* 7, 257–65.

— (1974b). Contact stimulation, androgenized females and accelerated sexual maturation in female mice. *Behav. Biol.* 12, 101–10.

— (1975a). Contact stimulation and accelerated sexual maturation of female mice. *Behav. Biol.* 15, 113–5.

— (1975b). Female mouse maturation: relative importance of social factors and daylength. *J. Reprod. Fert.* 44, 147–50.

— (1976). Effects of size and sex ratio of litter on the sexual maturation of female mice. *J. Reprod. Fert.* 46, 369–74.

— (1977). Delay of sexual maturation in female house mice by exposure to grouped females or urine from grouped females. *J. Reprod. Fert.* 51, 77–81.

— (1979). Acceleration and delay of first estrus in wild *Mus musculus*. *J. Mammal.* 60, 215–16.

— (1981a). Selection for age of sexual maturation in mice and the consequences for population regulation. *Behav. neural Biol.* 31, 82–9.

— (1981b). Acceleration and delay of sexual maturation in female house mice previously selected for early and late first vaginal oestrus. *J. Reprod. Fert.* 63, 325–9.

— (1982). Acceleration and delay of the first vaginal oestrus in female mice by urinary chemosignals: dose levels and mixing urine treatment sources. *Anim. Behav.* 30, 456–60.

— and Assmann, S. M. (1981). Acceleration and delay of puberty in female house mice: methods of delivery of the urinary stimulus. *Dev. Psychobiol.* 14, 487–97.

— and Hoover, J. E. (1979). Effects of urine from pregnant and lactating female house mice on sexual maturation of juvenile females. *Dev. Psychobiol.* 12, 545–51.

— and McIntosh, T. K. (1980). Effects of adrenalectomy on the presence of a maturation-delaying pheromone in the urine of female mice. *Horm. Behav.* 14, 146–52.

— — and Rose, E. A. (1978). Effects of ovariectomy on the presence of a maturation delaying pheromone in the urine of female mice. *Horm. Behav.* 11, 131–7.

— and Murphy, R. X. Jr (1978). Female mouse maturation: effects of excreted and bladder urine from juvenile and adult males. *Dev. Psychobiol.* 11, 63–72.

Drori, D. and Folman, Y. (1964). Effects of cohabitation on the reproductive system, kidneys and body composition of male rats. *J. Reprod. Fert.* 8, 351–9.

Eleftheriou, B. E., Bronson, F. H., and Zarrow, M. X. (1962). Interaction of olfactory and other environmental stimuli on implantation in the deermouse. *Science, NY* 137, 764.

— Christenson, C. M., and Zarrow, M. X. (1973). The influence of exteroceptive stimuli and pheromonal facilitation of ovulation in different strains of mice. *J. Endocr.* 57, 363–70.

Estes, R. D. (1972). The role of the vomeronasal organ in mammalian reproduction. *Mammalia* 36, 315–41.

Fairbairn, D. J. (1977). Why breed early? A study of reproductive tactics in *Peromyscus. Can. J. Zool.* 55, 862–71.

Fleming, A., Vaccarino, F., Tambosso, L., and Chee, P. (1979). Vomeronasal and olfactory system modulation of maternal behavior in the rat. *Science, NY* 203, 372–4.

Folman, Y. and Drori, D. (1966). Effects of social isolation and of female odours on the reproductive system, kidneys and adrenals of unmated male rats. *J. Reprod. Fert.* 11, 43–50.

Fox, K. A. (1968). Effects of prepubertal habitation conditions on the reproductive physiology of the male house mouse. *J. Reprod. Fert.* 17, 75–85.

Freeman, M. E., Smith, M. S., Nazian, S. J., and Neill, J. D. (1974). Ovarian and hypothalamic control of the daily surges of prolactin secretion during pseudopregnancy in the rat. *Endocrinology* 94, 875–82.

Fullerton, C. and Cowley, J. J. (1971). The differential effect of the presence of adult male and female mice on the growth and development of the young. *J. genet. Psychol.* 119, 89–98.

Fuxe, K., Ferland, L., Andersson, K., Eneroth, P., Gustafsson, J.-Å., and Skett, P. (1978). On the functional role of hypothalamic catecholamine neurons in control of the secretion of hormones from the anterior pituitary, particularly in the control of LH and prolactin secretion. In *Brain–endocrine interaction III: Neural hormones and reproduction* (ed. D. E. Scott, G. P. Koslawski, and A. Weindl) pp. 172–82. Karger, Basel.

Gandelman, R., Zarrow, M. X., and Denenberg, V. H. (1972). Reproductive and maternal performance in the mouse following removal of the olfactory bulbs. *J. Reprod. Fert.* 28, 453–6.

Graham, J. M. and Desjardins, C. (1979). Conditioned secretion of luteinizing hormone and testosterone in male rats. *Fedn. Proc.* 38, 1107. (Abstr.)

— — (1980). Classical conditioning: induction of luteinizing hormone and testosterone secretion in anticipation of sexual activity. *Science, NY* 210, 1039–41.

Gray, G. D., Davis, H. N., Zerylnick, M., and Dewsbury, D. A. (1974). Oestrus and induced ovulation in montane voles. *J. Reprod. Fert.* **38**, 193-6.

Graziadei, P. P. C. (1977). Functional anatomy of the mammalian chemo-receptor system. In *Chemical signals in vertebrates* (ed. D. Müller-Schwarze and M. M. Mozell) pp. 435-54. Plenum Press, New York.

Grosvenor, C. E. (1965). Evidence that exteroceptive stimuli can release prolactin from the pituitary gland of the lactating rat. *Endocrinology* **76**, 340-2.

— Maiweg, H., and Mena, F. (1970). A study of factors involved in the development of the exteroceptive release of prolactin in the lactating rat. *Horm. Behav.* **1**, 111-20.

— and Mena, F. (1973). Evidence that suckling pups, through an exteroceptive mechanism, inhibit the milk stimulatory effects of prolactin in the rat during late lactation. *Horm. Behav.* **4**, 209-22.

— — and Whitworth, N. S. (1977). Sensory stimuli from pups involved in inhibition of milk secretion in rats during late lactation. *Horm. Behav.* **8**, 287-96.

Grumbach, M. M. (1975). Onset of puberty. In *Puberty: biologic and psycho-social components* (ed. S. R. Berenberg) pp. 1-21. Stenfert Kroese, Leiden.

Handelmann, G., Ravizza, R., and Ray, W. J. (1980). Social dominance determines estrous entrainment among female hamsters. *Horm. Behav.* **14**, 107-15.

Harding, C. F. (1981). Social modulation of circulating hormone levels in the male. *Am. Zool.* **21**, 223-31.

Hardy, D. F. (1970). The effect of constant light on the estrous cycle and behavior of the female rat. *Physiol. Behav.* **5**, 421-5.

Harned, M. A. and Casida, L. E. (1972). Failure to obtain group synchrony of estrus in the guinea pig. *J. Mammal.* **53**, 223-5.

Hartung, T. G. and Dewsbury, D. A. (1979). Paternal behavior in six species of muroid rodents. *Behav. neural Biol.* **26**, 466-78.

Hasler, J. F. and Banks, E. M. (1975). The influence of mature males on sexual maturation in female collared lemmings (*Dicrostonyx groenlandicus*). *J. Reprod. Fert.* **42**, 583-6.

Hasler, M. J. and Conaway, C. H. (1973). The effect of males on the reproductive state of female *Microtus ochrogaster*. *Biol. Reprod.* **9**, 426-36.

— and Nalbandov, A. V. (1974). The effect of weanling and adult males on sexual maturation in female voles, *Microtus ochrogaster*. *Gen. comp. Endocr.* **23**, 237-8.

Henkin, R. L. (1975). Effects of ACTH, adrenocorticosteroids and thyroid hormone on sensory function. In *Anatomical neuroendocrinology* (ed. W. E. Stumpf and L. D. Grant) pp. 298-316. Karger, Basal.

Hennessy, M. B., Hollister, T. A., and Levine, S. (1981). Pituitary-adrenal responsiveness of mothers to pups in mice of two inbred strains. *Behav. neural Biol.* **31**, 304-13.

Herz, Z., Folman, Y., and Drori, D. (1969). The testosterone content of the testes of mated and unmated rats. *J. Endocr.* **44**, 127-8.

Hill, J. L. (1974). Peromyscus: effect of early pairing on reproduction. *Science, NY* **186**, 1042-4.

Ho, H. and Wilson, J. R. (1980). Genetic and hormonal aspects of female facilitation of PMSG-induced ovulation in immature mice. *J. Reprod. Fert.* **59**, 57-61.

Hoppe, P. C. (1975). Genetic and endocrine studies of the pregnancy-blocking pheromone of mice. *J. Reprod. Fert.* **45**, 109-15.

— and Whitten, W. K. (1972). Pregnancy block: imitation by administered gonadotropin. *Biol. Reprod.* 7, 254–9.

Hoover, J. E. and Drickamer, L. C. (1979). Effects of urine from pregnant and lactating female house mice on estrous cycles of adult females. *J. Reprod. Fert.* 55, 297–302.

Horton, L. and Shepherd, B. A. (1977). Olfactory bulbectomy and estrus induction in *Microtus ochrogaster*. *Am. Zool.* 17, 925. (Abstr.)

Hughes, R. L. (1964). Effect of changing cages, introduction of the male and other procedures on the oestrus cycle of the rat. *CSIRO wildl. Res.* 9, 115–22.

Jannett, F. J. (1979). Experimental laboratory studies on the interactions of sympatric voles (*Microtinae*). *Am. Zool.* 19, 966. (Abstr.)

Johns, M. A. (1980). The role of the vomeronasal system in mammalian reproductive physiology. In *Chemical signals: vertebrates and aquatic invertebrates* (ed. D. Müller-Schwarze and R. M. Silverstein) pp. 341–64. Plenum Press, New York.

— Feder, H. H., and Komisaruk, B. R. (1980). Reflex ovulation in light-induced persistent estrus (LLPE) rats: role of sensory stimuli and the adrenals. *Horm. Behav.* 14, 7–19.

— — — and Mayer, A. D. (1978). Urine-induced reflex ovulation in anovulatory rats may be a vomeronasal effect. *Nature, Lond.* 272, 446–8.

Kakihana, R., Ellis, L. B., Gerling, S. A., Blum, S. L., and Kessler, S. (1974). Bruce effect competence in yellow-lethal heterozygous mice. *J. Reprod. Fert.* 40, 483–6.

Kamel, F., Mock, E. J., Wright, W. W., and Frankel, A. I. (1975). Alterations in plasma concentrations of testosterone, LH and prolactin associated with mating in the male rat. *Horm. Behav.* 6, 277–88.

— Wright, W. W., Mock, E. J., and Frankel, A. I. (1977). The influence of mating and related stimuli on plasma levels of luteinizing hormone, follicle stimulating hormone, prolactin and testosterone in the male rat. *Endocrinology* 101, 421–9.

Kaneko, N., Debski, E. A., Wilson, M. C., and Whitten, W. K. (1980). Puberty acceleration in mice II. Evidence that the vomeronasal organ is a receptor for the primer pheromone in male mouse urine. *Biol. Reprod.* 22, 873–8.

Kennedy, J. M. and Brown, K. (1970). Effects of male odor during infancy on the maturation, behavior and reproduction of female mice. *Dev. Psychobiol.* 3, 179–89.

Kenney, A. McM., Evans, R. L., and Dewsbury, D. A. (1977). Postimplantation pregnancy disruption in *Microtus ochrogaster, M. pennsylvanicus* and *Peromyscus maniculatus. J. Reprod. Fert.* 49, 365–7.

Keverne, E. B. (1979). The dual olfactory projections and their significance for behaviour. In *Chemical ecology: odour communication in animals* (ed. F. J. Ritter) pp. 75–83. Elsevier, New York.

Kirchhof-Glazier, D. A. (1979). Absence of sexual imprinting in house mice cross-fostered to deermice. *Physiol. Behav.* 23, 1073–80.

Knowlton, N. (1979). Reproductive synchrony, parental investment, and the evolutionary dynamics of sexual selection. *Anim. Behav.* 28, 1022–33.

Kranz, L. K. and Berger, P. J. (1975). Pheromone maintenance of pregnancy in *Microtus montanus*. *Abstracts of the 55th Annual Meeting of the American Society of Mammalogists*, p. 66.

Labov, J. B. (1980). Factors influencing infanticidal behavior in wild male house mice (*Mus musculus*). *Behav. Ecol. Sociobiol.* 6, 297–303.

— (1981a). Male social status, physiology and ability to block pregnancies in female house mice (*Mus musculus*). *Behav. Ecol. Sociobiol.* 8, 287–91.

— (1981b). Pregnancy blocking in rodents: adaptive advantages for females. *Am. Nat.* 118, 361–71.

Lamond, D. R. (1959). Effect of stimulation derived from other animals of the same species on oestrous cycles in mice. *J. Endocr.* 18, 343–9.

Lawton, A. D. and Whitsett, J. M. (1979). Inhibition of sexual motivation by a urinary pheromone in male prairie deer mice. *Horm. Behav.* 13, 128–38.

Lecyk, M. (1967). The influence of crowded population stimuli on the reproduction of the common vole. *Acta theriol.* 12, 177–9.

Lee, S. Vander and Boot, L. M. (1955). Spontaneous pseudopregnancy in mice. *Acta physiol. pharmacol. neerl.* 4, 442–3.

— — (1956). Spontaneous pseudopregnancy in mice II. *Acta physiol. pharmacol. neerl.* 5, 203–4.

Leibich, H. M., Zlatkis, A., Bertsch, W., van Dahm, R., and Whitten, W. K. (1977). Identification of dihydrothiazoles in urine of male mice. *Biomed. Mass Spectro.* 4, 69–72.

Lisk, R. D., Reuter, L. A., and Raub, J. A. (1974). Effects of grouping on sexual receptivity in female hamsters. *J. exp. Zool.* 189, 1–6.

Lloyd, J. A. and Christian, J. J. (1969). Reproductive activity of individual females in three experimental freely growing populations of house mice (*Mus musculus*). *J. Mammal.* 50, 49–59.

Lombardi, J. R. and Vandenbergh, J. G. (1977). Pheromonally induced sexual maturation in females: regulation by the social environment of the male. *Science, NY* 196, 545–6.

— — and Whitsett, J. M. (1976). Androgen control of the sexual maturation pheromone in house mouse urine. *Biol. Reprod.* 15, 179–86.

— and Whitsett, J. M. (1980). Effects of urine from conspecifics on sexual maturation in female prairie deermice, *Peromyscus maniculatus bairdii*. *J. Mammal.* 61, 766–8.

Lott, D. F. and Hopwood, J. H. (1972). Olfactory pregnancy-block in mice (*Mus musculus*): an unusual response acquisition paradigm. *Anim. Behav.* 20, 263–7.

Louch, C. D. and Higginbotham, M. (1967). The relation between social rank and plasma corticosterone levels in mice. *Gen. com. Endocr.* 7, 441–4.

Lupo di Prisco, C., Lucarini, N., and Dessi-Fulgheri, F. (1978). Testosterone aromatization in rat brain is modulated by social environment. *Physiol. Behav.* 20, 345–8.

McClintock, M. K. (1978). Estrous synchrony and its mediation by airborne chemical communication (*Rattus norvegicus*). *Horm. Behav.* 10, 264–76.

— (1981). Social control of the ovarian cycle and the function of estrous synchrony. *Am. Zool.* 21, 243–56.

— and Adler, N. T. (1978). Induction of persistent estrus by airborne chemical communication among female rats. *Horm. Behav.* 11, 414–18.

McEwen, B. S., Denef, C. J., Gerlach, J. L., and Plapinger, L. (1974). Chemical studies of the brain as a steroid hormone target tissue. In *The neurosciences: third study program* (ed. F. O. Schmidt and F. G. Worden) pp. 599–620. MIT Press, Cambridge, Mass.

McGuire, M. R. and Getz, L. L. (1981). Incest taboo between sibling *Microtus ochrogaster*. *J. Mammal.* 62, 213–15.

McIntosh, T. K. and Drickamer, L. C. (1977). Excreted urine, bladder urine and the delay of sexual maturation in female house mice. *Anim. Behav.* 25, 999–1004.

McKinney, T. D. (1972). Estrous cycle in house mice: effects of grouping, preputial gland odors, and handling. *J. Mammal.* 53, 391–3.

— and Desjardins, C. (1973). Intermale stimuli and testicular function in adult and immature house mice. *Biol. Reprod.* 9, 370–8.

McNeilly, A. S., Cooper, K. J., and Crighton, D. B. (1970). Modification of the oestrous cycle of the under-fed rat induced by the proximity of a male. *J. Reprod. Fert.* 22, 359–61.

Macrides, F. (1976). Olfactory influences on neuroendocrine function in mammals. In *Mammalian olfaction, reproductive processes, and behavior* (ed. R. L. Doty) pp. 29–65. Academic Press, New York.

— Bartke, A., and Dalterio, S. (1975). Strange females increase plasma testosterone levels in mice. *Science, NY* 189, 1104–6.

— — Fernandez, F., and D'Angelo, W. (1974). Effects of exposure to vaginal odor and receptive females on plasma testosterone in the male hamster. *Neuroendocrinology* 15, 355–64.

— — and Svare, B. (1977). Interactions of olfactory stimuli and gonadal hormones in the regulation of rodent social behavior. In *Olfaction and taste VI* (ed. J. Le Magnen and P. MacLeod) pp. 143–7. Information Retrieval, London.

Mallory, F. F. and Brooks, R. J. (1978). Infanticide and other reproductive strategies in the collared lemming, *Dicrostonyx groenlandicus. Nature, Lond.* 273, 144–6.

— — (1980). Infanticide and pregnancy failure: reproductive strategies in the female collared lemming (*Dicrostonyx groenlandicus*). *Biol. Reprod.* 22, 192–6.

— and Clulow, F. V. (1977). Evidence of pregnancy failure in the wild meadow vole, *Microtus pennsylvanicus. Can. J. Zool.* 55, 1–17.

Marchlewska-Koj, A. (1977). Pregnancy block elicited by urinary proteins of male mice. *Biol. Reprod.* 17, 729–32.

— (1980*a*). Partial isolation of pregnancy block pheromone in mice. In *Chemical signals: vertebrates and aquatic invertebrates* (ed. D. Müller-Schwarze and R. M. Silverstein) pp. 413–14. Plenum Press, New York.

— (1980*b*). Male pheromone effect on the enzyme activity of the uterus and on the efficiency of pregnancy in mice. *Symp. zool. Soc. Lond.* 45, 277–88.

— (1981). Pregnancy block elicited by male urinary peptides in mice. *J. Reprod. Fert.* 61, 221–4.

— and Jemiolo, B. (1978). Evidence for the involvement of dopaminergic neurons in the pregnancy block effect. *Neuroendocrinology* 26, 186–92.

— — Wozniacka, J., and Kozlowski, K. (1980). Male-pheromone effect on the efficiency of pregnancy in female mice. *J. Reprod. Fert.* 58, 363–7.

Marsden, H. M. and Bronson, F. H. (1964). Estrous synchrony in mice: Alteration by exposure to male urine. *Science, NY* 144, 1469.

— — (1965*a*). The synchrony of oestrus in mice: relative roles of the male and female environments. *J. Endocr.* 32, 313–19.

— — (1965*b*). Strange male block to pregnancy: its absence in inbred mouse strains. *Nature, Lond.* 207, 878.

Maruniak, J. A. and Bronson, F. H. (1976). Gonadotropic responses of male mice to female urine. *Endocrinology* 99, 963–9.

— Coquelin, A., and Bronson, F. H. (1978). The release of LH in male mice in response to female urinary odors: characteristics of the response in young males. *Biol. Reprod.* 18, 251–5.

Massey, A. and Vandenbergh, J. G. (1980). Puberty delay by a urinary cue from female house mice in feral populations. *Science, NY* **209**, 821-2.

—— —— (1981). Puberty acceleration by a urinary cue from male mice in feral populations. *Biol. Reprod.* **24**, 523-7.

Mena, F. and Grosvenor, C. E. (1971). Release of prolactin in rats by exteroceptive stimulation: sensory stimuli involved. *Horm. Behav.* **2**, 107-16.

—— —— (1972). Effect of suckling and of exteroceptive stimulation upon prolactin release in the rat during late lactation. *J. Endocr.* **52**, 11-22.

Meredith, M. (1980). The vomeronasal organ and accessory olfactory system in the hamster. In *Chemical signals: vertebrates and aquatic invertebrates* (ed. D. Müller-Schwarze and R. M. Silverstein) pp. 303-26. Plenum Press, New York.

—— Marques, D. M., O'Connell, R. J., and Stern, F. L. (1980). Vomeronasal pump: significance for male hamster sexual behavior. *Science, NY* **207**, 1224-6.

—— and O'Connell, R. J. (1979). Efferent control of stimulus access to the hamster vomeronasal organ. *J. Physiol., Lond.* **286**, 301-16.

Miller, L. L., Whitsett, J. M., Vandenbergh, J. G., and Colby, D. R. (1977). Physical and behavioral aspects of sexual maturation in male golden hamsters. *J. comp. physiol. Psychol.* **91**, 245-59.

Milligan, S. R. (1974). Social environment and ovulation in the vole, *Microtus agrestis. J. Reprod. Fert.* **41**, 35-47.

—— (1975*a*). Mating, ovulation and corpus luteum function in the vole, *Microtus agrestis. J. Reprod. Fert.* **42**, 35-44.

—— (1975*b*). Further observations on the influence of the social environment on ovulation in the vole, *Microtus agrestis. J. Reprod. Fert.* **44**, 543-4.

—— (1976*a*). Pregnancy blocking in the vole, *Microtus agrestis* I. Effect of the social environment. *J. Reprod. Fert.* **46**, 91-5.

—— (1976*b*). Pregnancy blocking in the vole, *Microtus agrestis* II. Ovarian, uterine and vaginal changes. *J. Reprod. Fert.* **46**, 97-100.

—— (1979). Pregnancy blockage and the memory of the stud male in the vole (*Microtus agrestis*). *J. Reprod. Fert.* **57**, 223-5.

—— (1980). Pheromones and rodent reproductive physiology. *Symp. zool. Soc. Lond.* **45**, 251-75.

—— Charlton, H. M., and Versi, E. (1979). Evidence for a coitally induced 'mnemonic' involved in luteal function in the vole (*Microtus agrestis*). *J. Reprod. Fert.* **57**, 227-33.

—— and Mackinnon, C. B. (1976). Correlation of plasma LH and prolactin levels with the fate of the corpus luteum in the vole, *Microtus agrestis. J. Reprod. Fert.* **47**, 111-3.

Mody, J. K. (1963). Structural changes in the ovaries of IF mice due to age and various other states: demonstration of pseudopregnancy in grouped virgins. *Anat. Rec.* **145**, 439-47.

Moltz, H., Levin, R., and Leon, M. (1969). Prolactin in the postpartum rat: synthesis and release in the absence of suckling stimulation. *Science, NY* **163**, 1083-4.

Monder, H., Lee, C. T., Donovick, P. J., and Burright, R. G. (1978). Male mouse urine extract effects on pheromonally mediated reproductive functions of female mice. *Physiol. Behav.* **20**, 447-52.

Moss, R. L. (1971). Modification of copulatory behavior in the female rat following olfactory bulb removal. *J. comp. physiol. Psychol.* **74**, 374-82.

—— (1979). Actions of hypothalamic-hypophysiotropic hormones in the brain. *A. Rev. Physiol.* **41**, 617-31.

Moulton, D. G. and Beidler, L. M. (1967). Structure and function in the peripheral olfactory system. *Physiol. Rev.* 47, 1-52.

Murphy, M. R. (1976). Olfactory impairment, olfactory bulb removal, and mammalian reproduction. In *Mammalian olfaction, reproductive processes, and behavior* (ed. R. L. Doty) pp. 95-117. Academic Press, New York.

— and Schneider, G. E. (1970). Olfactory bulb removal eliminates mating behavior in the male golden hamster. *Science, NY* 167, 302-3.

Netto, G. and Pederson, V. (1976). The preputial gland as the source of the estrus inducing pheromone in *Microtus ochrogaster. Trans. Ill. St. Acad. Sci.* 69, 253. (Abstr.)

Nock, B. L. and Leshner, A. I. (1976). Hormonal mediation of the effects of defeat on agonistic responding in mice. *Physiol. Behav.* 17, 111-9.

Noirot, E. (1969). Selective priming of maternal responses by auditory and olfactory cues from mouse pups. *Dev. Psychobiol.* 2, 273-6.

Novotny, M., Jorgenson, J. W., Carmack, M., Wilson, S. R., Boyse, E. A., Yamazaki, K., Wilson, M., Beamer, W., and Whitten, W. K. (1980). Chemical studies of the primer mouse pheromones. In *Chemical signals: vertebrates and aquatic invertebrates* (ed. D. Müller-Schwarze and R. M. Silverstein) pp. 377-90. Plenum Press, New York.

Ojeda, S. R., Advis, J. P., and Andrews, W. W. (1980). Neuroendocrine control of the onset of puberty in the rat. *Fedn. Proc.* 39, 2365-71.

— and Ramirez, V. D. (1972). Plasma level of LH and FSH in maturing rats: response to hemigonadectomy. *Endocrinology* 90, 466-72.

Otte, D. (1974). Effects and functions in the evolution of signaling systems. *A. Rev. Ecol. Systemat.* 5, 385-417.

Parkes, A. S. and Bruce, H. M. (1961). Olfactory stimuli in mammalian reproduction. *Science, NY* 134, 1049-54.

— — (1962). Pregnancy-block in female mice placed in boxes soiled by males. *J. Reprod. Fert.* 4, 303-8.

Payman, B. C. and Swanson, H. H. (1980). Social influence on sexual maturation and breeding in the female mongolian gerbil (*Meriones unguiculatus*). *Anim. Behav.* 28, 528-35.

Petrusewicz, K. (1958). Investigation of experimentally induced population growth. *Ekol. Pol. A* 5, 281-309.

Pfaff, D. W. and Pfaffman, C. (1969). Behavioral and electrophysiological responses of male rats to female rate urine odors. In *Olfaction and taste III* (ed. C. Pfaffman) pp. 258-267. Rockefeller University Press, New York.

Phillips, H. S., Hostetter, G., Kerdelhue, B., and Kozlowski, G. P. (1980). Immunocytochemical localization of LHRH in central olfactory pathways of hamster. *Brain Res.* 193, 574-9.

Pietras, R. J. and Moulton, D. G. (1974). Hormonal influences on odor detection in rats: changes associated with estrous cycle, pseudopregnancy, ovariectomy and administration of testosterone propionate. *Physiol. Behav.* 12, 475-91.

Powers, J. B., Fields, R. B., and Winans, S. S. (1979). Olfactory and vomeronasal system participation in male hamsters' attraction to female vaginal secretions. *Physiol. Behav.* 22, 77-84.

— and Winans, S. S. (1975). Vomeronasal organ: critical role in mediating sexual behavior of the male hamster. *Science, NY* 187, 961-3.

Purvis, K., Cooper, K. J., and Haynes, N. B. (1971). The influence of male proximity and dietary restriction on the oestrous cycle of the rat. *J. Reprod. Fert.* 27, 167-76.

— and Haynes, N. B. (1972). The effect of female rat proximity on the reproductive system of male rats. *Physiol. Behav.* **9**, 401-7.

— — (1974). Short-term effects of copulation, human chorionic gonadotrophin injection and non-tactile association with a female on testosterone levels in the male rat. *J. Endocr.* **60**, 429-39.

— — (1978). Effect of odor of female rat urine on plasma testosterone concentrations in male rats. *J. Reprod. Fert.* **53**, 63-6.

Raisman, G. (1972). An experimental study of the projection of the amygdala to the accessory olfactory bulb and its relationship to the concept of dual olfactory system. *Exp. brain Res.* **14**, 395-408.

Raum, W. J., Glass, A. R., and Swerdloff, R. S. (1980). Changes in hypothalamic catecholamine neurotransmitters and pituitary gonadotropins in the immature female rat: relationships to the gonadostat theory of puberty onset. *Endocrinology* **106**, 1253-8.

Reynolds, J. and Keverne, E. B. (1979). The accessory olfactory system and its role in the pheromonally mediated suppression of oestrus in grouped mice. *J. Reprod. Fert.* **57**, 31-5.

Richmond, M. and Conaway, C. H. (1969). Induced ovulation and oestrus in *Microtus ochrogaster*. *J. Reprod. Fert.* Suppl. **6**, 357-76.

— and Stehn, R. (1976). Olfaction and reproductive behavior in microtine rodents. In *Mammalian olfaction, reproductive processes, and behavior* (ed. R. L. Doty) pp. 197-219. Academic Press, New York.

Roche, K. E. and Leshner, A. L. (1979). ACTH and vasopressin treatments immediately after a defeat increase future submissiveness in male mice. *Science, NY* **204**, 1343-4.

Rogers, C. H. (1971). Influence of copulation on ovulation in the cycling rat. *Endocrinology* **88**, 433-6.

Rogers, J. G. Jr and Beauchamp, G. K. (1976*a*). Influence of stimuli from populations of *Peromyscus leucopus* on maturation of young. *J. Mammal.* **57**, 320-30.

— — (1976*b*). Some ecological implications of primer chemical stimuli in rodents. In *Mammalian olfaction, reproductive processes, and behavior* (ed. R. L. Doty) pp. 181-95. Academic Press, New York.

Ropartz, P. (1966). Contribution à l'étude du déterminisme d'un effet de groupe chez les souris. *C.r. Acad. Sci. Paris D* **262**, 2070-2.

Ruf, K. B. (1973). How does the brain control the process of puberty? *Z. Neurol.* **204**, 95-105.

Ryan, K. D. and Schwartz, N. B. (1977). Grouped female mice: demonstration of pseudopregnancy. *Biol. Reprod.* **17**, 578-83.

Sar, M. and Stumpf, W. E. (1973). Autoradiographic localization of radioactivity in the rat brain after the injection of 1, 2- ^{3}H-testosterone. *Endocrinology* **92**, 251-6.

Scalia, F. and Winans, S. S. (1975). The differential projections of the olfactory bulb and accessory olfactory bulb in mammals. *J. comp. Neurol.* **161**, 31-56.

— — (1976). New perspectives on the morphology of the olfactory system: olfactory and vomeronasal pathways in mammals. In *Mammalian olfaction, reproductive processes, and behavior* (ed. R. L. Doty) pp. 7-28. Academic Press, New York.

Schwagmeyer, P. L. (1979). The Bruce effect: an evaluation of male–female advantages. *Am. Nat.* **114**, 932-8.

Scott, J. P., Stewart, J. M., and DeGhett, V. J. (1974). Critical periods in the organization of systems. *Dev. Psychobiol.* **7**, 489-513.

Sheridan, P. J. (1979). The nucleus interstitialis striatae terminalis and the nucleus amygdaloideus medialis: prime targets for androgen in the rat forebrain. *Endocrinology* **104**, 103–36.

Silver, W. L. and Moulton, D. G. (1982). Chemosensitivity of rat nasal trigeminal receptors. *Physiol. Behav.* **28**, 927–31.

Silverman, A. J. (1976). Distribution of luteinizing hormone-releasing hormone (LHRH) in the guinea pig brain. *Endocrinology* **99**, 30–41.

— Krey, L. C., and Zimmerman, E. A. (1979). A comparative study of the luteinizing hormone releasing hormone (LHRH) neuronal networks in mammals. *Biol. Reprod.* **20**, 98–110.

Singer, A. G., Macrides, F., and Agosta, W. C. (1980). Chemical studies of hamster reproductive pheromones. In *Chemical signals: vertebrates and aquatic invertebrates* (ed. D. Müller-Schwarze and R. M. Silverstein) pp. 365–75. Plenum Press, New York.

Skryja, D. D. (1975). Reproductive inhibition in female cactus mice. *J. Arizona Acad. Sci.* **10**, 23 (Abstr.).

— (1978). Reproductive inhibition in female cactus mice (*Peromyscus eremicus*). *J. Mammal.* **59**, 543–50.

Smotherman, W. P., Brown, C. P., and Levine, S. (1977). Maternal responsiveness following differential pup treatment and mother–pup interactions. *Horm. Behav.* **8**, 242–53.

— Wiener, S. G., Mendoza, S. P., and Levine, S. (1977). Maternal pituitary-adrenal responsiveness as a function of differential treatment of rat pups. *Dev. Psychobiol.* **10**, 113–22.

Snyder, R. L. and Taggart, N. E. (1967). Effects of adrenalectomy on male-induced pregnancy block in mice. *J. Reprod. Fert.* **14**, 451–5.

Stehn, R. A. and Jannett, F. J. Jr (1981). Male-induced abortion in various microtine rodents. *J. Mammal.* **62**, 369–72.

— and Richmond, M. E. (1975). Male-induced pregnancy termination in the prairie vole, *Microtus ochrogaster. Science, NY* **187**, 1211–13.

Steinach, E. (1936). Zur Geschichte des männlichen Sexualhormons und seiner Wirkungen am Säugetiere und beim Menschen. *Wien Klin. Wochenschr.* **49**, 161–72; 196–205.

Stiff, M. E., Bronson, F. H., and Stetson, M. H. (1974). Plasma gonadotropins in prenatal and prepubertal female mice: disorganization of pubertal cycles in the absence of a male. *Endocrinology* **94**, 492–6.

Stoddart, D, M. (1973). Preliminary characterisation of the caudal organ secretion of *Apodemus flavicollis. Nature, Lond.* **246**, 501–3.

Sung, K.-L. P., Bradley, E. L., and Terman, C. R. (1977). Serum corticosterone concentrations in reproductively mature and inhibited deermice (*Peromyscus maniculatus bairdii*). *J. Reprod. Fert.* **49**, 201–6.

Svare, B., Bartke, A., and Macrides, F. (1978). Juvenile male mice: an attempt to accelerate testes function by exposure to adult female stimuli. *Physiol. Behav.* **21**, 1009–13.

Swanson, H. H. (1980). Social and hormonal influences on scent marking in the Mongolian gerbil. *Physiol. Behav.* **24**, 839–42.

— and Lockley, M. R. (1978). Population growth and social structure of confined colonies of Mongolian gerbils: scent gland size and marking behaviour as indices of social status. *Aggress. Behav.* **4**, 57–89.

Teague, L. G. and Bradley, E. L. (1978). The existence of a puberty accelerating pheromone in the urine of the male prairie deermouse (*Peromyscus maniculatus bairdii*). *Biol. Reprod.* **19**, 314–17.

Terkel, J., Damassa, D. A., and Sawyer, C. H. (1979). Ultrasonic cries from infant rats stimulate prolactin release in lactating mothers. *Horm. Behav.* **12**, 95–102.

Terman, C. R. (1965). A study of population growth and control exhibited in the laboratory by deermice. *Ecology* **46**, 890–5.

— (1968). Inhibition of reproductive maturation and function in laboratory deermice: a test of pheromone influence. *Ecology* **49**, 1169–72.

— (1969a). Weights of selected organs of deermice (*Peromyscus maniculatus bairdii*) from asymptotic laboratory populations. *J. Mammal.* **50**, 311–20.

— (1969b). Pregnancy failure in female prairie deermice related to parity and social environment. *Anim. Behav.* **17**, 104–8.

— (1973). Reproductive inhibition in asymptotic populations of prairie deermice. *J. Reprod. Fert.* Suppl. **19**, 457–63.

Thomas, D. and Terman, C. R. (1975). The effects of differential prenatal and post-natal social environments on sexual maturation of young prairie deermice (*Peromyscus maniculatus bairdii*). *Anim. Behav.* **23**, 241–8.

Thomas, T. R. and Neiman, C. R. (1968). Aspects of copulatory behavior preventing atrophy in male rat's reproductive systems. *Endocrinology* **83**, 633–5.

Trivers, R. L. (1972). Parental investment and sexual selection. In *Sexual selection and the descent of man* (ed. B. Campbell) pp. 136–79. Heinemann, London.

— (1974). Parent–offspring conflict. *Am. Zool.* **14**, 249–64.

Vandenbergh, J. G. (1967). Effect of the presence of a male on the sexual maturation of female mice. *Endocrinology* **81**, 345–9.

— (1969). Male odor accelerates female sexual maturation in mice. *Endocrinology* **84**, 658–60.

— (1971). The influence of the social environment on sexual maturation in male mice. *J. Reprod. Fert.* **24**, 383–90.

— (1973a). Acceleration and inhibition of puberty in female mice by pheromones. *J. Reprod. Fert.* Suppl. **19**, 411–19.

— (1973b). Effects of central and peripheral anosmia on reproduction of female mice. *Physiol. Behav.* **10**, 257–67.

— (1975). Hormones, pheromones and behavior. In *Hormonal correlates of behavior*, Vol. 2 (ed. B. E. Eleftheriou and R. L. Sprott) pp. 551–84. Plenum Press, New York.

— (1976). Acceleration of sexual maturation in female rats by male stimulation. *J. Reprod. Fert.* **46**, 451–3.

— (1977). Reproductive coordination in the golden hamster: female influences on the male. *Horm. Behav.* **9**, 264–75.

— (1980). The influence of pheromones on puberty in rodents. In *Chemical signals: vertebrates and aquatic invertebrates* (ed. D. Müller-Schwarze and R. M. Silverstein) pp. 229–41. Plenum Press, New York.

— Drickamer, L. C., and Colby, D. R. (1972). Social and dietary factors in the sexual maturation of female mice. *J. Reprod. Fert.* **28**, 397–405.

— Finlayson, J. S., Dobrogosz, W. J., Dills, S. S., and Kost, T. A. (1976). Chromatographic separation of puberty accelerating pheromone from male mouse urine. *Biol. Reprod.* **15**, 260–5.

— Whitsett, J. M., and Lombardi, J. R. (1975). Partial isolation of a pheromone accelerating puberty in female mice. *J. Reprod. Fert.* **43**, 515–23.

Weir, B. J. (1971). The evocation of oestrus in the cuis, *Galea musteloides. J. Reprod. Fert.* **26**, 405–8.

— (1973). The role of the male in the evocation of oestrus in the cuis, *Galea musteloides* (Rodentia: Hystricomorpha). *J. Reprod. Fert.* Suppl. **19**, 421–32.

Weizenbaum, F., McClintock, M., and Adler, N. (1977). Decrease in vaginal acyclicity of rats when housed with female hamsters. *Horm. Behav.* **8**, 342–7.

Whitsett, J. M., Gray, L. E. Jr, and Bediz, G. M. (1979). Gonadal hormones and aggression toward juvenile conspecifics in prairie deermice. *Behav. Ecol. Sociobiol.* **6**, 165–8.

Whitten, W. K. (1956*a*). Modification of the oestrus cycle of the mouse by external stimuli associated with the male. *J. Endocr.* **13**, 399–404.

— (1956*b*). The effect of removal of the olfactory bulbs on the gonads of mice. *J. Endocr.* **14**, 160–3.

— (1957). Effect of exteroceptive factors on the oestrous cycle of mice. *Nature, Lond.* **180**, 1436.

— (1958). Modification of the oestrous cycle of the mouse by external stimuli associated with the male: changes in the oestrous cycle determined by vaginal smears. *J. Endocr.* **17**, 307–13.

— (1959). Occurrence of anoestrus in mice caged in groups. *J. Endocr.* **18**, 102–7.

— (1966). Pheromones and mammalian reproduction. In *Advances in reproductive physiology I* (ed. A. McLaren) pp. 155–78. Academic Press, New York.

— (1969). Mammalian pheromones. In *Olfaction and taste III* (ed. C. Pfaffman) pp. 252–7. Rockefeller University Press, New York.

— and Bronson, F. H. (1970). Role of pheromones in mammalian reproduction. In *Advances in chemoreception*, Vol. I (ed. J. W. Johnson Jr, D. G. Moulton, and A. Turk) pp. 309–26. Appleton-Century-Crofts, New York.

— — and Greenstein, J. A. (1968). Estrus-inducing pheromone of male mice: transport by movement of air. *Science, NY* **161**, 584–5.

— and Champlin, A. K. (1973). The role of olfaction in mammalian reproduction. In *Handbook of physiology*, Section 7, *Endocrinology*, Vol. II, Part 1 (ed. R. O. Greep) pp. 109–23. Americal Physiological Society, Washington, DC.

Wilson, E. O. (1975). *Sociobiology*. Belknap Press of Harvard University, Cambridge, Mass.

Wilson, M. C., Beamer, W. G., and Whitten, W. K. (1980). Puberty acceleration in mice I. Dose–response effects and lack of critical time following exposure to male mouse urine. *Biol. Reprod.* **22**, 864–72.

Winans, S. S. and Powers, J. B. (1977). Olfactory and vomeronasal deafferentiation of male hamsters: histological and behavioral analyses. *Brain Res.* **126**, 325–44.

Wuensch, K. L. (1979). Adrenal hypertrophy in mice following exposure to crowded males' odors. *Behav. neural Biol.* **27**, 222–6.

Wuttke, W., Honma, K., Lamberts, R., and Höhn, K. G. (1980). The role of monoamines in female puberty. *Fedn. Proc.* **39**, 2378–83.

Wysocki, C. J. (1979). Neurobehavioral evidence for the involvement of the vomeronasal system in mammalian reproduction. *Neurosci. Biobehav. Rev.* **3**, 301–41.

— Wellington, J. L., and Beauchamp, G. L. (1980). Access of urinary non-volatiles to the mammalian vomeronasal organ. *Science, NY* **207**, 781–3.

Young, W. C., Boling, J. C., and Blandau, R. J. (1941). The vaginal smear picture, sexual receptivity and time of ovulation in the albino rat. *Anat. Rec.* **80**, 37–45.

Zarrow, M. X., Christenson, C. M., and Eleftheriou, B. E. (1971). Strain differences in the ovulatory response of immature mice to PMS and to the pheromonal facilitation of PMS-induced ovulation. *Biol. Reprod.* 4, 52–6.

— Eleftheriou, B. E., and Denenberg, V. H. (1972). Pheromonal facilitation of HCG-induced ovulation in different strains of immature mice. *Biol. Reprod.* 6, 277–80.

— Estes, S. A., Denenberg, V. H., and Clark, J. H. (1970). Pheromonal facilitation of ovulation in the immature mouse. *J. Reprod. Fert.* 23, 357–60.

— Schlein, P. A., Denenberg, V. H., and Cohen, H. A. (1972). Sustained corticosterone release in lactating rats following olfactory stimulation from the pups. *Endocrinology* 91, 191–6.

9 The rodents II: suborder Myomorpha

RICHARD E. BROWN

There are 32 families of living rodents with over 350 genera (Walker 1975). Using the classical system of dividing the rodents into suborders (Simpson 1945; Eisenberg 1981), separate chapters have been assigned to the Sciuromorph (Chapter 10) and the Hystricomorph (Chapter 11) rodents. This chapter will focus on the Myomorpha with particular emphasis on the Cricetidae and Muridae, on which the majority of the research has been carried out.

Rodents use urine, faeces, specialized scent glands, and vaginal secretions for olfactory communication. They distribute these odours in a variety of ways including urine-marking, urine-spraying, anogenital drag, sandbathing, and scent-gland rubbing. The scent glands and marking behaviour often show age, sex, and seasonal differences and many rodent scent glands and odorous secretions are controlled by steroid hormones. Often the social function of these odours has only been implied. Table 9.1 attempts to summarize the information in the text, giving the odour sources, marking postures and implied functions of the social odours for representative species of rodents in each family.

There have been a number of extensive studies in the anatomy, physiology and histology of rodent scent glands which have not been concerned with the role of these glands in communication and social behaviour. These include Ortmann's (1960) and McColl's (1967) studies on anal glands in mammals; Lederer's (1949, 1950) summary of the chemistry of urine and scent-gland secretions in beaver, muskrat, and hyrax; Schaffer's (1940) compendium of mammalian skin glands; Quay's (1972) study of the evolution of vertebrate scent glands, and the numerous studies of Pocock (see Hindle (1948) for a bibliography). This chapter attempts to integrate information on the odour sources, marking behaviours, and odour-related behaviour in the rodents. For ease of presentation the chapter is organized by families and genera, with specific odour sources, marking, and odour-related behaviours discussed under each genus or species.

Aplondontidae

There is only one species of this family, *Aplodontia rufa*, the sewellel or 'mountain beaver', which is one of the most primitive rodents. According to Wandeler and Pilleri (1965), who observed a single male in captivity, a 'territory' is established and marked with urine. Apart from the identification of sebaceous glands in the oral angle and lips of a male *Aplondontia rufa* by Quay (1965b; see Plate 9.4), little seems to be known of the skin glands or the role of olfaction in social behaviour in this species.

Table 9.1. This table summarizes odour sources, scent-marking postures, and odour function in representative species of rodents from each genus. Each family is numbered to correspond to the proper section of the text. Odour sources may have been identified for males (M), females (F), both or without specifying the sex (?). The same marking posture is often given different names by different references. Refer to the text for details. To denote a function for each odour is often difficult and the implied functions are often questionable (?). Abbreviations: gl. = gland; secr. = secretion.

Common name and species	Odour source	Sex	Marking posture	Implied function	References
1. Aplodontidae					
Mountain beaver *Aplodontia rufa*	oral lips	M			Quay (1965b)
	urine	M		territory marking (?)	Wandeler and Pilleri (1965)
2. Sciuridae	See Table 10.1				
3. Geomyidae					
Plains pocket gopher *Geomys bursarius*	oral lips	M			Quay (1965b)
Pygmy pocket gopher *Thomomys umbrinus*	oral lips	M			Quay (1965b)
4. Heteromyidae					
California pocket mouse *Perognathus californicus*	oral lips	M, F	ventral rub		Quay (1965b)
	preputial gl.	M	perineal drag		Brown and Williams (1972)
	caudal gl.	M, F	sandbathing	sex, age, or rut identification	Eisenberg (1963, 1964)
	urine	M, F			
Pale kangaroo mouse *Microdipodops pallidus*	oral lips	M, F	side and ventral rub		Quay (1965b)
	ventral gl.	M, F	sandbathing	sex or rut identification	Eisenberg (1963, 1964)
	palmer gl.	M, F	perineal drag		Quay (1965c)
Merriam kangaroo rat *Dipodomys merriami*	oral lips	M, F	sandbathing	sex, rut, or individual identification	Quay (1965b)
	dorsal gl.	M, F			Eisenberg (1963, 1964)
			perineal drag		Eisenberg (1963, 1964)

Taxon	Gland/source	Sex	Behaviour	Function	Reference
Heerman kangaroo rat					
Dipodomys heermanni	oral lips	M, F	sandbathing	sex, rut, or individual identification	Quay (1965*b*)
	dorsal gl.	M, F	perineal drag ventral and side rub		Laine and Griswold (1976)
Spiny pocket mouse					
Liomys pictus	oral lips	M, F	ventral rub perineal drag		Quay (1965*b*) Eisenberg (1963, 1964)
Forest spiny pocket mice					
Heteromys desmarestianus	oral lips	M, F	perineal drag ventral rub 'belly rub'		Quay (1965*b*) Eisenberg (1963, 1964)
Heteromys anomalus					
Heteromys lepturus					
5. Castoridae					
Beaver	anal gl.	M, F	anal drag	colony bond; individual recognition; sign that a territory is occupied	Aleksiuk (1968)
Castor canadensis	castor gl. (preputial)	M, F	scent mounding		Butler and Butler (1979)
Castor fiber					Müller-Schwarze and Heckman (1980)
	urine (castoreum)	M, F	urine spraying		Svendsen (1978, 1980)
	pedal gl.	M, F			Pocock (1922)
6. Anomaluridae—nothing known					
7. Pedetidae					
Springhare	anal gl.	?			Schaffer (1940)
Pedetes caffer	perineal gl.	?			
8. Cricetidae					
Rice rats					
Oryzomys palustris	oral lips	M, F			Quay (1965*b*)
Oryzomys couesi					
Harvest mice					
Reithrodontomys humulis	oral lips	M, F			Quay (1965*b*)
Reithrodontomys megalotis					

Table 9.1 (*cont.*)

Common name and species	Odour source	Sex	Marking posture	Implied function	References
Cricetidae (*cont.*)					
Deer mouse *Peromyscus maniculatus*	saliva (?)	M, F			Aquadro and Patton (1980)
	oral lips	M, F			Quay (1965b)
	mid-ventral gl.	M, F	perineal drag	species	Richmond and Roslund (1952); Doty and Kart (1972)
	urine	M, F	urine marking	sex, dominance, rut, and individual recognition	Doty (1972, 1973)
	vaginal secr.	F			
Oldfield mouse *Peromyscus polionotus*	oral lips	M, F			Quay (1965b)
	mid-ventral gl.	M, F	perineal drag		Doty and Kart (1972)
White-footed mouse *Peromyscus leucopus*	saliva (?)	M, F			Aquadro and Patton (1980)
	oral lips	F			Quay (1965b)
	urine	M, F	urine marking	recognition of sex, species, and kin	Mazdzer et al. (1976)
	vaginal secr.	F		rut	Doty (1972, 1973)
California mouse *Peromyscus californicus*	preputial gl.	M, F			Brown and Williams (1972)
Cactus mouse *Peromyscus eremicus*	mid-ventral gl.	M, F			Doty and Kart (1972)
Grasshopper mouse *Onychomys leucogaster*	oral lips	M, F			Quay (1965b)
Burrowing mouse *Oxymycterus sp*	preputial gl.	M			Brown and Williams (1972)
	clitoral gl.	F			
Common rat *Sigmodon hispidus*	oral lips	M			Quay (1965b)
	preputial gl.	M only		sex, dominance (?)	Brown and Williams (1972)
Dusky-footed woodrat *Neotoma fuscipes*	oral lips	M, F			Quay (1965b)
	mid-ventral gl.	M, F	ventral rub	sex, dominance; aggressive display	Howell (1926)
	anal gl.	M, F			Linsdale and Tevis (1951)
	pedal gl.	M, F			Schaffer (1940)

Species	Source	Sex	Behaviour	Function	Reference
Bushy-tailed woodrat *Neotoma cinerea*	oral lips	M, F	'kiss'	individual recognition	Quay (1965*b*)
	mid-ventral gl.	M, F	ventral rub	species, individual or mother-pup recognition	Egoscue (1962); Bailey (1936)
			sandbathe	precopulatory or aggressive display	Howe (1977)
				territory occupied; alarm	
Eastern woodrat *Neotoma floridana*	urine	M, F	urine marking		Esherich (1981)
	anal gl.	M, F	anogenital rub		Quay (1965*c*)
	preputial gl.	M			Quay (1965*b*)
	pedal gl. (½)	M, F			Howe (1977)
					Clarke (1975)
Stephen's woodrat *Neotoma stephensi*	oral lips	M	ventral rub	aggressive display	Howe (1977)
	mid-ventral gl.	M, F	anal rub		
	anal gl.	M, F			
Southern plains woodrat *Neotoma micropus*	mid-ventral gl.	M, F	ventral rub	aggressive or dominance display	August (1978)
	urine	M, F	perineal drag		
	faeces (?)	M, F	sandbathing		
Dwarf hamster *Phodopus sungoris*	mid-ventral gl. (abdominal)	M, F		sex or rut identification	Vorontsov and Gurtovoi (1959)
Hamster *Cricetulus barabensis*	oral lips	M, F		sex or rut identification	Quay (1965*b*)
	mid-ventral gl.	M, F			Vorontsov and Gurtovoi (1959)
Chinese hamster *Cricetulus griseus*	flank	M, F	flank marking 'drum marking' perineal drag	threat or dominance display	Skirrow and Rysan (1976)
Common hamster *Cricetus cricetus*	oral lips	M, F	flank marking	territory marking	Schaffer (1940)
	ear gl.	M, F			
	mid-ventral gl.	M, F			
	flank gl.	M, F	flank marking	sex or rut identification	Vorontsov and Gurtovoi (1959)
	preputial gl.	M			Vrtis (1931)
	anal gl.	M, F			Brown and Williams (1972)
	pedal gl.	M, F			Schaffer (1940)

Table 9.1 (*cont.*)

Common name and species	Odour source	Sex	Marking posture	Implied function	References
Cricetidae (*cont.*)					
Golden hamster *Mesocricetus auratus*	oral lips	F		sex or rut identification	Quay (1965b)
	Harderian gl.	M, F		sex identification	Payne et al. (1979)
	ear gl.	M, F			Landauer et al. (1980)
	flank gl.	M, F	flank marking	sex and dominance identification; dominance display	Drickamer et al. (1973); Johnston (1977a, b)
	urine	M, F		fear or alarm	Payne (1974); Sherman (1974)
	vaginal secr.	F	vaginal marking	sex and rut identification; sexual solicitation	Johnston (1975a, b, 1981b); Darby et al. (1975); Macrides et al. (1977)
Turkish hamster *Mesocricetus brandti*	flank gl.	M, F	flank marking	aggressive display	Murphy (1977, 1978)
	vaginal secr.	F	vaginal marking	sex or rut identification; sexual solicitation	Frank and Johnston (1981)
Romanian hamster *Mesocricetus neutoni*	flank gl.	M, F		sex or rut identification	Murphy (1977, 1978)
	vaginal secr.	F	vaginal marking		
Crested rat *Lophiomys imhausi*	dorsal gl. (scapular)	M, F			Schaffer (1940)
	preputial gl.	M			Brown and Williams (1972)
Collared lemming *Dicrostonyx groenlandicus*	Meibomian gl.	M, F			Quay (1954b)
	oral lips	F			Quay (1965b)
	caudal (rump) gl.	M, F		individual recognition (?)	Huck and Banks (1979); Quay (1968)
	preputial gl.	M		species, sex, or rut identification	Huck and Banks (1980a); Quay (1960)
	clitoraal gl.	F			
	pedal gl.	M, F			Schaffer (1940)
Bog lemmings *Synaptomys cooperi* *Synaptomys borealis*	oral lips	F			Quay (1954b)
	Meibomian gl.	M, F		sex identification	Quay (1968)
	flank gl.	M, F			

Species	Source	Sex	Marking	Function	References
Norway lemming *Lemmus lemmus*	Meibomian gl.	M, F			Quay (1954b)
	ear gl.	M, F			Schaffer (1940)
	preputial gl.	M			Brown and Williams (1972)
	caudal (rump) gl.	M, F	'lavatory piles'	species, sex, or dominance recognition	Quay (1968); Huck et al. (1981)
Brown lemming *Lemmus trimucronatus*	pedal gl.	M, F			Myllymaki et al. (1962)
	faeces	M, F			
	oral lips	M			Quay (1965b)
	Meibomian gl.	M, F		dominance status (?)	Quay (1954b)
	rump (caudal) gl.	M, F			Quay (1968); Huck et al. (1981)
Bank vole *Clethrionomys glareolis*	oral lips	F		sex recognition	Quay (1965b)
	Meibomian gl.	M, F			Quay (1954b); Griffiths and Kendall (1980b)
	flank gl.	M		sex, dominance, or species recognition	Brown and Williams (1972)
	preputial gl.	M			
	urine	M, F	urine marking		Gustafsson et al. (1980); Christiansen (1980)
	plantar gl.	M, F			Johnson (1975); Griffiths and Kendall (1980a)
Pere David's voles *Eothenomys* sp.	Meibomian gl.	M, F			Quay (1954b)
	flank gl.	M, F		sex or rut identification	Quay (1968)
High mountain voles *Alticola* sp.	flank gl.	M, F			Quay (1968)
Water vole *Arvicola amphibius*	Meibomian gl.	M, F			Quay (1954b)
	oral lips	M, F			Quay (1965b)
	ear gl.	M, F			Schaffer (1940)
Ground vole *Arvicola terrestris*	pedal gl.	M, F			Quay (1968)
	flank gl.	M, F	'drum marking'	aggressive display; species, age, sex, or rut identification	Jannett and Jannett (1974); Stoddart (1972a); Stoddart, Aplin and Wood (1975)
	preputial gl.	M			
	anal gl.	M, F			Brown and Williams (1972)

Table 9.1 (cont.)

Common name and species	Odour source	Sex	Marking posture	Implied function	References
Cricetidae (cont.)					
Muskrat *Ondatra zibethicus*	Meibomian gl.	M, F			Quay (1954b)
	oral lips	M			Quay (1965b)
	preputial gl.	M, F			Brown and Williams (1972)
	anal gl.	M, F	anal drag		Ewer (1968)
Florida water rat *Neofiber alleni*	oral lips	F			Quay (1965b)
	flank gl.	M, F			Quay (1968)
Mountain phenacomys *Phenacomys intermedius*	oral lips	M			Quay (1965b)
	Meibomian	M, F			Quay (1954b)
	flank	M			Quay (1968)
Pine vole (N. America) *Pitymys (Microtus) pinetorum*	oral lips	F		species, sex, or individual identification (?)	Quay (1965b)
	Meibomian	M, F		sexual advertisement	Quay (1954b)
	anal	M, F			Hrabě (1973)
Pine vole (Europe) *Pitymys (Microtus) subterraneus*	Meibomian gl.	M, F			Quay (1954b)
	hip gl. (?)	M			Hrabě (1974); Quay (1968)
	anal gl.	M, F			Hrabě (1973)
Richardson vole *Microtus richardsonii*	oral lips	M		species, age, or sex identification (?)	Quay (1965b)
	Meibomian gl.	M, F			Quay (1954b)
	flank gl.	M, F			Quay (1968)
Short-tailed vole *Microtus agrestis*	Meibomian gl.	M, F		species, sex, or rut identification	Quay (1954b)
	hip gl.	M, F			Clarke and Frearson (1972)
	neck gl.	M, F			Wilson (1973)
	preputial gl.	M, F		play (?)	Jackson (1938)
Common vole *Microtus arvalis*	Meibomian gl.	M, F		sex, rut, aggressive display	Quay (1954b)
	hip gl.	M		sex, species	Quay (1968)
	pedal gl.	M, F		sex	Schaffer (1940)
	Preputial gl.	M, F		species	Vadasz (1975)
	urine	M, F			Hrabě (1973)
	anal gl.	M			

Species	Source	Sex	Behaviour	Function	Reference
Montane vole *Microtus montanus*	Meibomian gl.	M, F	hip marking	species, sex	Quay (1954b)
	hip gl.	M	'pelvic press'	rut identification	Quay (1968)
	preputial gl.	M	anal drag	territory marking	Jannett (1978)
	anal gl.	M	'scooting'		
	urine	M	urine marking		Jannett (1981b)
California vole *Microtus californicus*	oral lips	M, F			Quay (1965b)
	Meibomian gl.	M, F			Quay (1954b)
	hip gl.	M	hip marking	aggressive display	Quay (1968); Lidicker (1980)
Sagebrush vole *Lagurus curtatus*	oral lips	M			Quay (1965b)
	Meibomian gl.	M, F		species recognition	Dearden (1959)
	flank gl.	M	'drum marking'	sex or aggressive display	Quay (1968); Jannett (1981a)
Gerbil *Gerbillus gerbillus* *Gerbillus pyramidarum*	oral lips	F			Quay (1965b)
	ear gl.	M, F			Schaffer (1940)
	preputial gl.	M, F	perineal drag; side rub		Eisenberg (1967)
Mongolian gerbil *Meriones unguiculatus*	Harderian gl.	M, F		identification of sex or dominance status	Thiessen et al. (1976)
	saliva	M, F	'mouthing'	parents, littermates and sex recognition	Block et al. (1981)
	chin gl.				Thiessen et al. (1971)
	ventral gl.	M, F	ventral rub	species, sex, dominance, and individual recognition	Sokolov and Skurat (1966)
			perineal drag	territory marking	Halpin (1974)
	urine	M, F	sandbathing urine marking	sex, rut, and individual discrimination	Gregg and Thiessen (1981)
Israeli gerbil *Meriones tristrami*	oral lips	F		identification of sex and dominance	Quay (1965b)
	ventral gl.	M, F	ventral rub		Sokolov and Skurat (1966); Daly and Daly (1975a)
Sandrat *Psammomys obesus*	chin gl.	M, F	chin marking		
	ventral gl.	M, F	ventral rub	recognition of sex or rut	Daly and Daly (1975b)
	urine	M, F	'urine balls'		
Great gerbil *Rhombomys opimus*	ventral gl.	M, F	ventral rub		Holzman and Paskhina (1974)
	urine	M, F	urine marking		

Table 9.1 (*cont.*)

Common name and species	Odour source	Sex	Marking posture	Implied function	References
9. Spalacidae					
Mole rat					
Spalax ehrenbergi	urine (?)	M, F		species and sex recognition	Nevo and Heath (1976); Nevo *et al.* (1976)
10. Rhizomyidae—nothing known					
11. Muridae					
Old World field mice					
Apodemus flavicollis	oral lips	F		species, sex, and age discrimination (?)	Quay (1965b)
	caudal gl.	M			Stoddart (1972b)
	preputial gl.	M, F			Brown and Williams (1972)
Apodemus sylvaticus	oral lips	M		species, age	Quay (1965b)
	caudal gl.	M		sex and dominance identification (?)	Flowerdew (1971)
Polynesian rat					
Rattus exulans	ventral gl.	M, F			Egoscue (1970)
					Quay and Tomich (1963)
Rattus sabanus	ventral gl.	M, F			Rudd (1966)
Rattus canus	ventral gl.	M only			Rudd (1966)
Norway rat					
Rattus norvegicus	Harderian gl.	M, F			Venable and Grafflin (1940)
	Meibomian gl.	M, F			Schaffer (1940)
	pedal gl.	M, F			Munger and Brusilow (1971)
	anal gl.	M, F			Montagna and Noback (1947)
	preputial gl.	M, F		recognition of sex, rut, and dominance status	Hall (1949); Gawienowski (1977)
					Schaffer (1940)
	urine	M, F	urine marking	individual recognition, fear, alarm	Birke (1978); Brown (1977a)
					Mackay-Sim and Laing (1981b)

Species	Source	Sex	Behaviour	Function	Reference
House mouse *Mus musculus*	oral lips	M, F	mouthing	identification of sex or rut	Quay (1965b) Lee and Ingersoll (1979)
	saliva	M, F			Schaffer (1940)
	ear gl.	M, F		recognition of individuals or colony members	Ropartz (1977)
	pedal gl.	M, F			
	preputial gl.	M, F		sex or dominance recognition	Bronson and Caroom (1971) Bronson and Marsden (1973) Caroom and Bronson (1971)
	urine	M, F	urine marking	identification of sex, species individual or rut; fear or stress odour	Maruniak et al. (1975) Müller-Velten (1966) Hayashi and Kimura (1974) Breen and Leshner (1977)
	vaginal secr.	F		rut	
	faeces	F		maternal recognition	
Spiny mouse *Acomys cahirinus*	faeces (?)	F		maternal recognition	Porter and Doane (1976)
Pest-rat *Nesokia indica*	oral lips	F			Quay (1965b)
	preputial gl.	M			Brown and Williams (1972)
African giant rat *Cricetomys gambianus*	cheek gl. (?)	M, F	urine mark		Ewer (1967)
	urine	M, F	'hand-stand'		
	faeces	M, F			
Vlei rat *Otomys irroratus*	anal gl.	M, F	anal rub	alarm odours (?)	Davis (1972)
	urine	M, F			

Sciuridae

The Sciuridae mark with urine and a number of specialized scent glands. The role of olfaction in social behaviour of the Sciuridae is discussed in Chapter 10.

Geomyidae

Quay (1965*b*) found sebaceous glands in the oral angle of the male plains pocket gopher (*Geomys bursarius*) and pygmy pocket gopher (*Thomomys umbrinus*), but no information seems to be available on olfaction and social behaviour in the Geomyidae (see Eisenberg 1981; Walker 1975).

Heteromyidae

Glands

Kangaroo rats of at least seven species (*Dipodomys merriami, D. agilis, D. heermanni, D. deserti, D. ordi, D. spectabilis,* and *D. panamintinus*) have a dorsal sebaceous holocrine skin gland on the arch of the back (Quay 1953, 1954*a*, see Plate 9.1). This gland is found on both male and female *Dipodomys*, but not on other heteromyid rodents, so is considered to be generic character of *Dipodomys* (Quay 1953). Male *D. merriami* and *D. agilis* show increased glandular size and secretory activity during the breeding season (February–July) and during this period the male gland is larger than that of the female. During the rest of the year male and female glands do not differ in size. Male *D. deserti* show some increase in gland size from April to June, but their glands are not of much greater size than those of females. Male and female *D. heermanni* have dorsal glands of the same size which show little monthly fluctuation and, while male and female *D. ordi* also have glands of the same size, they show substantial monthly fluctuation in secretory activity, reaching a peak in November–December (Quay 1953).

The pocket mice (*Perognathus*) have a caudal sebaceous gland on the ventral surface of their tail about one-fourth of the distance from the base of the tail (Quay 1965*a*; see Fig. 9.1). This gland occurs in at least 15 species of *Perognathus* and is seen in both sexes. Such a caudal gland appears to be a characteristic of the genus *Perognathus* as it is not known in other heteromyid genera (Quay 1965*a*). These glands show great variability in size and may show age and sex differences, but this has not yet been investigated.

There are sebaceous glands in the oral angle and lips of both sexes in all five genera of the Heteromyidae (Quay 1965*b*) and many species have sweat glands on the soles of the feet (Quay 1965*c*; 1966).

Marking

Heteromyid rodents engage in 'sandbathing' which involves ventral and side-rubbing in the sand. Complete sandbathing, consisting of an integrated sequence

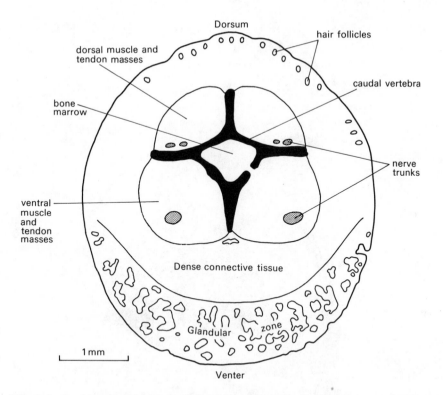

Fig. 9.1. A cross-section of the tail of an adult male *Perognathus artus* in the region of maximal development of the caudal gland. (From Quay (1965*a*) © The Southewestern Association of Naturalists.)

of ventral and side rubbing, occurs in *Perognathus*, *Microdipodops*, and *Dipodomys*, while ventral rubbing alone occurs in *Heteromys* and *Liomys*. The perineal drag occurs in all five genera of Heteromyidae and may be used to deposit urine, faeces, and glandular secretions from caudal and perineal glands on to the sandbathing site (Eisenberg 1963, 1967; see Fig. 9.2). According to Eisenberg (1967, p. 18) the perineal drag may have originated as a cleaning movement consisting of 'wiping the anal-genital area on the substrate after urination or defecation' and then gained a communicatory function.

Odour-related behaviour

Kangaroo rats (*Dipodomys agilis* and *D. merriami*) approach conspecific odours in the breeding season, but not during the non-breeding season. Oestrous females, for example, show preferences for male-scented traps whereas non-oestrous females prefer unscented traps (Daly, Wilson, and Behrends 1980).

Dorsal-gland secretions and sandbathing are important for grooming and pelage care in many species of *Perognathus* and *Dipodomys* which tend to

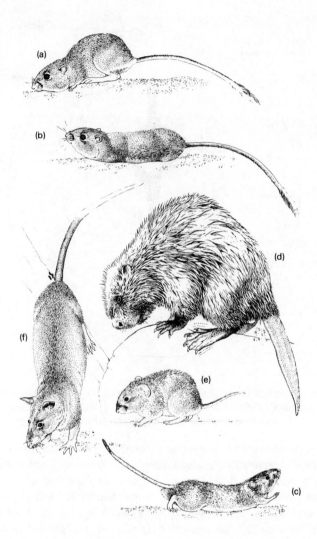

Fig. 9.2. Some rodent scent-marking behaviour. Sandbathing by *Dipodomys*: (a) flexion of the body followed by extension and (b) side rubbing with the side pressed into the substrate. (c) Ventral rub by *Liomys pictus*. (d) When a beaver (*Castor fiber*) deposits castoreum, it hunches its back and makes scratching movements with one hind leg as the castoreum is noisily sprayed out. (e) Urine-marking by a male Skomer vole (*Clethrionomys glareolus skomerensis*). (f) Hand-stand posture for defecation by a juvenile African giant rat (*Cricetomys gambianus* Waterhouse).

develop a matted, greasy pelage if not given access to sand in which to dress their fur (Eisenberg 1963). Deprivation of sand leads to an accumulation of glandular secretion on the fur and to increased sandbathing when sand is made available (Borchelt, Griswold, and Brancheck 1976). Quay (1953) suggested that the dorsal gland of *Dipodomys* may function to dress the hair or skin with any oil secretion or as a scent organ. Bailey (1936) suggested that this gland is used for individual recognition. Because the dorsal gland of *Dipodomys* is active only in adult breeding males of some species, but is active in both sexes throughout the year in other species, this gland may serve different functions in different species.

Perognathus inornatus have preferred sandbathing locations, and the sandbathing location of one animal elicits sandbathing in conspecifics (Eisenberg 1963). Over 90 per cent of the sandbathing bouts of male and female bannertail kangaroo rats (*Dipodomys spectabilis*) occur in places where conspecifics have sandbathed (Laine and Griswold 1976). This suggests a communicatory function for sandbathing, but does not indicate what function is served. Although some species of the Heteromyidae may use odours for species recognition, Martin (1977) found little or no preference by males or females for conspecifics of the opposite sex or their odours among three species of silky pocket mice (*Perognathus flavus, P. merriami*, and *P. flavescens*).

Castoridae

Glands

Both male and female North American beaver (*Castor canadensis*) and European beaver (*Castor fiber*) possess paired anal glands, paired castor or musk glands (see Fig. 9.3) and glands on the soles of the feet (Pocock 1922; Schaffer 1940; Svendsen 1978). The anal sebaceous glands possess ducts which open to the rump on each side of the cloaca. The secretory cells of the anal glands are richly vascularized and surrounded by a muscular sheath which constricts to force out drops of the secretion. The castor glands or 'castors' have often been called preputial glands (see Wilsson 1971). They open into the urethra where their secretion becomes mixed with urine and this compound is referred to as 'castoreum' (Svendsen 1978).

Castoreum is a mixture of many substances, including low-volatile metabolites in the urine and at least four compounds secreted by the castor gland. One of these, castoramine, is a pungent, highly volatile chemical which gives the beaver its characteristic odour (Svendsen and Jollick 1978). Bacteria do not appear to exist in the castor gland, but a bacterial flora is present in the anal gland. Whether these bacteria alter the odour of the anal gland secretion is unknown (Svendsen and Jollick 1978).

The size of both anal and castor glands is positively correlated with body weight. The glands increase in size as the beaver develops (Wilsson 1971).

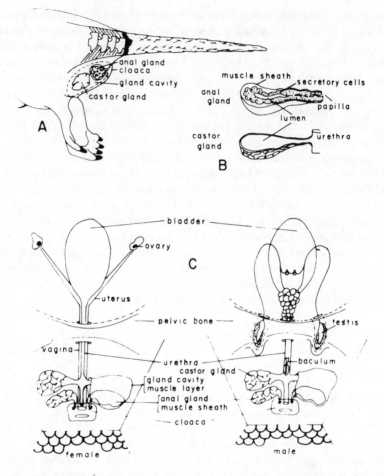

Fig. 9.3. The anal and castor glands of the beaver (*Castor canadensis*) showing (a) the position of the glands in the body; (b) a longitudinal section of the anal and castor glands; and (c) the anatomical relationships of the anal and castor glands to the other anogenital structures of the male and female beaver. (From Svendsen (1978).)

Whether or not the chemical composition of the secretions vary according to the age or sex of the animal is unknown.

Marking

Beaver mark with both castoreum and with the anal glands (see Fig. 9.2(d)). Castoreum is usually sprayed onto piles of earth, tree trunks, or protruding objects along with urine (Wilsson 1971). During scent-marking the beaver arches its body over the object to be marked and sprays castoreum over it. The cloaca is also dragged over the object, marking it with the protruding anal glands (Butler and Butler 1979; Wilsson 1971).

Beaver have 'sign places' at which they scratch up piles of earth and then deposit castoreum on these piles (Wilsson 1971). Such 'sign places', also called 'scent mounts', may be up to 60 cm in height. A colony may have up to seven scent mounds, most of which are at the edge of the territory. Castoreum is deposited on scent mounds in regular (possibly daily) visits by all age and sex classes of beavers (Aleksiuk 1968). Castoreum marking is elicited by the odour of a strange beaver (Aleksiuk 1968; Wilsson 1971) and although beaver avoid the territories of other beaver, the scent mounds do not appear to serve as territorial boundaries because they do not keep trespassers away. Beaver which intrude into a strange territory mark over the scent mounds of the resident beaver (Wilsson 1971). When beaver are stressed by handling they also release castoreum (Svendsen 1978).

Odour-related behaviour

Scent-mound construction in beaver is highest between April and June, low from July–October, and does not occur from November–March (Butler and Butler 1979; Svendsen 1980). Scent-mound construction is not correlated with the number of animals in a colony, nor the amount of shoreline around a colony (Butler and Butler 1979), but is related to the number of neighbours and occurs most frequently at the time when two-year-olds are dispersing from their home lodge and new litters are being born (Svendsen 1980). Beaver living in high-density areas or in areas where they have a high probability of encountering their neighbours build more scent mounds than beaver living farther away from their neighbours (Butler and Butler 1979; Müller-Schwarze and Heckman 1980). Beaver lodges with one or no scent mounds are likely to be 600 metres or more from their nearest neighbour (Müller-Schwarze and Heckman 1980).

When scent from an alien beaver is placed on a mound in a resident's territory, the resident often responds by hissing loudly, sniffing the mound, tailslapping, and/or marking over the alien odour (Aleksiuk 1968; Butler and Butler 1979; Müller-Schwarze and Heckman 1980). The odours of alien males are reacted to more often than the odours of alien females, and adult males respond to scent marks more frequently than females or juveniles (Butler and Butler 1979; Müller-Schwarze and Heckman 1980). When scent mounds are artificially constructed near unoccupied beaver lodges and artificially marked with castoreum, these lodges are avoided by beaver. Unoccupied lodges with no scent mounds often become occupied (Müller-Schwarze and Heckman 1980).

The most common interpretation of scent-mound function is that they serve to maintain a system of territorial rights (Aleksiuk 1968, p. 760) or act as a 'fence' between territories of adjacent colonies (Müller-Schwarze and Heckman 1980). Butler and Butler (1979, p. 451), on the other hand, state that 'no reports to date (including the present study) offer quantitative evidence to support the role of beaver scent marking in territoriality'. The finding that marking is most frequent when the number of transient two-year-old beaver is highest, and that unoccupied lodges with fresh scent mounds are less likely to become occupied by transients than unmarked lodges, however, suggests that the scent mounds may serve to indicate that a territory is occupied (Svendsen 1980).

Many other functions have been suggested for scent mounds in beaver, including orientation in the home-range (Müller-Schwarze and Heckman 1980); individual, age, or sex discrimination (Svendsen 1980); modification of the emotional state of residents and non-residents: 'increasing the confidence and reducing anxiety in residents smelling their own scent mound and decreasing the confidence and increasing the readiness to flee in trespassers encountering a strange scent mound' (Svendsen 1980, p. 146); and serving as a 'colony odour' to 'reinforce the intra-colony social bond' (Butler and Butler 1979, p. 452). Scent-marking in beaver does not appear to be related to sexual behaviour since scent glands are not sexually dimorphic and few scent mounds are constructed during the female's oestrus and mating period (Svendsen 1980).

Any changes in anal and castor-gland secretions at puberty or changes over the female's oestrous cycle are unknown in beaver, as are the abilities of beaver to use olfactory signals for age, sex, individual, colony, or social status discrimination. Many more behavioural studies are required in order to understand the role of castoreum and anal-gland secretions in the social organization of the beaver.

Anomaluridae

According to Kingdon (1974, p. 454) the role of smell in the social behaviour of Anomaluridae is 'difficult to assess'. Apart from the comments that *Anomalurus derbianus* gives off a 'very strong odour reminiscent of that of some monkeys' and that some females appear to have two small subcutaneous glands on their lower abdomen (Kingdon 1974), nothing appears to be known about the scent glands or the role of olfaction in the social behaviour of the four genera (nine species) of scaly-tailed squirrels.

Pedetidae

Although Schaffer (1940) has described anal and perineal glands of *P. caffer*, the role of these glands in the social behaviour of the two species of spring-hares (*Pedetes*) seems to be unknown (see Kingdon 1974).

Cricetidae

With 100 genera and over 728 species (Walker 1975), the Cricetidae (Old World rodents) are known almost world wide and some, such as *Peromyscus, Mesocricetus, Microtus,* and *Meriones* are common laboratory animals. Eisenberg (1981) has summarized the literature on the ecology and behaviour of the Cricetidae.

Oryzomys

With over 100 species of rice rats, it is odd that so little information exists on their scent glands and social behaviour. Quay (1965b) found sebaceous and sudoriferous glands in the oral angle and lips of both male and female *O. palustris* and *O. couesi,* but no information seems to exist on the social behaviour associated with these glands.

Reithrodontomys

Two of the 16 species of *Reithrodontomys* (*R. humulus* and *R. megalotis*) have glands of the oral angle and lips (Quay 1965b) but little else seems to be known about these animals.

Peromyscus

Many of the 55 species of *Peromyscus* are well known laboratory animals whose scent glands and social behaviour have been extensively studied.

Glands *Peromyscus* species possess sebaceous and sudoriferous glands in the oral angle and lips (Quay 1965b), a mid-ventral sebaceous gland (Doty and Kart 1972; Richmond and Roslund 1952) and use urine for marking (Dagg, Bell, and Windsor 1971; Eisenberg 1962; Maruniak, Desjardins, and Bronson 1975). Some species, such as *P. californicus,* have large preputial glands (Eisenberg (1962). Although *Peromyscus maniculatus* and *P. leucopus* can be differentiated by analysis of salivary amylase (Aquadro and Patton 1980), there is, so far, no evidence that saliva is used for olfactory communication in *Peromyscus.*

In a comparative study, Doty and Kart (1972) identified mid-ventral glands in males and females of four species of *Peromyscus,* and found eight species where no gland could be identified. All subspecies of *P. maniculatus* and two subspecies of *P. polionotus* (*P.p. rhoadsi,* and *P.p. subgriseus*) possess mid-ventral glands. Males have larger mid-ventral glands than females, and these glands are dependent on gonadal hormones for their development (see Plates 9.2 and 9.3). Castration causes atrophy of the mid-ventral gland of male deer mice and testosterone injections reinstate growth (Blum, Kakihana, and Kessler 1971; Doty and Kart 1972). Oestrogen and/or progesterone may inhibit the development of the mid-ventral gland of female deer mice (Doty and Kart 1972), but the hormonal control of female ventral glands does not seem to have been systematically investigated.

Marking   Four species of *Peromyscus* mark with a perineal drag (*P. maniculatus, P. californicus, P. eremicus, P. crinitus*), but *Peromyscus* do not sandbathe (Eisenberg 1962, 1967). Deer mice urine-mark (Eisenberg 1962; Maruniak, Desjardins, and Bronson 1975), and urine-marking is stimulated by conspecific odours (Dagg *et al.* 1971). Dominant male *P. maniculatus* urine-mark more frequently than subordinates; sexually experienced males urine-mark more than naive males; and dominant males mark more frequently after an agonistic encounter while subordinates show reduced urine-marking after an agonistic encounter (Farr, Andrews, and Klein 1978).

Odour-related behaviour Deer mice (*Peromyscus maniculatus*) may use urine and other odours for species, sex, and individual identification, but neither white-footed mice (*P. leucopus*) nor oldfield mice (*P. polionotus*) show preferences for conspecific odours in laboratory experiments (Doty 1972, 1973; Halpin 1980; Moore 1965). In field experiments, however, both male and female *P. leucopus* are attracted to traps baited with the odours from opposite sex conspecifics (Mazdzer, Capone, and Drickamer 1976). *P. maniculatus* males and females prefer to enter traps baited with the odours of conspecifics during the breeding season, but outside the breeding season odours are avoided and unscented traps are entered more often (Daly, Wilson, and Faux 1978; Daly *et al.* 1980).

Since adult deer mice attack juveniles (Whitsett, Gray, and Bediz 1979). olfactory cues may be used for age and sex identification, but no studies have demonstrated this. Some species of *Peromyscus* (e.g. *P. leucopus*) may be 'imprinted' to species odours during the pre-weaning period (McCarty and Southwick 1977) and may use odours for kinship recognition (Grau 1982).

Onychomys

The grasshopper mouse (*Onychomys leucogaster*) has sebaceous and sudoriferous glands of the oral lips and angle (Quay 1965b) and *Onychomys torridus* may use odours for species recognition (McCarty and Southwick 1977), but little work has been done on scent-marking or the role of odours in social behaviour in these species.

Oxymycterus

Both males and females of some species of *Oxymycterus* have preputial (clitoral) glands (Brown and Williams 1972) but their social function is unknown.

Sigmodon

At least one of the three species of cotton rats (*Sigmodon hispidus*) has sebaceous and sudoriferous glands in the oral lips and angle, although these glands are small (Quay 1965b). Male, but not female *Sigmodon* possess preputial glands (Brown and Williams 1972; Schaffer 1940).

Subordinate male cotton rats (*S. hispidus*) appear to avoid the odours of dominant males, particularly after social interactions, whereas dominant males approach the odours of other males (Summerlin and Wolfe 1973).

Neotoma

Glands At least four of the 22 species of *Neotoma* possess sebaceous and sudoriferous glands of the oral angle and lips (Quay 1965*b*; see Plate 9.4). *Neotoma* also possess eccrine sweat glands on the soles of the feet (Quay 1965*c*; Schaffer 1940) and adult males appear to have preputial glands (Escherich 1981). Some species of *Neotoma* including the Allegheny wood rat (*N. magister*), the desert wood rat (*N. lepida*), the eastern wood rat (*N. floridana*), the dusky-footed wood rat (*N. fuscipes*), and the bushy-tailed wood rat (*N. cinerea*), have mid-ventral and anal glands. The mid-ventral sebaceous gland occurs most frequently in adult and sub-adult males. The size and secretion of the mid-ventral glands differ among the different *Neotoma* species (Howe 1977), but in most species females have smaller glands than males and juveniles do not show mid-ventral glands (Escherich 1981; Howell 1926; Linsdale and Tevis 1951; Poole 1940). The mid-ventral gland is largest in the breeding season (October–February) and smaller during the rest of the year. These glands are reduced in size after castration and enlarged by testosterone, but not by oestrogen injections (Clarke 1975; Fleming and Tambosso 1980).

The anal gland surrounds the rectum and produces a thick, yellowish fluid which appears to cover the faeces (Howell 1926; Linsdale and Tevis 1951). The anal gland appears to be larger in males than in females (Escherich 1981; Linsdale and Tevis 1951) but this does not seem to have been systematically investigated.

Marking Wood rats scent-mark by rubbing the mid-ventral gland on rocks and the ground (ventral rubbing) (Escherich 1981; Howe 1977). *Neotoma* also sandbathe (Howe 1977) or sand-roll (Fleming and Tambosso 1980). Dominant *Neotoma* mark more than subordinates, and subordinates appear to avoid the odours of dominants (Howe 1977). Both male and female *Neotoma stephensi* and *N. cinerea* sandbathe during mating sequences (Fig. 9.4) (Escherich 1981; Howe 1977).

Male desert wood rats (*Neotoma lepida*) mark with the ventral rub more often than females but females sandbathe or sand-roll more often than males (Fleming and Tambosso 1980). Males ventral rub on stimuli with female odours, while females direct their rolling toward locations with male odours. Ventral rubbing occurs more frequently in dominant males and is increased by testosterone injections or the presence of an oestrous female. Ventral rubbing by females is increased by androgen injections, but dominant females do not mark more than subordinate females, nor do social interactions elevate the frequency of scent-marking in females (Fleming and Tambosso 1980).

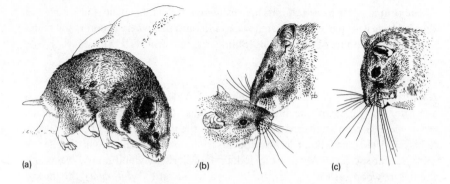

Fig. 9.4. (a) Female golden hamster beginning to flank-rub. (b) Bushy-tailed woodrats: 'kiss' between female (right) and young. (c) Gerbil (*Meriones unguiculatus*) grooming area of Harderian secretion.

The anal gland may be rubbed against sticks or other objects and its secretions may coat the faeces (Linsdale and Tevis 1951). Wood-rat faeces are piled in middens near their lodgings (Bailey 1936) and may be used to mark novel objects (Vestal 1938; Linsdale and Tevis 1951). Urine is used for marking specific objects in the environment, leaving visible white calcareous marks (Bailey 1936; Escherich 1981). Males and females of all ages urine-mark their home-areas.

Odour-related behaviour Male southern plains wood rats (*Neotoma micropus*) are attracted to female urine and faeces, and females are attracted to male urine and mid-ventral gland secretions (August 1978). Male desert wood rats (*Neotoma lepida*) prefer the urine and vaginal odours from oestrous females to those of non-oestrous females, and anosmic (olfactory bulbectomized) males fail to mate, indicating the importance of olfactory signals in sexual behaviour of this species (Fleming, Chee, and Vaccarino 1981).

Ventral rubbing in *Neotoma* appears to function as an agonistic signal, while sandbathing or sand-rolling appears to be a courtship or sexual activity (Escherich 1981). In male–male encounters, dominant males do almost all of the ventral marking (Howe 1977; Fleming and Tambosso 1980). Male wood rats aggressively defend and mark around their dens, and do not approach the odours of other males, so that ventral marks and urine marks may be used by males to denote occupation of a territory (Egoscue 1962; Escherich 1981; Linsdale and Tevis 1951). Since females are as aggressive as males (August 1978), one would also expect them to mark around their dens.

Wood rats are able to use urine and ventral-gland secretions to discriminate between the sexes, but little is known about species, age, colony, or individual recognition. Since females show increased rolling when in oestrus (Howe 1977), this may act to disseminate a rut odour. Whether or not olfactory cues act as reproductive isolating mechanisms in *Neotoma*, as suggested by Howe (1977) is yet to be determined. Anal- and ventral-gland odours may be used for home identification and trail-marking (Linsdale and Tevis 1951). Female ventral glands increase in size during lactation and appear to be used to mark pups (Escherich 1981). Urination occurs when *Neotoma* are chased and captured and may serve as an alarm odour (Escherich 1981).

Phodopus

Vorontsov and Gurtovoi (1959) have reported the existence of a mid-ventral or abdominal gland in *Phodopus sungoris* and *Ph. roborovskii*. This gland becomes enlarged during the reproductive season and is much smaller during the rest of the year. It may function to mark individual territories.

Cricetulus

Many species of *Cricetulus*, including *Cr. eversmanni*, *Cr. migratorius*, *Cr. barabensis*, *Cr. longicaudatus*, and *Cr. kamensis* possess a mid-ventral (abdominal) gland which secretes during the breeding season (Vorontsov and Gurtovoi 1959).

Both male and female striped hamsters (*Cricetulus barabensis*) have oral glands (Quay 1965b). The Chinese hamster (*Cricetulus griseus*) has flank glands which it occasionally uses to flank-mark. More often, however, the hindfoot is rubbed over the flank gland and the sebum is spread onto the substrate from the foot (Skirrow and Rysan 1976). Flank-marking is performed more often by females than males and may serve as a threat or dominance signal. Chinese hamsters also mark using a perineal drag (Skirrow and Rysan 1976).

Cricetus

The common hamster (*Cricetus cricetus*) possesses flank glands which are used to flank-mark objects and the substrate (Eibl-Eibesfeldt 1953; Schaffer 1940; Vrtis 1931). These hamsters also possess a ventral gland (Vorontsov and Gurtovoi 1959), anal glands, ear glands, glands of the oral lips, glands on the soles of the feet, and males possess preputial glands (Brown and Williams 1972; Schaffer 1940). The flank glands are larger in males than females, larger in the breeding season, and begin to secrete before puberty (Vrtis 1930b). In old hamsters the gland decreases in size but does not disappear (Vrtis 1930b). These findings indicate that sex hormones may enhance glandular activity but are not completely necessary for its maintenance.

Mesocricetus

The majority of the research in this section has been done using the golden hamster (*Mesocricetus auratus*) and a few comparative studies have examined Turkish hamsters (*M. brandti*) and Romanian hamsters (*M. neutoni*; see Table 9.1).

Glands   Hamsters have sebaceous glands on the ventral side of the ears (Plewig and Luderschmidt 1977), costovertebral of flank glands (Markel 1952), and Harderian glands in the orbit of the eye (Payne 1979), all of which may be used for olfactory communication. Glands of the oral angle have been found in the female (Quay 1965*b*). In addition to these glands, female hamster vaginal secretions are used in sexual communication (Singer, Macrides, and Agosta 1980). Urine, faeces, and preputial-gland secretions do not appear to be used for olfactory communication by hamsters (Johnston 1977*b*).

Flank gland   Flank glands have been identified on both sexes of golden, Turkish, and Romanian hamsters. Flank glands (see Plate 9.5) are larger in males than in females, and begin to develop around 40 days of age (Stölzer 1959). Pre- and postnatal development of the flank is described by Algard, Dodge, and Kirkman (1964; 1966). Castration or ovariectomy causes the flank glands to decrease in size and secretory rate, while androgen injections produce a rapid increase in size and secretion (Hamilton and Montagna 1950; Montagna and Hamilton 1949; see Plate 9.6). Oestradiol injections do not increase the size of the flank glands. In males, the flank gland shows a linear increase in size as testosterone doseage increases and the size of the male's gland is, therefore, a reliable index of testosterone level (Vandenbergh 1973). Dominant male and female hamsters have larger flank glands than subordinates (Drickamer, Vandenbergh, and Colby 1973; Drickamer and Vandenbergh 1973).

Flank-gland marking   Both male and female hamsters flank-mark by rubbing their flank glands against vertical surfaces or protruding objects (Dieterlen 1959; Johnston 1975*b*, 1977*a*) (Fig. 9.4). Flank-marking in male hamsters occurs at very low rates before puberty, and increases to adult levels at 45–50 days of age. Castration reduces, but does not eliminate flank-marking in male hamsters (Whitsett 1975; Johnston 1981*a*) and testosterone replacement reinstates high levels of flank-marking (Johnston 1981*a*). Removal of the male's flank gland does not reduce his marking frequency but the odour of a glandectomized male does not stimulate as much marking by other males as the odour of an intact male (Johnston 1975*c*).

In males, flank-marking occurs infrequently if the male is alone in his home-cage, and much more frequently in the presence of odours of other males or odours of dioestrous females, but flank-marking is inhibited by the odours of oestrous females (Johnston 1975*a*, 1980). Flank-marking occurs frequently during aggressive encounters between two males or a male and a dioestrous female (Payne and Swanson 1970; Johnston 1975*b*). In male–male encounters, the dominant male marks more than the subordinate male, but in male–female agonistic encounters, males mark more than females whether they have won or lost the encounter (Johnston 1975*b*). When paired males live in the same home-cage, the dominant male marks more frequently and over more of the living area than the subordinate male. Subordinates mark only around their home burrow (Johnston 1975*d*).

Flank-gland odours from male hamsters retain the ability to elicit investigation for over 25 days (Johnston and Lee 1976) and it is not until the flank mark is about 50 days old that it loses its stimulus quality (Johnston and Schmidt 1979). Whereas fresh scent marks (24 h old) are investigated more than older marks, male hamsters investigate 40-day-old flank marks significantly longer than clean areas (Johnston and Schmidt 1979).

Female hamsters flank-mark more in the presence of female hamster odours than in the presence of male odours or no odour. In encounters with males, dioestrous females flank-mark more than oestrous females (Carmichael 1980; Johnston 1977*a*). Flank-marking is associated with aggressive behaviour in female golden hamsters, and in female–female encounters, subordinate females mark less than dominant females (Johnston 1977*a*). Similarly, female Turkish hamsters flank-mark more when dioestrous and in this species flank-marking also appears to be an aggressive signal (Frank and Johnston 1981).

Harderian gland   There is a marked sex difference in the Harderian gland of hamsters (see Plates 9.7–9.9). Most striking is the colour difference: male Harderian glands are yellow-grey and female Harderian glands have black granules on a grey background (Christensen and Dam 1952; Woolley and Worley 1954). The black pigment (porphyrin) granules are not found in males (compare Plates 9.8 and 9.9). The cell structure of the Harderian gland in the golden hamster also shows a sex difference, with males having more large vacuole cells than females (Payne, McGadey, Moore, and Thompson 1975; Woolley and Worley 1954). Castration of males reduces the number of large vacuole cells and increases pigmentation (porphyrin content) of the Harderian gland, so that it resembles the female gland. Oestrogen treatment of males also increases pigmentation and produces an increase in large vacuole cells (Hoffman 1971; Payne *et al.* 1975, 1977; Woolley and Worley 1954).

Although Hoffman (1971) reported that ovariectomy had no effect on the structure of the Harderian gland in females, Payne *et al.* (1979) found that glandular weight and porphyrin content showed cyclical changes over the four-day oestrous cycle, being highest at oestrus and then decreasing at metoestrus. Pregnant and lactating females have large Harderian glands and a much greater secretion of photoporphyrin than cycling females (Payne *et al.* 1979). Thus, the Harderian gland secretions may provide information on the reproductive state of the female.

Male hamsters attack another male less frequently if that male has been smeared with female Harderian gland secretion (Payne 1977). In a preference test, male hamsters spend more time investigating Harderian glands from females than those from males, but do not investigate the Harderian glands of oestrus females more than those from dioestrus females (Payne 1979). Thus, these glands can be used to discriminate males from females (by males—females have not been used as subjects), but the Harderian gland secretion does not seem to be used to discriminate oestrous from non-oestrous females (although a preference test does not eliminate the possibility that there is a difference in odour of the female Harderian gland at different times in the oestrous cycle). Males may be conditioned to discriminate between these odours for some form of reward, but this experiment does not seem to have been done.

Ear gland   Golden hamsters have sebaceous glands on the ventral side of their ears. These glands are sexually dimorphic; males having much larger glands than females (Plewig and Luderschmidt 1977; see Plate 9.10) and are androgen dependent. Injections of testosterone increase the size of the ear gland in the female (Plewig and Luderschmidt 1977), and cyproterone acetate, an anti-androgen, reduces the size of the ear gland in the male (Luderschmidt and Plewig 1977). Sexually experienced male hamsters investigate the ear glands of females and castrated males more than those of intact males (Landauer, Liu, and Goldberg 1980) indicating that ear gland odours can be used for sexual discrimination.

Urine   Male hamsters urine-mark very infrequently in clean cages, whether placed into the cage alone, with another male, or with a female (Maruniak, Desjardins, and Bronson 1975), but hamsters will urinate over the urine mark of another hamster (Dagg *et al.* 1971).

The role of urine in hamster social behaviour has not often been investigated, and the results are inconsistent. Landauer (1975) reported that male hamsters did not show preferences for the urine of female hamsters or castrated male hamsters over the urine of intact males and Johnston (1974) found no preference for female urine odours over no odour by male hamsters. Evans and Brain (1974) reported that smearing castrated male hamsters with urine from intact males or intact females did not influence the aggressive behaviour of territorial male or female hamsters when the urine-smeared intruder was introduced into their home-cage. Payne (1974), however, found that castrated males smeared with the urine of intact males were attacked more often than castrates smeared with urine from female hamsters or castrated male hamsters. Whether or not hamsters are able to use urine odours to discriminate between male and female hamsters thus remains in doubt.

The urine of stressed hamsters is avoided by other hamsters and produces an increase in flight responses, suggesting that hamsters may release a fear odour in the urine (Sherman 1974).

Female hamster vaginal discharge (FHVD)   The female hamster has a four-day oestrous cycle during which the vaginal discharge shows characteristic changes in consistency and in quantity produced (Orsini 1961; O'Connell, Singer, Stern, Jesmajian, and Agosta 1981). Females have a vaginal pouch which stores the vaginal secretion and appear to have muscular control over the amount secreted, releasing small quantities when vaginal marking and large quantities before copulation (Johnston, Zahorik, Immler, and Zakon 1978; LaVelle 1951).

Male hamsters are attracted to the vaginal odour of females, whether the female is in oestrus or dioestrus (Carmichael 1980; Johnston 1974, 1980, 1981*b*; Landauer and Banks 1973; Murphy 1973) and this attraction does not depend on the male's sexual experience (Gregory, Engel, and Pfaff 1975; Landauer, Banks and Carter 1977; Macrides, Johnson, and Schneider 1977). Male hamsters do, however, discriminate between the bedding of oestrous and dioestrous females, flank-marking less in cages of oestrous females (Johnston 1980). Males can also discriminate between oestrous (cycling) females and pregnant and lactating (non-cycling) females, showing a greater preference for the oestrous females (Johnston 1980).

Male hamsters are less aggressive toward and attempt to mate with males which have been smeared with FHVD (Johnston 1975*a*; Murphy 1973). Placing FHVD on to inanimate objects such as clay or fur increases the time that males spend sniffing and licking these objects, but sexual responses are given only to anaesthetized male or female hamsters which have been smeared with FHVD; inanimate objects are not mounted (Darby, Devor, and Chorover 1975). The responses of males toward FHVD are the same whether the discharge is taken from oestrus, dioestrus, or ovariectomized females, indicating that the FHVD is attractive to males and stimulates mating during all phases of the oestrous cycle (Darby *et al.* 1975; Kwan and Johnston 1980; Landauer *et al.* 1978). Vaginectomized females are investigated less than intact females, but males mate with vaginectomized females as readily as with intact females, indicating that FHVD is not necessary to initiate copulation (Kwan and Johnston 1980).

Vaginal odours of females elicit sniffing and licking for at least 25 days (Johnston and Lee 1976) and significant investigation of vaginal odours occurs even when the odour is 100 days old (Johnston and Schmidt 1979), suggesting that the active components of this odour are non-volatile or bound to non-volatile substances.

Many attempts have been made to identify the chemical components of FHVD which are attractive to males. Singer, Agosta, O'Connell, Pfaffman, Bowen, and Field (1976) used gas chromatography to identify the volatile components of FHVD. Male hamsters were presented with fractions in an attraction test and the active components of the FHVD were identified by mass spectrometry as dimethyl-disulphide and dimethyl-trisulphide. Male hamsters spend almost as much time sniffing dimethyl-disulphide as they do sniffing FHVD in a preference test in which the odours are presented in bottles. When dimethyl-disulphide is smeared on anogenital areas of anaesthetized males, it is investigated more than no odour, but less than FHVD. The concentration of dimethyl-disulphide in the FHVD shows a cyclic fluctuation, having a peak at oestrus and the lowest concentration at dioestrus (O'Connell *et al.* 1981). Enough dimethyl-disulphide is produced at all stages of the oestrous cycle to explain why FHVD is attractive to males irrespective of the female's hormonal state and males are sensitive to very small amounts of dimethyl-disulphide (O'Connell *et al*. 1979).

While males attempt to mount those smeared with FHVD, almost no mounting is elicited by dimethyl-disulphide (Macrides *et al.* 1977). Thus, dimethyl-disulphide is as attractive to males as FHVD when presented in a preference test, but is not effective in eliciting sexual arousal when applied to an anaesthetized male. In an attempt to determine other chemicals which might act in conjunction with dimethyl-disulphide to produce attractiveness and sexual arousal, O'Connell *et al.* (1978) identified eight aliphatic acids and four alcohols in FHVD and used these in combination with dimethyl-disulphide. These acids and alcohols were not themselves attractive to males, but when mixed with dimethyl-disulphide, they slightly enhanced its attractiveness. This compound, however, was still far less attractive than FHVD. Rubbing the dimethyl-disulphide + alcohol + acid mixture over the anogenital area of anaesthetized males did not produce sexual arousal in test males.

The chemical in FHVD which stimulates mounting appears to be a high-molecular-weight compound, and the isolation of this chemical requires different analytical procedures than the isolation of volatile chemicals. Using ion-exchange chromatography, it is possible to separate out a fraction of FHVD which is as attractive as FHVD and stimulates the same degree of mounting by males (Singer *et al.* 1980). As yet the active chemicals in this fraction have not been identified.

Vaginal-marking Female hamsters vaginal-mark by rubbing their vagina on the ground using a posture that resembles the perineal drag (Dieterlen 1959; Johnston 1977*a*). Vaginal-marking is correlated with sexual behaviour and appears to be used for sexual solicitation. Vaginal-marking occurs more frequently in the presence of male odours than in the presence of female odours or clean areas (Johnston 1977*a,b*, 1979). During social encounters with males, females vaginal-mark most frequently when in pro-oestrous (Johnston 1977*a*, 1979). Like golden hamsters, female Turkish hamsters vaginal-mark most frequently during pro-oestrus and appear to use vaginal marks to attract males (Frank and Johnston 1981). Pregnant and early lactating females (7–9 days post-partum) show virtually no vaginal-marking, whereas females in late lactation (21–23 days post-partum) mark at high rates (Johnston 1979). Olfactory bulbectomy inhibits both flank-marking and vaginal-marking in female hamsters (Kairys, Magalhaes, and Floody 1980).

Odour-related behaviour   Olfaction is essential for mating behaviour in male hamsters, as anosmia produced by olfactory bulbectomy or by infusion of procaine hydrochloride eliminates male sexual behaviour (Devor and Murphy 1973; Doty and Anisko 1973; Murphy and Schneider 1970). Bilateral transection of the lateral olfactory tract inhibits both sexual behaviour and flank-marking in the male hamster (Macrides, Firl, Schneider, Bartke, and Stein 1976). This inhibition of sexual behaviour also occurs when the vomeronasal organ alone is deafferentated (Powers and Winans 1975). Olfactory bulbectomy does not affect sexual behaviour (lordosis) in female hamsters (Carter 1973).

Male hamsters are attracted to the odours of other males, as well as the odours of females, and prefer to investigate male odours to non-odourized stimuli, even if the odours are those of dominant males who have defeated them in aggressive encounters (Solomon and Glickman 1977). Male hamsters do, however, prefer the odours of castrated males to those of intact males, or castrated males with testosterone injections, indicating that the absence of testosterone increases the attractiveness of a male's odour for other males (Landauer, Banks, and Carter 1977). The scent of an alien male hamster increases the latency of a male to enter a novel environment from its home-cage (Alderson and Johnston 1975).

While FHVD elicits sniffing, licking, and mounting in sexually naive males, suggesting that the association of FHVD with sexual behaviour is not learned (Johnston 1975a), male hamsters will avoid FHVD if it is associated with illness through injections of lithium chloride (Johnston and Zahorik 1975; Zahorik and Johnston 1976; Johnston et al. 1978). When male hamsters are made ill after ingestion of a novel substance (phenylacetic acid), they do not avoid mating with a female which has been swabbed with this substance but do show longer latencies to investigate the female (Emmerick and Snowdon 1976). Males which have been poisoned after their first exposure to FHVD also show longer latencies to mount females, fewer mounts, and less time mounting than control males but they do not fail to mate (Johnston et al. 1978).

Female hamsters, whether sexually experienced or naive, prefer the odours of intact males to those of castrated males (Johnston 1981b). Female Turkish (*M. brandti*), Romanian (*M. neutoni*), and Syrian (*M. auratus*) hamsters spend more time investigating the odours of conspecific males than those of heterospecific males. Female Turkish hamsters also present to and mate with conspecific males, but attack heterospecific males (Murphy 1977, 1978), indicating that odours may be used for species identification.

Development of odour preferences   Hamsters are attracted to their home-cage odours as early as 7-8 days of age, and are also attracted to the odours of alien male and female hamsters, although the odours of lactating females and their pups, whether from their own or another litter, are preferred to other hamster odours (Devor and Schneider 1974; Gregory and Bishop 1975). Infant hamsters avoid certain odours as early as 3-4 days of age, but adapt to these odours after continuous exposure (Cornwell 1975, 1976). Female odour preferences seem to be more easily modified by odour exposure in infancy than male odour preferences (Cornwell-Jones and Holder 1979).

Both male and female hamsters begin to investigate female vaginal secretions when they are seven days of age, and show considerable interest in FHVD at 14 days of age, but this wanes by 17 days of age. Male, but not female hamsters show preferences for FHVD from 43 days of age and this preference is testosterone dependent; testosterone injections produce preferences for FHVD in 25-day-old male and female hamsters (Johnston and Coplin 1979).

Lophiomys

Male *Lophiomys* have preputial glands (Brown and Williams 1972) and a scapular (dorsal) gland (Schaffer 1940) or side glands (Kingdon 1974) but nothing is known of the social function of these glands.

Dicrostonyx

Glands   The collared lemming (*Dicrostonyx groenlandicus*) has small sebaceous glands of the oral angle and lips (identified on one female; Quay 1965*b*); a caudal sebaceous gland (rump gland) found on both males and females (Quay 1968; see Fig. 9.5); preputial or clitoral glands (Quay 1960) and glands on the soles of the feet (Schaffer 1940). Both *Dicrostonyx groenlandicus* and *D. hudsonius* have large Meibomian or tarsal glands in the eyelids (Quay 1954*b*). Lemmings deposit faeces on paths and at burrow entrances (Huck and Banks 1979), but it is not known whether or not these faeces have any communicative function.

Odour-related behaviour   Collared lemmings can discriminate between familiar and unfamiliar animals of the opposite sex by their odours, suggesting that odours can be used for individual recognition (Huck and Banks 1979). Lemmings can also discriminate between conspecifics of the opposite sex and those of another related species (*Lemmus trimucronatus*) by their odours (Huck and Banks 1980*a*). It is possible that lemmings learn to recognize their species through early olfactory experiences, as male *Dicrostonyx* cross-fostered to *Lemmus trimucronatus* dams at one day of age prefer the odours of female *Lemmus* in adulthood (Huck and Banks 1980*a*) and copulate with *Lemmus* females, whereas non-fostered *Dicrostonyx* males do not attempt to mate with *Lemmus* females (Huck and Banks 1980*b*). Cross-fostered female *Dicrostonyx*, however, do not show olfactory preferences for *Lemmus* males nor do they mate with male *Lemmus* (Huck and Banks 1980*a*,*b*).

Synaptomys

The southern bog lemming (*Synaptomys cooperi*) has sebaceous and sudoriferous glands of the oral angle and lip (identified on one female; Quay 1962, 1965*b*) and Meibomian glands around the eye (Quay 1954*b*). Both *S. cooperi* and *S. borealis* (the northern bog lemming) have flank glands which are sexually dimorphic, occurring on 95 per cent of adult males and 9 per cent of adult females (Quay 1968; see Fig. 9.5). The flank gland appears to secrete all year round, without seasonal fluctuation in size or secretory rate (Quay 1968). No observations seem to have been made on the behaviour associated with these glands.

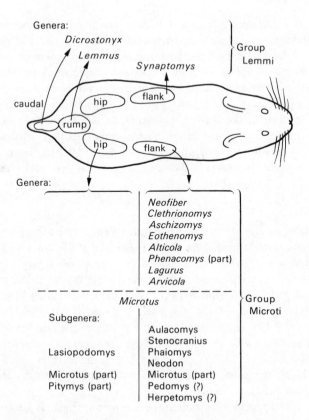

Genera:
Dicrostonyx
Lemmus
Synaptomys

Group
Lemmi

caudal
hip
flank
rump
hip
flank

Genera:

Neofiber
Clethrionomys
Aschizomys
Eothenomys
Alticola
Phenacomys (part)
Lagurus
Arvicola

Microtus

Group
Microti

Subgenera:

Aulacomys
Stenocranius
Lasiopodomys   Phaiomys
Neodon
Microtus (part)   Microtus (part)
Pitymys (part)   Pedomys (?)
Herpetomys (?)

Fig. 9.5. The locations of caudal, rump, hip, and flank glands of microtene rodents and a provisional taxonomic classification of these glands. (From Quay (1968) © American Society of Mammalogists.)

Lemmus

Glands   Male *Lemmus* have preputial glands (Brown and Williams 1972); sebaceous and sudoriferous glands of the oral angle and lips (Quay 1965*b*); ear glands; and glands on the soles of the feet (Schaffer 1940). Both male and female Norway lemmings (*L. lemmus*) and brown lemmings (*L. trimucronatus*) have Meibomian glands (Quay 1954*b*) and rump (caudal) glands (Quay 1968; see Fig. 9.5). Subspecies of *L. trimucronatus* living on the Pribilof Islands (*L. nigripes*) and in Siberia (*L. obensis*) also possess rump glands (Quay 1968). The Norway lemming defecates along trails and runways and these 'lavatory piles' may serve some communcation function (Myllymaki, Aho, Lind, and Tast 1962).

Odour-related behaviour   Both male and female brown lemmings are able to discriminate opposite sex conspecifics from *Dicrostonyx groenlandicus* by their odour (Huck and Banks 1980*a*) and females are able to discriminate between dominant and subordinate males by their odours (Huck, Banks, and Wang 1981) The source of the odours used for these discriminations is unknown.

Brown lemmings appear to learn the odour characteristics of their species by olfactory experience as males and females cross-fostered onto *Dicrostonyx* dams show a preference for *Dicrostonyx* of the opposite sex in adulthood (Huck and Banks 1980*a*). Unlike *Dicrostonyx* males, cross-fostered *Lemmus* males do not mate with females of the other species, while *Lemmus* females do allow *Dicrostonyx* males to mount them (Huck and Banks 1980*b*). Sexual behaviour between heterospecific animals, however, occurs far less often than among conspecifics.

Clethrionomys

Glands   Males of five species of *Clethrionomys* (*caesarius, gapperi, glareolus, rufocanus*, and *rutilus*) have flank glands (Lehmann 1962; Quay 1968; see Fig. 9.5). All species of *Clethrionomys* have Meibomian glands (Quay 1954*b*). California redback voles (*C. occidentalis*) (one male); Boreal redback voles (*C. gapperi*) (one male); and bank voles (*C. glareolis*) (one female) have sebaceous and sudoriferous glands of the oral angle and lips (Quay 1962, 1965*b*; see Plates 9.11 and 9.12). In the bank vole, the glands of the oral lips appear to be sexually dimorphic, with males having larger secretory tubules than females (Griffiths and Kendall 1980*b*). Male and female bank voles of all ages have sweat glands on the soles of the feet (plantar glands) (Griffiths and Kendall 1980*a*; see Plate 9.13).

Both male and female *Clethrionomys* have preputial glands which are larger in males than females, larger in adults than sub-adults, and show a seasonal fluctuation in size (Brown and Williams 1972; Christiansen, Wiger, and Eilertsen 1978; Jackson 1938; Kratochvil 1962; Paclt 1952; see Fig. 9.6 and Plate 9.14). Dominant male bank voles have larger preputial glands than subordinate males (Gustafsson, Andersson, and Meurling 1980), suggesting androgen control of glandular development and secretion of the preputial glands.

Marking   Although bank voles have flank glands, flank-marking does not seem to occur (Johnson 1975). Bank voles do, however, urine-mark, leaving urine trails (see Fig. 9.2(3)). Only sexually mature males urine-mark in trails, females and immature males leave 'puddles' (Christiansen 1980; Johnson 1975). Preputial gland secretion may be deposited along with urine during marking and these trails may serve as sex attractants or for the identification of dominance, territoriality, or aggressive motivation (Christiansen 1980).

Odour-related behaviour   Male bank voles can discriminate between females of a number of subspecies of *Clethrionomys* and prefer the odours of oestrous females of their own subspecies to those of other subspecies (Godfrey 1958). Male bank voles do not, however, prefer the odours of conspecific males to the odours of *Microtus*, nor do they avoid the odours of conspecific males (De Jonge 1980).

Eothenomys

Four species of *Eothenomys* have Meibomian glands (Quay 1954*b*) and some species have flank glands which appear to increase in size during the breeding season (Quay 1968; see Fig. 9.5). Little seems to be known about the behaviour associated with these glands.

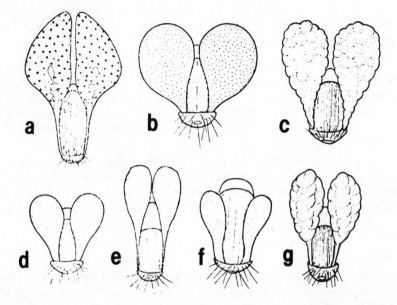

Fig. 9.6. Preputial glands of microtene rodents (a) *Clethrionomys glareolus*; (b) *Microtus (Chionomys) nivalis*; (c) *Microtus (Lasiopodmys) brandti*; (d) *Pitymys tatricus*; (e) *Microtus oeconomus*; (f) *Lagurus lagurus*; and (g) *Microtus middendorfi*. (From Kratochvil (1962) © Academia, Publishing House of the Czeckoslovak Academy of Sciences.)

Alticola

Adult males and some females of *Alticola montosa*, *A. stracheyi*, and *A. worthingtoni* have flank glands similar to those of *Clethrionomys* (Quay 1968; see Fig. 9.5), but the communicatory function of these glands is unknown.

Arvicola

Glands The water vole (*Arvicola amphibius*) and the ground vole (*Arvicola terrestris*) have Meibomian glands (Quay 1954*b*); preputial glands (Jackson 1938;

Brown and Williams 1972); anal glands, ear glands, glands of the oral lips and glands on the soles of the feet (Schaffer 1940). *Arvicola* of both sexes have highly developed flank glands which begin to develop immediately after birth (Fig. 9.5). In adulthood, males have larger flank glands than females, especially in the breeding season (Vrtis 1930*a*, 1931; Quay 1968; Schaffer 1940). Male *Arvicola terrestris*, for example, have the largest flank glands from June to August, and greatly reduced glands from November to February (Stoddart 1972*a*; see Plate 9.15). Castration reduces the size of the flank gland and testosterone injections increase flank gland size (Stoddart 1972*a*). Chemical analysis of the flank-gland secretions of *Arvicola terrestris* using gas–liquid chromatography indicates sex, age, and population differences in the chemical makeup of these secretions (Stoddart, Aplin, and Wood 1975).

Marking   Male *Arvicola terrestris* and *A. richardsoni* scent-mark by rubbing or scratching the hind-feet over the flank gland and stamping or drumming the feet on the ground. Such marking has been called 'drum-marking' (Frank 1956; Jannett and Jannett 1974). Males drum-mark most frequently in the presence of an alien male or his odour; seldom in a clean area. Females do not drum-mark at all (Jannett and Jannett 1974).

  Arvicola males, like hamsters, appear to use the odour of the flank gland in agonistic encounters. It also seems likely that *Arvicola* can use flank-gland odours for sex, age, rut, and subspecies identification, but no behavioural studies have been done on olfactory communication in these species, which appear ideally suited for this research.

Ondatra

Both male and female muskrats (*Ondatra zibethicus*) have preputial glands which are modified to form large scent glands (Jackson 1938; Brown and Williams 1972). Muskrats have Meibomian glands (Quay 1954*b*); and sebaceous glands in the oral angle (one male; Quay 1965*b*). Ewer (1968) reports that the muskrat has anal glands and marks with an anal drag, but there seem to be no other behavioural studies of these rodents.

Neofiber

The Florida water rat (*Neofiber alleni*) has large sebaceous glands in the oral angle and lip (one female; Quay 1965*b*) and Meibomian glands (Quay 1954*b*). Both males and females of all ages of *Neofiber* also possess flank glands but these glands may not be large (Quay 1968; Schaffer 1940; see Fig. 9.5) and their social function has not been studied.

Phenacomys

Tree phenacomys (*P. longicaudus*) have sebaceous glands in the oral angle (one male; Quay 1965*b*), and both tree phenacomys and mountain phenacomys (*Phenacomys intermedius*) have Meibomian glands (Quay 1954*b*). The adult male mountain phenacomys has flank glands (flank patches) which do not occur on *P. albipes* or *P. longicaudus* (Quay 1968; see Fig. 9.5).

Pitymys

The pine vole (*Pitymys (Microtus) pinetorum*) and *P. ibericus* both have sebaceous glands of the oral angle and lips (one female of each; Quay 1965*b*), and at least five species of *Pitymys* (*neomoralis, pinetorum, pelandorius, savi, subterraneus*) have Meibomian glands (Quay 1954*b*). Hrabě (1974) found that *P. subterraneus* has a greater number of tarsal (Meibomian) glands than *P. tatricus* and thus considers the number of tarsal glands to be a distinguishing feature of the two species. No sex differences were found in either species in number of tarsal glands. *P. tatricus, P. subterraneus*, and *P. pinetorum* also have anal glands which are enlarged in breeding males and may be used for communication during sexual activity (Benton 1955; Hrabě 1973). Some species of *Pitymys* have hip glands (Quay 1968; see Fig. 9.5) and some, including *Pitymys tatricus*, have preputial glands (Kratochvil 1962; see Fig. 9.6).

Pine voles may use olfactory cues to recognize the sex of conspecifics and to determine whether or not specific individuals are familiar (Geyer, Beauchamp, Seygal, and Rogers 1981). Infant pine voles prefer odours of their own home-cage to other vole odours at 16-20 days of age, and can discriminate between home-cage odours and clean bedding as early as six days of age. Infant pine voles also produce more ultrasounds when in the presence of home-cage odours than when in the presence of clean bedding (Geyer 1979).

Microtus

Glands Meibomian glands are found in at least 18 species and subspecies of *Microtus* (Quay 1954*b*) and at least six species (*californicus, oregoni, abbreviatus, nivalis, richardsoni*, and *mexicanus*) have sebaceous glands of the oral angle and lip (Quay 1962, 1965*b*). Males of a number of species of *Microtus* have preputial glands (Kratochvil 1962; Jackson 1938; Brown and Williams 1972; Jannett 1978) (see Fig. 9.6). Male *M. montanus, M. arvalis*, and *M. nivalis* have anal glands (Hrabě 1973; Jannett 1978; Schaffer 1940). *M. arvalis* also have glands on the soles of the feet (Schaffer 1940).

The most-studied cutaneous glands of *Microtus* are the glands on the hindquarters. Some species, such as *M. richardsoni* have flank glands; others, such as *M. koreni*, have hip glands; and some, such as *M. oregoni, M. umbrosus, M. breweri, M. longicaudus, M. mexicanus, M. nivalis*, and *M. pennsylvanicus* have no posterolateral glands (Lehmann 1966; Quay 1968; Schaffer 1940; see Fig. 9.5).

At least 11 species of *Microtus* have flank glands (*richardsoni, brandti, gregalis, major, muirus, abbreviatus, leucurus, irene, roberti, xanthognathus*, and *chrotorrhinus*) (Kratochvil 1962; Quay 1968). The flank glands are largest in breeding males but occur in immature males and in females of some species.

In *Microtus (Aulacomys) richardsoni*, for example, the flank gland is present in both sexes at all ages, while in *Microtus (Stenocranius) gregalis* and *M. (Steno-cranius) muirus* the flank gland is largest in adult males, and occasionally seen in lactating females, but seldom seen in non-lactating females or immature animals of either sex. In *Microtus (Phaiomys) leucurus*, the flank gland shows a seasonal fluctuation, being larger in the summer in the males, and sometimes detectable in females in the summer (Quay 1968).

At least 13 species of *Microtus* have hip glands (*agrestis, arvalis, californicus, fortis, guentheri, incertus, middendorfi, montanus, montebelli, oeconomus, orcadenis, townsendii*, and *ungurensis*) (Kratochvil 1962; Quay 1968). As with the flank gland, there are age and sex differences in the distribution of hip glands in different *Microtus* species. Some species such as *M. oeconomus* (*M. koreni*) have hip glands in adults of both sexes; other species such as *M. macfarlani* have hip glands in adult males, but in only 30–60 per cent of females, and other species such as *M. limnophilus, M. agrestis*, and *M. californicus* have hip glands in adult males but not adult females (Clarke and Frearson 1972; Lidicker 1980; Quay 1968; see Plate 9.16). The size of the hip glands on male voles is variable (Lidicker 1980) but hip glands are usually largest on the oldest males and their presence on females may be related to geographical variables, with females from more northerly climates (*M. koreni*) having hip glands while those from more southerly areas (*M. limnophilus*) have no hip glands (Quay 1968).

In the short-tailed vole (*Microtus agrestis*), Townsend vole (*M. townsendii*), and the montane vole (*M. montanus*) the hip gland of the male has been shown to be under androgen control. Castration reduces the size of the hip gland and testosterone replacement increases the size of the gland in proportion to the dose of the hormone administered (Clarke and Frearson 1972; Jannett 1978; MacIsaac 1977). Seasonal changes in testosterone levels in *M. agrestis* and *M. townsendii* result in large hip glands in the breeding season (May) and very small glands in the non-breeding season (December) (Clarke and Frearson 1972; MacIsaac 1977). Female *M. agrestis* given testosterone injections have large increases in gland size, but the largest female glands are only one-third the size of male glands (Clarke and Frearson 1972). The anal and preputial glands of *M. montanus* are also testosterone dependent (Jannett 1978).

Even though *M. pennsylvanicus* and *M. longicaudus* generally do not possess posterolateral glands, injection of testosterone stimulates the development of enlarged sebaceous gland units in the hip regions thence giving rise to true hip glands, suggesting that these species are closely related to those with hip glands naturally (Jannett 1975).

After observing olfactory investigations during social play in juvenile *M. agrestis*, Wilson (1973) suggested that juveniles may secrete an odour from skin at the back of the head which functions to stimulate play. Using thin-layer chromatography, she showed that skin from the back of the head of juvenile *M. agrestis* had a different pattern of components than skin from adults.

Marking   Montane voles mark with their preputial glands using a 'pelvic press' in which the vole depresses its anogenital area onto the substrate (Jannett 1978). Montane voles also mark with the hip gland. To do this the vole backs into a tunnel, raises the hindquarters, and rubs its hips on the sides of the tunnel (Jannett 1978; Jannett and Jannett 1974). Both male and female *M. townsendii* hip-mark in the same manner, moving the hips in 'figure-eight motions' to mark the sides of a tube (MacIsaac 1977). Male *M. californicus* hip-mark the sides of burrows by walking with a 'swagger' (Lidicker 1980). *M. montanus* anal-mark using a 'scooting' behaviour or anal drag in which the anus is rubbed against the substrate while the animal moves forward (Jannett 1978). Male and female meadow voles (*M. pennsylvanicus*) urine-mark in the presence of dried urine from conspecifics (Dagg *et al.* 1971) and montane voles urine-mark runways, stones, and twigs (Jannett 1981*b*). Dominant, territorial male montane voles leave faeces in conspecuous places and these males are more likely to scent-mark with the anal gland than subordinate or juvenile males (Jannett 1981*b*).

Odour-related behaviour   Voles are attracted to conspecific odours and traps containing urine and faeces are entered more often by Townsend voles (*M. townsendii*) than clean traps (Boonstra and Krebs 1976). Short-tailed voles (*M. agrestis*) are more likely to enter traps with their own odour or the odour of a conspecific of the opposite sex (Stoddart 1982).

   Female common voles (*M. arvalis*) can discriminate between the urine odours of male mice (*Mus*) and those of male *M. arvalis*, and show a preference for the odours of conspecific males (Vadasz 1975). Given a choice between odours from *M. agrestis* and *M. arvalis*, the *M. agrestis* prefer conspecific odours, while *M. arvalis* do not (De Jonge 1980). Both male and female *M. arvalis* and *M. oeconomus*, however, appear to be able to discriminate between conspecific and heterospecific odours (Skurat, Litvin, and Golubev 1976). Some species of voles (such as the montane vole) may learn to identify the odours of their own species through early exposure to conspecifics such as parents and siblings, while other species (such as they grey-tailed vole (*Microtus canicaudus*)) may prefer the odours of conspecifics irrespective of their early rearing experiences (McDonald and Forslund 1978).

   In many species of *Microtus*, olfactory communication plays a role in both inter- and intraspecific aggression. *M. agrestis*, for example, are aggressive toward conspecifics and will avoid the odours of conspecifics, particularly after agonistic encounters (De Jonge 1980). Attacks by male montane voles (*M. montanus*) are directed at the flank glands of conspecifics and heterospecifics more often than at the non-glandular tissue (Jannett 1981*a*), suggesting that males use flank-gland odours for species, sex, and/or age discrimination. Males of species without posterolateral glands such as *M. longicaudus* do not direct their attacks to the hip areas of other males more than to any other area (Jannett 1981*a*). The introduction of female vaginal and urine odours into groups of male prairie voles (*M. ochrogaster*) provokes an increase in the number of investigatory and aggressive interactions among the males (Stehn, Richmond, and Kollisch 1976).

Scent glands may be useful for the study of the taxonomic relationships among the sub-genera of *Microtus* (see Quay 1954*b*, 1968) but little is known of their behavioural function. Species, sex, age, and possibly dominance status may be determined from gland size in some species, as may reproductive state. The use of odours in aggressive and sexual encounters is, however, not well understood, nor is the motivational control of the different types of scent-marking.

Wilson (1973) observed that juvenile *M. agrestis* sniffed the rump, nose, mouth, and back-of-head of other juveniles. She suggests that the odour from the back of the head of juveniles stimulates play, that the odour from the glands of the oral lips indicates the sex of the animal and the odour from the flank-rump area gives animals an individual scent. More research is required before these hypotheses can be confirmed.

Lagurus

The sagebrush vole (*Lagurus curtatus*) has sebaceous glands of the oral angle and lips (one male; Quay 1965*b*) and *L. paurperrimus* have Meibomian glands (Quay 1954*b*). Dearden (1959) has carefully examined Meibomian glands from *L. lagurus* and six subspecies of *L. curtatus* and found species differences in the distribution of these glands. There are also species differences in flank glands. Adult *L. przewalskii* of both sexes have flank glands, but only adult male *L. curtatus* have flank glands (Quay 1968) and *L. lagurus* have no flank glands (Kratochvil 1962).

L. curtatus scent-mark by scraping the hind-feet over the flank glands and stamping or drumming their feet on the substrate (Jannett 1981*a*). These voles direct attacks toward the flank glands of conspecific males (Jannett 1981*a*) and may thus use the gland for sex discrimination.

Gerbillus

Gerbillus gerbillus has small sebaceous glands of the oral angle and lips (one female; Quay 1965*b*) and *G. pyramidarum* have preputial glands and ear glands (Schaffer 1940). *Gerbillus* scent-mark using a side rub and a perineal drag (Eisenberg 1967; Fiedler 1973).

Tatera

Tatera indica have small sebaceous glands of the oral angle and lips (Quay 1965*b*). These gerbils scent-mark using the perineal drag and show fully integrated patterns of sandbathing including ventral rubbing, side rubbing, and rolling over (Eisenberg 1967). Whether urine or other secretions are transferred to the substrate during the perineal drag and during sandbathing is unknown.

Meriones

Many species of *Meriones* have ventral glands (Sokolov and Skurat 1966). *M. unguiculatus* have Harderian glands (Thiessen, Clancy, and Goodwin, 1976; Thiessen and Yahr 1977), and may use urine (Dagg *et al.* 1971) and saliva (Block, Volpe, and Hayes 1981) for olfactory communication. Thiessen, Yahr, and Lindzey (1971) report that *M. unguiculatus* have a sebaceous gland complex around the neck and chin, but no histological studies seem to have been done to support this. *Meriones tristrami* has sebaceous glands of the oral angle and lip (Quay 1965*b*). Unless otherwise noted, the use of the term gerbil refers to *M. unguiculatus*.

Saliva   Saliva is secreted from the submaxillary and sublingual glands (see Fig. 9.7). Mouthing or 'mouth licking' occurs often during social behaviour of gerbils (Fiedler 1973) and, in a series of experiments, Block *et al.* (1981) have shown that 25–36-day-old gerbil pups are able to discriminate between the saliva of their mother and father when saliva samples are presented on an anaesthetized virgin female. Sub-adult (65–70-day-old) male and female gerbils prefer saliva odours of littermates to those of non-littermates, and adult males prefer the saliva odours of oestrous females to those of non-oestrous females. Thus, saliva appears to provide a maternal odour, a littermate odour, and sex odour in gerbils.

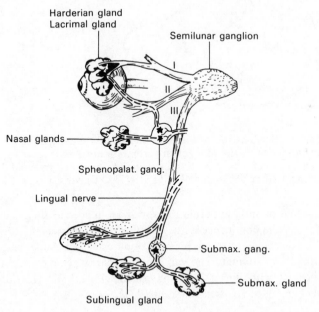

Fig. 9.7. The facial glands of the gerbil and their neural connections. These include the Harderian gland, nasal gland, and the salivary (sublingual and submaxillary) glands. (From Thiessen (1977) © Plenum Press.)

Harderian gland   Secretions from the Harderian glands, as well as saliva, are spread over the gerbil's face during grooming (Fig. 9.4). The relationship between the Harderian glands and the salivary glands is shown in Fig. 9.7. Figure 9.8 shows the location of the Harderian glands and the Harder–lacrimal canal. Males have larger glands than females, but both sexes have the same gland:body weight ratio (Thiessen and Yahr 1977). Male gerbils have high rates of Harderian gland secretion after grooming their face and are investigated more by other males after completing a bout of grooming (Thiessen *et al.* 1976). Male gerbils with their Harderian glands removed are investigated less than intact males and intact males are dominant to Harderianectomized males (Thiessen and Yahr 1977).

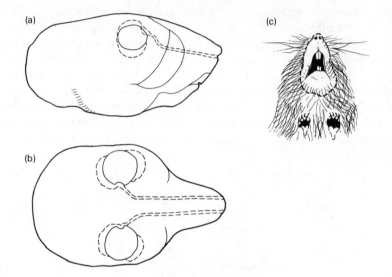

Fig. 9.8. (a, b) The location of the Harderian gland and the Harder–lacrimal canal leading to the nares of the Mongolian gerbil, *Meriones unguiculatus* and (c) the areas of the face over which the Harderian gland secretion is smeared during facial grooming. (From Theissen and Yahr (1977) © University of Texas Press.)

Harderian-gland odours thus seem to provide an important social cue, and may be used for sex and dominance status identification (Thiessen and Rice 1976).

Male gerbils can learn to bar-press for food using the Harderian-gland odour as a discriminative stimulus, and will avoid Harderian-gland odours if these have been associated with illness by being paired with lithium chloride injections (Thiessen *et al.* 1976; Thiessen and Yahr 1977).

Urine   Both male and female gerbils urine-mark in the presence of urine from other gerbils (Dagg *et al.* 1971), but males urine-mark more often than females (Thiessen, Blum, and Lindzey 1970). Male gerbils urine-mark at low rates when isolated (once every six minutes) and more often when paired with another male or female (Maruniak, Desjardins, and Bronson 1975).

Male gerbils are attracted to the urine odours of females, preferring the urine of oestrous females to that of non-oestrous females, but show little attraction to male urine (Pettijohn 1974, 1977). Pettijohn (1977) found that female gerbils do not show preferences for investigating male urine odours but, using rearing as a measure of preference, Gregg and Thiessen (1981) found that oestrous female gerbils showed a preference for urine of intact males over that of castrated males while males did not show a preference for the urine of oestrous over that of non-oestrous females. Thus it is possible that urine odours may differentiate sexually active gerbils of both sexes from sexually inactive gerbils, but different measures of preference give different results. Urine can be used by gerbils to discriminate between males and females and to identify individual males (Dagg and Windsor 1971).

Ventral gland Mid-ventral abdominal glands occur in both sexes of at least nine species of *Meriones (unguiculatus, hurrianae, tristrami, persicus, meridianus, erythrourus, crassus, vinogradovi*, and *tamariscinus*; Sokolov and Skurat 1966; Fiedler 1973; Kumari and Prakash 1981). Males have larger ventral glands than females in all species, and adults have larger glands than juveniles (Kumari and Prakash 1981; Sokolov and Skurat 1966; see Plate 9.17). In a morphological study, Feldman and Mitchell (1968) identified the ventral gland of male *M. unguiculatus* in a 10-day-old embryo and a ventral gland composed of about 10 cells on one-day-old males. At 28 days of age the gland hypertrophies and continues to enlarge and thicken until 86 days of age when it has up to 300 tubulo-alveolar holocrine glands (Glenn and Gray 1965). In female gerbils the ventral gland develops later than in the male and is not observable until 16–32 weeks of age (Glenn and Gray 1965).

Using thin-layer chromatography and gas chromatography, Thiessen, Regnier, Rice, Goodwin, Isaacks, and Lawson (1974) fractionated the ventral gland secretion of male *M. unguiculatus* and identified the components by mass spectrometry. Using a conditioning paradigm and a preference test, the active component was identified as phenylacetic acid.

The development of the ventral gland is affected by hormones and the social conditions under which the animals are reared. Young gerbils of both sexes kept with their parents have delayed ventral-gland development (Payman and Swanson 1980; Swanson and Lockley 1978). Females are more sensitive than males to inhibition of ventral-gland development. Juveniles reared with parents whose ventral gland has been excised do not, however, reach puberty earlier, nor have larger ventral glands than those reared with intact parents (Blum, Balsiger, Ricci, and Spiegel 1975), indicating that it is the presence of the parents or some odour other than that of the ventral gland that affects puberty.

Male gerbils castrated as juveniles do not develop ventral glands and males castrated in adulthood show progressive decreases in gland size until complete involution occurs eight weeks after castration (Arluk 1968; Blum and Thiessen 1971; Glenn and Gray 1965; Lindzey, Thiessen, and Tucker 1968; Mitchell 1965; Thiessen, Owen, and Lindzey 1971). Testosterone implants in juvenile and adult males increase the glandular size and secretory rate within two weeks (Arluk 1968; Glenn and Gray 1965; Lindzey, Thiessen, and Tucker 1968; Mitchell 1965; Nyby and Thiessen 1971; Thiessen, Friend, and Lindzey 1978). Oestrogen and oestrogen plus progesterone alone have no effect (Arluk 1968; Nyby and Thiessen 1971).

Ovariectomy reduces the size of the ventral gland in adult females (Thiessen, Owen, and Lindzey 1971) and testosterone replacement increases the size of the female gland (Glenn and Gray 1965; Owen and Thiessen 1973; Thiessen 1968; Thiessen and Lindzey 1970). Oestradiol and oestradiol plus progresterone also increase ventral-gland size in ovariectomized females, but progesterone alone has no effect (Owen and Thiessen 1973; Yahr and Thiessen 1975).

Marking　　*Meriones unguiculatus* and *M. hurrianae* scent-mark using a perineal drag and sandbathe using a side rub and rolling over (Eisenberg 1967). *M. unguiculatus, M. tristrami, M. libycus,* and other species mark low-lying objects with their ventral gland using a ventral rub (Daly and Daly 1975a; Fiedler 1973; Thiessen 1968; Thiessen, Wallace, and Yahr 1973; Thiessen, Yahr, and Lindzey 1971; Thiessen, Blum, and Lindzey 1970).

Male *M. unguiculatus* ventral-mark about twice as often as females (Thiessen, Blum, and Lindzey 1970) and this sex difference in marking frequency develops at about 16 weeks of age (Lee and Estep 1971). An intact ventral gland is not necessary for ventral-marking, as adult males whose gland has been surgically removed mark almost as often as unoperated males (Blum and Thiessen 1970; Baran and Glickman 1970; Swanson and Norman 1978).

Extensive research has been done on the hormonal control of ventral-gland marking in both male and female *M. unguiculatus.* In male gerbils, castration reduces scent-marking and testosterone, oestradiol, and oestradiol plus progesterone reinstate marking in castrates (Nyby and Thiessen 1971; Swanson and Norman 1978; Thiessen, Friend, and Lindzey 1968; Turner 1979). The response of male gerbils to androgens is dose dependent and large doses will produce 'super markers' which mark up to twice as often as intact males (Blum and Thiessen 1971; Thiessen, Owen, and Lindzey 1971; Thiessen and Rice 1976; Turner 1979; Yahr 1981). Injections of testosterone into the preoptic area of the hypothalamus will also reinstate scent-marking in castrated males (Thiessen and Yahr 1970; Thiessen, Yahr, and Owen 1973).

The results of ovariectomy and hormone replacement on ventral-marking in females are often contradictory. Scent-marking frequency is often so low in females that ovariectomy has little effect (Swanson and Norman 1978; Thiessen, Owen, and Lindzey 1971) and less than 30 per cent of female gerbils display moderate to high levels of scent-marking (Yahr and Thiessen 1975). Low-marking females differ from low-marking males in that massive hormone doses do not increase marking. However, in females which show higher levels of ventral-marking ovariectomy reduces marking frequency (Wallace, Owen, and Thiessen 1973). On the other hand, ovariectomized gerbils develop ventral-marking behaviour at the same rate as intact females, even though their ventral gland is completely involuted (Whitsett and Thiessen 1972).

Testosterone injections increase ventral-marking in both intact and ovariectomized females (Owen and Thiessen 1973; Swanson and Norman 1978; Thiessen 1968; Thiessen and Lindzey 1970). Oestradiol injections increase marking in ovariectomized females but decrease it in intact females (Thiessen and Lindzey 1970; Owen and Thiessen 1974). Oestradiol followed by progesterone increases scent-marking in ovariectomized females but the amount of scent-marking is very sensitive to the dose of oestradiol and progesterone used (Owen and Thiessen 1974). Testosterone, oestrogen, and oestrogen plus progesterone implanted into the preoptic area and anterior hypothalamus of the female gerbil increase scent-marking (Owen, Wallace, and Thiessen 1974).

Ventral-marking in female gerbils is modified by hormonal changes during the oestrous cycle, pregnancy, and lactation. Female Indian desert gerbils (*M. hurrianae*) scent-mark more during pro-oestrus and oestrus than during di-oestrus (Kumari and Prakash 1981). During pregnancy, the female's ventral gland increases in size, reaching a maximum size two weeks post-partum and then decreases in size. Scent-marking frequency in female *M. unguiculatus* parallels gland growth, reaching a maximum 1-2 weeks post-partum and declining during weaning (Wallace *et al.* 1973). Even females which seldom mark increase their rate of marking during lactation (Yahr 1976). Lactating females mark clean pups, indicating that the ventral-gland secretion may be used for pup identification (Wallace *et al.* 1973).

Social, environmental, and sensory factors as well as hormones affect scent-marking in gerbils. Juveniles which cohabit with their parents have scent-marking delayed or inhibited; sons mark less than half as often than their fathers and daughters mark less often than their mothers (Swanson 1980; Swanson and Lockley 1978). Removal of the father disinhibits marking by the sons, increasing the number which scent-mark. If juvenile males which are cohabiting with their father are injected with testosterone they show a slight increase in marking but still mark less than their brothers living alone (Swanson 1980).

In mixed-sex groups of adult gerbils, males have a dominance hierarchy consisting of sequential alpha-males. In this situation, only the alpha-male ventral-marks and most marking is directed at novel objects (Gallup and Waite 1970; Thiessen, Owen, and Lindzey 1971; Thiessen, Lindzey, Blum, and Wallace 1971). Males which become submissive after being paired with another male show a decrease in scent-marking when compared to the pre-pairing levels (Yahr

1977), while those that become dominant show an increase in marking, especially in the 'territory' of the submissive male (Thiessen, Owen, and Lindzey 1971).

When males and females are paired in a novel arena, males mark frequently but females seldom mark. In male–male encounters, males mark far less than in male–female encounters (Swanson 1974). After repeated male–male encounters, males decrease their marking in areas containing the odours of other males, but mark at higher frequencies in clean (unscented) areas (Nyby, Thiessen, and Wallace 1970; Yahr 1977).

Scent-marking is also influenced by the sensory information received by the gerbil. Visual cues may be important as less marking is done in a dark chamber than in a lighted chamber (Baran and Glickman 1970). Subordinate males decrease their scent-marking in a familiar, but clean area in which they have been defeated by a dominant gerbil (Yahr 1977). Olfactory cues are extremely important in regulating scent-marking in gerbils. Olfactory bulbectomy virtually eliminates ventral-marking in male gerbils (Baran and Glickman 1970; Thiessen, Lindzey, and Nyby 1970; Lumia, Raskin, and Eckhert 1977; Lumia, Westervelt, and Rieder 1975). After massive doses of testosterone, but not 'normal' doses, low-frequency marking is reinstated in some bulbectomized males (Thiessen, Lindzey, and Nyby 1970; Lumia *et al.* 1975). Testosterone does not increase marking in bulbectomized females, however (Lumia, Canino, and Drozdowski 1976).

Male gerbils mark more often in the presence of ventral-gland odours from unknown males than in a 'clean' environment but mark the most in the presence of female gerbil odours (Daly 1977). Males housed singly in the same room as females scent-mark more often than males housed singly in a room containing only males, when they are paired in a neutral arena (Rieder and Reynierse 1971).

Odour-related behaviour Male and female gerbils (*M. unguiculatus*) are able to discriminate between the odours of male gerbils and males of other species (rats, mice, hamsters, and lemmings); between male and female gerbils using body odours or urine; and between two male gerbils using their body odours or urine (Dagg and Windsor 1971). The easiest discrimination is between species odours; discrimination of sex odours and the odours of individual males are more difficult. Male gerbils can also discriminate between individual male gerbils using odours from ventral-gland secretions and soiled cage bedding, but not using faeces odours (Halpin 1974, 1980). Anosmic gerbils are not able to discriminate between familiar and unfamiliar conspecifics and this alters the animals social behaviour (Halpin 1976).

Both male and female gerbils respond to odours of the same and the opposite sex with ultrasonic vocalizations. The highest rate of vocalization occurs when males are presented with the odours of other males. Body hair from male gerbils contains an olfactory stimulus sufficient to elicit ultrasonic vocalizations from other gerbils (Thiessen, Graham, and Davenport 1978).

Male gerbils are attracted to the ventral gland odour of both male and female conspecifics (Baran and Glickman 1970; Baran 1973; Thiessen, Lindzey, Blum, and Wallace 1971), but female gerbils are not attracted to the ventral gland odour of male nor female conspecifics (Thiessen, Lindzey, Blum, and Wallace 1971). Similar results have been found in studies with the Israeli gerbil (*M. tristrami*) in which males are attracted to the odours of conspecific males, but females are not (Thiessen, Wallace, and Yahr 1973). Although male gerbils may avoid the odours of males which have defeated them in an agonistic encounter (Nyby *et al.* 1970; Thiessen and Dawber 1972), this may depend on the test situation, as male gerbils may also show preferences for the odours of male conspecifics after agonistic interactions (Halpin 1978). Gerbils will, however, avoid the soiled bedding odours of conspecifics and the odour of phenylacetic acid (the active component of the ventral-gland secretion) if these have been associated with illness by injection of lithium chloride (Pettijohn 1979; Pettijohn and Jamora 1980).

Infant gerbils (12–20 days old) are attracted to their home-cage bedding odours and to the ventral-gland odours of their mother and father, if he cohabits with them (Cornwell-Jones and Azar 1982; Gerling and Yahr 1982) and this preference is modified by diet: infants reared on 'Purina lab chow' prefer the odour of adults fed this food more than the odour of adults fed on 'Pooch dog food' (Skeen and Thiessen 1977). The ventral-gland odour of lactating female gerbils is preferred to the odour of clean bedding or to the ventral-gland odour of non-lactating females but not to the odour of familiar males (fathers) (Gerling and Yahr 1982).

Olfactory experience may be important in the development of odour preferences. Gerbils reared with mothers whose ventral glands had been excised do not prefer the ventral gland odours of lactating females to those of virgin females and are less attracted to the ventral-gland odours of adult males (Blum *et al.* 1975; Gerling and Yahr 1982). Pups reared with both parents prefer their own nest odour to that of other litters, but those reared with the mother alone do not discriminate between their own nest odour and that of other litters (Gerling and Yahr 1982).

Gerbil odours can be used for individual, sex, rut, species, and parent–offspring recognition. It seems unlikely that odours serve a 'territorial' function, but they may serve to identify particular age and sex classes of individuals and their home-areas. Glandectomized animals seem to lose some of their identity and have a reduction in social attention. There is some evidence that ventral-gland size and scent-marking are related to dominance, but no indication that the ventral gland is necessary for sexual activity (Mitchell 1967). Olfactory bulbectomy does not inhibit mating in male gerbils (Cheal and Domesick 1979).

Psammomys

The 'sandrat' (*Psammomys obesus*) has a mid-ventral sebaceous gland, and a glandular patch under the chin (Daly and Daly 1975*b*). Both of these glands are used for scent-marking objects in the environment. The sandrat also shows the side-rubbing component of sandbathing, but does not show fully integrated sandbathing. Sandrats urine-mark frequently and produce 'urine balls'. These are made by digging up sand, urinating on this and then making it into a 'ball' which is investigated by other sandrats. Males are particularly interested in investigating the urine balls of oestrous females (Daly and Daly 1975*b*).

Rhombomys

The great gerbil (*Rhombomys opimus*) has a mid-ventral abdominal gland and marks using a ventral rub. Urine-marking may also be used for communication in this species (Holzman and Paskhina 1974).

Spalacidae

The fossorial mole rat (*Spalax ehrenbergi*) has different chromosomal forms ($2n = 52, 54, 58$ or 60). Aggression levels are high between different chromosomal forms (karotypes) and mating occurs primarily between animals of the same form (Nevo 1969; Nevo, Naftali, and Guttman 1975). These mole rats are blind and females prefer males of their own karotype and can discriminate between these males by the odours of their soiled bedding and urine (Nevo and Heth 1976; Nevo, Bodmer, and Heth 1976). Little seems to be known about the location and role of skin glands in communication in *Spalax*.

Rhizomyidae

Nothing appears to be known about the scent glands nor the role of odours in social communication of the three genera (18 species) of this family (see Kingdon 1974; Walker 1975).

Muridae

This family of 100 genera and over 840 species (Walker 1975) has not been well represented in studies of scent glands and social behaviour. The majority of research has been done with *Rattus* and *Mus* and even within these genera the majority of species are unknown. Of the 570 species of *Rattus*, for example, the majority of research has centred around *Rattus norvegicus* and of the 15 species of *Mus*, research has been directed primarily at *Mus musculus*. Eisenberg (1981) has summarized the literature on the ecology and behaviour of the Muridae.

Apodemus

At least two species of Old World field mice (*Apodemus flavicollis* and *A. sylvaticus*) have sebaceous glands in the oral angle and lips (Quay 1965*b*). Both male and female *Apodemus* have preputial (clitoral) glands (Brown and Williams 1972; Jackson 1938).

Males of at least four species of *Apodemus* (*agrarius, flavicollis, microps* and *sylvaticus*) have a sebaceous caudal gland (see Plate 9.18). This gland is larger in *A. flavicollis* and *A. sylvaticus* than in the other species and does not occur in the females of any species. Larger (older) males have larger caudal glands than small males (Stoddart 1972*b*; Flowerdew 1971). Using gas-liquid chromatography, Stoddart (1973) has shown that the secretion of the caudal gland of the adult male *A. flavicollis* has 26 principal components. The secretions of immature males and of adult and immature females have very few components.

The caudal secretions of *Apodemus* may be used to discriminate between the sexes and between adults and juveniles, but no behavioural studies have been done to confirm this. Male *Apodemus sylvaticus* have a dominance hierarchy, live in well-defined territories, and appear to recognize the dominance status of other males (Brown 1966), but there have been no studies on the role of scent-marking or olfactory communication within this social system. Similarly, while it has been suggested that the caudal gland may function to mark territories or nests, or used as a self-advertisement or for mate recognition (Stoddart 1972*b*, 1973) there is no evidence to support these suggestions.

Both male and female *Apodemus sylvaticus* recognize their own soiled bedding odours and those from alien conspecifics, and produce different amounts of ultrasonic vocalizations in different olfactory environments. Females produce more ultrasounds than males and the most ultrasounds are produced in areas soiled with their own home-cage bedding; fewest ultrasounds are made in environments with bedding soiled by alien conspecifics (Schenk 1978).

Rattus

Glands Some species of *Rattus*, such as *R. exulans*, have a mid-ventral sebaceous gland (see Plate 9.19). This gland occurs primarily in adult males, but is also visible on pregnant and lactating females. Immature males have no visible ventral gland. There have been no observations of ventral-gland marking in these rats (Egoscue 1970; Quay and Tomich 1963). Rudd (1966) examined 12 species of Malaysian *Rattus* and found evidence for ventral glands on 11 of these. Nine species (*annandalei, muelleri, sabanus, rajah, sunifer, cremoriventer, bowersi, edwardsi*, and *rattus jajorensis*) showed heavy staining of the pelage around the gland; two species (*R. canus* and *Chiropodomys gliroides*) had light stains, indicating smaller glands, and no gland was evident in a male *R. whiteheadi*. Though only a few specimens were examined, there appear to be sex differences in the ventral glands of some species but not others. In *R. sabanus* and *R. rajah*, for example, the ventral gland occurs in both adult males and adult females while in other species such as *R. cremoriventer* and *R. canus* the ventral gland is present in adult males, but not in females (Rudd 1966).

Although black rats (*Rattus rattus*) appear to have no ventral glands, they scent-mark by rubbing their cheeks and the ventral surface of their body, from the throat to the anogenital region, on objects such as tree branches (Ewer 1971), thus they may have some specialized glands in these areas.

A number of scent glands have been identified on *Rattus norvegicus*, including a lateral nasal gland (Warshawsky 1963; Schaffer 1940); holocrine anal glands (Montagna and Noback 1947; Schaffer 1940); Harderian glands, Meibomian glands, and lacrimal glands around the eye (Ebling, Ebling, Randall, and Skinner 1975; Greene 1935; Schaffer 1940; Venable and Grafflin 1940; see Fig. 9.9); sebaceous glands on the palmar and plantar soles of the feet (Munger and Brusilow 1971; Quay 1965c; Ring and Randall 1947; Schaffer 1940; see Plate 9.20) and preputial glands (Beaver 1960; Brown and Williams 1972; Hall 1949; Jackson 1938; Montagna and Noback 1946; Schaffer 1940; Stanley and Powell 1941).

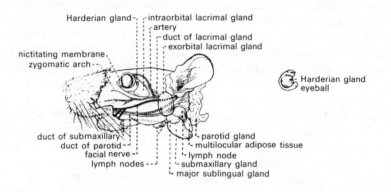

Fig. 9.9. The facial glands of the rat, *Rattus norvegicus*, including the Harderian gland, lachrymal glands, and salivary (submaxillary, sublingual, and partotid) glands. (From Greene (1935).)

As well as these specialized glandular areas, there are also simple sebaceous glands in the skin of the body of *R. norvegicus* as in probably nearly all rodents. In a series of studies Ebling has shown that castration reduces the size of the sebaceous glands in the dorsal skin of male rats and testosterone increases gland size and rate of proliferation of gland cells (Ebling 1957*a,b*). Oestradiol reduces the size of the male's dorsal sebaceous glands (Ebling 1957*b*). In the female rat, sebaceous glands on the dorsal skin show hypertrophy after testosterone injections (Ebling 1948, 1951, 1961). The sebaceous glands on the dorsal skin of the adult female rat show fluctuations in size during the oestrous cycle, being largest at pro-oestrus, and reduced at oestrus and metoestrus (Ebling 1954).

Preputial gland   Different species and strains of rats have different size pre-putial glands. The Alexandrine rat, for example, has much larger preputial glands than the Norway rat and the wild Norway rat has much larger preputial glands than domestic (Wistar albino) rats (Hall 1949). Male domestic rats have larger preputial glands than females (see Fig. 9.10).

Although Noble and Collip (1941) found no effects of castration on the male preputial gland, the majority of work suggests that castration reduces the size of the preputial gland in male rats and testosterone replacement produces hyper-trophy of the preputial gland. Oestradiol has little effect on the preputial gland of the castrated male, nor does progesterone at low doses, but high doses of progesterone increase the weight of the preputial gland in castrated males (Beaver 1960; Ebling, Ebling, and Skinner 1969*b*). Hypophysectomy causes severe atrophy of the male preputial gland and neither testosterone nor pro-gesterone reinstate preputial-gland development in hypophysectomized rats. Testosterone given with growth hormone, however, does restore the preputial gland in hypophysectomized males (Ebling, Ebling, and Skinner 1969*a*; Ebling *et al.* 1975; Thody and Shuster 1975).

Some authors report that ovariectomy has little effect on the size of the preputial glands of the female (Beaver 1960; Noble and Collip 1941) while others report a significant decrease in the size of the female preputial gland after ovariectomy (Thody and Dijkstra 1978). No differences in the preputial gland size have been observed to occur over the female oestrous cycle (Beaver 1960; Thody and Dijkstra 1978). Testosterone and progesterone treatments increase the size and secretion of the female preputial gland, but oestradiol has little effect (Beaver 1960; De Groot, Lely, and Kooij 1965; Thody and Dijkstra 1978). Hypophysectomy produces atrophy of the preputial gland and testosterone alone does not reinstate gland size (Beaver 1960; Noble and Collip 1941). Detailed summaries of hormonal control of the preputial glands of the rat are found in Strauss and Ebling (1970) and Brown and Williams (1972).

Preputial gland and behaviour   In preference tests, female rats spend more time investigating the preputial gland tissue of male rats than they spend investigating submaxillary glands, foot-pad tissue, or coagulating glands and males show a similar preference for the odours of female preputial glands. Males are not, how-ever, attracted to male preputial gland secretions nor are females attracted to female preputial-gland secretions (Gawienowski 1977; Gawienowski, Orsulak, Stacewicz-Sapuntzakis, and Joseph 1975; Gawienowski, Orsulak, Stacewicz-Sapuntzakis, and Pratt 1976; Orsulak and Gawienowski 1972).

Female rats prefer the preputial-gland secretions of intact males to those of castrated males and male rats prefer the odours of preputial glands from intact females to those of ovariectomized females (Gawienowski *et al.* 1975; Gawienow-ski 1977; Thody and Dijkstra 1978). Similarly, males prefer the preputial gland odours from oestrous and pro-oestrous females to those from dioestrous

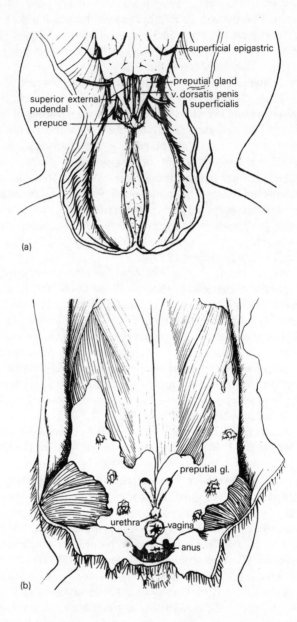

Fig. 9.10. The preputial glands of (a) a male and (b) a female rat (*Rattus novegicus*). (From Greene (1935).)

females (Gawienowski 1977; Lucas, Donohoe, and Thody 1982; Thody and Dijkstra 1978).

Using gas–liquid chromatography, Gawienowski *et al.* (1975) and Gawienowski, Orsulak, Stacewicz-Sapuntzakis, and Pratt (1976) fractioned male and female rat preputial-gland secretions and presented each fraction to animals of the opposite sex in order to discover the active components in the secretions. These studies demonstrate differences in the chemical components of male and female preputial glands and suggest that the active components are lipids. Using mass spectrometry and gas–liquid chromatography, Stacewicz-Sapuntzakis and Gawienowski (1977) suggested that the active components of the male preputial gland were a number of aliphatic acetates. Female rats were attracted to the odour of these acetates, while males were indifferent to the majority of the acetates and avoided some others (Gawienowski 1977).

The preputial-gland odour can thus be used to discriminate between males and females and between sexually active and inactive rats. Since preputial-gland development is dependent on gonadal hormones, the preputial odours may also be used to discriminate pre- and postpubertal animals. Since whole-body odours of females are more attractive to male rats than preputial odours alone; however, other odour sources such as urine, vaginal secretions or skin odours as well as preputial gland odours may be involved in sexual discrimination (Thody and Dijkstra 1978).

Urine-marking Urine odours provide a great deal of information about rats and are used for many types of communication. Both wild and domestic rats urine-mark novel objects in their environment, particularly if these objects contain the odours of other rats (Adams 1976; Brown 1973, 1975; Price 1977; Reiff 1951; Richards and Stevens 1974). Both sexes use anogenital drag and leg-lift urine-marking postures and males urine-mark more than females (Brown 1975; Price 1977). Male rats urine-mark over female odours more than male odours and mark over male odours more than no odour. While Brown (1977*a,b*) found that female rats showed no preferences for marking over male odours, Taylor, Haller, and Regan (1982) found that females urine-mark more over the odours of intact males than over the odours of castrates. Isolation rearing reduces urine-marking by both male and female rats and isolated males mark most over their own odours, avoiding odours of conspecifics (Brown 1982*a*).

Castration eliminates urine-marking in male rats and testosterone, oestrogen, and oestrogen plus progesterone reinstate marking (Brown 1977*a*, 1978; Price 1975). Testosterone also increases the frequency of urine-marking in ovariectomized female rats, but oestradiol and oestradiol plus progesterone have little effect (Brown 1978). Oestrous females urine-mark more than dioestrous females and ovariectomized females virtually never urine-mark (Brown 1977*a,b*; Birke 1978).

During male–male interactions, both intruders and 'territorial' males urine-mark, but the amount of marking does not appear to be correlated with aggression. Intruders do, however, show a decrease in urine-marking after they have been attacked by the resident male (Adams 1976). Sexual arousal may be more important than agonistic arousal for motivating urine-marking, as male rats mark more after encounters with females than after male encounters (Brown 1982*b*). During sexual encounters, male rats urine-mark most before their first mount and after ejaculation, and urine may thus be used as an indication of sexual arousal or function along with the 22-kHz ultrasounds to mediate male–female contact during the refractory period between bouts of sexual activity (Anisko, Adler, and Suer 1979; McIntosh, Davis, and Barfield 1979). Male urine odours in combination with ultrasonic vocalizations facilitate female sexual arousal, but urine odours alone have little effect (Geyer, McIntosh, and Barfield 1978).

Urine-marking may also occur in stressful situations. Steiniger (1950) reported that wild Norway rats urine-mark and deposit faeces on poison baits but this response may be due to the novelty of the stimuli rather than their poisonous nature. Stressed rats produce urine odours which delay running of other rats in runways, suggesting that the urine may carry 'fear odours' (Mackay-Sim and Laing 1981*b*).

Urine and behaviour   Urine odours can be used by rats to discriminate males from females and sexually active from sexually inactive adults. Males prefer the urine odours of female rats to those of males and prefer the urine odours of oestrous females to those of non-oestrous females (Brown 1977*a,b*; Lydell and Doty 1972). Female rats prefer the urine odours of males to those of females and prefer the urine odours of intact males to those of castrated males (Brown 1977*a, b*; Lucas *et al.* 1982). Although there is some evidence that male rats may avoid the odours of other males (Gawienowski, DeNicola, and Stacewicz-Sapuntzakis 1976), other studies have indicated that male rats are attracted to the urine odours of males, preferring male urine odours to no odour and to their own odour (Brown 1975, 1977*a*; Fass, Gutermann and Stevens 1978).

General body odour   Many studies have examined the responses of rats to the general body odours of conspecifics without attempting to identify the source of the odours. These include studies of the role of olfaction in sexual, aggressive, fear, and maternal behaviour. The odours involved include a combination of urine, faeces, preputial gland, and general skin-gland secretions.

Male rats prefer body odours of female rats to those of males (LeMagnen (1952) and prefer the odours of oestrous females to dioestrous females, but these preferences depend on the age, hormonal condition and sexual experience of the test subjects. Sexually experienced intact males show the greatest preference for odours of oestrous females, while sexually naive adults and immature males show no preferences for oestrous over dioestrous females, and castrates show no preference at all (Carr 1974; Carr, Loeb, and Dissinger 1965; Carr, Loeb, and

Wylie 1966; Carr, Wylie, and Loeb 1970; Landauer, Wiese, and Carr 1977; Stern 1970; Wylie 1968). Male rats can discriminate between the odours of individual females, but their preferences depend on their previous sexual experiences and state of arousal (Carr, Krames, and Costanzo 1970; Carr, Hirsch and Balazs 1980; Krames and Mastromatteo 1973).

Olfactory stimulation is not necessary for sexual behaviour in male rats, but anosmia induced before puberty and anosmia combined with social isolation greatly reduce sexual behaviour (Larsson 1971, 1975; Thor and Flannelly 1977; Wilhelmsson and Larsson 1973).

Male rats can learn to discriminate between the body odours of two males when the odour of one male is associated with food in an operant conditioning paradigm (Husted and McKenna 1966), but preferences of male rats for the odours of other males have been difficult to determine. While Krames and Shaw (1973) found that males preferred the odours of their cage-mates to the odours of novel males, Carr, Yee, Gable, and Marasco (1976) found that the odours of novel adult males were preferred to those of familiar males and Brown (1977*a*) found no preference between odours of alien and familiar males. Similarly, while Carr, Wylie, and Loeb (1970) found that male rats showed no preference between the odours of castrated and intact males, Brown (1977*a*) found that male rats preferred the odours of castrated males to those of intact males. Male rats can discriminate between the odours of dominant and submissive males and prefer the odour of the submissive male (Krames, Carr, and Bergman 1969). The ability to recognize individual colony members and estimate dominance status from odour cues may allow groups of rats to be integrated into a social unit (Thor 1979).

Female rats prefer the odour of males to females and prefer the odour of intact males to castrated males (Brown 1977*a,b*). Sexual experience does not seem to effect female odour preferences, but the female's hormonal state does. Oestrous females show greater preferences for male odours than dioestrous females and ovariectomized females show no odour preferences (Brown 1977*a,b*; Carr 1974; Carr *et al.* 1965; Carr, Wylie, and Loeb 1970; LeMagnen 1952). Females rats can discriminate between the odours of oestrous and dioestrous females, but preferences for one type of female odour or the other depend on hormonal state and sexual experience (Carr, Wylie, and Loeb 1970). Females can also discriminate between the odours of individual males (Krames 1970), but their preferences for familiar versus novel males depend on their sexual experience and state of sexual arousal (Carr, Hirsch, and Balazs 1980; Carr, Demesquita-Wander, Sachs, and Maconi 1979; Carr, Krames, and Costanzo 1970; Krames and Mastromatteo 1973).

Olfactory signals from other rats appear to be important cues in the establishment of social and sexual relationships, including the maintenance of colonies and recognition of non-colony members (Calhoun 1962; Barnett 1967; Thor 1979). The importance of olfactory information to the rat is supported by results of odour discrimination studies using non-rat odours. In these experiments rats have been shown to acquire learning sets at a rapid rate, and olfaction seems to be the most important sensory modality for learning in rats (Nigrosh, Slotnick, and Nevin 1975; Slotnick and Katz 1974).

Maternal odour   Faeces of lactating female rats carry an odour which is attractive to pre-weanling rats. This odour is produced from 14 to 27 days post-partum and is more attractive to pups than the odour of non-lactating females (Holinka and Carlson 1976; Leon 1978b; Leon and Moltz 1971, 1972). Leon (1974) believes that the maternal odour is produced by bacteria in the caecum of the mother and the resulting caecotrophe coats the faeces. The production of caecotrophe is stimulated by the developing pups and is dependent on high prolactin levels (Leon 1978b; Leon and Moltz 1973; Moltz and Leon 1973). Nulliparous females, but not males, which are housed with pups and show maternal behaviour, produce maternal odours and the production of this odour also depends on prolactin secretion (Leidahl and Moltz 1975, 1977).

There is, however, a second theory of how the maternal odour is produced. According to this theory prolactin acts on the liver of the lactating females, altering the production of bile, and this bile may be important in the production of the maternal odour (Lee, Lee, and Moltz 1982). Although Leon (1978a) has argued that bile acts merely as a laxative, the injection of bile from lactating females into adult males causes these males to produce a 'maternal odour' which is attractive to pups (Moltz and Leidahl 1977). Bile acid levels increase in 12-21-day post-partum females, and nulliparous females fed choic acid emit a maternal odour, suggesting that cholic acid in the bile is essential for the production of the maternal odour (Moltz and Kilpatrick 1978; Kilpatrick, Bolt, and Moltz 1980). Cholic acid appears to be the primary precursor of the maternal odour and is converted to an odorous additive to the faeces through the action of caecal bacteria (Moltz and Lee 1981).

The control of the timing of the release of the maternal odour appears to be through a mother–pup feedback loop. Growth of the bacteria in the mother's gut is controlled by the pH of her intestines. As the mother licks her pups and stimulates their elimination, she consumes their faeces. When the pups are young and feeding only on milk, these faeces are acidic, and a highly acidic gut inhibits bacterial growth, thus the early lactating female produces no maternal odour. When the pups begin to eat solid food, their faeces are no longer acidic, and when eaten by the mother produce a more alkaline gut environment which allows the growth of caecal bacteria. These bacteria break down the cholic acid, thus producing a maternal odour (Lee and Moltz 1980; Moltz and Lee 1981). As the infant rat produces its own odour, the mother's odour becomes less attractive (Leon and Behse 1977).

There may be strain differences in the production of maternal odours. While Leon's and Moltz's work has been done with albino rats, studies with Long-

Evans hooded rats indicate that these pups are attracted to the odours of non-lactating females and are attracted to the whole-body odours of lactating females more than to the faeces alone (Galef and Heiber 1976; Galef and Muskus 1979). Maternal odours can be modified by the female's diet (Leon 1975; Galef 1981*b*), suggesting that the individual females may acquire specific odours by eating variable diets, but different diets are not necessary for the production of individual differences in maternal odours. Sixteen- to 20-day-old Long–Evans hooded rats prefer faeces from their own mother to those of strange lactating females or non-lactating females on the same diet (Brown and Elrick 1983).

As well as approaching the maternal odour, pups ingest the mother's faeces (Leon 1974). Both mother's milk and the mother's faeces appear to provide the pups with an immunity to necrotizing enterocolitis, a lethal disease of the gastro-intestinal tract (Moltz and Kilpatrick 1978). The bile acids in the mother's faeces may protect the pups from *E. coli* bacteria which cause the gastro-intestinal disease. A second advantage to pups which ingest maternal faeces may be that the bile acids absorb fatty acids from foods and use them for myelinization of brain cells (Moltz and Lee 1981).

Olfaction and development   The maternal odour appears to be a 'prepotent' stimulus for infant rats. Rearing rats with artificial odours such as perfume, peppermint or lemon produces a preference for these odours and this preference appears to be caused by mere exposure to the odour; association of the odours with the mother, food, or warmth are not necessary (Marr and Lilliston 1969; Leon 1980; Leon, Galef, and Behse 1977; Galef and Kaner 1980). The mother's odour, however, is preferred to lemon odour even when the lemon odour is as familiar as the mother's odour (Schumacher and Moltz 1982). While Leon (1975) and Leon *et al.* (1977) claim that the infants must learn to approach the maternal odour, Schumacher and Moltz (1982) claim that the attraction of the infants to the maternal odour is unlearned.

As well as being attracted to the maternal odour, infant rats respond to olfactory cues in a number of social situations. Newborn rats require an olfactory stimulus to direct them to the mother's nipples. Rats whose mother's ventral surface has been washed fail to locate the nipples. The odour source which directs the pups appears to be amniotic fluid and saliva spread over the nipples as the lactating female grooms (Blass and Teicher 1980; Blass, Teicher, Cramer, Bruno, and Hall 1977; Hofer, Shair, and Singh 1976; Teicher and Blass 1977). The odour which attracts pups to the nipples may depend on the hormone oxytocin for its release (Singh and Hofer 1978). Rat pups which have been made anosmic by olfactory bulbectomy or zinc sulphate irrigation do not attach to the nipples, nurse less often than controls, and lose weight (Hofer 1976; Singh, Tucker, and Hofer 1976; Rouger and Schneirla 1977; Singh 1977; Tobach 1977).

Olfactory cues from the mother accelerate heart rate in infant rats, inhibit activity, and reduce the amount of ultrasonic vocalizations (Compton, Koch, and Arnold 1977; Geyer 1979; Hofer and Shair 1978; Schapiro and Salas 1970). Contact with littermates (huddling) depends on temperature cues in very young rats, but by 15 days of age olfactory cues become more important and rats choose to huddle with conspecifics having a familiar odour (Alberts 1978; Alberts and Brunjes 1978; Brunjes and Alberts 1979).

Infant rats attend to the odour of the nest (which contains odours from the mother and littermates); the odour of the bedding material; and the odour of the food consumed by the parents. As early as 5-6 days of age, rats are attracted to the odours of their mothers, their own nest, and the odour of the wood shavings used as bedding (Carr, Marasco, and Landauer 1979; Cornwell-Jones and Holder 1979; Cornwell-Jones and Sobrian 1977; Goldblatt 1978a; Gregory and Pfaff 1971; Nyakas and Endröczi (1970). Nest odours also reduce ultrasonic vocalizations of infant rats while unfamiliar odours increase the frequency of ultrasonic vocalizations (Conely and Bell 1978; Oswalt and Meier 1975). When the behaviour of neonatal rats is examined in detail, differences in head movements, pivoting and locomotion are seen in the presence of familiar nest odours versus unfamiliar odours in rats as young as three days of age (Singh, Lederhendler, Desantis, and Beckhorn 1977; Tobach 1977). Sixteen- to 20-day-old Long-Evans rats prefer their own nest odour to the nests of other litters, and rats reared with both parents appear to discriminate between own and strange nests more than rats reared with their mother alone (Brown 1982c).

The social odours experienced by rats during infancy may influence olfactory peceptions at later ages. The maternal odour may be the basis for group or colony odours in adulthood (Leon 1975). Rats reared with cologne odours, for example, show preferences for cologne-smelling rats in adulthood (Marr and Gardner 1965). The exposure of rats to odours in infancy may result in 'olfactory imprinting' (see Brown 1979). It has been hypothesized that early exposure to odours may somehow modify the development of the olfactory bulb, but there is, as yet, no evidence for the neural basis of olfactory imprinting (Leon 1980) even though olfactory deprivation in infancy appears to inhibit olfactory bulb development (Meisami 1976).

The attraction of pups to the odours of adults facilitates the choice and consumption of solid foods. Rat pups prefer to eat the food which their mother eats and select food which is near the faeces or body odours of other rats (Bronstein and Crockett 1976; Galef 1981a), indicating social mediation of food selection. Mere exposure to an odour will also facilitate eating food which has the familiar odour (Hennessy, Smotherman and Levine 1977; Galef 1981a).

The importance of the sense of olfaction for infant rats extends to the ability of rats to associate olfactory cues with illness as young as two days of age and remember this association at eight days of age by avoiding the odour which was paired with illness (Rudy and Cheatle 1977). Between seven and 10 days of age, rats are able to learn a more complex second-order odour aversion (Cheatle and Rudy 1978, 1979).

Olfaction and maternal behaviour Since the amniotic fluid is attractive to pups and directs them to the mother's nipples, it is not surprising that the presence of the placenta on the pups accelerates the onset of maternal behaviour in virgin rats (Kristal, Whitney, and Peters 1981). Lactating females are able to discriminate between male and female pups on the basis of their odours and direct more licking toward male anogenital areas than toward females (Moore 1981; Moore and Morelli 1979). While lactating female rats will retrieve both their own and alien pups, pups with unfamiliar odours are retrieved more slowly than pups with a familiar odour (Beach and Jaynes 1965a,b). Smotherman *et al.* (1974) found that olfactory cues were necessary for Long-Evans female rats to retrieve pups, and responses to these olfactory cues were facilitated by infant ultra-sounds. Olfactory bulbectomized lactating females and zinc-sulphate-infused females show deficits in pup retrieval and pup care (Benuck and Rowe 1975). In virgin female rats, however, olfactory bulbectomy and vomeronasal nerve cuts facilitate the onset of maternal behaviour (Fleming and Rosenblatt 1974a,b; Fleming *et al.* 1979). The role of olfaction in maternal behaviour of the rat has been reviewed by Rosenblatt, Siegel and Mayer (1979).

Fear odour Rats avoid the odours of dead conspecifics (Carr, Landauer, and Sonsino 1981) and show fright reactions (freezing, elimination, or retreat-ing) to conspecific muscle and blood but not to rat brain tissue nor to guinea-pig muscle and blood (Mackay-Sim and Laing 1981a; Stevens and Gerzog-Thomas 1977; Stevens and Saplikoski 1973).

Whether the urine, faeces, or glandular secretions of stressed rats act as alarm odours, eliciting fright reactions, is unclear. Stevens and Koster (1972) found no evidence that rats avoided the odours of stressed conspecifics, but Mackay-Sim and Laing (1980, 1981a,b) have indicated that the urine and general body odour of stressed rats is aversive, and the greater the stress a rat undergoes, the more aversive an odour it produces. Rats do not appear to show a stereotyped response to the odours of stressed rats; any fear response depends on the environmental situation (e.g. runway, conditioning apparatus, open field) and the experience of the test rat in that environment. The odour of the stressed rat appears to be a salient cue on which a rat can base its response (Courtney, Reid, and Wasden 1968; King 1969; King, Pfister, and DiGuisto 1975; Davis and Tapp 1972; Minor and Lolordo 1981; Valenta and Rigby 1968).

Although faeces have been suggested as a source of fear odours in rats (King 1969), there seems to be little evidence to support this suggestion. Faeces from stressed rats do not produce a delay in running which is caused by urine from stressed rats (Mackay-Sim and Laing 1981b) and rats do not avoid food or novel environments which have the faeces of stressed rats on them (Brown and Hartley 1982).

As well as alarm or fear odours, rats produce odours when frustrated or non-rewarded in a runway in which they have normally been rewarded. Such 'frustration odours' can be perceived by other rats and used as discriminative stimuli (Bloom and Phillips 1973; Burns, Thomas, and Davis 1981; Collerain 1978; Collerain and Ludvigson 1972, 1977; Davis, Whiteside, Bramlett, and Petersen 1981; Eslinger and Ludvigson 1980; Means, Bates, and Cahoon 1973; Mellgren, Fouts, and Martin 1973; Morrison and Ludvigson 1970; Taylor and Ludvigson 1980*a,b*). Frustration odours produce different responses in the mitral cells of the olfactory bulbs of female rats than urine or food odours, suggesting a chemical specificity (Cattarelli, Vernet-Maury, and Chanel 1977; Cattarelli, Vernet-Maury, Chanel, MacLeod, and Brandon 1975; Voorhees and Remley 1981). Frustration odours do not appear to be produced in the urine, preputial glands, accessory sex glands, or foot-pad glands of the rat (Weaver, Whiteside, Janzen, Moore, and Davis 1982) and the source of these odours remains unknown.

Mus

Using body odours, mice (*Mus*) are able to discriminate between members of their own species and another species (*Peromyscus*), between males and females and between two males (Bowers and Alexander 1967; Hahn and Tumolo 1971; Hahn and Simmel 1968; Halpin 1980). The odours which are used for these discriminations and for other forms of social communication have a number of glandular and non-glandular sources. These include the preputial glands, coagulating glands, submaxillary glands, urine, and faeces. Mice also have glands of the oral angle and lips (Quay 1965*b*; Schaffer 1940), ear glands, plantar glands on the soles of the feet (Ortmann 1956; Quay 1965*c*; Schaffer 1940), and anal glands (Schaffer 1940). The discussion of 'mice' below refers primarily to *Mus musculus*.

Salivary glands   The submaxillary glands of mice are sexually dimorphic in histology, enzyme content, and histochemistry. Males have larger glands, greater tubule diameter and more intense acidophilic granule production than females (see Plate 9.21). This sexual dimorphism does not appear until after puberty and is influenced by gonadal hormones (Harvey 1952). Castration and ovariectomy reduce the size of the submaxillary glands. Testosterone injections increase the size and weight of these glands in gonadectomized mice of both sexes, while oestrogens increase the submaxillary gland size in females (Harvey 1952). Protease activity in the submaxillary gland is inhibited by gonadectomy and increased by testosterone or oestrogen injection (Junqueira, Fajer, Rabinovitch, and Frankenthal 1949).

Saliva appears to be used for social communication, as male mice investigate the mouth areas of castrates more than the mouth areas of intact males. Castrated males smeared with the saliva of testosterone-injected castrates are attacked more than castrates smeared with saliva from oil-injected castrates, suggesting that males can discriminate between intact and castrated males on the basis of salivary odours (Lee and Ingersoll 1979). This salivary odour may be from the secretions of the submaxillary glands, other salivary glands, the glands of the oral angle and lips, or the Harderian glands, none of which have been thoroughly investigated.

Plantar glands   The sudoriferous plantar glands are located in the foot pads (see Fig. 9.11) and produce a secretion which may be used for recognizing individuals or colony members (see Ropartz 1977).

Fig. 9.11. The plantar surface of the hindfoot of the mouse (*Mus musculus*) showing the glandular areas. (From Ortmann (1956), after Schaffer (1940).)

Preputial gland   Both male and female *Mus musculus* have large preputial or clitoral glands in relation to their body size (Jackson 1938; Brown and Williams 1972). The preputial gland secretion of male mice is attractive to females, who prefer preputial homogenate to male urine and prefer the urine of intact males to the urine of preputialectomized males (Bronson 1966; Bronson and Caroom 1971; Vadasz 1975). This preference for male preputial-gland odour depends on the female's hormonal state and sexual experience: sexually-experienced females and oestrogen-injected sexually-naive females prefer male preputial glands over muscle and fat tissue, but sexually-naive females, pregnant females, and progesterone-injected females show no preferences (Caroom and Bronson 1971).

Dominant male mice have larger preputial glands than isolated or subordinate males suggesting that as males achieve dominance, their preputial glands increase in size (Bronson and Marsden 1973; Hucklebridge, Nowell, and Wouters 1972; see also McKinney and Christian 1970). Lane-Petter (1967) suggested that male mice release their preputial gland secretion in clean environments in order to mark a territory or to establish dominance.

The odour of the preputial gland may be important in the definition of 'maleness'. Preputialectomized males are no less aggressive than intact males, but when paired with an intact male, the preputialectomized male is more likely to attack first (McKinney and Christian 1970). Similarly, male mice will attack testosterone-injected females faster than control females and the removal of the preputial gland from the testosterone-injected females decreases the intensity of the male's attack (Mugford and Nowell 1971c).

While the male preputial gland appears to produce an odour which is attractive to females and stimulates aggression in males, the odours from the female preputial gland appear to have little social function. Male mice are not able to use preputial-gland odours to discriminate between oestrous and non-oestrous females (Hayashi and Kimura 1974). Whether or not mice can use preputial gland secretions to discriminate between males and females is unknown.

Vaginal odour   The vaginal odour of female mice changes over the oestrous cycle and male mice can discriminate between oestrous and non-oestrous females on the basis of their vaginal odours (Hayashi and Kimura 1974; Sokolov, Skurat, and Kotenkova 1976).

Urine   Both male and female mice mark with urine (Dagg *et al.* 1971) and urine odours serve many social functions. In the wild, house mice form and regularly mark 'urinating posts' along trails (Reiff 1951; Welch 1953). Male house mice urine-mark novel environments and objects at a high rate, depositing urine every 15–24 seconds in a two-hour test (Bronson 1976; Maruniak, Desjardins, and Bronson 1975; Maruniak, Owen, Bronson, and Desjardins 1974; Matthews 1980; Van Abeelen and De Vries 1978). Males mark more in the presence of a female than in the presence of another male (Bronson 1973, 1976; Desjardins, Maruniak, and Bronson 1973; Matthews 1981; Reynolds 1971) and urine-mark more frequently in the presence of odours from female mice or castrated males than in the presence of odours from intact male mice (Maruniak *et al.* 1974). The male mouse's urine-marking response to a novel environment depends primarily on olfactory cues; anosmic males urine-mark less than one-third as frequently as control males (Maruniak, Darney, and Bronson 1975).

Female mice urine-mark at a rate of 25–95 marks/h (versus 100–200 marks/h by males) and urine-mark at higher frequencies in the presence of a male than in the presence of a female or when in isolation. Females also urine-mark more frequently in the presence of urine from intact males than in the presence of urine from castrated males (Maruniak, Owen, Bronson, and Desjardins 1975). Although Maruniak *et al.* (1975) found no fluctuation in rates of urine-marking during different periods of the oestrous cycle in female *CF-1* mice, Wolff and Powell (1979) found that female *C3M/Me-Mg* self-chocolate mice marked more often during pro-oestrus and oestrus than during metoestrus and dioestrus.

Both male and female mice are attracted to the odours from mice of the opposite sex and, in trapping studies, mice are most likely to be caught in traps baited with odours from the opposite sex (Davies and Bellamy 1972; Rowe 1970). In preference tests, female mice prefer male urine odours over no odour, male mouse urine over urine from male field voles (*M. arvalis*), and urine from intact males over urine from castrated males (Scott and Pfaff 1970; Vadasz 1975). Oestrous, but not dioestrous female mice are more attracted to the urine odour of dominant males than to the urine odour of subordinate males (Jones and Nowell 1974a). Olfactory bulbectomized female mice are not receptive to males, but females made anosmic by zinc sulphate infusion show only slight decrease in receptivity, suggesting that olfactory sensitivity is not essential for female sexual behaviour (Edwards and Burge 1973).

Female mice prefer the odours of unrelated males of the same strain to males of another strain (Gilder and Slater 1978). This indicates that females which were reared with their father and then housed in isolation from weaning, may have the ability to distinguish between relatives and non-relatives by their odours.

Male mice can discriminate between male and female mouse odours (Bowers and Alexander 1967) and prefer the urine odours of oestrous females to those of dioestrous females (Rose and Drickamer 1975; Sokolov *et al.* 1976). While both sexually naive and experienced intact males prefer the odours of oestrous female urine, castrated males show no preference (Rose and Drickamer 1975). Male mice which have observed pairs of mice copulating also show a preference for oestrous female odours over those of dioestrous females (Hayashi and Kimura 1976).

Urine and ultrasonic communication   Monitoring the male's ultrasounds provides a sensitive bioassay for the attractiveness of female odours. Male mice produce ultrasonic vocalizations in the presence of a female or her urine and the oestrous state of the female does not affect ultrasound production, although there are fewer ultrasounds produced in the presence of the urine from ovariectomized females (Nyby, Wysocki, Whitney, Dizinno, and Schneider 1979; Whitney and Nyby 1979; Whitney, Alpern, Dizinno, and Horowitz 1974). Males do, however, produce ultrasounds in the presence of prepubertal females, castrated males, hypophysectomized males, and hypophysectomized females even though urine from these animals does not elicit ultrasounds (Nyby, Wysocki, Whitney, Dizinno, Schneider, and Nunez 1981). Males produce ultrasounds in the presence of voided urine and bladder urine from females, and urine stored for up to 31 days still effects ultrasounds (Nyby and Zakeski 1980). In addition to female urine, female facial odours and vaginal secretions will elicit ultrasounds from male mice, but male odours do not (Nyby, Wysocki, Whitney, and Dizinno 1977). Similarly, the urine from testosterone-injected ovariectomized females fails to elicit ultrasounds (Whitney and Nyby 1979) as does urine from hypophysectomized females (Nyby *et al.* 1979).

The male's production of ultrasonic vocalizations in the presence of female odours depends on the hormonal state and the sociosexual experience of the male (Nyby and Whitney 1980). Castration increases the latency to produce ultrasounds in the presence of females and injection of testosterone into castrated males reduces this latency and increases the frequency of ultrasounds (Dizinno and Whitney 1977; Nyby, Dizinno, and Whitney 1977). Similarly, injection of testosterone into ovariectomized females increases their production of ultrasounds in the presence of a stimulus female (Nyby, Dizinno, and Whitney 1977). Male mice with no sociosexual experience in adulthood do not emit ultrasounds in the presence of female urine and males which have mated with perfume-covered females will emit ultrasounds in the presence of perfume alone (Dizinno, Whitney, and Nyby 1978; Nyby, Whitney, Schmitz, and Dizinno 1978; Nyby and Whitney 1980). The vomeronasal organ is the receptor which determines the male's ultrasonic vocalizations to female urine. Damage to the vomeronasal receptors inhibits ultrasonic vocalizations whereas damage to the main olfactory system has little effect (Bean 1982).

Female urine, sex and aggression   Exposing males to female urine or rubbing one member of a pair of unfamiliar male mice with female urine reduces the amount of aggression shown by the other male and increases his social investigation and sexual behaviour (Connor 1972; Davies and Bellamy 1974; Dixon and Mackintosh 1971, 1976; Mugford and Nowell 1970*b*, 1971*a*). Female urine is effective in reducing aggression whether it is taken from oestrous or dioestrous females (Dixon and Mackintosh 1975; Mugford and Nowell 1971*a*). The effects of ovariectomy are unclear. Whereas Mugford and Nowell (1971*a*) found that urine from ovariectomized females did not reduce aggression, Dixon and Mackintosh (1975) found that urine from ovariectomized females did reduce aggression. The urine from ovariectomized females injected with testosterone or oestradiol does not reduce aggression according to Mugford and Nowell (1971*a*) but Lee (1976) suggests that the urine of ovariectomized females injected with either oestrogen or progesterone does inhibit aggression.

Only urine from oestrous females increases sexual behaviour (Dixon and Mackintosh 1976), suggesting that inhibition of aggression and excitation of sexual behaviour are two independent responses of males to female urine. Bladder urine from female mice inhibits aggression and increases sexual behaviour in males as does female urine which has been exposed to air at room temperature for seven days (Evans, Mackintosh, Kennedy, and Robertson 1978). Attempts to isolate the active components from female urine have not been successful. One can conclude only that the active components are non-volatile and stable (Evans *et al.* 1978).

Although female mice are not usually aggressive toward males (see Ropartz 1977), they are aggressive against juveniles and other females, particularly when living in groups (Gray, Whitsett, and Ziesenis 1978; Haug 1972; Ropartz and Haug 1975). The aggressive behaviour of grouped females is higher toward lactating females or females smeared with urine from lactating females and lower toward ovariectomized females, suggesting hormonal control of a female odour which elicits aggression from grouped females (Haug 1972; Ropartz and Haug 1975). For their part, lactating females are more aggressive against males scented with the urine from strange, non-lactating females than against unscented males or males smeared with water (Lynds 1976).

Male urine and aversion Male mice avoid traps scented with urine from other males (Davies and Bellamy 1972) and, in a forced-choice preference test, males avoid the areas scented with the urine of intact males (Jones and Nowell 1973*a*). In addition to avoiding male urine, male mice reduce their exploratory behaviour, rear less, groom less, and take longer to approach food in the presence of urine or soiled bedding from other males (Jones and Nowell 1974*c,e*, 1977*a*). Male urine retains its aversive properties for about two days; urine marks three or more days old are not avoided (Jones and Nowell 1977*b*). Male mice appear to avoid only the urine of males of their own strain; urine of rats, hamsters, or other strains of mice is not avoided (Jones and Nowell 1974*e*). That male mice avoid the urine of other males suggests that urine marks may be used as 'territorial boundaries'. Urine marks, however, seem less salient as boundary marks than visual cues (Mackintosh 1973) and when used may only be recognized by dominant males (Harrington 1976).

The urine of castrated mice is not avoided by other males, nor is the urine of males whose testosterone levels are inhibited by injection of cyproterone acetate, but the urine of castrates injected with testosterone is avoided and, as the dosage of testosterone is increased, the aversiveness of the urine increases (Jones and Nowell 1973*a*, 1974*b,d*, 1977*a*; Sawyer 1978). Urine from isolated males is avoided by other males, but urine from group-housed males, which have lower testosterone levels, is not avoided (Jones and Nowell 1974*e*, 1977*a*).

The effect of social status on the production of the urinary aversive odour is unclear. While Jones and Nowell (1973*b*, 1974*a*) have shown that only the urine of a dominant male is aversive, other authors (Carr, Martorano, and Krames 1970; Sawyer 1978) have found that the urine of subordinate males is avoided as well. It is possible that there is a difference between the odour of subordinate males immediately after a fight and the odour of subordinate males 24 h later. Jones and Nowell, for example, did not collect urine from subordinate males until at least 24 hours after they lost a fight, but Sawyer collected urine immediately after a fight, and Carr, Martorano, and Krames collected urine within 2 h of the aggressive encounter. Urine samples collected soon after a fight may have contained 'stress odours' (see below). Also, since Carr, Martorano, and Krames (1970) used a free-choice test rather than the forced-choice test used by Jones and Nowell and by Sawyer, the results may not be comparable.

Odour exposure in infancy or post-weaning may alter responses to the odours of dominant males. Submissive males exposed to urine from submissive males when 25-34 days of age prefer the odour of a dominant male at 73 days of age whereas submissive males exposed to the odour of dominant males and dominant males, whether exposed to urine from dominant or submissive males, avoid the odours of dominant males (although not significantly so) (Hennessy 1980).

The responses of males to the urine odours of other males may depend on the hormonal state and social experience of the test subjects, and the relationship between the urine donor and test subject (Fass and Stevens 1977). While Jones and Nowell (1977a) found that both group-housed and isolated males avoided the urine of isolated males; Sawyer (1978) found that group-housed, but not isolated males avoided urine of dominant males, thus the effects of housing are unclear. Both dominant and subordinate males avoid the urine odour of dominant males (Jones and Nowell 1974a), but castrated males do not avoid the odour of dominant males (Sawyer 1980). Testosterone-injected castrate males do, however, avoid the urine of dominant males (Sawyer 1980) and castrates can be trained to avoid urine from intact males in order to avoid shock (Sawyer 1981). Thus the reason that castrates do not avoid male urine odours is motivational rather than sensory. Finally, whether or not the test subject has interacted with the odour donor may alter its behaviour. Subordinate males may avoid the urine odour of the particular dominant male that defeated them, but not the odour of unknown dominant males (Carr, Martorano, and Krames 1970).

The source of the urinary aversive odour may involve the coagulating glands. The bladder urine of isolated mice is not aversive, nor is the coagulating-gland secretion alone, but urine plus coagulating-gland secretion is aversive (Jones and Nowell 1973c).

Male urine and aggression   Male mice are more aggressive toward intact males than toward juvenile males, castrated males or females, but are just as aggressive toward testosterone-injected castrates or females, suggesting that the hormonal status of an opponent determines its ability to elicit aggressive behaviour (Brain and Evans 1974*b*; Lee and Brake 1971, 1972; Lee and Griffo 1974; Mugford and Nowell 1970*a*, 1971*b*; Taylor 1982). Painting urine from dominant males or isolated males onto juveniles, castrated males or females increases the aggressive behaviour shown by intact males towards these mice (Connor 1972; Mugford and Nowell 1970*b*; Taylor 1982). Urine from subordinate males or castrated males, however, does not elicit aggression (Brain and Evans 1974*b*; Jones and Nowell 1973*b*) and urine from juvenile males reduces aggression (Taylor 1982). The behaviour of the urine-coated castrated male does not seem to influence the olfactory stimulation of aggressive behaviour (Lee and Crump 1980).

That the aggression eliciting odour is under hormonal control has been shown in many studies (see Lee 1976; Fass and Stevens 1977). Females masculinized by testosterone injections in infancy are attacked more than control females and when urine from these masculinized females is rubbed on castrated males, they are attacked more frequently (Lee and Griffo 1973). Progesterone inhibits the production of the aggression eliciting odour in testosterone-injected castrates (Lee and Griffo 1974; Lee, Griffo, Braunstein, Mars, and Stein 1976). Since males attack juveniles, however, there must be some factor other than an androgen-dependent odour which elicits aggression (Gray *et al.* 1978).

There are strain differences in the ability of male urine to elicit aggression. Urine from *DBA* males elicits more aggression from other *DBA* males than urine from *CBA* or *C57/BL* males (Kessler, Harmatz, and Gerling 1975). Although this has been explained as a genetic difference in the production of the aggression eliciting odour and it is known that *C57/BL* males are low producers of urinary odours (Marchlewska-Koj 1977), it is possible that the *DBA* mice are able to discriminate among the odours of different strains and direct more attacks toward males of their own strain (Jones and Nowell 1974*e*). Brain and Evans (1974*a*), on the other hand, found that hormones were more important than strain in determining aggression-eliciting properties of different strains of mice and suggested that strain differences in the production of attack-eliciting odours are due to strain differences in the metabolization of androgens. The distinction between strain differences in male odours and differences in the production of male odours (i.e. hormonal versus genetic differences) does not seem to have been thoroughly investigated.

Familiarity influences aggression in mice   Mice living in small family groups attack unrelated males and females which are introduced into the group and will attack family members if they have been removed from the group for two or three weeks (Rowe and Redfern 1969). Familiarity need not mean relatedness. Unrelated males which have lived in pairs for long periods of time show little aggression towards each other, but if one member of the pair is scented with the urine of a strange male, it is attacked more frequently by its cagemate. Similarly, if a strange male is covered with urine from a familiar cagemate or has its odour altered by being covered with perfume, the strange male is attacked less frequently (Mackintosh and Grant 1966; Ropartz 1968).

Whether or not the odour of a strange male must be painted on to another male in order to increase aggression is a disputed point. Some researchers report that placing cohabiting pairs of male mice or an intact male and a castrated male into cages having the urine odours of strange males produces an increase in aggression (Archer 1968; Mugford 1973). Other researchers, however, have found that placing mice into cages soiled by strange males has either no effect or reduces aggression. Haug (1970) and Harmatz, Boelkins, and Kessler (1975) found odours from grouped males inhibited aggression; Mugford (1973) found an increase in aggression and Jones and Nowell (1975) found no effect. Similarly, while Mugford (1973) found that the odours of isolated males increased aggression, Jones and Nowell (1975) found a decrease in aggression.

The resolution of these differences does not seem to be easy because a number of factors interact. First, the home-cage odour of a male produces a rapid increase in aggression toward castrates (Jones and Nowell 1973d, 1975) and therefore, the more similar the foreign male odour to the test male's own odour, the more aggression may occur. Second, male urine placed onto a castrate increases aggression, so if any urine from the bedding becomes rubbed on the castrate, it will be attacked more often. Third, the olfactory environment of the test male may alter his aggressiveness as males exposed to the odours of their opponents or of other strange mice do not show as much aggression when they meet these strange mice (Kimelman and Lubow 1974; Connor and Lynds 1977). Although it has been suggested that males attack familiar males less because they habituate to their odours (Archer 1975; Kimelman and Lubow 1974), the results of Connor and Lynds (1977) indicate that there are many dimensions to the odour of a male mouse, such as a strain odour, a sex odour, and an individual odour and all of these plus the factors mentioned above will affect aggressiveness. Thus, the concepts of 'aggression-eliciting' and 'aggression-inhibiting' odours on substrate material are difficult to substantiate (see Ropartz 1977; Fass and Stevens 1977).

Whatever the resolution of this problem, there have been a number of sources suggested for the urine odours which affect aggression in male mice, but again the evidence is contradictory. Although preputial gland odours have been suggested as the source of aggression-eliciting odours (Mugford and Nowell 1971c; Nowell and Wouters 1975), the urine of preputialectomized, testosterone-injected females or castrated males will elicit aggression (Jones and Nowell 1973b; Mugford and Nowell 1971c) thus there must be a second source of this odour, which may be the bladder urine (Lee 1976). An attempt to identify the component of the urine of male mice which elicits aggression using a head-space distillation technique has not been successful and has shown only that the active component of the male's urine is in the volatile portion (Lee, Lukton, Bobotas, and Ingersoll 1980). The source of the aggression-inhibiting odour in male urine appears to be the coagulating glands (Haug 1971; Jones and Nowell 1973c).

Maternal odour Infant mice are attracted to the body odours and excreta of lactating females and prefer these to the odours of non-lactating females, but do not prefer the odour of their own mother to that of another lactating female (Breen and Leshner 1977; Hennessy, Li, and Levine 1980). Sixteen- to 20-day-old albino mice do, however, prefer the excreta of lactating females on the same diet as their own dam to the excreta of lactating females on a different diet (Brown 1982d). Mice begin to show a preference for the odours of lactating females at about 10 days of age (Koski, Dixon, and Fahrion 1977). Rearing mice with artificially odourized mothers produces a preference for the artificial odour in juvenile mice, suggesting that the mice learn to associate particular odours with lactating females (Goldblatt 1978b).

Olfaction and development The responses of mice to sex odours and odours associated with aggression appear to be dependent on the olfactory experiences of the infant and juvenile mice. Male and female *Mus musculus* reared with lactating female pygmy mice (*Baiomys*) and their litters from birth to weaning show a preference for the odours of female *Baiomys* over those of female *Mus* when tested as adults, and spend more time in social interactions with *Baiomys*, but still mate with conspecifics (Quadagno and Banks 1970). Females show a stronger preference for the odours of the foster species than males (Quadagno and Banks 1970). Similar results were found by Mainardi, Marsan, and Pasquali (1965) who reared mice with perfumed mothers. Males reared with perfumed mothers showed only a slight preference for perfumed females in adulthood but females showed a highly significant preference for the odours of perfumed males. Female house mice reared with deer mice (*Peromyscus*) litters, however, show no evidence of acquiring a preference for the odours of deer mice, nor do they respond physiologically to the urine of deer mice (i.e. puberty acceleration) (Kirchhof-Glazier 1979). These females were kept in groups of same-sex conspecifics from weaning to adulthood and this experience may have affected odour preferences.

Wuensch (1981) found that rearing *Mus* males with *Peromyscus* females produces some attraction to *Peromyscus* odours in adult males that had been housed alone from the time of weaning, but since *Mus* which had not been cross-fostered showed just as great an attraction toward *Peromyscus* odours, the cross-fostering seems to have had little effect. Mice avoid rat odours and cross-fostering *Mus* onto rat mothers produces a preference for rat odours in adulthood, indicating that early experience with these odours produced a preference, or eliminated the aversion (Wuensch 1981).

Female mice reared with preputialectomized mothers show a preference for the odours of preputialectomized females and males in adulthood, while females reared with intact mothers prefer the odours of intact adult males and females, even though they were reared in groups from weaning to 90 days of age (Hayashi 1979). The attraction to the preputial gland odour of the lactating female may therefore be generalized to other mice in adulthood.

Using ultrasonic vocalizations as a measure of preference, Nyby *et al.* (1978) found that males reared with perfumed mothers from birth to weaning, and then reared with same-sex litter-mates and experiencing sexual behaviour with non-perfumed females showed very few ultrasonic vocalizations in the presence of a perfume odour. Males which had been reared with non-perfumed mothers but had copulated with perfumed females produced more ultrasounds in the presence of perfume, but males exposed to perfumed mothers and perfumed sex partners produced the most ultrasounds to perfume odours. This would suggest that olfactory experiences after puberty are more important in determining odour preferences of adult male mice than olfactory experiences ('imprinting') in infancy, but the two olfactory experiences (infancy and adult) interact to produce a greater preference than that produced by olfactory experience at only one age (Nyby and Whitney 1980). Similar results were found by Denenberg, Hudgens, and Zarrow (1964) who reared mice with rats and found that exposure to rat mothers and siblings as well as rats after weaning produced a greater preference for rats than exposure to rats after weaning alone. Mice which were never exposed to rats avoided rat odours.

Odours of adult conspecifics other than the mother may also influence odour preferences. Female mice reared in housing rooms with male odours present prefer the odours of intact males over those of castrated males, while females reared in rooms with no males prefer the odours of castrated males (Hayashi and Kimura 1978). The olfactory environment of infant mice may affect their aggressive behaviour in adulthood as well as their sex-odour preferences. Mice reared with rats are less aggressive than those reared with their own parents (Denenberg *et al.* 1964). Male mice reared in rooms housing *Peromyscus* and thus exposed to odours (and ultrasounds) of both *Mus* and *Peromyscus*, however, are more aggressive toward *Peromyscus* males than *Mus* reared with no *Peromyscus* odours (Stark and Hazlett 1972). The mice exposed to *Peromyscus* odours may have overcome their aversion to *Peromyscus* odours. Similarly, male mice reared with their mother and father or the odour of their father (father behind wire mesh) are more aggressive toward conspecific males than mice reared without fathers (Mugford and Nowell 1972; Wuensch and Cooper 1981). On the other hand, mice reared as single males in a litter are more aggressive in adulthood than males reared in all male litters (Namikas and Wehmer 1978).

Maternal behaviour   Olfactory cues from mouse pups interact with infant ultrasounds to stimulate cleaning and retrieval of pups by lactating females (Noirot 1969; Smotherman, Bell, Starzec, Elias, and Zachman 1974). Olfactory bulb lesions made before parturition eliminate maternal behaviour in female mice (Gandelman, Zarrow, and Denenberg 1972; Gandelman, Zarrow, Denenberg, and

Myers 1971) while olfactory bulbectomy after parturition reduces maternal behaviour, but does not eliminate it (Cowley and Cooper 1977).

Olfactory recognition of individuals and kin   Mice can discriminate between the odours of individuals of the same sex (Bowers and Alexander 1967) and when mice differing in a single gene locus (the H-2 locus) are given mating preference tests, they choose to mate with females of the different H-2 type (Yamazaki, Boyse, Miké, Thaler, Mathieson, Abbott, Boyse, Zayas, and Thomas 1976). It appears that genetic differences at this locus produce urine odours which can be used to identify individual mice (Yamazaki, Yamaguchi, Boyse, and Thomas 1980; Yamazaki, Yamaguchi, Baranoski, Bard, Boyse, and Thomas 1979).

Mate selection in mice appears to involve preferences for animals of the same species which are not kin. Female mice of both Porton and Steel strains prefer the odours of strange males of their own strain to males of the other strain or the odours of their brothers (Gilder and Slater 1978). Female *C57BL6/J* mice, however, prefer the odours of sibling males to non-siblings whereas *SEC1ReJ* females show no preferences for siblings (D'Udine and Partridge 1981). Kareem and Barnard (1982) have shown that both male and female *CFLP* mice associate more with siblings and investigate non-siblings more than siblings. Mice may, therefore, use the odours of kin as 'reference odours' for choosing mates which are of the same species, but not kin and thus 'strike an optimal balance between inbreeding and outbreeding' (Bateson 1978, p. 660).

Fear odour   When a mouse is stressed or frightened by rough handling, electric shock, or saline injections, it produces a fear or stress odour which is avoided by other mice (Carr, Martorano, and Krames 1970; Rottman and Snowdon 1972; Sprott 1969). This fear odour or 'Angstgeruch' may be secreted from the skin glands or in the faeces, but can be shown to occur in the urine, which is avoided by other mice (Müller-Velten 1966). Anosmia produced by infusion of zinc sulphate into the nares eliminates aversion to fear odours (Rottman and Snowden 1972). The urine retains the fear odour for 8–24 h, after which it is no longer avoided. Test mice soon adapt to the fear odour (Müller-Velten 1966). The fear odour appears to be species specific; mice (*Mus musculus*) do not avoid the odours of stressed field mice (*Apodemus sylvaticus*) nor the odours of stressed fat dormice (*Glis glis*) (Müller-Velten 1966).

Müller-Velten (1966) found that there was a sex difference in the response to fear odours. Both males and females avoided fear odours of males, but only females avoided urine odours of stressed females; males showed no fright reaction to the urine of stressed females. This finding, however, was not supported by Carr, Roth, and Amore (1971) who found that males avoided fear odours from both males and females.

Although Rottman and Snowden (1972) found that mice reared in social isolation did not avoid odours of stress, Carr *et al.* (1971) did find that males isolated from puberty avoided the odours of stressed conspecifics, so it is difficult to know the effects of social experience. More recently, Carr, Zunino, and Landauer (1980) examined the development of responsiveness to fear odours. Neither males nor females housed with their parents and littermates avoided odours of stressed mice at 24 days of age, and only females showed a significant avoidance at 48 days of age. Thus, it is possible that adult social interactions or aggressive encounters may influence the development of responsiveness to fear odours, but the definitive study has yet to be completed.

Testosterone levels do not appear to influence the production or the responses of males to fear odours; both castrated and intact males avoid the urine odours of stressed castrates (Sawyer 1980). The results of this study indicate that fear odours differ from the 'aversive odours' of isolated males which are not avoided by castrated males.

Acomys

Apart from Quay's (1965*b*) report that *Acomys dimidiatus* have small sebaceous glands of the oral angle and lip (one female) little is known of the scent glands of *Acomys*. All of the research has been directed at the responses of infant and juvenile *Acomys* toward maternal odours and the role of odours in learning in infant *Acomys*.

Maternal odour   Lactating female spiny mice (*Acomys cahirinus*) produce maternal odours which are attractive to the precocial pups. As early as one day of age, *Acomys* pups show preferences for bedding soiled by lactating females over clean bedding and bedding soiled by non-lactating females, bedding soiled by newborn *Acomys* pups or bedding soiled by adult males (Porter and Doane 1976; Porter and Ruttle 1975). *Acomys* pups prefer the odours of lactating females over those of non-lactating females until they are 25 days of age and then begin to prefer the odours of non-lactating females. Lactating females, however, appear to produce odours which are attractive to newborn *Acomys* until 38 days post-partum (Porter, Doane, and Cavallaro 1978).

The maternal odour of *Acomys* is influenced by the mother's diet. Pups prefer the odour of lactating females eating the same food as their mother over the odour of lactating females eating another diet. *Acomys* pups also prefer the odour of lactating female *Mus* eating a familiar diet to the odour of lactating female *Acomys* eating an unfamiliar diet, indicating that diet is more important than genetics in the production of the maternal odour of *Acomys* (Porter and Doane 1977; Porter, Deni, and Doane 1977).

While normally reared *Acomys* pups are attracted to odours of both lactating female *Mus* and lactating female *Acomys*, they show a preference for maternal odours of their own species. *Acomys* pups reared by *Mus* females, however, prefer the maternal odours from lactating *Mus* (Porter *et al.* 1977).

Olfaction and development   Juvenile spiny mice prefer to huddle with their litter-mates and this preference is reduced by housing non-litter-mates together for five days or making pups anosmic via zinc sulphate infusion (Porter, Wyrick, and Pankey 1978). *Acomys* pups kept in isolation for five days still prefer to huddle with siblings but eight days of isolation eliminates this preference (Porter and Wyrick 1979). Sibling preferences begin at 14–16 days of age (Porter and Wyrick 1979) which may be when *Acomys* develop their own individual odours. The preference for litter-mates appears to result from familiarity rather than recognition of some specific 'family odour' and there is, as yet, no evidence that *Acomys* prefer the odours of litter-mates over the odours of strange conspecifics. The only studies done have used whole animals, so behavioural differences, visual cues or ultrasounds may be used as well as odours.

Acomys, however, are very sensitive to olfactory stimuli and develop olfactory preferences at 1–2 days of age. As little as 1 h of exposure to artificial odours (cumin or cinnamon) is sufficient to produce a preference for that odour in newborn *Acomys* (Porter and Etscorn 1974). *Acomys* pups are more sensitive to odours during the first two days post-partum than during post-partum days 3–5 as exposure to an artificial odour on days 1 or 2 produces a greater preference for that odour in six-day-old *Acomys* than exposure on days 3–5 (Porter and Etscorn 1976). If *Acomys* are exposed to two different odours, for 1 h each, the first at one day of age and the second at two days of age, they prefer the first odour to which they were exposed. This primacy effect is eliminated if the pups are exposed to the second odour for much longer than they are exposed to the first odour (Porter and Etscorn 1975).

Bandicota

Apart from the report that male and female *Bandicota* of a number of species have sebaceous glands on the oral angle and lips (Quay 1965*b*), nothing appears to be known about the role of olfaction in the social behaviour of this genus.

Nesokia

The pest rat (*Nesokia indica*) has sebaceous glands of the oral angle and lip (one female; Quay 1965*b*) and males, but not females, have preputial glands (Brown and Williams 1972; Schaffer 1940). Nothing seems to be known about the role of these glands in behaviour, however.

Cricetomys

Adult male African giant rats (*Cricetomys gambianus*) cheek-rub females and objects in their environment, suggesting that they possess scent glands of the oral angle or other facial glands. These rats also urine-mark by dribbling urine in a strange environment. Defecation normally occurs in lavatory areas but faeces may be deposited in prominent places using a peculiar 'hand-stand' position, particularly when the animal is disturbed, suggesting that faeces may be used as alarm signals (Ewer 1967) (see Fig. 9.2(f)).

Chiropodomys

Both males and females of various species of *Chiropodomys* have preputial glands (Brown and Williams 1972; Schaffer 1940), but nothing is known about the role of these glands in social behaviour.

Otomys

The Vlei rat (*Otomys irroratus*) has paired anal glands and scent-marks by backing up to an object or wall and rubbing the anal region against the object by raising and lowering the hind-legs. Urination occurs in specific locations, but it is unclear whether these excretions are used for communication (Davis 1972).

Discussion

Rodent social odours provide information about the identity and emotional state of the animal. While much research during the last decade has identified the sources of odours and some of their social functions, there are a number of questions about the social odours of rodents which may provide the basis for future research. Eight such questions are examined here.

1. Can scent glands be used as taxonomic traits?

Glands which have been suggested as taxonomic traits include the dorsal gland of *Dipodomys* (Quay 1953); the caudal gland of *Perognathus* (Quay 1965a); and caudal gland of *Apodemus* (Stoddart 1972b); the rump gland of *Lemmus* and *Dicrostonyx* (Quay 1968); the flank glands of *Cricetus* and *Mesocricetus* (Vrtis 1931; Stolzer 1959); the mid-ventral glands of *Peromyscus* (Doty and Kart 1972); *Neotoma* (Howe 1977); *Meriones* (Sokolov and Skurat 1966) and *Rattus* (Rudd 1966); and the hip, flank, rump, and caudal glands of *Microtus* (Quay 1968; see Fig. 9.5).

Many other glands such as salivary, preputial, nasal, Harderian, and Meibomian glands and those of the oral angle and lips occur so frequently among the rodents that their absence may be of more taxonomic interest than their presence. These glands may, however, show sex differences in some species, but not others and this difference may be used taxonomically.

Unfortunately, it is difficult to use scent glands as fixed taxonomic traits because many of these glands vary with the age, sex, dominance status, or reproductive state of the animal. Because of this added physiological dimension in variability, large samples of animals must be examined systematically at different times of the year. Most of the taxonomic studies have not done this, but have used museum collections, examined animals for 'staining' of the pelage or examined only one animal of a species.

Thus, while there is some evidence that scent glands may be used as taxonomic traits, much systematic work is required before secure use of scent glands in taxonomy is possible.

2. What information is contained in the animal's odorous secretions?

While preference and discrimination tests have shown that many rodents can identify the sex, species, reproductive status, age, dominance status, and even individuality of another animal by their odours alone, there is little information on exactly which secretions provide the information. Chemical analyses have identified vast numbers of components in many mammalian secretions. The chemistry of the castor gland of the beaver (Lederer 1949); the flank glands of *Arvicola* (Stoddart *et al.* 1975); the ventral gland of the gerbil (Thiessen *et al.* 1974); the caudal gland of *Apodemus* (Stoddart 1973); the preputial gland of the rat (Gawienowski *et al.* 1975, 1976); the vaginal secretion of the female hamster (O'Connell *et al.* 1978; Singer *et al.* 1976, 1980) have all been examined, but there is still little information on which chemicals in the secretions from these glands code which information.

It is apparent from many behavioural bioassays that an animal's response to a complex chemical such as a glandular secretion cannot be elicited by only a few chemical components from the secretion. Thus, while dimethyl disulphide appears to be the chemical in the vaginal secretion of the female hamster which elicits the most investigation from males, it does not elicit mounting (Macrides *et al.* 1977). While information about the active components in the glandular secretions are important, it appears that the concept of an 'active component' may be oversimplified and it may require a 'recipe' of ingredients in order to replicate the information in the whole gland secretion. Thus, a combination of dimethyl disulphide + alcohol + acids is more attractive to male hamsters than dimethyl disulphide alone (O'Connell *et al.* 1978), but some more complex chemical compound appears to be necessary to elicit mounting (Singer *et al.* 1980).

The principal component of the ventral-gland secretion of the Mongolian gerbil has been identified as phenylacetic acid (Thiessen *et al.* 1974), but the information that gerbils gain from this chemical is unknown. Gerbils will avoid phenylacetic acid if it is paired with illness (Pettijohn 1979), but whether or not the phenylacetic acid content in the ventral-gland secretion is related to the age, sex, dominance status, or level of stress of an individual, and whether the Israeli gerbil also secretes phenylacetic acid from its ventral gland, or if this chemical is specific to the Mongolian gerbil is unknown.

Thus, while a start has been made in the identification of the chemical components of the glandular secretions of rodents, the relationship of the chemical components of the scent glands to the information carried in the glandular secretions and to the behaviour shown toward the odours of these secretions is yet to be understood.

3. Do the 'neglected glands' have a communicatory function?

Just as the secretion from a whole gland is a more powerful stimulus than one chemical isolated from the secretion, the odour of a whole animal is often a more attractive stimulus than the odour from only one gland. Gerbils, for example, spend more time investigating bedding soiled by adult male gerbils than they spend investigating male ventral gland or urine odours alone (Halpin 1974). The 'whole animal' odour includes urine, faeces, and secretions from the specialized sebaceous glands (mid-ventral, dorsal, or flank) as well as secretions from feet, eyes, mouth, ears and from smaller, more widely dispersed and often neglected glands of the general skin.

While some of these 'neglected' glands have been shown to have a communicatory function, many remain unevaluated. The Harderian gland secretions of hamsters and gerbils can be used to identify the sex of the animal (Payne *et al.* 1975, 1977; Thiessen and Rice 1976); but the Harderian- and lachrymal-gland secretions of other rodents have yet to be examined for a communicative function. Similarly, the secretion of the ear gland of the hamster can be used for sex discrimination (Landauer *et al.* 1980) and saliva can be used to indicate sex and possibly individuality in gerbils and mice (Block *et al.* 1981; Lee and Ingersoll 1979), but remain neglected in other species.

The pedal and plantar glands, anal glands and the small sebaceous glands of the general body surface have only rarely been examined for communicatory function in rodents. Investigation of the social function of these glands may provide information about the sources of odours which contribute to the 'body odour' of rodents.

4. Are enough different animals being studied?

The majority of the research on olfaction and social behaviour in rodents has been carried out on four species: the Mongolian gerbil, the golden hamster, the Norway rat, and the house mouse. Many species which appear to be ideally suited for olfactory research remain neglected.

Some species such as the mole rats (*Spalax*) are blind and may prove to be important species for research on olfactory communication. Other species such as *Microtus* have an abundance of scent glands and enough variability to examine species, sex, age, and seasonal differences in the function of the odours from these glands. Still other species such as *Arvicola* have a flank gland which differs in chemical composition between males and females, young and adult, and breeding and non-breeding season, but no behavioural studies appear to be directed at the communicative functions of this gland.

Of the species of *Rattus*, those with ventral glands have not been examined more than superficially, nor have the rice rats (*Oryzomys*), muskrats (*Ondatra*) or water rats (*Neofiber*) been studied for olfactory communication. Research on olfactory communication in these neglected species may answer some of the questions of the function of scent glands.

5. What is the function of scent-marking?

The understanding of scent-marking has advanced greatly since the time when all marking was considered to be 'territorial' (see Ralls 1971), but the functions of scent-marking are not yet fully understood. While Ralls (1971) pointed out that not all scent-marking is territorial, she focused on the aggressive motivation for marking. Dominance, the appearance of a strange conspecific of the same sex and aggressive encounters stimulate marking in many species, but these are only some of the factors which must be considered in understanding the function of rodent scent marks. Johnson (1973) has reviewed the role of scent marks as sex attractants, orientation marks, responses to novelty, indicators of individual identity and alarm signals as well as their role in aggressive behaviour. Marking mates, colony members and offspring for recognition, marking as a sexual advertisement and marking as a maternal behaviour can also be added to the list of scent-marking functions (see Brown 1979).

Species such as rats and mice tend to scent-mark novel environments and, as suggested by Eisenberg and Kleiman (1972), this may provide an animal with an 'optimum odour field'. Such an odour field may serve many functions, including a territorial one. Beaver scent mounds, for example, appear to indicate that a lodge is occupied. If a lodge does not have a scent mound, it is 'vacant' and may be taken over by alien beaver (Svendsen 1980). The scent mounds may also act to orient beaver within their home-range, to increase the 'confidence' of the residents (Svendsen 1980) or to 'bond' colony members together (Butler and Butler 1979).

Unfortunately, many of the functional explanations for scent-marking remain untested. Aggressive and sexual motivations have been most carefully defined, both behaviourally and hormonally, and maternal motivation can be defined in the same ways, but 'confidence' 'colony bonds', and 'optimum odour fields' are more difficult to define operationally and thus provide weaker explanations for the motivation of scent-marking.

The reason for using different postures for scent-marking has also not been explained. Some, such as sandbathing in the Heteromyidae may have evolved from toilet behaviours, and others, such as the 'hip-swagger' of Microtus may depend on the location of the objects to be marked. The peculiar hand-stand posture of Cricetomys (Ewer 1967) may involve a visual as well as an olfactory signal (see Eisenberg and Kleiman 1972).

Different types of scent-marking may occur under different motivational states and have different functions. The desert woodrat (Neotoma lepida) ventral marks when aggressively motivated and sandbathes or sand rolls when sexually motivated (Fleming and Tambosso 1980). Female hamsters flank mark in agonistic encounters with other females, but vaginal mark in the presence of males, indicating sexual motivation (Johnson 1977a). Female Mongolian gerbils seldom ventral-mark except while lactating and then ventral-marking reaches a peak from 1–2 weeks post-partum, suggesting that the ventral-gland marking is motivated by 'maternal' arousal (Yahr 1976).

Examination of the motivational and situational factors involved in scent-marking in a wide range of rodents and examination of different marking patterns in species with more than one gland may therefore help to determine the functions of scent-marking.

6. What factors determine the development and secretory patterns of rodent scent glands?

The development and secretory patterns of scent glands appear to be controlled by two main factors: genetic differences and hormonal secretions. In addition, hormonal secretions can be altered by environmental variables such as breeding season and social variables such as dominance status. Whether or not the odours of rodents are altered by bacterial action seems not to have been investigated as extensively as it has in carnivores (see Chapter 15). One might expect bacterial action to modify vaginal secretions and the secretions from glands with sacs such as the anal glands. The anal gland of the beaver has a bacterial flora, but whether these bacteria influence odour production is unknown (Svendson and Jollick 1978).

Genetic factors   The different chromosomal forms of the mole rat (*Spalax*) produce different odours and these odours direct aggression and sexual behaviour toward members of the same karotype (Nevo and Heth 1976). In *Mus*, it appears that olfactory differences between strains or subspecies, between kin and even individual differences in odours may be controlled genetically (Grau 1982; Kareem and Barnard 1982; Yamazaki *et al.* 1979, 1980). In many of these cases the genetic differences are known, but the mechanism for converting genetic information into odour differences remains unknown.

Differences in the presence or absence of a scent gland between closely related species, or the location of scent glands also suggest genetic control of scent production. Thus, some species of *Peromyscus* (Doty and Kart 1972) and *Neotoma* (Howe 1977) have ventral glands while others do not. *Microtus* species may have flank glands, hip glands, or no glands and those that do have glands may have them on both sexes or only one sex, and the glands may vary with the breeding season or secrete all year round (Quay 1968).

Differences in glands among closely related species may serve to allow animals to discriminate between conspecifics and closely related species as has been found in *Clethrionomys* (Godfrey 1958), but little appears to be known of the genetic factors controlling scent glands or the linkage between glands and other features.

Hormones   Gonadal, pituitary, and adrenal hormones have been implicated in the development of scent glands and control of odorous secretions in rodents. The sex differences seen in many glands have been shown to be controlled by the gonadal hormones, primarily testosterone. This is true for the ventral gland of *Peromyscus* (Doty and Kart 1972); the ventral gland of *Neo-*

toma (Howe 1977); the flank gland of *Mesocricetus* (Vandenbergh 1973); the flank gland of *Arvicola* (Stoddart 1972); the anal, preputial, and hip glands of *Microtus montanus* (Jannett 1978); the ventral gland of *Meriones* (Glenn and Gray 1965); and the preputial gland of *Rattus* (Ebling *et al.* 1969*a,b*). In all of these examples, castration inhibits and testosterone injections increase the size and secretory rate of the glands.

Much less research has been directed toward the hormonal control of glandular secretions in female rodents. Vaginal secretions change over the oestrous cycle of *Mesocricetus* (Orsini 1961) and *Mus* (Hayashi and Kimura 1974) but, where studied, the skin glands seem difficult to control using oestrogen and progesterone. The multitude of studies on the hormonal control of the ventral gland of the female gerbil (see pp. 385-6) have shown only minor effects of ovariectomy or oestrogen and progesterone injection.

During lactation, on the other hand, the female gerbil's ventral gland does increase in size (Yahr 1976) and the pituitary hormone prolactin, or some other hormonal change during pregnancy may cause this increase. The flank gland of female *Microtus gregalis* also secretes more during lactation (Quay 1968). The maternal odour secreted in the faeces of lactating rats has been shown to be controlled by prolactin (Leon and Moltz 1973). While few attempts have been made to examine the control of scent glands by pituitary hormones, the rat preputial gland appears to be controlled by both prolactin and MSH as well as testosterone (Thody and Shuster 1975).

The adrenal gland may be involved in the secretion of fear or alarm odours in mice (Archer 1969) and may also be involved in the decrease in size of the preputial glands in subordinate mice and in crowded mice (Brown and Williams 1972).

In many species skin glands show increased activity during the 'breeding season' during which the gonadal hormone levels are at a maximum. Whether stimulated via changes in the light cycle or some other climatic change (see Gwinner 1981), the dorsal gland of *Dipodomys* (Quay 1953); the mid-ventral gland of *Neotoma* (Clarke 1975); the mid-ventral gland of *Cricetulus* (Vorontsov and Gurtovoi, 1959); the preputial gland of *Clethrionomys* (Christiansen *et al.* 1978); and the flank glands of *Arvicola* (Stoddart 1971) become enlarged during the breeding season.

In terms of social factors affecting hormone levels and scent-gland secretion, dominance appears to be the most important; the flank glands of *Mesocricetus* (Drickamer and Vandenbergh 1973; Drickamer *et al.* 1973); the preputial glands of *Clethrionomys* (Gustafsson *et al.* 1980) and the preputial glands of *Mus* (Bronson and Marsden 1973) are larger in dominant than in subordinate animals.

Delayed puberty (see Chapter 8) also inhibits the development of scent glands. For example, gerbils kept with their parents have delayed puberty and delayed development of the ventral gland (Payman and Swanson 1980).

7. What variables affect the reception and use of olfactory information by rodents?

The presence of an odour is not sufficient for communication. There must be a recipient whose behaviour or physiological processes are altered when the odour is perceived. Many variables affect the responses of rodents to social odours. These include age, sex, hormonal state, dominance status, and social experience (see Brown 1979). It is important to consider these receiver variables because of the necessity for using behavioural bioassays for evaluating the 'active components' of odorous secretions, determining the information content of odorous secretions and determining the 'function' of scent glands and scent-marking.

The age of rodents affects their response to maternal odours (e.g. *Rattus*, Leon and Behse 1977), to sex odours (e.g. *Rattus*; Carr, Wylie, and Loeb 1970; and to stress odours (e.g. *Mus*, Carr, Zunino, and Landauer 1980).

The sex of the recipient is important because the odours of conspecifics of the same sex are usually less attractive than the odours of the opposite sex. In some species, only one sex may be attracted to conspecific odours. This occurs in *Meriones* in which males are attracted to the ventral gland odour of conspecifics but females are not (Thiessen *et al.* 1971).

The effect of hormonal state on responses to odours is most pronounced with respect to odours from conspecifics of the opposite sex. Castration eliminates attraction to conspecific odours in *Rattus* (Brown 1977*a*) and *Mus* (Rose and Drickamer 1975). Rodents which have seasonal breeding such as *Peromyscus* are attracted to odours of conspecifics of the opposite sex during the breeding season, but not outside of this season (Daly *et al.* 1978, 1980).

Dominance status is important because dominant and subordinate animals may show different responses to odours. Subordinate male cotton rats (*Sigmodon*), for example, avoid the odours of dominant males, but dominant males approach these odours (Summerlin and Wolfe 1973).

Social experiences alter odour-directed behaviour in many different ways. In *Rattus*, sexual experience increases the attractiveness of female odours to males but has little effect on females (Brown 1977*a*; Carr 1974). Sexual experience, however, does not seem to affect the attraction of male *Mesocricetus* to female hamster vaginal discharge (Gregory *et al.* 1975). Aggressive experiences may lead to the averson of the odours of particular conspecifics. Male *Microtus agrestis* (De Jonge 1980) and male *Mus* (Carr, Martorano, and Krames 1970) avoid the odours of males which have defeated them in agonistic encounters. Responses of *Rattus* and *Mus* to fear and stress odours may depend on the experience of the test animals (Mckay-Sim and Laing 1980). Social isolation may alter the responses of rodents to stress odours (Rottman and Snowdon 1972; Carr, Roth, and Amore 1971) and to sex odours. Rats reared in isolation from weaning to adulthood show different patterns of scent marking than socially experienced rats (Brown 1982*b*). Familiarity with particular individuals through close association leads to changes in odour investigation preferences (e.g. in *Rattus*, Carr *et al.* 1976; Krames and Shaw 1973) and to reduced aggressiveness (e.g. in *Mus*, Mackintosh and Grant 1966; Ropartz 1968).

The mechanism through which experience alters responsiveness to odours is unclear. In some cases associative learning appears to occur; in other cases learning appears to be by mere exposure to an odour, and in still other cases, experience seems to have little effect on odour preference suggesting some genetic mechanism of odour identification.

The associative learning studies with the most dramatic effects have paired animal odours with lithium chloride injections to produce sickness. Male *Mesocricetus* associate female hamster vaginal odour with sickness and avoid this odour for a brief period (Johnston and Zahorick 1975) and male *Meriones* will avoid the odour of other gerbils when these were associated with sickness (Pettijohn 1979).

In other situations, particularly the attraction of infant rodents to odours of their nest or odours of a lactating female, the odours may gain their attractive value through mere exposure during some 'critical period'. Such exposure learning may occur in infant *Mesocricetus* (Cornwell 1975, 1976); *Rattus* (Leon, Galef, and Behse 1977; Galef and Kaner 1980); *Mus* (Goldblatt 1978*a*) and *Acomys* (Porter and Etscorn 1974). Schumacher and Moltz (1982), on the other hand, have argued that infant rats are attracted more to maternal odours than other odours which are equally familiar and that the preference for maternal odours may be 'innate'. Whether infants are innately attracted to maternal odours or acquire this preference through 'imprinting', or some form of associative learning during a critical period are questions to which much future research will be directed.

The development of preferences for conspecific odours over the odours of other species remains unknown. Cross-fostering studies suggest that some species may develop preferences for the odours of a foster species, but do not direct social and sexual behaviour toward members of the foster species. Other species do not show preferences for the odours of the foster species nor direct sociosexual behaviour toward the foster species. *Dicrostonyx* males cross-fostered to *Lemmus* prefer the odours of *Lemmus* in adulthood and copulate with *Lemmus* females, but cross-fostered female *Dicrostonyx* show no such preferences for their foster species (Huck and Banks 1980*a,b*). Both male and female *Lemmus* cross-fostered to *Dicrostonyx* prefer the odours of *Dicrostonyx*, but only the females mate with their foster species (Huck and Banks 1980*a,b*). *Mus* reared with *Baiomys* prefer the odours of *Baiomys*, but mate only with *Mus* (Quadagno and Banks 1970) and *Mus* reared with *Peromyscus* sometimes show a preference for their foster species (Wuensch 1981) and sometimes do not (Kirchhof-Glazier 1979). While *Acomys* reared with *Mus* show a preference for the maternal odours of *Mus* (Porter, Deni, and Doane 1977) there have been no studies on whether this preference lasts into adulthood. So far, no species of rodent has shown a complete shift of all odour preferences and social behaviour toward a foster species, suggesting that there is some genetic basis for species attraction which is resistent to experimental modification.

8. What is the function of olfactory communication in the evolution of rodent social behaviour?

Social behaviour has evolved to benefit individuals for interacting with conspecifics. The benefits of social behaviour include predator avoidance, increased nutrition, reduced likelihood of disease and increased ability for reproduction. The social interactions which evolve are those that are genotypically selfish and these include reciprocal altruism, nepotism, and parental manipulation of progeny (see Alexander 1974).

Olfactory communication may play an important role in the evolution of social behaviour because social odours provide the basis for the recognition of other animals, and the mediation of social interactions including kin recognition, reproductive competition, mate selection, parental care, territory and home-range identification, and food selection.

Kin selection   To be genetically selfish means that altruistic behaviour should be directed towards kin. The ability to recognize kin should therefore be an essential feature of animal communication systems and social odours may provide a basis for kin recognition in rodents as they appear to in the social insects (Getz 1981). For some rodents, kin recognition appears to be genetically based (Grau 1982), while for others, familiarity may be important and familiar animals may be regarded as kin (Kareem and Barnard 1982). Since animals prefer to have social contact with kin, are less aggressive toward kin, and tolerate kin in their territory, kin recognition could be an important function of social odours.

Aggressiveness   Since rodents direct more agonistic responses toward unfamiliar adult conspecifics of the same sex, odours may be used for recognition of species, age, sex, and familiarity of an opponent. Once an opponent has been identified, odours may be used to assess potential aggressiveness (see Maynard Smith 1974; Parker and Rubenstein 1981).

Since dominant males of many species have larger scent glands than subordinates, an animal's odour may provide a valid indicator of its potential aggressiveness as well as its dominance status. Dominant *Sigmodon* (Summerlin and Wolfe 1973); *Neotoma* (Howe 1977); *Mesocricetus* (Drickamer *et al.* 1973); *Lemmus* (Huck *et al.* 1981); *Meriones* (Gallup and Waite 1970); *Rattus* (Krames, Carr, and Bergman 1969); and *Mus* (Bronson and Marsden 1973) all can be discriminated from subordinates by olfactory cues. Thor (1979) has suggested that these differences in odour allow for interactions to be carried out with a minimum of aggressive behaviour.

Mate selection   Whether mating is monogamous, polygamous, or promiscuous (Wittenberger 1981) animals must select mates which are adult conspecifics of the opposite sex and are in a reproductive state. Odours may be used to identify the species, age, sex, and reproductive state of prospective mates (Daly, Wilson, and Behrends 1980; D'Udine and Partridge 1981; Godfrey 1958; Vadasz 1975). In many cases odours may also be used to select unrelated conspecifics as mates (Bateson 1978; Gilder and Slater 1978). As well as using odours for mate selection, scent marking in many species of rodents may be used for mate solicitation or sexual advertisement (Brown 1979; Doty 1974; Johnston 1977*a*).

Rodents which are monogamous may use different cues to select mates than those that are promiscuous. Monogamous species may use odours to recognize their mates and thus attend to the odours of specific individuals of the opposite sex while promiscuous species are attracted to the odours of novel females or males (Carr, Hirsch, and Balazs 1980). For promiscuous species, selecting oestrous females or sexually active males may be the primary function of olfactory communication. The odours of sexually active conspecifics are preferred to castrates, juveniles or those that are not sexually active in *Rattus* (Orsulak and Gawienowski 1972; Lydell and Doty 1972); *Mus* (Caroom and Bronson 1971; Rose and Drickamer 1975); *Dipodomys* (Daly, Wilson, and Behrends 1980); and *Neotoma* (Fleming, Chee, and Vaccarino 1981).

Parental care   All female and some male rodents show parental care. Odours appear to be important stimuli for offspring to attach to the nipples of the mother (e.g. in *Rattus*, Teicher and Blass 1977) and may be used to differentiate family members from non-family members. In species in which the parents discriminate between own and alien offspring (e.g. goats, Klopfer and Gamble 1966) the mother may mark the offspring for recognition and reject alien infants. Rodents do not seem to form attachments to their young and will accept foster young (Gubernick 1981) but the increase in the size of the scent glands, and in scent-marking frequency and the changes in odours of faeces in lactating females of many species suggest that odours play an important role in maternal care in rodents.

Male *Meriones* (Elwood 1975), *Microtus* (Hartung and Dewsbury 1979), *Peromyscus* (Dudley 1974) and *Rattus* (Brown 1981) show paternal care, but must first have their tendencies toward infanticide inhibited. In these species cohabitation with a female and her litter inhibits infanticide (Elwood 1980; Labov 1980; Brown 1981). The odour of the female on the pups, or the aggressive behaviour of the female towards the male in the presence of the pups may inhibit infanticide by males.

Odours may be used to recognize siblings as well as offspring. *Acomys* prefer to huddle with siblings rather than non-siblings (Porter, Wyrick, and Pankey 1978) but this 'sibling identification' may depend more on experimental variables such as exposure to siblings than to 'genetic' variables (see Bekoff 1981). Kin selection and altruism depend on the recognition of relatives and there may be an olfactory basis for this, but only if the relatives are reared together and live continuously in the same group.

Home-range and nest marking   One important function of scent-marking may be for place recognition or attachment to home-areas (see Brown 1979). Many rodents are more aggressive in their home-area and may use scent marks for trail and nest identification.

Mouse 'runways' (Welch 1953) are regularly marked and animals which have elaborate tunnel systems such as kangaroo rats (Seton 1901) and wood rats (Rainey 1956; see Fig. 9.12) may mark tunnels and their entrances. Since voles are more likely to enter traps with their own odours (Stoddart 1982), they may be more likely to enter a tunnel system with their own odour than one with no odour and least likely to enter tunnels with the odours of strange animals. Animals might also run more rapidly through tunnels which they had previously scent-marked and thus be able to escape predators more easily than in unmarked runways.

No matter what sort of nest a rodent makes, the nest will probably absorb the animal's odour so that an individual rodent should be able to recognize its own nest by its odour alone. The research on nest identification in rodents has stressed the ability of infant spiny mice (Porter and Ruttle 1975), hamsters (Devor and Schneider 1974), and rats (Brown 1982c; Carr, Marasco, and Landauer 1979) to locate their nests and no research seems to have been done on the ability of adults to identify nests.

Olfaction in foraging and feeding   Foraging takes an animal away from its home-area and thus makes it more vulnerable to predation. One would expect rodents to mark unfamiliar areas more often than those with their own odour. Mice (Bronson 1976); voles (Johnson 1975), and rats (Brown 1975, 1977a) urine-mark novel, 'clean' areas and marking over clean objects may represent the marking of an unfamiliar trail. Nocturnal species may scent-mark trails more than diurnal species and species which store food in caches may scent-mark the cache sites, but no evidence of this has been presented for rodents.

Carnivorous species such as *Onychomys* and omnivorous species such as *Rattus* and *Mus* may be more likely to cannibalize infants than graniverous or herbivorous species of rodents, so might be more likely to mark their pups for recognition, but again there is no evidence for this.

Little is known about the role of scent-marking in determining feeding patterns. Galef and Heiber (1976) have shown that food choice in young rats is directed toward food eaten by adults and having the odours of adults nearby. Faeces may function as guides to food selection, so that examination of faeces may direct an animal to choose foods eaten by a conspecific and feeding areas having conspecific odours may be chosen in preference to 'clean' feeding areas.

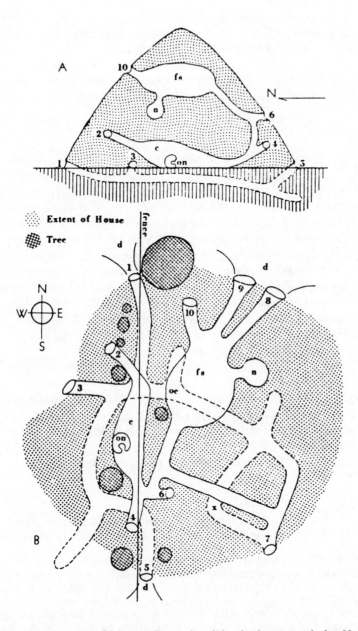

Fig. 9.12. The side view (a) and floor plan (b) of a house made by *Neotoma floridana* drawn to a scale of 1 : 18. fs = food store; d = debris from inside the house; n = nest chamber; on = unused nest chamber; c = chamber; oc = osage orange cuttings; x = site of opened coffee beans and walnuts. Numbers refer to entrances. (From Rainey (1956) © The University of Kansas.)

Since areas having the odours of dominant mice are avoided by subordinates (Jones and Nowell 1974c), the scent marks of a dominant animal may inhibit other animals from encroaching on the food supply. Rodents may also use odours to mark poisonous foods or foraging areas which are to be avoided. Steiniger (1950) reported that rats defecated on poisonous baits which were then avoided by other rats, but no experimental studies have replicated this.

Summary

Odours play an important role in the social behaviour of rodents. This chapter has attempted to examine the known relationships between odour sources, odour reception, and odour function in rodent social behaviour. Scent glands, scent-marking postures, and the information content of social odours were examined and some questions for further research suggested. While some species have been studied extensively, there are many species which have a wide variety of odour sources and which display sexual, seasonal, age, and individual differences in their odorous secretions which have not been extensively studied.

The importance of social odours in mate selection, aggressive interactions and parent-offspring interactions have been examined as well as the role of odours as conditioned stimuli. Finally, the importance of the animal's olfactory environment during development and the effects of early olfactory experience on later behaviour has been examined.

Acknowledgements

I would like to thank Diane Elrick, Patti Lynch, Mary MacConnachie, and the librarians in the Macdonald Science Library at Dalhousie University for their assistance in the preparation of this chapter and Dr W. B. Quay for his comments on the manuscript. Financial assistance was provided by grant A-7441 from the National Science and Engineering Research Council of Canada.

References

Adams, D. B. (1976). The relation of scent-marking, olfactory investigation, and specific postures in the isolation-induced fighting of rats. *Behaviour* **56**, 3-4; 286-97.

Alberts, J. R. (1978). Huddling by rat pups: multisensory control of contact behaviour. *J. comp. physiol. Psychol.* **92**, 220-30.

— and Brunjes, P. C. (1978). Ontogeny of thermal and olfactory determinants of huddling in the rat. *J. comp. physiol. Psychol.* **92**, 897-906.

Alderson, J. and Johnston, R. E. (1975). Responses of male golden hamsters (*Mesocricetus auratus*) to clean and male scented areas. *Behav. Biol.* **15**, 505-10.

Aleksiuk, M. (1968). Scent-mound communication, territoriality, and population regulation in beaver (*Castor canadensis* Kuhl.). *J. Mammal.* **49**, 759-62.

Alexander, R. D. (1974). The evolution of social behavior. *A. Rev. Ecol. System.* **5**, 325-83.

Algard, F. T., Dodge, A. H., and Kirkman, H. (1964). Development of the flank organ (scent gland) of the Syrian hamster. I. Embryology. *Am. J. Anat.* **114**, 435–55.

— — — (1966). Development of the flank organ (scent gland) of the Syrian hamster. II. Postnatal development. *Am. J. Anat.* **118**, 317–26.

Anisko, J. J., Adler, N. T., and Suer, S. (1979). Pattern of postejaculatory urination and sociosexual behavior in the rat. *Behav. neural Biol.* **26**, 169–76.

Aquadro, C. F. and Patton, J. C. (1980). Salivary amylase variation in *Peromyscus*: use in species identification. *J. Mammal.* **61**, 703–7.

Archer, J. E. (1968). The effects of strange male odour on aggressive behaviour in male mice. *J. Mammal.* **49**, 572–5.

— (1969). Adrenocortical responses to olfactory social stimuli in male mice. *J. Mammal.* **50**, 839–41.

— (1975). Comment on Harmatz, Boelkins and Kessler, 1975: 'Habituation not primer pheromone reduces attack in odor exposed mice'. *Behav. Biol.* **15**, 519–20.

Arluk, D. J. (1968). The hormonal regulation of the ventral (sebaceous) gland of the Mongolian gerbil. *Diss. Abstr. Int.* **30B**, 29.

August, P. V. (1978). Scent communication in the southern plains wood rat, *Neotoma micropus. Am. mid. Nat.* **99**, 206–18.

Bailey, V. (1936). The mammals and life zones of Oregon. *N. Am. Fauna* **55**, US Department of Agriculture, Washington, DC.

Baran, D. (1973). Responses of male Mongolian gerbils to male gerbil odours. *J. comp. physiol. Psychol.* **84**, 63–72.

— and Glickman, S. E. (1970). 'Territorial marking' in the Mongolian gerbil: a study of sensory control and function. *J. comp. physiol. Psychol.* **71**, 237–45.

Barnett, S. A. (1967). *A study in behaviour.* Methuen, London.

Bateson, P. (1978). Sexual imprinting and optimal outbreeding. *Nature, Lond.* **273**, 659–60.

Beach, F. A. and Jaynes, J. (1956*a*). Studies of maternal retrieving in rats I: recognition of young. *J. Mammal.* **37**, 177–80.

— — (1956*b*). Studies of maternal retrieving in rats III: Sensory cues involved in the lactating female's response to her young. *Behaviour* **10**, 104–25.

Bean, N. J. (1982). Olfactory and vomeronasal mediation of ultrasonic vocalizations in male mice. *Physiol. Behav.* **28**, 31–7.

Beaver, D. L. (1960). A re-evaluation of the rat preputial gland as a 'dicrine' organ from the standpoint of its morphology, histochemistry and physiology. *J. exp. Zool.* **143**, 153–73.

Bekoff, M. (1981). Mammalian sibling interactions: genes, facilitative environments, and the coefficient of familiarity. In *Parental care in mammals* (ed. D. J. Gubernick and P. H. Klopfer) pp. 307–46. Plenum Press, New York.

Benton, A. H. (1955). Observations on the life history of the northern pine mouse. *J. Mammal.* **36**, 52–62.

Benuck, I. and Rowe, F. A. (1975). Centrally and peripherally induced anosmia: influences on maternal behavior in lactating female rats. *Physiol. Behav.* **14**, 439–47.

Birke, L. I. A. (1978). Scent marking and the oestrous cycle of the female rat. *Anim. Behav.* **26**, 1165–6.

Blass, E. M. and Teicher, M. H. (1980). Suckling. *Science, NY* **210**, 15–22.

— — Cramer, C. P., Bruno, J. P., and Hall, W. G. (1977). Olfactory, thermal and tactile controls of suckling in preauditory and previsual rats. *J. comp. physiol. Psychol.* **91**, 1248–60.

Block, M. L., Volpe, L. C., and Hayes, M. J. (1981). Saliva as a chemical cue in the development of social behavior. *Science, NY* **211**, 1062–4.

Bloom, J. M. and Phillips, J. M. (1973). Conspecific odours as discriminative stimuli in the rat. *Behav. Biol.* **8**, 279–83.

Blum, S. L., Balsiger, D., Ricci, J. S., and Spiegel, D. K. (1975). Effects of early exposure to ventral gland odor on physical and behavioral development and adult social behavior in Mongolian gerbils. *J. comp. physiol. Psychol.* **89**, 1210–19.

— Kakihana, R., and Kessler, S. (1971). Atrophy of the ventral gland of the male deermouse after castration. *J. Endocr.* **49**, 695–6.

— and Thiessen, D. D. (1970). Effect of ventral gland excision on scent marking in the male Mongolian gerbil (*Meriones unguiculatus*). *J. comp. physiol. Psychol.* **73**, 461–4.

— — (1971). The effect of different amounts of androgen on scent marking in the male Mongolian gerbil. *Horm. Behav.* **2**, 93–105.

Boonstra, R. and Krebs, C. J. (1976). The effect of odour on trap responses in *Microtus townsendii. J. Zool., Lond.* **180**, 467–76.

Borchelt, P. L., Griswold, J. G., and Branchek, R. S. (1976). An analysis of sand-bathing and grooming in the kangaroo rat (*Dipodomys merriami*). *Anim. Behav.* **24**, 347–53.

Bowers, J. M. and Alexander, B. K. (1967). Mice: individual recognition by olfactory cues. *Science, NY* **158**, 1208–10.

Brain, P. F. and Evans, C. M. (1974*a*). Influences of two naturally occurring androgens on the attack directed by 'trained fighter' TO strain mice towards castrated mice of three different strains. *IRCS (Res. Endocr. Syst. Neurobiol. Neurophysiol. Physiol. Psychol.)* **2**, 1672.

— — (1974*b*). Effects of androgens on the attack ability of gonadectomized mice by TO trained fighter individuals: confirmatory experiments. *IRCS (Res. Endocr. System., Neurobiol. Neurophysiol. Physiol. Psychol.)* **2**, 1730.

Breen, M. F. and Leshner, A. I. (1977). Maternal pheromone: a demonstration of its existence in the mouse (*Mus musculus*). *Physiol. Behav.* **18**, 527–9.

Bronson, F. H. (1966). A sex attractant function for mouse preputial glands. *Am. Zool.* **6**, 535. (Abstr.)

— (1973). Behavioral and physiological effectiveness of reproductive pheromones in mice. *Int. J. Chronobiol.* **1**, 320–1. (Abstr.)

— (1976). Urine marking in mice: causes and effects. In *Mammalian olfaction, reproductive processes, and behavior* (ed. R. L. Doty) pp. 119–43. Academic Press, New York.

— and Caroom, D. (1971). Preputial gland of the male mouse: attractant function. *J. Reprod. Fert.* **25**, 279–82.

— and Marsden, H. M. (1973). The preputial gland as an indicator of social dominance in male mice. *Behav. Biol.* **9**, 625–8.

Bronstein, P. M. and Crockett, D. P. (1976). Exposure to the odor of food determines the eating preferences of rat pups. *Behav. Biol.* **18**, 387–92.

Brown, J. C. and Williams, J. D. (1972). The rodent preputial gland. *Mamm. Rev.* **2**, 105–47.

Brown, L. E. (1966). Home range and movements of small animals. *Symp. zool. Soc. Lond.* **18**, 111–42.

Brown, R. E. (1973). The fire hydrant effect: stimuli eliciting urine marking in the rat (*Rattus norvegicus*). *Bull. ecol. Soc. Am.* **54**, 44. (Abstr.)

— (1975). Object-directed urine-marking by male rats (*Rattus norvegicus*). *Behav. Biol.* **15**, 251–4.

— (1977a). Odour preference and urine-marking scales in male and female rats: effects of gonadectomy and sexual experience on responses to conspecific odors. *J. comp. physiol. Psychol.* **91**, 1190–206.

— (1977b). Odour preference scales in rats. In *Proc. 6th Int. Symp. Olfaction and Taste* (ed. J. LeMagnen and P. Macleod) pp. 188. (Abstr.) Information Retrieval, London.

— (1978). Hormonal control of odor preferences and urine-marking in male and female rats. *Physiol. Behav.* **20**, 21–4.

— (1979). Mammalian social odors: a critical review. In *Advances in the study of behavior* (ed. J. S. Rosenblatt, R. A. Hinde, C. Beer, and M.-C. Busnel) Vol. 10, pp. 103–62. Academic Press, New York.

— (1981). Paternal behaviour in rats. Paper presented at the 42nd annual meeting of the Canadian Psychological Association, Toronto, Ontario, June.

— (1982a). Effects of social isolation on odour preferences and urine-marking in male and female rats. Unpublished manuscript.

— (1982b). Effects of prior social encounters on odor preference and urine-marking scales of male rats. Unpublished manuscript.

— (1982c). Preferences of pre- and post-weaning Long-Evans rats for nest odours. *Physiol. Behav.* **29**, 865–74.

— (1982d). Diet influences preference for maternal odours in mice (*Mus musculus*). Unpublished manuscript.

— and Elrick, D. (1983). Preferences of pre-weaning Long-Evans rats for anal excreta of adult males and females. *Physiol. Behav.* **30**, 567–71.

— and Hartley, C. (1982). Investigation of rat faeces as the source of a fear odour. Unpublished manuscript.

Brunjes, P. C. and Alberts, J. R. (1979). Olfactory stimulation induces filial preferences for huddling in rat pups. *J. comp. physiol. Psychol.* **93**, 548–55.

Burns, R. A., Thomas, R. L., and Davis, S. F. (1981). Three-trial sequences of reward magnitudes with odor cues maximized. *Bull. psychonom. Soc.* **17**, 266–8.

Butler, R. B. and Butler, L. A. (1979). Toward a functional interpretation of scent marking in the beaver (*Castor canadensis*). *Behav. neural Biol.* **26**, 442–54.

Calhoun, J. B. (1962). *The ecology and sociology of the Norway rat.* US Department of Health, Education and Welfare, Public Health Services Publication No. 1008, Washington.

Carmichael, M. S. (1980). Sexual discrimination by golden hamsters (*Mesocricetus auratus*). *Behav. neural Biol.* **29**, 73–90.

Caroom, D. and Bronson, F. H. (1971). Responsiveness of female mice to preputial attractant: effects of sexual experience and ovarian hormones. *Physiol. Behav.* **7**, 659–62.

Carr, W. J. (1974). Pheromonal sex attractants in the Norway rat. In *Advances in the study of communication and affect*, Vol. I *Nonverbal communication* (ed. L. Krames, P. Pliner, and T. Alloway). Plenum Press, New York.

— Demesquita-Wander, M., Sachs, S. R., and Maconi, P. (1979). Responses of female rats to odors from familiar vs. novel males. *Bull. psychonom. Soc.* **14**, 118–20.

— Hirsch, J. T., and Balazs, J. M. (1980). Responses of male rats to odors from familiar vs. novel females. *Behav. neural Biol.* **29**, 331–7.

— Krames, L., and Costanzo, D. J. (1970). Previous sexual experience and olfactory preference for novel versus original sex partners in rats. *J. comp. physiol. Psychol.* **71**, 216–22.

— Landauer, M. R., and Sonsino, R. (1981). Responses by rats to odors from living versus dead conspecifics. *Behav. neural Biol.* **31**, 67–72.

— Loeb, L. S., and Dissinger, M. L. (1965). Responses of rats to sex odors. *J. comp. physiol. Psychol.* **59**, 370–7.

— — and Wylie, N. R. (1966). Responses to feminine odors in normal and castrated male rats. *J. comp. physiol. Psychol.* **62**, 336–8.

— Marasco, E., and Landauer, M. R. (1979). Responses by rat pups to their own nest versus a strange conspecific nest. *Physiol. Behav.* **23**, 1149–51.

— Martorano, R. D., and Krames, L. (1970). Responses of mice to odors associated with stress. *J. comp. physiol. Psychol.* **71**, 223–8.

— Roth, P., and Amore, M. (1971). Responses of male mice to odors from stressed vs. nonstressed males and females. *Psychonom. Sci.* **25**, 275–6.

— Wylie, N. R., and Loeb, L. S. (1970). Responses of adult and immature rats to sex odors. *J. comp. physiol. Psychol.* **72**, 51–9.

— Yee, L., Gable, D., and Marasco, E. (1976). Olfactory recognition of conspecifics by domestic Norway rats. *J. comp. physiol. Psychol.* **90**, 821–8.

— Zunino, P. A., and Landauer, M. R. (1980). Responses by young house mice (*Mus musculus*) to odors from stressed vs. nonstressed adult conspecifics. *Bull. psychonom. Soc.* **15**, 419–21.

Carter, C. S. (1973). Olfaction and sexual receptivity in the female golden hamster. *Physiol. Behav.* **10**, 47–51.

Cattarelli, M., Vernet-Maury, E., and Chanel, J. (1977). Control of the rat olfactory bulb activity induced by biologically significant odors. *Physiol. Behav.* **19**, 381–7.

— — — Macleod, P., and Brandon, A. M. (1975). Olfactory bulb and integration of some odorous signals in the rat: behavioral and electro-physiological study. In *Olfaction and taste V* (ed. D. A. Denton and J. P. Coghlan). pp. 235–7. Academic Press, New York.

Cheal, M. and Domesick, V. (1979). Mating in male Mongolian gerbils after olfactory bulbectomy. *Physiol. Behav.* **22**, 199–202.

Cheatle, M. D. and Rudy, J. W. (1978). Analysis of second-order odor aversion conditioning in neo-natal rats: implications for Kamin's blocking effect. *J. exp. Psychol. Anim. Behav. Processes* **4**, 237–49.

— — (1979). Ontogeny of second-order odor aversion conditioning in neo-natal rats. *J. exp. Psychol. Anim. Behav. Processes* **5**, 142–51.

Christiansen, E. (1980). Urinary marking in wild bank voles, *Clethrionomys glateolus* in relation to season and sexual status. *Behav. neural Biol.* **28**, 123–7.

— Wiger, R., and Eilertsen, E. (1978). Morphological variations in the preputial gland of wild bank voles, *Clethrionomys glareolus. Holarct. Ecol.* **1**, 321–5.

Christensen, F. and Dam, H. (1952). A sexual dimorphism of the Harderian gland in hamsters. *Acta physiol. scand.* **27**, 333–6.

Clarke, J. R. and Frearson, S. (1972). Sebaceous glands on the hindquarters of the vole, *Microtus agrestis. J. Reprod. Fert.* **31**, 477–81.

Clarke, J. W. (1975). Androgen control of the ventral scent gland in *Neotoma floridana. J. Endocr.* **64**, 393–4.

Collerain, I. (1978). Frustration odor of rats receiving small numbers of prior rewarded running trials. *J. exp. Psychol. Anim. Behav. Processes* **4**, 120–30.

— and Ludvigson, H. W. (1972). Aversion of conspecific odor of frustrative nonreward in rats. *Psychonom. Sci.* **27**, 54-6.

— — (1977). Hurdle-jump responding in the rat as a function of conspecific odor of reward and nonreward. *Anim. Learn. Behav.* **5**, 177-83.

Compton, R. P., Koch, M. D., and Arnold, W. J. (1977). Effect of maternal odor on the cardiac rate of maternally separated infant rats. *Physiol. Behav.* **18**, 769-3.

Conely, L. and Bell, R. W. (1978). Neonatal ultrasounds elicited by odor cues. *Dev. Psychobiol.* **11**, 193-8.

Connor, J. (1972). Olfactory control of aggressive and sexual behavior in the mouse (*Mus musculus L.*). *Psychonom. Sci.* **27**, 1-3.

— and Lynds, P. G. (1977). Mouse aggression and the intruder-familiarity effect: evidence for multiple factor determination. *J. comp. physiol. Psychol.* **91**, 270-80.

Cornwell, C. A. (1975). Golden hamster pups adapt to complex rearing odors. *Behav. Biol.* **14**, 175-88.

— (1976). Selective olfactory exposure alters social and plant odor preferences of immature hamsters. *Behav. Biol.* **17**, 131-7.

Cornwell-Jones, C. A. and Azar, L. M. (1982). Olfactory development in gerbil pups. *Dev. Psychobiol.* **15**, 131-7.

— and Holder, C. L. (1979). Early olfactory learning is influenced by sex in hamsters but not rats. *Physiol. Behav.* **23**, 1035-40.

— and Sobrian, S. K. (1977). Development of odor-guided behavior in Wistar and Sprague-Dawley rat pups. *Physiol. Behav.* **19**, 685-8.

Courtney, R. J., Reid, L. D., and Wasden, R. E. (1968). Suppression of running times by olfactory stimuli. *Psychonom. Sci.* **12**, 315-16.

Cowley, J. J. and Cooper, A. J. (1977). The effects of olfactory bulb lesions on the maternal behavior of the mouse. *Bull. psychonom. Soc.* **9**, 55-7.

Dagg, A. I., Bell, W. L. and Windsor, D. E. (1971). Urine marking of cages and visual isolation as possible sources of error in behavioral studies of small mammals. *Lab. Anim.* **5**, 163-7.

— and Windsor, D. E. (1971). Olfactory discrimination limits in gerbils. *Can. J. Zool.* **49**, 283-5.

Daly, M. (1977). Some experimental tests of the functional significance of scent marking by gerbils *Meriones unguiculatus*. *J. comp. physiol. Psychol.* **91**, 1082-94.

— and Daly, S. (1975a). Socio-ecology of Saharan gerbils, especially *Meriones libycus*. *Mammalia* **39**, 289-311.

— — (1975b). Behavior of *Psammomys obesus* (Rodentia: Gerbillinae) in the Algerian Sahara. *Z. Tierpsychol.* **37**, 298-321.

— Wilson, M. I., and Behrends, P. (1980). Factors affecting rodents' responses to odours of strangers encountered in the field: experiments with odour-baited traps. *Behav. Ecol. Sociobiol.* **6**, 323-9.

— — and Faux, S. F. (1978). Seasonally variable effects of conspecific odors upon capture of deermice, *Peromyscus maniculatus gambelii*. *Behav. Biol.* **23**, 254-9.

Darby, E. M., Devor, M., and Chorover, S. L. (1975). A presumptive sex pheromone in the hamster: some behavioral effects. *J. comp. physiol. Psychol.* **88**, 496-502.

Davies, V. J. and Bellamy, D. (1972). The olfactory response of mice to urine and effects of gonadectomy. *J. Endocr.* **55**, 11-20.

— — (1974). Effects of female urine on social investigation in male mice. *Anim. Behav.* **22**, 239-41.

Davis, R. G. and Tapp, J. T. (1972). Odor artifacts of an olfactometer evoke a conditioned emotional response in the rat. *Percept. motor Skills* **35**, 931–6.

Davis, R. M. (1972). Behavior of the Vlei rat, *Otomys irroratus* (Brants, 1827). *Zool. Afr.* **7**, 119–40.

Davis, S. F., Whiteside, D. A., Bramlett, J. A., and Petersen, H. (1981). Odor production and utilization under conditions of nonreward and small reward. *Learn. Motiv.* **12**, 364–82.

Dearden, L. C. (1959). Meibomian glands in *Lagurus*. *J. Mammal.* **40**, 20–5.

DeGroot, C. A., Lely, M. A. V. D., and Kooij, R. (1965). The effect of progesterone on the sebaceous glands of the rat. *Brit. J. Derm.* **77**, 617–21.

De Jonge, G. (1980). Response to con- and hetero-specific male odours by the voles *Microtus agrestis, M. arvalis* and *Clethrionomys glareolus* with respect to competition for space. *Behaviour* **73**, 277–302.

Denenberg, V. H., Hudgens, G. A., and Zarrow, M. X. (1964). Mice reared with rats: modification of behavior by early experience with another species. *Science, NY* **143**, 380–1.

Desjardins, C., Maruniak, J. A., and Bronson, F. H. (1973). Social rank in house mice: differentiation revealed by ultraviolet visualization of urinary marking patterns. *Science, NY* **182**, 939–41.

Devor, M. and Murphy, M. R. (1973). The effect of peripheral olfactory blockade on the social behavior of the male golden hamster. *Behav. Biol.* **9**, 31–42.

— and Schneider, G. E. (1974). Attraction to home-cage odor in hamster pups: specificity and changes with age. *Behav. Biol.* **10**, 211–21.

Dieterlen, F. (1959). Das Verhalten des syrischen Goldhamsters (*Mesocricetus auratus* Waterhouse): Untersuchungen zur Frage seiner Entwicklung und seiner angeborenen Anteile durch geruchsisolierte Aufzuchten. *Z. Tierpsychol.* **16**, 47–103.

Dixon, A. K. and Mackintosh, J. H. (1971). Effects of female urine upon the social behaviour of adult male mice. *Anim. Behav.* **19**, 138–40.

— — (1975). The relationship between the physiological condition of female mice and the effects of their urine on the social behaviour of adult males. *Anim. Behav.* **23**, 513–20.

— — (1976). Olfactory mechanisms affording protection from attack to juvenile mice (*Mus musculus* L.). *Z. Tierpsychol.* **41**, 225–34.

Dizzino, G. and Whitney, G. (1977). Androgen influence on male mouse ultrasounds during courtship. *Horm. Behav.* **8**, 188–92.

— — and Nyby, J. (1978). Ultrasonic vocalizations by male mice *Mus musculus* to female sex pheromone: experiential determinants. *Behav. Biol.* **22**, 104–13.

Doty, R. L. (1972). Odor preferences of female *Peromyscus maniculatus bairdi* for male mouse odors of *P.m. bairdi* and *P. leucopus noveboracensis* as a function of estrous state. *J. comp. physiol. Psychol.* **81**, 191–7.

— (1973). Reactions of deermice (*Peromyscus maniculatus*) and white-footed mice. (*Peromyscus leucopus*) to homospecific and heterospecific urine odors. *J. comp. physiol. Psychol.* **84**, 296–303.

— (1974). A cry for the liberation of the female rodent: courtship and copulation in *Rodentia. Psych. Bull.* **81**, 159–72.

— and Anisko, J. J. (1973). Procaine hydrochloride olfactory block eliminates mounting in the male golden hamster. *Physiol. Behav.* **10**, 385–97.

— and Kart, R. (1972). A comparative and developmental analysis of the midventral sebaceous glands of 18 taxa of *Peromyscus*, with an examination of gonadal-steroid influences in *Peromyscus maniculatus bairdii. J. Mammal.* **53**, 83–99.

Drickamer, L. C. and Vandenbergh, J. G. (1973). Predictors of social dominance in the adult female golden hamster (*Mesocricetus auratus*). *Anim. Behav.* **21**, 564–70.

— — and Colby, D. R. (1973). Predictors of dominance in the male golden hamster (*Mesocricetus auratus*). *Anim. Behav.* **21**, 557–63.

Dudley, D. (1974). Paternal behavior in the California mouse, *Peromyscus californicus. Behav. Biol.* **11**, 247–52.

D'Udine, B. and Partridge, L. (1981). Olfactory preferences of inbred mice (*Mus musculus*) for their own strain and for siblings: effects of strain, sex and cross-fostering. *Behaviour* **78**, 314–23.

Ebling, F. J. (1948). Sebaceous glands. I. The effect of sex hormones on the sebaceous glands of the female albino rat. *J. Endocr.* **5**, 297–302.

— (1951). Sebaceous glands. II. Changes in the sebaceous glands following the implantation of oestradiol benzoate in the female albino rat. *J. Endocr.* **7**, 288–98.

— (1954). Changes in the sebaceous glands and epidermis during the oestrus cycle of the albino rat. *J. Endocr.* **10**, 147–54.

— (1957a). The action of testosterone on the sebaceous glands and epidermis in castrated and hypophysectomized male rats. *J. Endocr.* **15**, 297–306.

— (1957b). The action of testosterone and estradiol on the sebaceous glands and epidermis of the rat. *J. Embryol. exp. Morph.* **5**, 74–82.

— (1961). Failure of progesterone to enlarge sebaceous glands in the female rat. *Br. J. Derm.* **73**, 65–8.

— Ebling, E., Randall, V., and Skinner, J. (1975). The effects of hypophysectomy and of bovine growth hormone on the responses to testosterone of prostate, preputial, Harderian and lachrymal glands and of brown adipose tissue in the rat. *J. Endocr.* **66**, 401–6.

— — and Skinner, J. (1969a). The influence of pituitary hormones on the response of the sebaceous glands of the male rat to testosterone. *J. Endocr.* **45**, 245–56.

— — — (1969b). The influence of the pituitary on the response of the sebaceous and preputial glands of the rat to progesterone. *J. Endocr.* **45**, 257–63.

Edwards, D. A. and Burge, K. G. (1973). Olfactory control of the sexual behavior of male and female mice. *Physiol. Behav.* **11**, 867–72.

Egoscue, H. J. (1962). The bushy-tailed wood rat: a laboratory colony. *J. Mammal.* **43**, 328–37.

— (1970). A laboratory colony of the Polynesian rat, *Rattus exulans. J. Mammal.* **51**, 261–6.

Eibl-Eibesfeldt, I. (1953). Zur ethologie des Hamsters (*Cricetus cricetus L.*). *Z. Tierpsychol.* **10**, 204–54.

Eisenberg, J. F. (1962). Studies on the behavior of *Peromyscus maniculatus gambelii* and *Peromyscus californicus parasiticus. Behaviour* **19**, 177–207.

— (1963). The behavior of heteromyid rodents. *Univ. Calif. Publ. Zool.* **69**, 1–100.

— (1964). A comparative study of sandbathing behavior in heteromyid rodents. *Behaviour* **22**, 16–23.

— (1967). A comparative study of rodent ethology with emphasis on evolution of social behavior. *Proc. US nat. Mus.* **122**, 1–51.

— (1981). *The mammalian radiations.* University of Chicago Press, Chicago.

— and Kleiman, D. G. (1972). Olfactory communication in mammals. *A. Rev. Ecol. System.* **3**, 1–32.

Elwood, R. W. (1975). Paternal and maternal behaviour in the Mongolian gerbil. *Anim. Behav.* **23**, 766–72.

— (1980). The development, inhibition and disinhibition of pup-cannibalism in the Mongolian gerbil. *Anim. Behav.* **28**, 1188–94.

Emmerick, J. J. and Snowdon, C. T. (1976). Failure to show modification of male golden hamster mating behavior through taste/odor aversion learning. *J. comp. physiol. Psychol.* **90**, 857–69.

Escherich, P. C. (1981). Social biology of the bushy-tailed woodrat, *Neotoma cinerea. Univ. Calif. Publ. Zool.* **110**, 1–126.

Eslinger, P. J. and Ludvigson, H. W. (1980). Are there constraints on learned responses to odors from rewarded and nonrewarded rats? *Anim. Learn. Behav.* **8**, 452–6.

Evans, C. M. and Brain, P. F. (1974). Some influences of sex steroids on the aggressiveness directed towards golden hamsters (*Mesocricetus auratus* Waterhouse) of both sexes by 'trained fighter' individuals. *J. Endocr.* **61**, xlvi–xlvii.

— Mackintosh, J. H., Kennedy, J. F., and Robertson, S. M. (1978). Attempts to characterise and isolate aggression reducing olfactory signals from the urine of female mice *Mus musculus L. Physiol. Behav.* **20**, 129–34.

Ewer, R. F. (1967). The behavior of the African giant rat (*Cricetomys gambianus* Waterhouse). *Z. Tierpsychol.* **24**, 6–79.

— (1968). *The ethology of mammals.* Plenum Press, New York.

— (1971). The biology and behavior of a free living population of black rats (*Rattus rattus*). *Anim. Behav. Monogr.* **4**, 127–74.

Farr, L. A., Andrews, R. V., and Kline, M. R. (1978). Comparison of methods for estimating social rank of deer mice. *Behav. Biol,* **23**, 399–404.

Fass, B., Gutermann, P. E., and Stevens, D. A. (1978). Evidence that rats discriminate between familiar and unfamiliar putative urinary odorants of adult male conspecifics. *Aggress. Behav.* **4**, 231–6.

— and Stevens, D. A. (1977). Pheromonal influences on rodent agonistic behavior. In *Chemical signals in vertebrates* (ed. D. Müller-Schwarze and M. M. Mozell) pp. 185–206. Plenum Press, New York.

Feldman, M. and Mitchell, O. G. (1968). The postnatal development of the pelage and ventral gland of the male gerbil. *J. Morph.* **125**, 303–13.

Fiedler, U. (1973). Beobachtungen zur Biologie einiger Gerbillinen, insbesondere *Gerbillus (Dipodillus) dasyurus,* (Myomorpha, Rodentia) in Gefangenschaft. *Z. Säugetierek.* **38**, 321–40.

Fleming, A. S., Chee, P., and Vaccarino, F. (1981). Sexual behaviour and its olfactory control in the desert woodrat (*Neotoma lepida lepida*). *Anim. Behav.* **29**, 727–45.

— and Rosenblatt, J. S. (1974a). Olfactory regulation of maternal behavior in rats. I. Effects of olfactory bulb removal in experienced and inexperienced lactating and cycling females. *J. comp. physiol. Psychol.* **86**, 221–32.

— — (1974b). Olfactory regulation of maternal behavior in rats. II. Effects of peripherally induced anosmia and lesions of the lateral olfactory tract in pup-induced virgins. *J. comp. physiol. Psychol.* **86**, 233–46.

— and Tambosso, L. (1980). Hormonal and sensory control of scent-marking in the desert woodrat (*Neotoma lepida lepida*). *J. comp. physiol. Psychol.* **94**, 564–78.

— Vaccarino, F., Tambosso, L., and Chee, P. (1979). Vomeronasal and olfactory system modulation of maternal behavior in the rat. *Science, NY* **203**, 372-4.

Flowerdew, J. R. (1971). The subcaudal glandular area of *Apodemus sylvaticus. J. Zool.,Lond.* **165**, 525-7.

Frank, D. H. and Johnston, R. E. (1981). Determinants of scent marking and ultrasonic calling by female Turkish hamsters, *Mesocricetus brandti. Behav. neural Biol.* **33**, 514-18.

Frank, F. (1956). Das Duftmarkieren der grossen Wühlmaus, *Arvicola terrestris* (L). *Z. Säugetierek.* **21**, 172-5.

Galef, B. G. Jr (1981*a*). Development of olfactory control of feeding-site selection in rat pups. *J. comp. physiol. Psychl.* **95**, 615-22.

— (1981*b*). Preference for natural odors in rat pups: implications of a failure to replicate. *Physiol. Behav.* **26**, 783-6.

— and Heiber, L. (1976). Role of residual olfactory cues in the determination of feeding site selection and exploration patterns of domestic rats. *J. comp. physiol. Psychol.* **90**, 727-39.

— and Kaner, H. C. (1980). Establishment and maintenance of preference for natural and artificial olfactory stimuli in juvenile rats. *J. comp. physiol. Psychol.* **94**, 588-95.

— and Muskus, P. A. (1979). Olfactory mediation of mother–young contact in Long-Evans rats. *J. comp. physiol. Psychol.* **93**, 708-16.

Gallup, G. G. and Waite, M. S. (1970). Some preliminary observations on the behavior of Mongolian gerbils (*Meriones unguiculatus*) under seminatural conditions. *Psychonom. Sci.* **20**, 25-6.

Gandelman, R., Zarrow, M. X., and Denenberg, V. H. (1972). Reproductive and maternal performance in the mouse following removal of the olfactory bulbs. *J. Reprod. Fert.* **28**, 453-6.

— — — and Myers, M. (1971). Olfactory bulb removal eliminates maternal behavior in the mouse. *Science, NY* **171**, 210-11.

Gawienowski, A. M. (1977). Chemical attractants of the rat preputial gland. In *Chemical signals in vertebrates* (ed. D. Müller-Schwarze and M. M. Mozell) pp. 45-60. Plenum Press, New York.

— DeNicola, D. B., and Stacewicz-Sapuntzakis, M. (1976). Androgen dependence of a marking pheromone in rat urine. *Horm. Behav.* **7**, 401-5.

— Orsulak, P. J., Stacewicz-Sapuntzakis, M., and Joseph, B. (1975). Presence of sex pheromone in preputial gland of male rats. *J. Endocr.* **67**, 283-8.

— — — and Pratt, J. J. Jr (1976). Attractant effect of female preputial gland extracts on the male rat. *Psychoneuroendocrinology* **1**, 411-18.

Gerling, S. and Yahr, P. (1982). Maternal and paternal pheromones in gerbils. *Physiol. Behav.* **29**, 667-73.

Getz, W. M. (1981). Genetically based kin recognition systems. *J. theoret. Biol.* **92**, 209-26.

Geyer, L. A. (1979). Olfactory and thermal influences on ultrasonic vocalization during development in rodents. *Am. Zool.* **19**, 420-31.

— Beauchamp, G. K., Seygal, G., and Rogers, J. G. Jr (1981). Social behaviour of pine voles, *Microtus pinetorum*: effects of gender, familiarity, and isolation. *Behav. neural Biol.* **31**, 331-41.

— McIntosh, T. K., and Barfield, R. J. (1978). Effects of ultrasonic vocalizations and male's urine on female rat readiness to mate. *J. comp. physiol. Psychol.* **92**, 457-62.

Gilder, P. M. and Slater, P. J. B. (1978). Interest of mice in conspecific male odors is influenced by degree of kinship. *Nature, Lond.* **274**, 364-5.

Glenn, E. M. and Gray, J. (1965). Effect of various hormones on the growth and histology of the gerbil (*Meriones unguiculatus*) abdominal sebaceous gland pad. *Endocrinology* **76**, 1115-23.

Godfrey, J. (1958). The origin of sexual isolation between bank voles. *Proc. R. phys. Soc. Edinb.* **27**, 47-55.

Goldblatt, A. (1978*a*). Effects of early olfactory experience on a tilt table odor preference. *Behav. Biol.* **22**, 269-73.

— (1978*b*). Does early olfactory exposure result in an odor preference or the loss of avoidance? *Percept. motor Skills* **47**, 196-8.

Grau, H. J. (1982). Kin recognition in white-footed deermice (*Peromyscus leucopus*). *Anim. Behav.* **30**, 497-505.

Gray, L. E., Whitsett, J. M., and Ziesenis, J. S. (1978). Hormonal regulation of aggression toward juveniles in female house mice. *Horm. Behav.* **11**, 310-22.

Greene, E. C. (1935). *The anatomy of the rat.* Hafner, New York.

Gregg, B. and Thiessen, D. D. (1981). A simple method of olfactory discrimination of urines for the Mongolian gerbil, *Meriones unguiculatus. Physiol. Behav.* **26**, 1133-6.

Gregory, E. H. and Bishop, A. (1975). Development of olfactory guided behavior in the golden hamster. *Physiol. Behav.* **15**, 373-6.

— Engel, K., and Pfaff, D. (1975). Male hamster preference for odors of female hamster vaginal discharges: studies of experiential and hormonal determinants. *J. comp. physiol. Psychol.* **89**, 442-6.

— and Pfaff, D. W. (1971). Development of olfactory-guided behavior in infant rats. *Physiol. Behav.* **6**, 573-6.

Griffiths, C. and Kendall, M. D. (1980*a*). Structure of the plantar sweat glands of the bank vole (*Clethrionomys glareolus*). *J. Zool., Lond.* **191**, 1-10.

— — (1980*b*). The structure of glands in the angulus oris of the bank vole (*Clethrionomys glareolus*). *J. Zool., Lond.* **192**, 311-22.

Gubernick, D. J. (1981). Parent and infant attachment in mammals. In *Parental care in mammals* (ed. D. J. Gubernick and P. H. Klopfer) pp. 243-305. Plenum Press, New York.

Gustafsson, T., Andersson, B., and Meurling, P. (1980). Effect of social rank on the growth of the preputial glands in male bank voles, *Clethrionomys glareolus. Physiol. Behav.* **24**, 689-92.

Gwinner, E. (1981). Circannual systems. In *Handbook of behavioral neurobiology*, Vol. 4 *Biological rhythms* (ed. J. Aschoff) pp. 391-410. Plenum Press, New York.

Hahn, M. E. Jr and Simmel, E. C. (1968). Individual recognition by natural concentration of olfactory cues in mice. *Psychonom. Sci.* **12**, 183-94.

— and Tumolo, P.(1971). Individual recognition in mice: how is it mediated. *Am. Zool.* **11**, 634-5. (Abstr.)

Hall, C. E. (1949). Comparison of the preputial glands in the Alexandrine, the wild, and the domestic Norway rat. *Proc. Soc. exp. Biol.* **69**, 233-7.

Halpin, Z. T. (1974). Individual differences in the biological odors of the Mongolian gerbil (*Meriones unguiculatus*). *Behav. Biol.* **11**, 253-9.

— (1976). The role of individual recognition by odors in the social interactions of the Mongolian gerbil (*Meriones unguiculatus*). *Behaviour* **58**, 117-30.

— (1978). The effects of social experience on the odour preferences of the Mongolian gerbil (*Meriones unguiculatus*). *Biol. Behav.* **3**, 169-79.

— (1980). Individual odors and individual recognition: review and commentary. *Biol. Behav.* **5**, 233-48.

Hamilton, J. B. and Montagna, W. (1950). The sebaceous glands of the hamster.

I. Morphological effects of androgens on integumentary structures. *Am. J. Anat.* **86**, 191–233.

Harmatz, P., Boelkins, R. C. and Kessler, S. (1975). Post isolation aggression and olfactory cues. *Behav. Biol.* **13**, 219–24.

Harrington, J. E. (1976). Recognition of territorial boundaries by olfactory cues in mice (*Mus musculus* L.). *Z. Tierpsychol.* **41**, 295–306.

Hartung, T. G. and Dewsbury, D. A. (1979). Paternal behavior in six species of muroid rodents. *Behav. neural Biol.* **26**, 466–78.

Harvey, H. (1952). Sexual dimorphism of submaxillary glands in mice in relation to reproductive maturity and sex hormones. *Physiol. Zool.* **25**, 205–22.

Haug, M. (1970). Mise en évidence de deux odeurs aux effects opposés de facilitation et d'inhibition des conduites aggressives chez la souris male. *C. r. hebd. Séanc. Acad. Sci.* **271**, 1567–70.

— (1971). Rôle probable des vésicules séminales et des glandes coagulantes dans la production d'une phéromone inhibitrice du comportement agressif chez la souris. *C. r. hebd. Séanc. Acad. Sci.* **273**, 1509–10.

— (1972). Effet de l'urine d'une femelle étrangère sur le compotement agressif d'un groupe de souris femelles. *C. r. hebd. Séanc. Acad. Sci.* **275**, 995–8.

Hayashi, S. (1979). A role of female preputial glands in social behavior of mice. *Physiol. Behav.* **23**, 967–9.

— and Kimura, T. (1974). Sex-attractant emitted by female mice. *Physiol. Behav.* **13**, 563–7.

— — (1976). Sexual behavior of the naive male mouse as affected by the presence of a male and a female performing mating behavior. *Physiol. Behav.* **17**, 807–10.

— — (1978). Effects of exposure to males on sexual preference in female mice. *Anim. Behav.* **26**, 290–5.

Hennessy, D. F. (1980). Early olfactory determinants of adult responsiveness to social status odors in *Mus musculus*. *J. Mammal.* **61**, 520–4.

Hennessy, M. B., Li, J., and Levine, S. (1980). Infant responsiveness to maternal cues in mice of 2 inbred lines. *Dev. Psychobiol.* **13**, 77–84.

— Smotherman, W. P., and Levine, S. (1977). Early olfactory enrichment enhances later consumption of novel substances. *Physiol. Behav.* **19**, 481–4.

Hindle, E. (1948). Reginald Innes Pocock. *Obit. Not. Fell. R. Soc.* **6**, 189–211.

Hofer, M. A. (1976). Olfactory denervation: its biological and behavioral effects in infant rats. *J. comp. physiol. Psychol.* **90**, 829–38.

— and Shair, H. (1978). Ultrasonic vocalization during social interaction and isolation in 2 week-old rats. *Dev. Psychobiol.* **11**, 495–504.

— — and Singh, P. (1976). Evidence that maternal ventral skin substances promote suckling in infant rats. *Physiol. Behav.* **17**, 131–6.

Hoffman, R. A. (1971). Influence of some endocrine glands, hormones and blinding on the histology and porphyrins of the Harderian glands of golden hamsters. *Am. J. Anat.* **132**, 463–78.

Holinka, C. F. and Carlson, A. D. (1976). Pup attraction to lactating Sprague-Dawley rats. *Behav. Biol.* **16**, 489–505.

Holzman, M. E. and Paskhina, N. M. (1974). Elements of social behaviour in *Rhombomys opimus Licht. Mosk. Obshch. Ispytat. Pr. Biull. Otel Biol.* **79**, 29–38.

Howe, R. J. (1977). Scent-marking behavior in three species of woodrats (*Neotoma*) in captivity. *J. Mammal.* **58**, 685–8.

Howell, A. B. (1926). *Anatomy of the wood rat.* Williams and Wilkins, Baltimore.

Hrabě, V. (1973). Die Analdrüsen dreier Arten der Familie Microtidae (*Rodentia*) aus dem Tatra-Gebirge. *Zool. Listy* **22**, 145–54.
— (1974). Tarsal glands in *Pitymys subterraneus* (de Sél. -Long.) and *Pitymys tatricus* Krat. (Microtidae, Mammalia). *Zool. Listy* **23**, 97–105.
Huck, U. W. and Banks, E. M. (1979). Behavioral components of individual recognition in the collared lemming, (*Dicrostonyx groenlandicus*). *Behav. Ecol. Sociobiol.* **6**, 85–90.
— — (1980a). The effects of cross-fostering on the behaviour of two species of North American lemmings, *Dicrostonyx groenlandicus* and *Lemmus trimucronatus*. I. Olfactory preferences. *Anim. Behav.* **28**, 1046–52.
— — (1980b). The effects of cross-fostering on the behaviour of two species of North American lemmings, *Dicrostonyx groenlandicus* and *Lemmus trimucronatus*. II. Sexual behaviour. *Anim. Behav.* **28**, 1053–62.
— — and Wang, S. -C. (1981). Olfactory discrimination of social status in the brown lemming. *Behav. neural Biol.* **33**, 364-71.
Hucklebridge, F. H., Nowell, N. W., and Wouters, A. (1972). A relationship between social experience and preputial gland function in the albino mouse. *J. Endocr.* **55**, 449–50.
Husted, J. R. and McKenna, F. S. (1966). The use of rats as discriminative stimuli, *J. exp. Anal. Behav.* **9**, 677–9.
Jackson, L. H. (1938). The preputial glands of British Muridae. *J. Anat.* **72**, 458–61.
Jannett, F. J. Jr (1975). 'Hip glands' of *Microtus pennsylvanicus* and *M. longicaudus* (Rodentia: Muridae), voles 'without' hip glands. *Syst. Zool.* **24**, 171–5.
— (1978). Dosage response of the vesicular, preputial, anal and hip glands of the male vole, *Microtus montanus*, to testosterone propionate. *J. Mammal.* **59**, 772–9.
— (1981a). Scent mediation of intraspecific, interspecific, and intergeneric agonistic behavior among sympatric species of voles (*Microtinae*). *Behav. Ecol. Sociobiol.* **8**, 293–6.
— (1981b). Sex ratios in high-density populations of the montane vole, *Microtus montanus*, and the behavior of territorial males. *Behav. Ecol. Sociobiol.* **8**, 297–307.
— and Jannett, L. Z. (1974). Drum marking by *Arvicola richardsoni* and its taxonomic significance. *Am. midl. Nat.* **92**, 230–4.
Johnson, R. P. (1973). Scent marking in mammals. *Anim. Behav.* **21**, 521–35.
— (1975). Scent marking with urine in two races of the bank vole (*Clethrionomys glareolus*). *Behaviour* **55**, 81–93.
Johnston, R. E. (1974). Sexual attraction function of golden hamster vaginal secretion. *Behav. Biol.* **12**, 111–17.
— (1975a). Sexual excitation function of hamster vaginal secretion. *Anim. Learn. Behav.* **3**, 161–6.
— (1975b). Scent marking by male golden hamsters (*Mesocricetus auratus*). I. Effects of odors and social encounters. *Z. Tierpsychol.* **37**, 75–98.
— (1975c). Scent marking by male golden hamsters (*Mesocricetus auratus*). II. The role of the flank gland scent in the causation of marking. *Z. Tierpsychol.* **37**, 138–44.
— (1975d). Scent marking by male golden hamsters (*Mesocricetus auratus*). III. Behavior in a seminatural environment. *Z. Tierpsychol.* **37**, 213–21.
— (1977a). The causation of two scent marking behavior patterns in female hamsters (*Mesocricetus auratus*). *Anim. Behav.* **25**, 317–27.

— (1977*b*). Sex pheromones in golden hamsters. In *Chemical signals in vertebrates* (ed. D. Müller-Schwarze and M. M. Mozell) pp. 225-49. Plenum Press, New York.

— (1979). Olfactory preferences, scent marking, and 'proceptivity' in female hamsters. *Horm. Behav.* **13**, 12-39.

— (1980). Responses of male hamsters to odors of females in different reproductive states. *J. comp. physiol. Psychol.* **94**, 894-904.

— (1981*a*). Testosterone dependence of scent marking by male hamsters (*Mesocricetus auratus*). *Behav. neural Biol.* **31**, 96-9.

— (1981*b*). Attraction to odors in hamsters: an evaluation of methods. *J. comp. physiol. Psychol.* **95**, 951-60.

— and Coplin, B. (1979). Development of responses to vaginal secretion and other substances in golden hamsters. *Behav. neural Biol.* **25**, 473-89.

— and Lee, N. A. (1976). Persistence of the odor deposited by two functionally distinct scent marking behaviors of golden hamsters. *Behav. Biol.* **16**, 199-210.

— and Schmidt, T. (1979). Responses of hamsters to scent marks of different ages. *Behav. neural Biol.* **26**, 64-75.

— and Zahorik, D. M. (1975). Taste aversions to sexual attractants. *Science, NY* **189**, 893-4.

— — Immler, K., and Zakon, H. (1978). Alterations of male sexual behavior by learned aversions to hamster vaginal secretion. *J. comp. physiol. Psychol.* **92**, 85-93.

Jones, R. B. and Nowell, N. W. (1973*a*). The effect of urine on the investigatory behaviour of male albino mice. *Physiol. Behav.* **11**, 35-8.

— — (1973*b*). Aversive and aggression-promoting properties of urine from dominant and subordinate male mice. *Anim. Learn. Behav.* **1**, 207-210.

— — (1973*c*). The coagulating glands as a source of aversive and aggression inhibiting pheromone(s) in the male albino mouse. *Physiol. Behav.* **11**, 455-62.

— — (1973*d*). The effect of familiar visual and olfactory cues on the aggressive behavior of mice. *Physiol. Behav.* **10**, 221-3.

— — (1974*a*). A comparison of the aversive and female attractant properties of urine from dominant and subordinate male mice. *Anim. Learn. Behav.* **2**, 141-4.

— — (1974*b*). Effects of androgen on the aversive properties of male mouse urine. *J. Endocr.* **60**, 19-25.

— — (1974*c*). Latency to approach food in male mice: effects of clean and soiled sawdust substrates. *Behav. Biol.* **12**, 409-12.

— — (1974*d*). Effects of cyproterone acetete upon urinary aversive cues and accessory sex glands in male albino mice. *J. Endocr.* **62**, 167-8.

— — (1974*e*). The urinary aversive pheromone of mice: species, strain and grouping effects. *Anim. Behav.* **22**, 187-91.

— — (1975). Effects of clean and soiled sawdust substrates and of different urine types upon aggressive behavior in male mice. *Aggress. Behav.* **1**, 111-21.

— — (1977*a*). The emotional responses of male laboratory mice to various urinary odors. *Behav. Biol.* **19**, 98-107.

— — (1977*b*). Aversive potency of male mouse urine: a temporal study. *Behav. Biol.* **19**, 523-6.

Junqueira, L. C., Fajer, A., Rabinovitch, M., and Frankenthal, L. (1949). Biochemical and histochemical observations on the sexual dimorphism of mice submaxillary glands. *J. cell. comp. Physiol.* **34**, 129-58.

Kairys, D. J., Magalhaes, H., and Floody, O. R. (1980). Olfactory bulbectomy depresses ultrasound production and scent marking by female hamsters. *Physiol. Behav.* **25**, 143–6.

Kareem, A. M., and Barnard, C. J. (1982). The importance of kinship and familiarity in social interactions between mice. *Anim. Behav.* **30**, 594–601.

Kessler, S., Harmatz, P., and Gerling, S. A. (1975). The genetics of pheromonally mediated aggression in mice. Part 1. Strain differences in the capacity of male urinary odors to elicit aggression. *Behav. Genet.* **5**, 233–8.

Kilpatrick, S. J., Bolt, M., and Moltz, H. (1980). The maternal pheromone and bile acids in the lactating rat. *Pharmacol. Biochem. Behav.* **12**, 555–8.

Kimelman, B. R. and Lubow, R. E. (1974). The inhibitory effect of pre-exposed olfactory cues on intermale aggression in mice. *Physiol. Behav.* **12**, 919–22.

King, M. G. (1969). Stimulus generalization of conditioned fear in rats over time: olfactory cues and adrenal activity. *J. comp. physiol. Psychol.* **69**, 590–600.

— Pfister, H. P., and DiGuisto, E. L. (1975). Differential preference for and activation by the odoriferous compartment of a shuttlebox in fear-conditioned and naive rats. *Behav. Biol.* **13**, 175–81.

Kingdon, J. (1974). *East African mammals.* Vol II B. Academic Press, London.

Kirchhof-Glazier, D. A. (1979). Absence of sexual imprinting in house mice cross-fostered to deermice. *Physiol. Behav.* **23**, 1073–80.

Klopfer, P. H. and Gamble, J. (1966). Maternal 'imprinting' in goats: the role of chemical senses. *Z. Tierpsychol.* **23**, 588–92.

Koski, M. A., Dixon, L. K., and Fahrion, N. (1977). Olfactory mediated choice behavior in mice: developmental and genetic aspects. *Behav. Biol.* **19**, 324–32.

Krames, L. (1970). Responses of female rats to the individual body odors of male rats. *Psychonom. Sci.* **20**, 274–5.

— Carr, W. J., and Bergman, B. (1969). A pheromone associated with social dominance among male rats. *Psychonom. Sci.* **16**, 11–12.

— and Mastromatteo, L. A. (1973). Role of olfactory stimuli during copulation in male and female rats. *J. comp. physiol. Psychol.* **85**, 528–35.

— and Shaw, B. (1973). Role of previous experience in the male rat's reaction to odors from group and alien conspecifics. *J. comp. physiol. Psychol.* **82**, 444–8.

Kratochvil, J. (1962). Sexualdrüsen bei den Säugetieren mit Rücksicht auf Taxonomie. Symposium Theriologicum (*Proc. Int. Symp. Methods of Mammalogical Investigation*), Praha, pp. 175–87.

Kristal, M. B., Whitney, J. F., and Peters, L. C. (1981). Placenta on pups' skin accelerates onset of maternal behaviour in non-pregnant rats. *Anim. Behav.* **29**, 81–5.

Kumari, S. and Prakash, I. (1981). Scent marking behaviour of *Meriones hurrianae* during oestrus. *Anim. Behav.* **29**, 1269–71.

Kwan, M. and Johnston, R. E. (1980). The role of vaginal secretion in hamster sexual behavior: males' responses to normal and vaginectomized females and their odors. *J. comp. physiol. Psychol.* **94**, 905–13.

Labov, J. B. (1980). Factors influencing infanticidal behavior in wild male house mice (*Mus musculus*). *Behav. Ecol. Sociobiol.* **6**, 297–303.

Laine, H. and Griswold, J. G. (1976). Sandbathing in kangaroo rats, *Dipodomys spectabilis. J. Mammal.* **57**, 408–10.

Landauer, M. R. (1975). Sexual and olfactory preferences of male hamsters (*Mesocricetus auratus* Waterhouse) for conspecifics in different hormonal conditions. *Diss. Abstr.* **36**, 2106B–7B.

— and Banks, E. M. (1973). Olfactory preferences of male and female golden hamsters, *Mesocricetus auratus. Bull. ecol. Soc. Am.* **54**, 44. (Abstr.)

— — and Carter, C. S. (1977). Sexual preferences of male hamsters (*Mesocricetus auratus*) for conspecifics in different endocrine conditions. *Horm. Behav.* **9**, 193–202.

— — — (1978). Sexual and olfactory preferences of naive and experienced male hamsters. *Anim. Behav.* **26**, 611–21.

— Liu, S., and Goldberg, N. (1980). Responses of male hamsters to the ear gland secretions of conspecifics. *Physiol. Behav.* **24**, 1023–6.

— Wiese, R. E., and Carr, W. J. (1977). Responses of sexually experienced and naive male rats to cues from receptive vs. nonreceptive females. *Anim. Learn. Behav.* **5**, 398–402.

Lane-Petter, W. (1967). Odour in mice. *Nature, Lond.* **216**, 794.

Larsson, K. (1971). Impaired mating performances in male rats after anosmia induced peripherally or centrally. *Brain Behav. Evol.* **4**, 463–71.

— (1975). Sexual impairment of inexperienced male rats following pre- and postpubertal olfactory bulbectomy. *Physiol. Behav.* **14**, 195–9.

LaVelle, F. W. (1951). A study of hormonal factors in the early sex development of the golden hamster. *Contr. Embryol.* **34**, 19–53.

Lederer, E. (1949). Chemistry and biochemistry of some mammalian secretions and excretions. *J. chem. Soc.* 2115–25.

— (1950). Odeurs et parfums des animaux. *Fortschr. Chem. org. NatStoffe* **6**, 87–153.

Lee, C. T. (1976). Agonistic behavior, sexual attraction, and olfaction in mice. In *Mammalian olfaction, reproductive processes, and behavior* (ed. R. L. Doty) pp. 161–80. Academic Press, New York.

— and Brake, S. C. (1971). Reactions of male fighters to male and female mice untreated or de-odorized. *Psychonom. Sci.* **24**, 209–11.

— — (1972). Reaction of male mouse fighters to male castrates treated with testosterone propionate or oil. *Psychonom. Sci.* **27**, 287–8.

— and Crump, M. (1980). A possible confound and the role of olfaction in mouse aggressive interactions. *Aggress. Behav.* **6**, 131–8.

— and Estep, D. (1971). The developmental aspect of marking and nesting behaviors in Mongolian gerbils (*Meriones unguiculatus*). *Psychonom. Sci.* **22**, 312–13.

— and Griffo, W. (1973). Early androgenization and aggression pheromone in inbred mice. *Horm. Behav.* **4**, 181–9.

— — (1974). Progesterone antagonism of androgen-dependent aggression-promoting pheromone in inbred mice (*Mus musculus*). *J. comp. physiol. Psychol.* **87**, 150–5.

— — Braunstein, A. Mars, H., and Stein, K. (1976). Progesterone antagonism of aggression promoting olfactory signals: a time-dependent pheromone. *Physiol. Behav.* **17**, 319–24.

— and Ingersoll, D. W. (1979). Salivary cues in the mouse: a preliminary study. *Horm. Behav.* **12**, 20–9.

— Lukton, A., Bobotas, G., and Ingersoll, D. W. (1980). Partial purification of male *Mus musculus* urinary aggression-promoting chemosignal. *Aggress. Behav.* **6**, 149–60.

Lee, T. M., Lee, C., and Moltz, H. (1982). Prolactin in liver cytosol and pheromonal emission in the rat. *Physiol. Behav.* **28**, 631–3.

— and Moltz, H. (1980). How rat young govern the release of a maternal pheromone. *Physiol. Behav.* **24**, 983–9.

Lehmann, E. v. (1962). Über die Seitendrüssen der mitteleuropäischen Rötelmaus (*Clethrionomys glareolus* Schreber.) *Z. Morph. Ökol. Tiere* **51**, 335–44.

— (1966). Über die Seitendrüssen der mitteleuropäischen Wühlmäuse der Gattung *Microtus* Schrnak. *Z. Morph. Ökol. Tiere* **56**, 436–43.

Leidahl, L. C. and Moltz, H. (1975). Emission of the maternal pheromone in the nulliparous female and failure of emission in the adult male. *Physiol. Behav.* **14**, 421–4.

— — (1977). Emission of the maternal pheromone in nulliparous and lactating females. *Physiol. Behav.* **18**, 399–402.

LeMagnen, J. (1952). Les phénomènes olfacto-sexuels chez le rat blanc. *Arch. Sci. Physiol.* **6**, 295–331.

Leon, M. (1974). Maternal pheromone. *Physiol. Behav.* **13**, 441–53.

— (1975). Dietary control of maternal pheromone in the lactating rat. *Physiol. Behav.* **14**, 311–19.

— (1978*a*). Filial responsiveness to olfactory cues in the laboratory rat. In *Advances in the study of behavior* (ed. J. S. Rosenblatt, R. A. Hinde, C. Beer, and M.-C. Busnel) Vol. 9, pp. 117–53. Academic Press, New York.

— (1978*b*). Emission of maternal pheromone. *Science, NY* **201**, 938–9.

— (1980). Development of olfactory attraction by young Norway rats. In *Chemical signals: vertebrates and aquatic invertebrates* (ed. D. Müller-Schwarze and R. M. Silverstein) pp. 193–209. Plenum Press, New York.

— and Behse, J. H. (1977). Dissolution of the pheromonal bond: waning of approach response by weanling rats. *Physiol. Behav.* **18**, 393–7.

— Galef, B. G. Jr, and Behse, J. H. (1977). Establishment of pheromonal bonds and diet choice in young rats by odor pre-exposure. *Physiol. Behav.* **18**, 387–91.

— and Moltz, H. (1971). Maternal pheromone: discrimination by pre-weanling albino rats. *Physiol. Behav.* **7**, 265–7.

— — (1972). The development of the pheromonal bond in the albino rat. *Physiol. Behav.* **8**, 683–6.

— — (1973). Endocrine control of the maternal pheromone in the postpartum female rat. *Physiol. Behav.* **10**, 65–7.

Lidicker, W. J. Jr (1980). The social biology of the California vole. *Biologist* **62**, 46–55.

Lindzey, G., Thiessen, D. D., and Tucker, A. (1968). Development and hormonal control of territorial marking in the male Mongolian gerbil (*Meriones unguiculatus*). *Dev. Psychobiol.* **1**, 97–9.

Linsdale, J. M. and Tevis, L. P. Jr (1951). *The dusky-footed woodrat*. University of California Press, Berkeley.

Lucas, P. D., Donohoe, S. M., and Thody, A. J. (1982). The role of estrogen and progesterone in the control of preputial gland sex attractant odors in the female rat. *Physiol. Behav.* **28**, 601–7.

Luderschmidt, C. and Plewig, G. (1977). Effects of cyproterone acetate and carboxylic acid derivitives on the sebaceous glands of the Syrian hamster. *Arch. derm. Res.* **258**, 185–91.

Lumia, A. R., Canino, G., and Drozdowski, B. (1976). Effects of androgen on scent marking and aggressive behavior of neonatally bulbectomized female Mongolian gerbils (*Meriones unguiculatus*). *Horm. Behav.* **7**, 461–71.

— Raskin, L. A., and Eckhert, S. (1977). Effects of androgen on marking and aggressive behavior of neonatally and prepubertally bulbectomized and castrated male gerbils. *J. comp. physiol. Psychol.* **91**, 1377–89.

— Westervelt, M. O., and Rieder, C. A. (1975). Effects of olfactory bulb ablation and androgen on marking and agonistic behavior in male Mongolian gerbils, *Meriones unguiculatus. J. comp. physiol. Psychol.* **89**, 1091–9.

Lydell, K. and Doty, R. L. (1972). Male rat odor preferences for female urine as a function of sexual experience, urine age, and urine source. *Horm. Behav.* **3**, 205–12.

Lynds, P. G. (1976). Olfactory control of aggression in lactating female house mice. *Physiol. Behav.* **17**, 157–9.

McCarty, R. and Southwick, C. H. (1977). Cross-species fostering: effects on the olfactory preferences of *Onychomys torridus* and *Peromyscus leucopus. Behav. Biol.* **19**, 255–60.

McColl, I. (1967). The comparative anatomy and pathology of anal glands. *Ann. R. Coll. Surg. Engl.* **40**, 36–67.

McDonald, D. L. and Forslund, L. G. (1978). The development of social preferences in the voles *Microtus montanus* and *Microtus canicaudus*: Effects of cross-fostering. *Behav. Biol.* **22**, 497–508.

McIntosh, T. K., Davis, P. G., and Barfield, R. J. (1979). Urine marking and sexual behavior in the rat (*Rattus norvegicus*). *Behav. neural Biol.* **26**, 161–8.

MacIsaac, G. L. (1977). Reproductive correlates of the hip gland in voles (*Microtus townsendii*). *Can. J. Zool.* **55**, 939–41.

Mackay-Sim, A. and Laing, D. G. (1980). Discrimination of odours from stressed rats by non-stressed rats. *Physiol. Behav.* **24**, 699–704.

— — (1981*a*). Rats' responses to blood and body odors of stressed and non-stressed conspecifics. *Physiol. Behav.* **27**, 503–10.

— — (1981*b*). The sources of odors from stressed rats. *Physiol. Behav.* **27**, 511–13.

McKinney, T. D. and Christian, J. J. (1970). Effect of preputialectomy on fighting behavior in mice. *Proc. Soc. exp. Biol. Med.* **134**, 291–3.

Mackintosh, J. H. (1973). Factors affecting the recognition of territory boundaries by mice, *Mus musculus. Anim. Behav.* **21**, 464–70.

— and Grant, E. C. (1966). The effect of olfactory stimuli on the agonistic behaviour of laboratory mice. *Z. Tierpsychol.* **23**, 584–7.

Macrides, F., Firl, A. C. Jr, Schneider, S. P., Bartke, A., and Stein, D. G. (1976). Effects of one-stage or serial transections of the lateral olfactory tracts on behavior and plasma testosterone levels in male hamsters. *Brain Res.* **109**, 97–109.

— Johnson, P. A., and Schneider, S. P. (1977). Responses of the male golden hamster to vaginal secretion and dimethyl disulfide: attraction versus sexual behavior. *Behav. Biol.* **20**, 377–86.

Mainardi, D., Marsan, M., and Pasquali, A. (1965). Causation of sexual preferences in the house mouse: the behaviour of mice reared by parents whose odor was artificially altered. Att. *Soc. Ital. Sci. nat.* **104**, 325–38.

Markel, K. (1952). Zur Kenntnis der Seitendrüsen des Goldhamsters (*Mesocricetus auratus*). *Zool. Anz.* **149**, 216–25.

Marchlewska-Koj, A. (1977). Pregnancy block elicited by urinary proteins in male mice. *Biol. Reprod.* **17**, 729–32.

Marr, J. N. and Gardner, L. E. (1965). Early olfactory experience and later social behavior in the rat: preferences, sexual responsiveness, and care of young. *J. genet. Psychol.* **107**, 167–74.

— and Lilliston, L. G. (1969). Social attachment in rats by odor and age. *Behaviour* **33**, 277–82.

Martin, R. E. (1977). Species preferences of allopatric and sympatric populations of silky pocket mice, genus *Peroganthus* (*Rodentia: Heteromyidae*). *Am. midl. Nat.* **98**, 124–36.

Maruniak, J. A., Darney, K. T. Jr, and Bronson, F. H. (1975). Olfactory perception of the nonsocial environment by male housemice. *Behav. Biol.* **14**, 237–40.

— Desjardins, C., and Bronson, F. H. (1975). Adaptations for urinary marking in rodents: prepuce length and morphology. *J. Reprod. Fert.* **44**, 567–70.

— Owen, K., Bronson, F. H., and Desjardins, C. (1974). Urinary marking in male house mice: responses to novel environmental and social stimuli. *Physiol. Behav.* **12**, 1035–9.

— — — — (1975). Urinary marking in female house mice: effects of ovarian steroids, sex experience, and type of stimulus. *Behav. Biol.* **13**, 211–17.

Matthews, M. K. Jr (1980). Urinary marking and tendency to investigate novelty in *Mus musculus*. *Behav. neural Biol.* **28**, 501–6.

— (1981). Short- and long-term effects of aggressive behavior on urinary marking in *Mus musculus*. *Behav. neural Biol.* **32**, 104–10.

Maynard Smith, J. (1974). The theory of games and the evolution of animal conflicts. *J. theoret. Biol.* **47**, 209–22.

Mazdzer, E., Capone, M. R., and Drickamer, L. C. (1976). Conspecific odor and trappability of deermice, *Peromyscus leucopus noveboracensis*. *J. Mammal.* **57**, 607–9.

Means, L. W., Bates, T. W., and Cahoon, R. U. (1973). The effects of nonreinforcement on the cue value of the odor trails of rats. *Behav. Biol.* **8**, 545–50.

Meisami, E. (1976). Effects of olfactory deprivation on postnatal growth of the rat olfactory bulb utilizing a new method for production of neonatal unilateral anosmia. *Brain Res.* **107**, 437–44.

Mellgren, R. L., Fouts, R. S., and Martin, J. W. (1973). Approach and escape to conspecific odors of reward and nonreward in rats. *Anim. Learn. Behav.* **1**, 129–32.

Minor, T. R. and Lolordo, V. M. (1981). Odor cues and the transfer of learned helplessness in the rat. Unpublished manuscript.

Mitchell, O. G. (1965). Effect of castration and transplantation on ventral gland of the gerbil. *Proc. Soc. exp. Biol. Med.* **119**, 953–5.

— (1967) The supposed role of the gerbil ventral gland in reproduction. *J. Mammal.* **48**, 142.

Moltz, H. and Kilpatrick, S. J. (1978). Response to the maternal pheromone in the rat as protection against necrotizing enterocolitis. *Neurosci. biobehav. Rev.* **2**, 277–80.

— and Lee, T. M. (1981). The maternal pheromone of the rat: identity and functional significance. *Physiol. Behav.* **26**, 301–6.

— and Leidahl, L. C. (1977). Bile, prolactin, and the maternal pheromone. *Science, NY* **196**, 81–3.

— and Leon, M. (1973). Stimulus control of the maternal pheromone in the lactating rat. *Physiol. Behav.* **10**, 69–71.

Montagna, W. and Hamilton, J. B. (1949). The sebaceous glands of the hamster. II. Some cytochemical studies in normal and experimental animals. *Amer. J. Anat.* **84**, 365–95.

— and Noback, C. R. (1946). The histochemistry of the preputial gland of the rat. *Anat. Rec.* **96**, 111–27.

— — (1947). Histochemical observations on the sebaceous glands of the rat. *Amer. J. Anat.* **81**, 39–62.

Moore, C. L. (1981). An olfactory basis for maternal discrimination of sex of offspring in rats (*Rattus norvegicus*). *Amer. Behav.* **29**, 383–6.

— and Morelli, G. A. (1979). Mother rats interact differently with male and female offspring. *J. comp. physiol. Psychol.* **93**, 677–84.

Moore, R. E. (1965). Olfactory discrimination as an isolating mechanism between *Peromyscus maniculatus* and *Peromyscus polionotus*. *Am. midl. Nat.* **73**, 85–100.

Morrison, R. R. and Ludvigson, H. W. (1970). Discrimination by rats of conspecific odors of reward and nonreward. *Science, NY* **167**, 904–5.

Mugford, R. A. (1973). Intermale fighting affected by home-cage odors of male and female mice. *J. comp. physiol. Psychol.* **84**, 289–95.

— and Nowell, N. W. (1970a). The aggression of male mice against androgenized females. *Psychonom. Sci.* **20**, 191–12.

— — (1970b). Pheromones and their effects on aggression in mice. *Nature, Lond.* **226**, 967–98.

— — (1971a). Endocrine control over production and activity of the anti-aggression pheromone from female mice. *J. Endocr.* **49**, 225–32.

— — (1971b). The relationship between endocrine status of female opponents and aggressive behaviour of male mice. *Anim. Behav.* **19**, 153–5.

— — (1971c). The preputial gland as a source of aggression-promoting odors in mice. *Physiol. Behav.* **6**, 247–9.

— — (1972). Paternal stimulation during infancy: effects upon aggression and open field performance of mice. *J. comp. physiol. Psychol.* **79**, 30–6.

Müller-Schwarze, D. and Heckman, S. (1980). The social role of scent marking in beaver (*Castor canadensis*). *J. chem. Ecol.* **6**, 81–95.

Müller-Velten, H. (1966). Über den Angstgeruch bei der Hausmaus (*Mus musculus* L.). *Z. Vergleich. Physiol.* **52**, 401–29.

Munger, B. L. and Brusilow, S. W. (1971). The histophysiology of rat plantar sweat glands. *Anat. Rec.* **169**, 1–22.

Murphy, M. R. (1973). Effects of female hamster vaginal discharge on the behavior of male hamsters. *Behav. Biol.* **9**, 367–75.

— (1977). Intraspecific sexual preferences of female hamsters. *J. comp. physiol. Psychol.* **91**, 1337–46.

— (1978). Oestrous Turkish hamsters display lordosis toward conspecific males, but attack heterospecific males. *Anim. Behav.* **26**, 311–12.

— and Schneider, G. E. (1970). Olfactory bulb removal eliminates mating behavior in the male golden hamster. *Science, NY* **167**, 302–4.

Myllymaki, A., Aho, J., Lind, E. A. and Tast, J. (1962). Behavior and daily activity of the Norwegian lemming (*Lemmus lemmus*) during autumn migration. *Ann. Zool. Soc. zool. bot. fenn. 'Vanamo'* **24**, 1–31.

Namikas, J. and Wehmer, F. (1978). Gender composition of the litter affects behavior of male mice. *Behav. Biol.* **23**, 219–24.

Nevo, E. (1969). Mole rat *Spalax ehrenbergi*: mating behavior and its evolutionary significance. *Science NY* **163**, 484–6.

— Bodmer, M., and Heth, G. (1976). Olfactory discrimination as an isolating mechanism in speciating mole rats. *Experientia* **32**, 1511–12.

— and Heth, G. (1976). Assortative mating between chromosome forms of the mole rat, *Spalax ehrenbergi*. *Experientia* **32**, 1509–10.

— Naftali, G., and Guttman, R. (1975). Aggression patterns and speciation. *Proc. nat. Acad. Sci.* **72**, 3250–4.

Nigrosh, B. J., and Slotnick, B. M., and Nevin, J. A. (1975). Olfactory discrimination reversal learning and stimulus control in rats. *J. comp. physiol. Psychol.* **89**, 285–94.

Noble, R. L. and Collip, J. B. (1941). A possible direct control of the preputial glands of the female rat by the pituitary gland and indirect effects produced through the adrenals and gonads by augmented pituitary extracts. *Endocrinology* **29**, 943–51.

Noirot, E. (1969). Selective priming of maternal responses by auditory and olfactory cues from mouse pups. *Dev. Psychobiol.* **2**, 273–6.

Nowell, N. W. and Wouters, A. (1975). Release of aggression-promoting pheromone by male mice treated with α-melanocyte-stimulating hormone. *J. Endocr.* **65**, 36P–7P.

Nyakas, C. and Endröczi, E. (1970). Olfaction guided approaching behaviour of infantile rats to the mother in maze box. *Acta physiol. acad. sci. hung.* **38**, 59–65.

Nyby, J., Dizinno, G., and Whitney, G. (1977). Sexual dimorphism in ultrasonic vocalizations of mice (*Mus musculus*): gonadal hormone regulation. *J. comp. physiol. Psychol.* **91**, 1424–31.

— and Thiessen, D. D. (1971). Singular and interactive effects of testosterone and estrogen on territorial marking in castrated male Mongolian gerbils (*Meriones unguiculatus*). *Horm. Behav.* **2**, 279–85.

— — and Wallace, P. (1970). Social inhibition of territorial marking in the Mongolian gerbil (*Meriones unguiculatus*). *Psychonom. Sci.* **21**, 310–12.

— and Whitney, G. (1980). Experience affects behavioral responses to sex odors. In *Chemical signals: vertebrates and aquatic invertebrates* (ed. D. Müller-Schwarze and R. M. Silverstein). pp. 173–92. Plenum Press, New York.

— — Schmitz, S., and Dizinno, G. (1978). Postpuberal experience establishes signal value of mammalian sex odor. *Behav. Biol.* **22**, 545–52.

— Wysocki, C. J., Whitney, G., and Dizinno, G. (1977). Pheromonal regulation of male mouse ultrasonic courtship, *Mus musculus. Anim. Behav.* **25**, 333–41.

— — — — and Schneider, J. (1979). Elicitation of male mouse ultrasonic vocalizations: I. Urinary cues. *J. comp. physiol. Psychol.* **93**, 957–75.

— — — — and Nunez, A. A. (1981). Stimuli for male mouse (*Mus musculus*) ultrasonic courtship vocalizations: presence of female chemosignals and/or absence of male chemosignals. *J. comp. physiol. Psychol.* **95**, 623–9.

— and Zakeski, D. (1980). Elicitation of male mouse ultrasounds: bladder urine and aged urine from females. *Physiol. Behav.* **24**, 737–40.

O'Connell, R. J., Singer, A. G., Macrides, F., Pfaffmann, C., and Agosta, W. C. (1978). Responses of the male golden hamster to mixtures of odorants identified from vaginal discharge. *Behav. Biol.* **24**, 244–55.

— — Pfaffmann, C., and Agosta, W. C. (1979). Pheromones of hamster vaginal discharge: attraction to femtogram amounts of dimethyl disulfide and to mixtures of volatile components. *J. chem. Ecol.* **5**, 575–85.

— — Stern, F. L., Jesmajian, S., and Agosta, W. C. (1981). Cyclic variations in the concentration of sex attractant pheromone in hamster vaginal discharge. *Behav. neural Biol.* **31**, 457–64.

Orsini, M. (1961). The external vaginal phenomena characterizing the stages of the estrous cycle, pregnancy, pseudopregnancy, lactation and the anestrous hamster,*Mesocricetus auratus*Waterhouse.*Proc. Anim. Care Panel* **11**, 193–206.

Orsulak, P. J. and Gawienowski, A. M. (1972). Olfactory preferences for the rat preputial gland. *Biol. Reprod.* **6**, 219–23.

Ortmann, R. (1956). Über die Musterbildung von Düftdrüsen in der Sohlenhaut der weissen Hausmaus (*Mus musculus* alba). *Z. Säugetierk* **21**, 138–41.

— (1960). Die Analregion der Säugetiere. *Handb. Zool.* **8**:26, 3(7), 1–68.

Oswalt, G. L. and Meier, G. W. (1975). Olfactory, thermal and tactile influences on infantile ultrasonic vocalization in rats. *Dev. Psychobiol.* 8, 129–35.

Owen, K. and Thiessen, D. D. (1973). Regulation of scent marking in the female Mongolian gerbil, *Meriones unguiculatus. Physiol. Behav.* 11, 441–5.

—— —— (1974). Estrogen and progesterone interaction in the regulation of scent marking in the female Mongolian gerbil, *Meriones unguiculatus. Physiol. Behav.* 12, 351–5.

—— Wallace, P., and Thiessen, D. D. (1974). Effects of intracerebral implants of steroid hormones on scent marking in the ovariectomized female gerbil, *Meriones unguiculatus. Physiol. Behav.* 12, 755–60.

Paclt, J. (1952). Scent glands in the bank vole. *Experientia* 8, 464.

Parker, G. A. and Rubenstein, D. I. (1981). Role assessment, reserve strategy, and acquisition of information in asymmetric animal conficts. *Anim. Behav.* 29, 221–40.

Payman, B. C. and Swanson, H. (1980). Social influence on sexual maturation and breeding in the female Mongolian gerbil (*Meriones unguiculatus*). *Anim. Behav.* 28, 528–35.

Payne, A. P. (1974). The effects of urine on aggressive responses by male golden hamsters. *Aggress. Behav.* 1, 71–9.

—— (1977). Pheromonal effects of Harderian gland homogenates on aggressive behavior in the hamster. *J. Endocr.* 73, 191–2.

—— (1979). The attractiveness of Harderian gland smears to sexually naive and experienced male golden hamsters. *Anim. Behav.* 27, 897–904.

—— McGadey, J., Moore, N. R., and Thompson, G. (1975). The effects of androgen manipulation on cell types and porphyrin content of the Harderian gland in the male golden hamster. *J. Anat.* 120, 615–16. (Abstr.)

—— —— —— —— (1977). Androgenic control of the Harderian gland in the male golden hamster. *J. Endocr.* 75, 73–82.

—— —— —— —— (1979). Changes in porphyrin content of the Harderian gland during the oestrus cycle, pregnancy and lactation. *Biochem. J.* 178, 597–604.

—— and Swanson, H. H. (1970). Agonistic behaviour between pairs of hamsters of the same and opposite sex in a neutral observation area. *Behaviour* 36, 259–69.

Pettijohn, T. F. (1974). Attractiveness of conspecific urine to the male Mongolian gerbil. Paper presented at the Animal Behavior Society, Champaign, Illinois, May.

—— (1977). Reactions of Mongolian gerbils in the presence of urine stimuli. *Anim. Learn. Behav.* 5, 370–2.

—— (1979). Conditioned olfactory aversion in the male Mongolian gerbil. *Physiol. Psychol.* 7, 299–302.

—— and Jamora, C. M. (1980). Learned aversion to soiled bedding in male Mongolian gerbils. *Physiol. Behav.* 24, 1031–4.

Plewig, G. and Luderschmidt, C. (1977). Hamster ear model for sebaceous glands. *J. invest. Derm.* 68, 171–6.

Pocock, R. I. (1922). On the external characters of the beaver (*Castoridae*) and of some squirrels (*Sciuridae*). *Proc. zool. Soc., Lond.* 1171–212.

Poole, E. L. (1940). A life history sketch of the Allegheny woodrat. *J. Mammal.* 21, 249–70.

Porter, R. H., Deni, R., and Doane, H. M. (1977). Responses of *Acomys cahirinus* pups to chemical cues produced by a foster species. *Behav. Biol.* 20, 244–51.

—— and Doane, H. M. (1976). Maternal pheromone in the spiny mouse (*Acomys cahirinus*). *Physiol. Behav.* 16, 75–8.

—— (1977). Dietary dependent cross species similarities in maternal chemical cues. *Physiol. Behav.* **19**, 129–31.

—— and Cavallaro, S. A. (1978). Temporal parameters of responsiveness to maternal pheromone in *Acomys cahirinus. Physiol. Behav.* **21**, 563–6.

— and Etscorn, F. (1974). Olfactory imprinting resulting from brief exposure in *Acomys cahirinus. Nature, Lond.* **250**, 732–3.

—— (1975). A primacy effect for olfactory imprinting in spiny mice (*Acomys cahirinus*). *Behav. Biol.* **15**, 511–17.

—— (1976). A sensitive period for the development of olfactory preference in *Acomys cahirinus. Physiol. Behav.* **17**, 127–30.

— and Ruttle, K. (1975). The responses of one-day-old *Acomys cahirinus* pups to naturally occurring chemical stimuli. *Z. Tierpsychol.* **38**, 154–62.

— and Wyrick, M. (1979). Sibling recognition by spiny mice (*Acomys cahirinus*): influence of age and isolation. *Anim. Behav.* **27**, 761–6.

—— and Pankey, J. (1978). Sibling recognition in spiny mice (*Acomys cahirinus*). *Behav. Ecol. Sociobiol.* **3**, 61–8.

Powers, J. B. and Winans, S. S. (1975). Vomeronasal organ: critical role in mediating sexual behavior of the male hamster. *Science, NY* **187**, 961–3.

Price, E. O. (1975). Hormonal control of urine-making in wild and domestic Norway rats. *Horm. Behav.* **6**, 393–7.

— (1977). Urine marking and the response to fresh versus aged urine in wild and domestic Norway rats. *J. chem. Ecol.* **3**, 9–25.

Quadagno, D. M. and Banks, E. M. (1970). The effect of reciprocal cross fostering on the behaviour of two species of rodents, *Mus musculus* and *Baiomys taylori ater. Anim. Behav.* **18**, 379–90.

Quay, W. B. (1953). Seasonal and sexual differences in the dorsal skin gland of the kangaroo rat (*Dipodomys*). *J. Mammal.* **34**, 1–14.

— (1954*a*). The dorsal holocrine skin gland of the kangaroo rat (*Dipodomys*). *Anat. Rec.* **119**, 161–75.

— (1954*b*). The Meibomian glands of voles and lemmings (*Microtinae*). *Misc. Publ. Mus. Zool. Univ. Mich.* **82**, 1–17.

— (1960). The reproductive organs of the collared lemming under diverse temperature and light conditions. *J. Mammal.* **41**, 74–8.

— (1962). Apocrine sweat glands in the angulus oris of microtine rodents. *J. Mammal.* **43**, 303–10.

— (1965*a*). Variation and taxonomic significance in the sebaceous caudal glands of pocket mice (*Rodentia: Heteromyidae*). *S West. Nat.* **10**, 282–7.

— (1965*b*). Comparative survey of sebaceous and sudoriferous glands of the oral lips and angle in rodents. *J. Mammal.* **46**, 23–37.

— (1965*c*). Integumentary modifications of North American desert rodents. In *Biology of the skin and hair growth* (ed. A. G. Lyne and B. F. Short) pp. 59–74. Elsevier, New York.

— (1966). Stimulation of the palmar sweat glands of the Kangaroo rat (*Dipodomys heermanni*) with acetylcholine. *Am. Zool.* **6**, 351. (Abstr.)

— (1968). The specialized posterolateral sebaceous glandular regions in microtine rodents. *J. Mammal.* **49**, 427–45.

— (1972). Integument and the environment: glandular composition, function, and evolution. *Am. Zool.* **12**, 95–108.

— and Tomich, P. Q. (1963). A specialized midventral sebaceous glandular area in *Rattus exulans. J. Mammal.* **44**, 537–42.

Rainey, D. G. (1956). Eastern woodrat, *Neotoma floridana:* life history and ecology. *Univ. Kansas Publ. Mus. nat. Hist.* **8**, 535–646.

Ralls, K. (1971). Mammalian scent marking. *Science, NY* **171**, 443–9.

Reiff, M. (1951). Territoriums markierung bei Hausratten und Hausmausen. *Schweiz. naturforsch. ges. Verhandl.* **13**, 150–1.

Reynolds, E. (1971). Urination as a social response in mice. *Nature, Lond.* **234**, 481–3.

Richards, D. B. and Stevens, D. A. (1974). Evidence for marking with urine by rats. *Behav. Biol.* **12**, 517–23.

Richmond, N. D. and Roslund, H. R. (1952). A mid-ventral dermal gland in *Peromyscus maniculatus. J. Mammal.* **33**, 103–4.

Rieder, C. A. and Reynierse, J. H. (1971). Effects of maintenance condition on aggression and marking behavior of the Mongolian gerbil (*Meriones unguiculatus*). *J. comp. physiol. Psychol.* **73**, 471–5.

Ring, J. R. and Randall, W. C. (1947). The distribution and histological structure of sweat glands in the albino rat and their response to prolonged nervous stimulation. *Anat. Rec.* **99**, 7–19.

Ropartz, P. (1968). The relation between olfactory stimulation and aggressive behavior in mice. *Anim. Behav.* **16**, 97–100.

— (1977). Chemical signals in agonistic and social behavior of rodents. In *Chemical signals in vertebrates* (ed. D. Müller-Schwarze and M. M. Mozell) pp. 169–84. Plenum Press, New York.

— and Haug, M. (1975). Olfaction and aggressive behavior in female mice. In *Olfaction and taste V* (ed. D. A. Denton and J. P. Coghlan). Academic Press, New York.

Rose, E. and Drickamer, L. C. (1975). Castration, sexual experience, and female urine odor preferences in adult BDF_1 male mice. *Bull. psychonom. Soc.* **5**, 84–6.

Rosenblatt, J. S., Siegel, H. I., and Mayer, A. D. (1979). Progress in the study of maternal behavior in the rat: hormonal, nonhormonal, sensory, and developmental aspects. In *Advances in the study of behavior* (ed. J. S. Rosenblatt, R. A. Hinde, C. Beer, and M.-C. Busnel) Vol. 10, pp. 225–311. Academic Press, New York.

Rottman, S. J. and Snowdon, C. T. (1972). Demonstration and analysis of an alarm pheromone in mice. *J. comp. physiol. Psychol.* **81**, 483–90.

Rouger, Y. and Schneirla, T. C. (1977). Effects of pre-weanling olfactory bulbectomy on the survival of Wistar (DAB) rat pups. *Ann. NY Acad. Sci.* **290**, 227–31.

Rowe, F. P. (1970). The response of wild house mice (*Mus musculus*) to live traps marked by their own and by a foreign mouse odour. *J. Zool.* **162**, 517–48.

— and Redfern, R. (1969). Aggressive behaviour in related and unrelated wild house mice (*Mus musculus* L.). *Ann. appl. Biol.* **64**, 425–31.

Rudd, R. L. (1966). The midventral gland in Malaysian murid rodents. *J. Mammal.* **47**, 331–2.

Rudy, J. W. and Cheatle, M. D. (1977). Odor-aversion learning in neonatal rats. *Science, NY* **198**, 845–6.

Sawyer, T. F. (1978). Aversive odors of male mice: experiential and castration effects, and the predictability of the outcomes of agonistic encounters. *Aggress. Behav.* **4**, 263–75.

— (1980). Androgen effects on responsiveness to aggression and stress-related odors of male mice. *Physiol. Behav.* **25**, 183–7.

— (1981). Learned aversion to the odors of male mice: effects of agonistic behavior. *Physiol. Behav.* **27**, 19–25.

Schaffer, J. (1940). *Die Hautdrüsenorgane der Säugetiere*. Urban & Schwarzenberg, Berlin.

Schapiro, S. and Salas, M. (1970). Behavioral response of infant rats to maternal odor. *Physiol. Behav.* 5, 815-17.

Schenk, F. (1978). Ultrasound production from the isolated adult wood mouse in various surroundings. *Behav. Biol.* 22, 219-29.

Schumacher, S. K. and Moltz, H. (1982). The maternal pheromone of the rat as an innate stimulus for pre-weanling young. *Physiol. Behav.* 28, 67-71.

Scott, J. W. and Pfaff, D. W. (1970). Behavioral and electrophysiological responses of female mice to male urine odors. *Physiol. Behav.* 5, 407-11.

Seton, E. T. (1901). *Lives of the hunted*. David Nutt, London.

Sherman, R. A. W. (1974). Demonstration of the hamster alarm pheromone. *Diss. Abstr. Int. B* 34, 3656.

Simpson, G. G. (1945). The principles of classification and a classification of the mammals. *Bull. Am. Mus. Nat. Hist.* 85, 1-350.

Singer, A. G., Agosta, W. C., O'Connell, R. J., Pfaffmann, C., Bowen, D. V., and Field, F. H. (1976). Dimethyl disulfide: an attractant pheromone in hamster vaginal secretion. *Science, NY* 191, 948-50.

— Macrides, F., and Agosta, W. C. (1980). Chemical studies of hamster reproductive pheromones. In *Chemical signals: vertebrates and aquatic invertebrates* (ed. D. Müller-Schwarze and R. M. Silverstein) pp. 365-75. Plenum Press, New York.

Singh, P. J. (1977). Effects of complete and partial olfactory bulbectomy on nursing behavior. *Ann. NY Acad. Sci.* 290, 237-9.

— and Hofer, M. A. (1978). Oxytocin reinstates maternal olfactory cues for nipple orientation and attachment in rat pups. *Physiol. Behav.* 20, 385-90.

— Lederhendler, I., Desantis, J., and Beckhorn, G. (1977). Responses of Wistar (DAB) rat pups to familiar and unfamiliar substrate odors. *Ann. NY Acad. Sci.* 290, 239-59.

— Tucker, A. M., and Hofer, M. A. (1976). Effects of nasal $ZnSo_4$ irritation and olfactory bulbectomy on rat pups. *Physiol. Behav.* 17, 373-82.

Skeen, J. T. and Thiessen, D. D. (1977). Scent of gerbil cuisine. *Physiol. Behav.* 19, 11-14.

Skirrow, M. H. and Rysan, M. (1976). Observations on the social behavior of the Chinese hamster, *Cricetulus griseus*. *Can. J. Zool.* 54, 361-8.

Skurat, L. N., Litvin, V. Yu, and Golubev, M. V. (1976). Reactions of two species of common vole to smells of their life activity. *Zool. Zhur.* 55, 1897-902.

Slotnick, B. M. and Katz, H. M. (1974). Olfactory learning set formation in rats. *Science, NY* 185, 796-8.

Smotherman, W. P., Bell, R. W., Starzec, J., Elias, J., and Zachman, T. A. (1974). Maternal responses to infant vocalizations and olfactory cues in rats and mice. *Behav. Biol.* 12, 55-66.

Sokolov, W. and Skurat, L. (1966). A specific mid-ventral gland of gerbils. *Nature, Lond.* 211, 544-5.

— — and Kotenkova, E. V. (1976). Patterns of smell signalization in house mice (*Mus musculus*). *Zool. Zhur.* 55, 1710-14.

Solomon, J. A. and Glickman, S. E. (1977). Attraction of male golden hamsters (*Mesocricetus auratus*) to the odors of male conspecifics. *Behav. Biol.* 20, 367-76.

Sprott, R. L. (1969). Fear communication via odor in inbred mice. *Psychol. Rep.* 25, 263-8.

Stacewicz-Sapuntzakis, M. and Gawienowski, A. M. (1977). Rat olfactory response to aliphatic acetates. *J. chem. Ecol.* 3, 411–17.

Stanley, A. J. and Powell, R. A. (1941). Studies on the preputial gland of the white rat. *Proc. Louisana Acad. Sci.* 5, 28–9. (Abstr.)

Stark, B. and Hazlett, B. A. (1972). Effects of olfactory experience on aggression in *Mus musculus* and *Peromyscus maniculatus*. *Behav. Biol.* 7, 265–9.

Stehn, R. A., Richmond, M. E., and Kollisch, N. (1976). Female odors and aggression among male *Microtus*. *Behav. Biol.* 17, 43–50.

Steiniger, F. (1950). Über Duftmarkierung bei der Wanderratte. *Z. ang. Zool.* 38, 357–61.

Stern, J. J. (1970). Responses of male rats to sex odors. *Physiol. Behav.* 5, 519–24.

Stevens, D. A. and Gerzog-Thomas, D. A. (1977). Fright reactions in rats to conspecific tissue. *Physiol. Behav.* 18, 47–51.

— and Koster, E. P. (1972). Open field responses of rats to odors from stressed and nonstressed predecessors. *Behav. Biol.* 7, 519–25.

— and Saplikoski, M. J. (1973). Rats' reactions to conspecific muscle and blood: evidence for an alarm substance. *Behav. Biol.* 8, 75–82.

Stoddart, D. M. (1972a). The lateral scent organs of *Arvicola terrestris* (Rodentia: Microtinae). *J. Zool., Lond.* 166, 49–54.

— (1972b). An examination of the caudal organ of *Apodemus agrarius, A. flavicollis, A. microps*, and *A. sylvaticus* from Moravia. *Zool. Listy* 21, 39–42.

— (1973). Preliminary characterisation of the caudal organ secretion of *Apodemus flavicollis. Nature, Lond.* 246, 501–3.

— (1982). Demonstration of olfactory discrimination by the short-tailed vole, *Microtus agrestis* L. *Anim. Behav.* 30, 293–301.

— Aplin, R. T., and Wood, M. J. (1975). Evidence for social difference in the flank organ secretion of *Arvicola terrestris* (Rodentia: Microtinae). *J. Zool., Lond.* 177, 529–40.

Stolzer, E. (1959). Beiträge zur Kenntnis der Seitendrüsen des syrischen Goldhamsters (*Mesocricetus auratus* Waterhouse). *Z. Säugetierk.* 23, 182–97.

Strauss, J. S. and Ebling, F. J. (1970). Control and function of skin glands in mammals. *Mem. Soc. Endocr.* 18, 341–71.

Summerlin, C. T. and Wolfe, J. L. (1973). Social influences on trap response of the cotton rat, *Sigmodon hispidus. Ecology* 54, 1156–9.

Svendsen, G. E. (1978). Castor and anal glands of the beaver (*Castor canadensis*). *J. Mammal.* 59, 618–20.

— (1980). Patterns of scent mounding in a population of beaver (*Castor canadensis*). *J. chem. Ecol.* 6, 133–48.

— and Jollick, J. D. (1978). Bacterial contents of the anal and castor glands of beaver *Castor canadensis. J. chem. Ecol.* 4, 563–70.

Swanson, H. H. (1974). Sex differences in behaviour of the Mongolian gerbil (*Meriones unguiculatus*) in encounters between pairs of same or opposite sex. *Anim. Behav.* 22, 638–44.

— (1980). Social and hormonal influences of scent marking in the Mongolian gerbil. *Physiol. Behav.* 24, 839–42.

— and Lockley, M. R. (1978). Population growth and social structure of confined colonies of Mongolian gerbils: scent gland size and marking behaviour as indices of social status. *Aggress. Behav.* 4, 57–89.

— and Norman, M. E. (1978). Central and peripheral action of testosterone propionate on scent gland morphology and marking behaviour in the Mongolian gerbil. *Behav. Process.* 3, 9–19.

Taylor, G. T. (1982). Urinary odors and size protect juvenile laboratory mice from adult male attack. *Dev. Psychobiol.* **15**, 171–86.

— Haller, J., and Regan, D. (1982). Female rats prefer an area vacated by a high testosterone male. *Physiol. Behav.* **28**, 953–8.

Taylor, R. D. and Ludvigson, H. W. (1980*a*). Selective removal of reward and nonreward odors to assess their control of patterned responding in rats. *Bull. psychon. Soc.* **16**, 101–4.

— — (1980*b*). Selective removal of alleyway paper flooring to air to assess locus of nonreward odor. *Bull. psychonom. Soc.* **16**, 105–8.

Teicher, M. H. and Blass, E. M. (1977). First suckling response of the newborn albino rat: the roles of olfaction and amniotic fluid. *Science, NY* **198**, 635–6.

Thiessen, D. D. (1968). The roots of territorial marking in the Mongolian gerbil: a problem of species-common topography. *Behav. Res. Meth. Instrum.* **1**, 70–6.

— (1977). Methodology and strategies in the laboratory. In *Chemical signals in vertebrates* (ed. D. Müller-Schwarze and M. M. Mozell) pp. 391–412. Plenum Press, New York.

— Blum, S. L., and Lindzey, G. (1970). A scent marking response associated with the ventral sebaceous gland of the Mongolian gerbil (*Meriones unguiculatus*). *Anim. Behav.* **18**, 26-30.

— Clancy, A., and Goodwin, M. (1976). Harderian gland pheromone in the Mongolian gerbil, *Meriones unguiculatus*. *J. chem. Ecol.* **2**, 231–8.

— and Dawber, M. (1972). Territorial exclusion and reproductive isolation. *Psychonom. Sci.* **28**, 159–60.

— Friend, H. C., and Lindzey, G. (1968). Androgen control of territorial marking in the Mongolian gerbil (*Meriones unguiculatus*). *Science, NY* **160**, 432–4.

— Graham, M., and Davenport, R. (1978). Ultrasonic signaling in the gerbil (*Meriones unguiculatus*): social interaction and olfaction. *J. comp. physiol. Psychol.* **92**, 1041–9.

— and Lindzey, G. (1970). Territorial marking in the female Mongolian gerbil: short-term reactions to hormones. *Horm. Behav.* **1**, 157–60.

— — Blum, S. L., and Wallace, P. (1971). Social interactions and scent marking in the Mongolian gerbil (*Meriones unguiculatus*). *Anim. Behav.* **19**, 505–13.

— — and Nyby, J. (1970). The effects of olfactory deprivation and hormones on territorial marking in the Mongolian gerbil (*Meriones unguiculatus*). *Horm. Behav.* **1**, 315–25.

— Owen, K., and Lindzey, G. (1971). Mechanisms of territorial marking in the male and female Mongolian gerbil (*Meriones unguiculatus*). *J. comp. physiol. Psychol.* **77**, 38–47.

— Regnier, F. E., Rice, M., Goodwin, M., Isaacks, N., and Lawson, N. (1974). Identification of a ventral scent marking pheromone in the male Mongolian gerbil (*Meriones unguiculatus*). *Science, NY* **184**, 83–5.

— and Rice, M. (1976). Mammalian scent gland marking and social behavior. *Psychol. Bull.* **83**, 505–39.

— Wallace, P., and Yahr, P. (1973). Comparative studies of glandular scent marking in *Meriones tristrami*, an Israeli gerbil. *Horm. Behav.* **4**, 143–7.

— and Yahr, P. (1970). Central control of territorial marking in the Mongolian gerbil. *Physiol. Behav.* **5**, 275–8.

— — (1977). *The gerbil in behavioral investigation: mechanisms of territoriality and olfactory communication.* University of Texas Press, Austin.

— — and Lindzey, G. (1971). Ventral and chin gland marking in the Mongolian gerbil (*Meriones unguiculatus*). *Forma Funct.* **4**, 171–5.

— — and Owen, K. (1973). Regulatory mechanisms of territorial marking in the Mongolian gerbil. *J. comp. physiol. Psychol.* **82**, 382–93.

Thody, A. J. and Dijkstra, H. (1978). Effect of ovarian steroids on preputial gland odors in the female rat. *J. Endocr.* **77**, 397–404.

— and Shuster, S. (1975). Control of sebaceous gland function in the rat by α-melanocyte-stimulating hormone. *J. Endocr.* **64**, 503–10.

Thor, D. H. (1979). Olfactory perception and inclusive fitness. *Physiol. Psychol.* **7**, 303–6.

— and Flannelly, K. J. (1977). Social-olfactory experience and initiation of copulation in the virgin male rat. *Physiol. Behav.* **19**, 411–17.

Tobach, E. (1977). Developmental aspects of chemoreception in the Wistar DAB rat: tonic processes. *Ann. NY Acad. Sci.* **290**, 226–69.

Turner, J. W. (1979). Effects of sustained release testosterone on marking behavior in the Mongolian gerbil. *Physiol. Behav.* **23**, 845–9.

Vadasz, C. (1975). Sex preference and species specificity of rodent (*Mus musculus* and *Microtus arvalis*) pheromones. *Acta biol. acad. Sci. hung.* **26**, 9–14.

Valenta, J. G. and Rigby, M. K. (1968). Discrimination of the odor of stressed rats. *Science, NY* **161**, 599–601.

Van Abeelen, J. H. F. and De Vries, A. E. (1978). Genetic control of sniffing and marking responses in mice. *Behav. Genet.* **8**, 219–22.

Vandenbergh, J. G. (1973). Effects of gonadal hormones on the flank gland of the golden hamster. *Horm. Res.* **4**, 28–33.

Venable, J. H. and Grafflin, A. L. (1940). Cross anatomy of the orbital glands of the albino rat. *J. Mammal.* **21**, 66–71.

Vestal, E. H. (1938). Biotic relations of the wood rat (*Neotoma fuscipes*) in the Berkeley hills. *J. Mammal.* **19**, 1–36.

Voorhees, J. W. and Remley, N. R. (1981). Mitral cell responses to the odors of reward and nonreward. *Physiol. Psychol.* **9**, 164–70.

Vorontsov, N. N. and Gurtovoi, N. N. (1959). Structure of the midabdominal gland of the true hamster (*Cricetini, Cricetinae, Rodentia, Mammalia*). *Acad. Sci. USSR Proc. biol. Sci.* **125**, 385–8.

Vrtis, V. (1930*a*). Glandular organ on the flancs of the water rat, their development and changes during breeding season. *Biol. Spis. Acad. vet. Brno* **9**(4), 1–51.

— (1930*b*). Glandular organ on the flancs of the hamster *Cricetus cricetus* (L). *Biol. Spis. Acad. vet. Brno* **9**(13–14), 1–31.

— (1931). Ueber die s.g. Seitendrüsen der Wasserratte 'Arvicola' und des Hamsters 'Cricetus'. *Arch. zool. ital.* **16**, 790–5.

Walker, E. P. (1975). *Mammals of the world.* 3rd edn. John Hopkins University Press, Baltimore.

Wallace, P., Owen, K., and Thiessen, D. D. (1973). The control and function of maternal scent marking in the Mongolian gerbil. *Physiol. Behav.* **10**, 463–6.

Wandeler, I. and Pilleri, G. (1965). Weitere Beobachtungen zum Verhalten von *Aplodontia rufa* Rafinesque (Rodential, Aplodontoidea) in Gefangenschaft. *Z. Tierpsychol.* **22**, 570–83.

Warshawsky, H. (1963). Investigations on the lateral nasal gland of the rat. *Anat. Rec.* **147**, 443–56.

Weaver, M. S., Whiteside, D. A., Janzen, W. C., Moore, S. A., and Davis, S. F. (1982). A preliminary investigation into the source of odor-cue production. *Bull. psychonom. Soc.* **19**, 284–6.

Welch, J. F. (1953). Formation of urinating 'posts' by house mice (*Mus*) held under restricted conditions. *J. Mammal.* **34**, 502–3.

Whitney, G., Alpern, M., Dizinno, G., and Horowitz, G. (1974). Female odors evoke ultrasounds from male mice. *Anim. Learn. Behav.* **2**, 13-18.

— and Nyby, J. (1979). Cues that elicit ultrasounds from adult male mice. *Am. Zool.* **19**, 457-63.

Whitsett, J. M. (1975). The development of aggressive and marking behavior in intact and castrated male hamsters. *Horm. Behav.* **6**, 47-57.

— Gray, L. E., and Bediz, G. M. (1979). Gonadal hormones and aggression toward juvenile conspecifics in prairie deer mice. *Behav. Ecol. Sociobiol.* **6**, 165-8.

— and Thiessen, D. D. (1972). Sex differences in the control of scent marking behavior in the Mongolian gerbil (*Meriones unguiculatus*). *J. comp. physiol. Psychol.* **78**, 381-5.

Wilhelmsson, M. and Larsson, K. (1973). The development of sexual behavior in anosmic male rats reared under various social conditions. *Physiol. Behav.* **11**, 227-32.

Wilson, S. (1973). The development of social behavior in the vole (*Microtus agrestis*). *Zool. J. Linn. Soc.* **52**, 45-62.

Wilsson, L. (1971). Observations and experiments on the ethology of the European beaver (*Castor fiber* L.). *Viltrevy, Stockh.* **8**, 1-266.

Wittenberger, J. F. (1981). The evolution of mating systems in birds and mammals. In *Handbook of behavioral neurobiology*, Vol. 3. *Social behavior and communication* (ed. P. Marler and J. G. Vandenbergh) pp. 271-349. Plenum Press, New York.

Wolff, P. R. and Powell, A. J. (1979). Urination patterns and estrous cycling in mice. *Behav. neural Biol.* **27**, 379-83.

Woolley, G. M. and Worley, J. (1954). Sexual dimorphism in the Harderian gland of the hamster (*Cricetus auratus*). *Anat. Rec.* **118**, 416-17. (Abstr.)

Wuensch, K. L. (1981). Cross-species fostering and investigation of biologically scented tunnels by house mice (*Mus musculus*). Paper presented at the annual meeting of the Animal Behavior Society, University of Tennessee, Knoxville, June.

— and Cooper, A. J. (1981). Preweaning paternal presence and later aggressiveness in male *Mus musculus. Behav. neural Biol.* **32**, 510-15.

Wylie, N. R. (1968). Neonatal castration modifies responses to sex odors. Proceedings, 76th Annual Convention, American Psychological Association, pp. 291-2.

Yahr, P. (1976). Effects of hormones and lactation on gerbils that seldom scent mark spontaneously. *Physiol. Behav.* **16**, 395-9.

— (1977). Social subordination and scent marking in male Mongolian gerbils (*Meriones unguiculatus*). *Anim. Behav.* **25**, 292-7.

— (1981). Scent marking, sexual behavior and aggression in male gerbils: comparative analysis of endocrine control. *Am. Zool.* **21**, 143-51.

— and Thiessen, D. D. (1975). Estrogen control of scent marking in female Mongolian gerbils (*Meriones unguiculatus*). *Behav. Biol.* **13**, 95-101.

Yamazaki, K., Boyse, E. A., Miké, V., Thaler, H. T., Mathieson, B. J., Abbot, J., Boyse, J., Zayas, Z. A. and Thomas, L. (1976). Control of mating preferences in mice by genes in the major histocompatability complex. *J. exp. Med.* **144**, 1324-35.

— Yamaguchi, M., Baranoski, L., Bard, J., Boyse, E. A., and Thomas, L. (1979). Recognition among mice: evidence from the use of a Y-maze differentially scented with congenic mice of different major histocompatability types. *J. exp. Med.* **150**, 755-60.

— — Boyse, E. A., and Thomas, L. (1980). The major histocompatability complex as a source of odors imparting individuality among mice. In *Chemical signals: vertebrates and aquatic invertebrates* (ed. D. Müller-Schwarze and R. M. Silverstein) pp. 267–73. Plenum Press, New York.

Zahorik, D. M. and Johnston, R. E. (1976). Taste aversions to food flavors and vaginal secretion in golden hamsters. *J. comp. physiol. Psychol.* **90**, 57–66.

10 The rodents III: suborder Sciuromorpha

ZULEYMA TANG HALPIN

Introduction

The importance of chemical communication in the social behaviour of many species of rodents is widely recognized (see Brown 1979 for a review). The Sciuridae is a successful family of rodents which includes approximately 51 genera and 260 species distributed over most regions of the world (Vaughan 1972). Levels of social organization among sciurids are diverse. The family includes not only many species which have a solitary social structure, but also some species, such as the black-tailed prairie dog (*Cynomys ludovicianus*) and the Olympic marmot (*Marmota olympus*), which are among the most social of all rodents (Wilson 1975).

Members of the family Sciuridae can be subdivided into two general groups, ground-dwelling species and arboreal species. Michener (1980) has classified the ground-dwelling sciurids into five levels of sociality:

1. Solitary—males and females maintain separate territories or home-ranges and, outside of the breeding season, social interactions are mostly agonistic. Species included in this level are the eastern chipmunk, *Tamias striatus*; the woodchuck, *Marmota monax*; and Franklin's ground squirrel, *Spermophilus franklinii*.

2. Single-family female kin clusters—males and females maintain separate home-ranges but female offspring remain within or near the home-ranges of their mothers. Social interactions within the female kin groups are friendly, but agonistic interactions predominate between groups. Outside the breeding season, males do not interact with the females, and there is no male parental care. Richardson's ground squirrels, *S. richardsonii*; Belding ground squirrels, *S. beldingi*; and white-tailed prairie dogs, *Cynomys leucurus*, belong in this category.

3. Female kin clusters with male territoriality—males maintain territories which overlap with the home-ranges of several related, adult females and their offspring. Friendly interactions are more common than in the preceding level, but females still defend their natal burrows against all other adults of both sexes. The arctic ground squirrel, *S. parryii* (*undulatus*) and the Columbian ground squirrel, *S. columbianus* are representative of this level.

4. Polygynous harems with male dominance—an adult male maintains a territory within which a number of females and their offspring live. The male is dominant to all other members of the group. This type of social structure is exemplified by the yellow-bellied marmot, *M. flaviventris*.

5. Egalitarian polygynous harems—one male shares a common home range or territory with several females and their offspring, and there is no male dominance.

Interactions within the group are egalitarian and friendly. This highest level of sociality is represented by *M. olympus* and *C. ludovicianus*.

Even among the ground-dwelling sciurids, however, the social organization of many species is not well understood. Ferron (1977), for example, reports that the golden-mantled ground squirrel, *S. lateralis* is less social than the Columbian ground squirrel, the arctic ground squirrel, and Richardson's ground squirrel, but more social than Franklin's ground squirrel. Social interactions are rare but may be either agonistic, friendly, or neutral. There is overlap of home-range use but with temporal separation, so that two animals do not occupy the same area at the same time. Thus, from this description it is not clear if *S. lateralis* should be placed in level (1) or (2) above. Similar problems of classification exist in other terrestrial sciurids.

Among arboreal sciurids (e.g. *Sciurus, Tamiasciurus, Glaucomys*), even less is known about social organization. Both male and female red squirrels, *T. hudsonicus*, and Douglas squirrels (chickarees), *T. douglasii*, are territorial and social interactions are predominantly agonistic, except during the breeding season (Smith 1968). Grey squirrels, *S. carolinensis*; tassel-eared squirrels, *S. aberti*; and fox squirrels, *S. niger*, are non-territorial, have overlapping home-ranges, and show a hierarchical social organization (Benson 1975). Social interactions in these species appear to be primarily agonistic or neutral. The social organization of other species of tree squirrels is not well understood.

In spite of their world-wide distribution, and the diversity of their social structures, relatively little is known about the importance of chemical communication in the social behaviour of sciurid rodents. To date, much of the available information has been limited to descriptive studies of various scent glands and scent-marking behaviours. Experimental studies of function have been rare and most discussions of the functional significance of odours have tended to be highly speculative. This chapter will review the currently available data on chemical communication among sciurid rodents (summarized in Table 10.1).

Distribution of glands

Oral glands

Sebaceous, apocrine sudoriferous, and mucous glands located in the oral lips and oral angle (mouth corners) have been described in a number of sciurid genera and species. Sebaceous glands are typically associated with, and secrete into the hair follicles of, the oral lips and angle. Quay (1965) examined 11 sciurid species, including ground squirrels (*Spermophilus* spp), antelope squirrels (*Ammospermophilus*), chipmunks (*Tamias, Eutamias*), and tree squirrels (*Tamiasciurus, Funambulus, Glaucomys*), and found that they all possess moderately developed sebaceous glands. More recently, Benson (1975, 1980) has described sebaceous glands in the lip skin of the fox squirrel, *Sciurus niger*. Well-developed apocrine sudoriferous glands, which also secrete into the hair follicles of the oral angle, have been found

Table 10.1. Presence of scent glands and scent-marking behaviours in sciurid species

	Oral sebaceous	Oral sudoriferous	Dorsal	Anal	Scent marking	Reference
Ammospermophilus leucurus	+	+				14
Tamias striatus	+	−	−	+		14
Eutamias minimus	+	−				14
E. sibiricus	+				O, Ur	5
E. speciosus	+	−				14
E. townsendii	+	+				14
Tamiasciurus hudsonicus	+	+				14
Sciurus carolinensis		+			Ur	17, 18
S. niger	−	+			O	3, 4
S. vulgaris					O, Ur	6, 8
Sundasciurus tenuis					PR	20
Funambulus palmarum	+	−			O, PR	14, 19
Paraxerus palliatus	+	−				14
P. cepapi	+	+			U, O, PR	21
Glaucomys volans	+	+	+		U, O, PR	21
Spermophilus armatus	+	+	+	+		14
S. columbianus	+	+		+	O, D	11

					References
S. franklinii	+	+	+	–	11
S. lateralis	+	+	+	O, D	7, 11, 14
S. richardsonii	+	+	+	O, D	11, 14
S. tridecemlineatus	+	+	+	O, D	11
S. undulatus				O, D	11
Marmota broweri				O	15
M. flaviventris			+	O	1
M. marmota				O	12, 13
M. monax			+		16
M. olympus				O	2
Cynomys ludovicianus			+		9, 10

+ means that the gland is present; – means the species has been examined for that particular gland and the gland has not been found. Blank spaces mean the species has not been examined for that particular gland. Only species in which glands have been described morphologically or histologically are included. Under scent marking behaviour, O = marking with oral glands, D = dorsal glands, Ur = urine, PR = perineal region. In some species, marking behaviour has been reported but the gland has never been described morphologically or histologically; in such cases the notation for the gland has been left blank but the behaviour is noted.

References: 1, Armitage (1976); 2, Barash (1973); 3, Benson (1975); 4, Benson (1980); 5, Dobroruka (1972); 6, Eibl-Eibesfeldt (1951); 7, Ferron (1977); 8, Figulla (1933); 9, Jones and Plakke (1981); 10, King (1955); 11, Kivett (1975); 12, Koenig (1957); 13, Münch (1958); 14, Quay (1965); 15, Rausch and Rausch (1971); 16, Smith and Hearn (1979); 17, Taylor (1968); 18, Taylor (1977); 19, Thomas and Alexander (1980); 20, Cranbrook (personal communication); 21, Viljoen (personal communication).

in all species of *Spermophilus* which have been studied to date (Quay 1965; Kivett, Murie, and Steiner 1976; Kivett 1978). Quay (1965), however, found no evidence of sudoriferous oral glands in *Tamias*, *Eutamias*, or *Funambulus*, and only poorly developed glands in *Ammospermophilus*, *Tamiasciurus*, and *Glaucomys*. Sudoriferous glands of a moderate degree of development are also present in the oral lips of *Sciurus niger* (Benson 1975, 1980). Mucous glands have been found in at least some species of all sciurid genera which have been examined (Quay 1965; Benson 1975), but there is much variation in their degree of development in different species. Lip plates (thickened areas of lip skin associated with secretory glands of the oral region) have been described in the European red squirrel, *Sciurus vulgaris* (Schumacher 1924, Figulla 1933). S. Viljoen (personal communication) has found no extensive glandular development in the oral angles of the African tree squirrels, *Paraxerus palliatus* and *P. cepapi*, but sebaceous glands do occur.

Kivett *et al*. (1976) and Kivett (1978) have provided detailed descriptions of the apocrine sudoriferous gland (which they designate the oral gland) of several species of ground squirrels (*Spermophilus* spp). Although there is some variation from one species to another, the morphology they describe for *S. columbianus* is similar to that of other sciurid species (Fig. 10.1). The apocrine sudoriferous

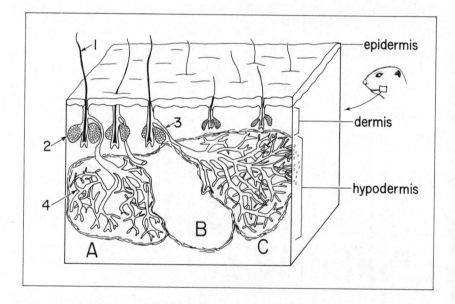

Fig. 10.1. Diagram of the morphology of the oral glands. 1. Hair; 2. Sebaceous gland; 3. Duct or oral gland (apocrine sudoriferous); 4. Branched tubules of oral gland. A, B, C represent the three lobes of the gland. The number of lobes varies in different species. (Based on Kivett *et al*. (1976) and Kivett (1978).)

gland consists of one or more lobes and is located in the lower dermis and hypo-dermis of the oral angle. Each lobe, composed of coiled, branched tubules lined by secretory epithelium, has a separate duct which originates in a small sinus and empties into a hair follicle of the oral angle. The secretions of the oral glands have an oily texture and a strong 'musky' odour which is easily detectable by man. Large sebaceous glands are also associated with each hair follicle.

Dorsal glands

Dorsal-gland fields, consisting of approximately 60 individual apocrine glands, have been described only in the genus *Spermophilus* (Hatt 1926; Kivett *et al.* 1976; Kivett 1978). While there is much variation in the location of the dorsal glands in different species, generally a large dorsal-gland field extends down the back from behind the scapular region. In some species (e.g. *S. columbianus* and *S. undulatus*), the field also extends anteriorly in two narrow bands which terminate in a small concentration of glands ventral to the ears. In *S. lateralis* the ventral gland field is limited to the area between the shoulders, and in *S. richardsonii* there are no narrow bands connecting the glands located on the back to those located ventral to the ear. In all species, dorsal glands are larger in males than in females, and in those species which have glands ventral to the ears, these glands are larger than those located on the back.

Kivett *et al.* (1976) and Kivett (1978) have described the morphology of the dorsal glands. Each dorsal gland is located in the lower dermis and upper hypo-dermis. Branched, coiled tubules lined by secretory epithelium terminate in a collecting sinus. From this sinus, a short, straight duct opens onto the free surface of the skin. The secretions of the dorsal glands are oily to the touch but do not have a strong odour as detectable by man (Kivett 1978).

Anal glands

Anal glands have been reported in all sciurid species which have been examined (Sleggs 1926; King 1955; Kivett *et al.* 1976; Armitage 1976; Kivett 1978; Yahner, Allen, and Peterson 1979). In all cases, three glandular masses, one of which is medial ventral and the other two lateral to the anal opening, are located within the folds of the anal canal. When these glands are in a retracted state, each mass of glandular tissue discharges its secretions into a channel which opens onto the anal wall. Thus, when the animal is in a relaxed state, the three openings of the gland are hidden within the anal canal. However, when the animal is frightened or alarmed, the channels of the three glands can be everted forming three ex-trusible, finger-like papillae which protrude from the anus and which can be made to pulsate (Fig. 10.2(a) and (b)).

Sleggs (1926), Kivett (1978), Smith and Hearn (1979), and Jones and Plakke (1981) have described the morphology of the anal glands. Each glandular mass which makes up an individual anal gland is composed of both sebaceous and apocrine elements (Fig. 10.2(c)). Very large (i.e. 700 μm in diameter in *Spermo-*

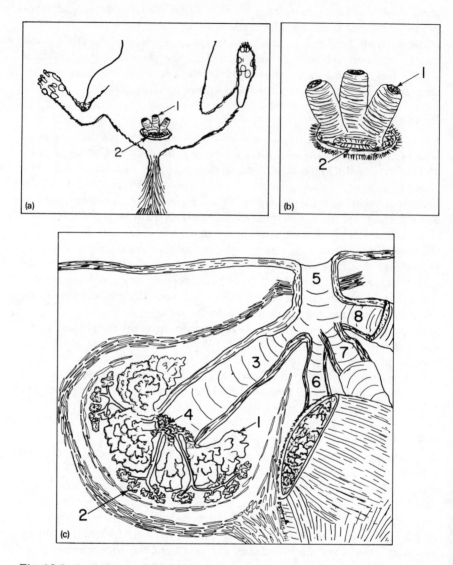

Fig. 10.2. Anal glands. (a) Diagram of a ground squirrel showing the position of the extruded anal glands. 1. Anal gland papilla; 2. Anal aperture. (b) Close-up of the three extruded anal papillae. 1. Anal papilla showing secretory product on tip; 2. Anal aperture. (c) Diagram of the morphology of an anal gland papilla. 1. Sebaceous glands; 2. Apocrine glands; 3. Channel of left gland; 4. Secretory mass; 5. Anal aperture; 6. Anal canal; 7. Channel of median gland; 8. Channel of right gland. (Based on Sleggs (1926).)

philus columbianus; Kivett 1978) sebaceous glands, which occupy most of each glandular mass, are comprised of dense concentrations of lobules located around the base of each channel. The apocrine glands are smaller and less numerous than the sebaceous glands, and are located between the latter and the muscle sheath surrounding each glandular mass. The lobules of each apocrine gland are composed of branched tubules which are lined by secretory epithelium. Ducts from both sebaceous and apocrine elements secrete directly into the channel of each anal gland. Striated muscle layers surround both the sebaceous and apocrine glands and may be responsible for forcing secretions from the extruded papillae. Eversion of the papillae is accomplished by striated muscle layers, while retraction is mediated by unstriated muscle fibres. The strong-smelling, thick, yellowish substance secreted by the anal glands results from a combination of the secretory products of both apocrine and sebaceous elements, but its chemical composition is not known.

Other glands

Kivett (1978) has described sweat glands consisting of coiled, unbranched tubules located in the foot pads of *Spermophilus columbianus*. A single duct from each gland empties on to the surface of each foot-pad. In the Indian palm squirrel, *Funambulus palmarum*, Thomas and Alexander (1980) have reported that there are specialized, integumentary skin glands in the orbital and perineal regions. S. Viljoen (personal communication) reports the presence of well-developed sudoriferous glands in the foot-soles of the African tree squirrels, *Paraxerus palliatus* and *P. cepapi*. Salivary glands which may be important in scent marking are also found in these squirrels. In addition, Viljoen believes that two small glands located on the side of the anus of females, and the bulbo-urethral gland of males, may also play a role in the scent marking behaviours of these species. Figure 10.3 shows the location of the primary scent glands and odour-producing areas of a generalized sciurid.

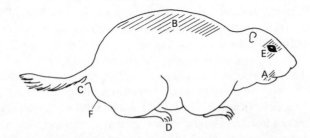

Fig. 10.3. Diagrammatical sketch of a generalized sciurid showing the location of all scent glands and odour-producing areas which have been reported. A = Oral or cheek glands; B = Dorsal glands; C = Anal glands; D = Pedal glands (reported only in *Spermophilus columbianus*); E = Orbital glands (reported only in *Funambulus palmarum*); F = Urine and glands of the perineal region.

Scent-marking behaviour

Ground squirrels

In ground squirrels belonging to the genus *Spermophilus*, scent marking may be active or passive. Steiner (1974), Kivett *et al.* (1976), Ferron (1977), and Kivett (1978) have described three patterns of active scent-marking:

1. Mouth–cheek rubbing. This behaviour involves the squirrel pressing its chin against the substrate then rapidly moving the head forward while at the same time twisting it sideways. As a result of this movement, the lips and mouth corners (oral glands) of the animal are pressed and rubbed against the surface to be marked. This behaviour is usually performed against a flat surface or against a sharp protruding object, such as a rock or stick, and it is generally preceded and followed by sniffing of the substrate.

2. Twist marking. The squirrel performs a rapid, forward, spiral or twisting movement of the body during which the oral glands, cheek glands (extension of dorsal glands as described above), and dorsal glands are rubbed against the substrate in quick succession. This behaviour is complex and stereotyped and is most frequently performed against vertical or sloping surfaces, or against objects such as rocks, tree stumps, or mounds of dirt. The behaviour frequently starts and ends with sniffing of the substrate, and in some cases the animal may vigorously scratch or claw at the substrate before marking it.

3. Ventral drag. The squirrel presses the hindquarters against the substrate then, while retaining this position, moves forward slowly, dragging and rubbing the ventral region against the substrate. This behaviour, which is generally preceded by very thorough sniffing of the substrate, results in the deposition of a liquid trail with a strong musky odour (probably urine but possibly anal-gland secretions—see Steiner (1974) and Kivett (1975). Although Steiner (1974) refers to this behaviour as an anal drag, J. Murie (personal communication) believes that in *S. columbianus* there is no distinctive behaviour pattern which can be classified as an anal drag because the squirrels do not specifically press or rub the anogenital regional against the substrate. He suggests that, if scent is deposited, it is likely to be done so passively (see below), as a result of a comfort behaviour. Furthermore, while urine is often deposited during the ventral drag, according to Murie it is less likely that anal-gland secretions are also involved.

Passive scent-marking is also common and occurs whenever scent transfer results from a squirrel's normal routine activities rather than from an active, specialized behaviour (Kivett *et al.* 1976). For example, as a squirrel runs along a burrow or under a low, overhanging object, such as a branch or rock outcropping, its back and shoulder (dorsal gland field) may passively come into contact with the substrate, leaving behind secretions. Self-grooming behaviours and comfort movements, such as dust-bathing and rubbing the body against the substrate, may also result in the passive deposition of scents, in self-marking, or in group-scent sharing (Steiner 1974). In fact, the potential for passive scent marking appears to

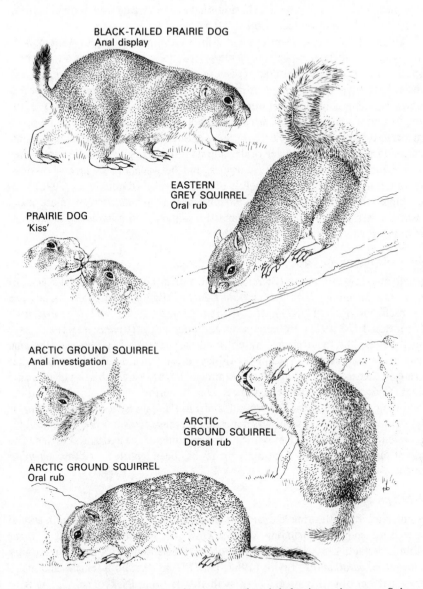

BLACK-TAILED PRAIRIE DOG
Anal display

EASTERN
GREY SQUIRREL
Oral rub

PRAIRIE DOG
'Kiss'

ARCTIC GROUND SQUIRREL
Anal investigation

ARCTIC
GROUND SQUIRREL
Dorsal rub

ARCTIC GROUND SQUIRREL
Oral rub

Fig. 10.4. Scent-marking and other scent-related behaviours in some Sciuromorph rodents. See text for explanation.

be so widespread in these squirrels, that Kivett *et al*. (1976) consider that most complex, active scent-marking behaviours have been derived from simpler patterns of passive scent transfer (see also Fig. 10.5).

Although the development of scent glands and the frequency of scent marking are variable in different species of *Spermophilus* (Kivett *et al*.1976), among those species which show sexual dimorphism (e.g. *S. columbianus* and *S. undulatus*), males have larger glands and scent mark more than females. Among other species, however, there is little sexual dimorphism either in gland development or in the frequency of marking (e.g. *S. lateralis*), and a few species show little or no scent-marking behaviours at all (e.g. *S. franklinii*). In general, there appears to be some correlation between the degree of development of scent glands and the frequency with which active scent marking occurs, and between the frequency of scent marking and the degree of sociality of each species (Kivett *et al*. 1976). For example, the more social *S. columbianus* and *S. undulatus* scent mark more than the less social *S. richardsonii*, and the solitary *S. franklinii* shows no scent marking at all.

Marmots and prairie dogs

Scent marking with cheek (oral?) glands has been reported for several species of marmots, including *Marmota marmota* (Koenig 1957, Münch 1958), *M. broweri* (Rausch and Rausch 1971), *M. olympus* (Barash 1973), and *M. flaviventris* (Armitage 1974, 1976). Both marmots and prairie dogs (*Cynomys* spp) also have anal glands, but these glands are not used for scent marking. Instead, the anal glands are used as part of agonistic displays during territorial disputes and other aggressive encounters (King 1955; Armitage 1976). Rarely, black-tailed prairie dogs (*Cynomys ludovicianus*) have been seen to perform behaviour similar to the anal drag described above for ground squirrels (Halpin, personal observation). It is not clear, however, if scent was deposited on these occasions or if the behaviour represents only a comfort movement or scratching of the anogenital region. The use of faeces or urine for scent marking has not been reported for either *Marmota* or *Cynomys*.

Tree squirrels

Scent marking with urine and/or with cheek glands has been described in several species of squirrels, including *Sciurus vulgaris*, *S. carolinensis*, and *S. niger* (Eibl-Eibesfeldt 1951, 1958; Taylor 1968, 1977; Benson 1975, 1980). In the grey squirrel, *S. carolinensis*, Taylor (1968, 1977) found that both females and males gnaw patches of bark on trees and spray them with urine. Peaks of marking activity occur both during and outside of the breeding seasons. Benson (1975, 1980) has described a scent-marking behaviour in the fox squirrel (*S. niger*) which involves wiping or rubbing the lips and face over the substrate. Generally, both sides of the face are wiped in quick succession. This behaviour, which probably involves the oral glands, is performed by both males and females and by squirrels of all ages. A more detailed analysis revealed that older and more dominant squirrels face wiped more than younger, lower-ranking individuals, but that there was no significant difference in the frequency of marking between males and females.

The palm squirrel, *Funambulus palmarum*, actively marks the substrate (ground or branch) with oral, chin, and perineal glands (Thomas and Alexander 1980). Marking behaviour is sexually dimorphic with males exhibiting a higher frequency of marking than females. Chinning and marking with oral glands is performed most frequently in locations which had been previously occupied by other squirrels. Perineal marking, which includes both an anal drag and scrotal marking, is associated with the release of urine and may involve the deposition of a urine trail (Thomas and Alexander 1980). Cranbrook (personal communication) has reported an apparent case of scent marking in *Sundasciurus tenuis*. An unsexed individual was seen to move about in a tree apparently rubbing its inguinal region on small branches. It is not clear if this behaviour involved scent-marking or was only a comfort movement. In the African squirrels, *Paraxerus palliatus* and *P. cepapi*, the salivary glands are used for scent-marking by means of mouthwiping (S. Viljoen, personal communication). As a result of mouthwiping, the sides of the face often become thoroughly wet with saliva. In addition, these squirrels also anal drag, leaving behind an odorous, wet smear. This smear most probably contains secretions from the perianal glands, as well as urine.

Many species of tree squirrels show mating behaviours which involve the prolonged following of oestrous females by males. During such chases, numerous males are attracted to and aggregate near the females. Layne (1954) described the swelling and other changes which occur in the vagina of oestrous female red squirrels (*Tamiasciurus hudsonicus*) and suggested that anal gland secretions may be involved in communicating the oestrous condition of the females. More recently, Smith (1968) has suggested that in *T. hudsonicus* and in *T. douglasii* vaginal secretions change as oestrus approaches and that males use these vaginal secretions to determine the day on which a female comes into oestrus. There have, however, been no experimental studies on the vaginal secretions of *Tamiasciurus* females or on the responses of males to female odours.

Chipmunks

Scent-marking behaviours have been described only in the Siberian chipmunk, *Eutamias (Tamias) sibiricus*. In this species, Dobroruka (1972) reports that scent marking with urine is the most frequently observed form of marking but that scent marking with cheek (oral) glands is also known to occur. Urine marking is performed by both male and female chipmunks, but the sites used for marking are different in the two sexes. Males drop urine on prominent places (e.g. stones, branches) within their home-ranges, and sometimes also mark by rubbing the urine-wet scrotum along branches. Females, on the other hand, generally urinate on the ground at constant marking sites, but may also drop urine on places that have been marked by males. Marking places are not individual but, instead, may be used by all male and female chipmunks in a population.

The mating behaviour of the eastern chipmunk (*Tamias striatus*) suggests that females may produce a vaginal secretion which signals oestrous condition to the males (Yahner 1978; Elliot 1978). However, specialized scent-marking behaviours have never been observed in this species (Yahner 1978).

Other scent-related behaviour

In addition to active and passive scent marking, sciurid rodents engage in a variety of other activities which may be related to scent production. Some of these activities may not involve communication with conspecifics, but may represent evolutionary antecedents of active scent marking (see Fig. 10.5).

Scent-related behaviours have been studied most extensively in the genus *Spermophilus* (Steiner 1974; Kivett *et al.* 1976; Ferron 1977). In these squirrels, both self- and allogrooming may result in the transfer of scents from the oral and pedal glands. During self-grooming, for example, the fore-paws are usually wiped over the mouth corners before being used to groom the rest of the body. In addition, squirrels often nibble and scratch their fur as a part of grooming. Both of these behaviours may spread scents from the oral glands (and possibly pedal glands) to other areas of the body, thereby resulting in self-marking or self-anointing (self-anointing is known to occur in other mammals; see for example Burton (1957) and Brockie (1976)). During allogrooming the groomer nibbles, pats, and scratches the fur of the conspecific. Thus, scent transfer from one animal to another may also occur as a result of these activities. Since patterns of self- and allogrooming are similar in other sciurid species (e.g. self-grooming in *Tamias striatus*; Halpin, personal observation), it is possible that self-marking, as well as scent transfer to conspecifics, may be common phenomena among sciurids. It should be noted, however, that the transfer of scents as a result of grooming has never been confirmed experimentally for any sciurid species.

Odour-producing areas appear to be extremely important in the social interactions of sciurids. Frequently, for example, amicable and investigatory behaviours (e.g. nibbling, licking, and sniffing) are directed towards the scent-producing body areas such as the mouth corners, cheeks, and anogenital region (see Steiner 1974). In addition, hostile behaviours (e.g. biting) are also directed to these same areas. The interest which sciurids show in the scent-bearing areas of conspecifics suggests that the odours produced by these areas may be related to individual or group identification (see pp. 474-5).

Another scent-related behaviour shown by sciurids is the food begging and oral investigation seen between mothers and their young and also between adults of social species. Muul (1970) and Ferron (1974) have reported that food marking occurs in both flying squirrels (*Glaucomys* spp) and in red squirrels (*Tamiasciurus hudsonicus*). It is likely that in other sciurids also, food marking may occur during

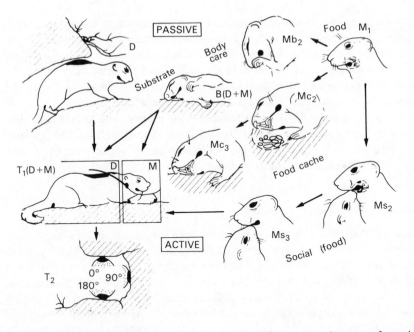

Fig. 10.5. Hypothetical derivation of scent-transfer system in some fossorial Sciuridae (diagrammatical). Original situation: passive transfer from strategically located dorsal (D) gland, burrow restricted. Later: transfer to larger areas and more variable landmarks favoured more active, oriented behavior = twist-marking (T_1(D + M)); combines scent transfer from dorsal gland (D), which is more substrate oriented, and from mouth-corner glands (M), which is more conspecific oriented. Three major derivation lines for mouth-gland marking (considering food marking M_1 as primordial, since mouth-corner glands are strategically located for this) are as follows: first, *body care line* (Mb_2)— mouth wiping leading to self-anointing, allogrooming (not represented), and mouth + body-rubbing substrate (B(D + M)); second, *food caching + marking* (Mc_2), and later substrate marking only (Mc_3); third, *social-feeding line* (Ms_2 to Ms_3), with scent transferred to food-begging young or conspecific (Ms_2), later ritualized into 'greeting'-identification (Ms_2), also involving scent sharing. T_2 shows flexibility of active-oriented marking of variable substrate. Scent-gland location indicated in black (larger than scale). (From Kivett *et al.* (1976).)

the storage of food in cheek pouches and during food caching. Given the position of the cheek pouches in the mouth corners, Steiner (1975), Kivett *et al.* (1976), and Ferron (1977) suggest that, in *Spermophilus*, food begging may also involve olfactory investigation and scent transfer, and that this behaviour may have given rise to the nose to mouth greeting which is common to all social sciurids (but note that not all sciurids have cheek pouches; also see discussion on nose-to-mouth greeting on pages 472-3). Furthermore, they also suggest that mouth–cheek rubbing may have been derived directly from marking the substrate and food at food caches with the oral glands.

In summary, scent-related behaviours include substrate-oriented behaviours, conspecific-oriented behaviours, and self-oriented behaviours (Kivett *et al*. 1976). In addition, these behaviours may be active as in the case of scent-marking or passive, as a by-product of the general activities of the animal. Unfortunately, since only the genus *Spermophilus* has been studied in detail, and since the discussions of scent-related behaviours (other than active scent marking) have tended to be speculative, it is not possible at the present time to draw definitive conclusions which can be applied to all sciurids. The studies on *Spermophilus* (e.g. Steiner 1974; Kivett 1975, 1978; Kivett *et al*. 1976), however, have been highly suggestive and have highlighted the need for additional studies.

Hormonal control of scent marking

The hormonal control of scent marking has been investigated in only one species, *Spermophilus columbianus*. Kivett (1975) found that, in castrated males, injections with testosterone propionate increased the size and secretory activity of both oral and dorsal glands, and also the frequency of scent marking with these glands. These results are compatible with the observation that among many (but not all) sciurid species scent-marking behaviours increase during the reproductive period (Steiner 1974).

The role of hormones in gland development and scent-marking behaviours of female sciurids has never been examined experimentally. However, the finding that female scent-marking often increases during pregnancy and lactation (e.g. *Spermophilus columbianus*, Kivett 1975) suggests that progesterone, oestrogen, or possibly prolactin may be the hormones responsible.

Functional significance of biological odours

Greeting behaviour

Almost all social or gregarious sciurids, particularly among the ground squirrels (*Spermophilus*), marmots (*Marmota*), and prairie dogs (*Cynomys*), perform a greeting behaviour in which one animal approaches and sniffs the mouth corners of another (see, for example, King 1955; Armitage 1962; Balph and Stokes 1963; Quanstrom 1971; Barash 1973; Watton and Keenleyside 1974; also Steiner 1975 for a review). Sometimes the olfactory investigation of the mouth areas is mutual, and both animals may open their mouths, apparently to facilitate investigation. More detailed observations (Steiner 1975) revealed that in many cases actual contact is made between the nose of the sniffer and the mouth corners of its partner. Furthermore, after the initial contact is made, the sniffer frequently pushes its nose and muzzle further into the mouth corners of the partner. It is almost certain that the oral glands located in the oral angle are involved in this behaviour.

Among social sciurids, nose-to-mouth greetings are performed by all individuals of all ages, but appear to be more common between well-acquainted members of the same social group. In some species (e.g. *Spermophilus columbianus, S. undulatus, Marmota olympus*), the more dominant individuals appear to initiate greetings more often than subordinate individuals (Steiner 1975). Between members of different social groups, greetings are rare and when they do occur they are usually performed hesitantly and are likely to lead to agonistic behaviours (i.e. avoidance or attack). The frequency of nose to mouth greetings appears to be related to degree of sociality; greetings are much more common in social/ gregarious species (e.g. *Cynomys ludovicianus, Spermophilus columbianus*) than in the less social sciurids (e.g. *C. leucurus, S. franklinii*) (Steiner 1975).

The functional significance of nose-to-mouth greetings is not completely understood. However, the prevalence of this behaviour among social sciurids and the contexts in which it occurs, suggest that the identification of individuals or of social groups may be a primary function (Steiner 1975; but see also Watton and Keenleyside 1974). Steiner (1975) has also suggested that during the investigation of the mouth corners scent sharing may occur. The sniffer, for example, may pick up on its muzzle the scent of its partner; this scent, as well as the sniffer's own scent, could then be passed on to the next animal which is investigated. Thus, if all group members engage in nose-to-mouth greetings, the oral gland scents of individual group members can be passed around and mixed, thereby creating a unique group odour. This idea, while intriguing and intuitively appealing, needs to be examined experimentally.

Correlational studies

Sciurid patterns of scent marking are so varied, and so few species have been studied, that it is difficult to draw any general conclusions regarding the function of social odours in this family. Correlational studies have concentrated primarily on the relationship between scent marking, dominance, and aggression; most of these studies have shown that in a number of species (e.g. *Spermophilus columbianus, S. undulatus, Sciurus niger, Marmota flaviventris*) scent marking with oral (cheek) and/or dorsal glands is more common among dominant, more aggressive individuals and among territory owners (Steiner 1974; Kivett 1975; Kivett *et al.* 1976; Benson 1975; Armitage 1976). This association between aggression and scent-marking is further strengthened by the observation that scent-marking frequently occurs either immediately before or immediately after aggressive interactions (Steiner 1974; Armitage 1976). Furthermore, levels of marking are generally higher during the breeding season when animals are more likely to be aggressive and/or to show territorial behaviour (Steiner 1974; Kivett 1975). It should also be noted, however, that some species (e.g. *Sciurus carolinensis*; Taylor 1968, 1977) show peaks of marking even outside the reproductive season, and most species show some marking activity throughout the year.

Halpin (1980) has emphasized the importance of distinguishing between message and function when discussing the behavioural significance of social odours. Although correlational studies do not allow us to reach definitive conclusions, they do provide evidence that the odours from oral and dorsal glands may carry information on individual, sex, age, and species identity (see also Kivett 1975). The correlation between aggression and scent-marking further suggests that the odours of these glands serve a threat function. This would be in accord with the view proposed by Ralls (1971) that, in many cases, mammalian odours and scent marking may function primarily as threat displays during aggressive encounters. Furthermore, scent marking may serve not only as a threat to conspecifics but, as Ewer (1968) has suggested, as a means of providing reassurance to the marker by saturating its environment with its own familiar odour. In the case of the more social sciurids, a group odour or the odour of the dominant animal within the group may have this same effect of reassuring group members.

It is also possible that different glands present within the same species could produce combinations of messages, resulting in more complex functions. As an example, Kivett (1975) suggests that in *Spermophilus columbianus*, the oral glands may have a threat function while the dorsal glands may convey information on occupancy (i.e. this area or burrow is occupied); twist marking might then have the dual function of alerting conspecifics that the area is occupied by another animal, and of threatening the conspecific if it enters into the occupied area. This idea is interesting and deserves further study.

Experimental studies

There have been few experimental studies on the functional significance of scent-marking and/or biological odours among sciurids. Haslett (1973) found that secretions from the anal glands of either male or female eastern woodchucks (*Marmota monax*) inhibit the activity of other woodchucks. Furthermore, in paired homosexual and heterosexual encounters, those animals which were least likely to extrude the anal papillae were, generally, more likely to become dominant. As a result of these findings, Haslett (1973) concluded that the anal glands of woodchucks produce an alarm or fear pheromone. Previous, correlational studies on other species had determined that sciurids are most likely to release anal gland scents either when they are frightened or alarmed (Steiner 1974) or during agonistic encounters (King 1955; Balph and Stokes 1963; Armitage 1976). In view of Haslett's (1973) findings, it seems crucial to determine whether during hostile encounters it is the dominant or the subordinate animal (or perhaps both) who releases anal-gland scent. The use of anal glands by the dominant individual would suggest a threat function, while use by the subordinate would be consistent with a fear or alarm function. The complex and stereotyped 'anal displays' used by some sciurids during agonistic encounters (e.g. *Cynomys ludovicianus*; King 1955) also need to be examined in more detail to determine the role of the anal glands and the patterns of release of anal-gland scents.

In an interesting field experiment, Harris and Murie (1979) studied the responses of male and female Columbian ground squirrels (*Spermophilus columbianus*) to the oral-gland secretions of male conspecifics. When presented with plastic cubes containing oral-gland scents from other squirrels, both male and female subjects spent more time investigating the odours of strange and neighbouring males than their own odours (odour of the resident male in the case of female subjects), and more time investigating the odours of strange males than of neighbouring males. These results are compatible with the hypothesis that the oral-gland scents of ground squirrels contain information on individual identity. Thus, the secretions of the oral glands may function in individual recognition and/or in the recognition of familiar versus unfamiliar conspecifics (see Barrows, Bell, and Michener (1975) and Halpin (1980) for a discussion of the need to distinguish between individual recognition and recognition of familiar versus unfamiliar animals).

In a more recent study, Keevin, Halpin, and McCurdy (1981) have determined that male and female eastern chipmunks (*Tamias striatus*) are able to distinguish both individual and sex-specific differences in the odours of conspecifics. Odours were presented in soiled bedding materials (wood shavings and cotton) which contained urine, faeces, and also probably secretions from the oral and/or anal glands. To test for the ability to discriminate individual differences in conspecific odours, test animals were allowed to habituate to the odour of one conspecific (odour A) and then simultaneously presented with odour from a second conspecific (odour B). The amount of time that the test animals spent sniffing the two odours was recorded; it was predicted that if the two odours were different the test animal should spend more time investigating odour B. Figure 10.6 shows that the test animals did, in fact, spend significantly more time investigating the second, unfamiliar odour. In the sexual recognition study, a simple preference test was used to determine if male and female subjects could distinguish between the odours of male versus female conspecifics. Table 10.2 shows that eastern chipmunks have the ability to distinguish sex-specific differences in conspecific odours; male subjects spent more time investigating female odours while female subjects spent more time investigating male odours. Thus, the results of this study suggest that the biological odours of the eastern chipmunk may function in individual recognition (or in the recognition of familiar versus unfamiliar animals) and in sexual recognition.

Although *Tamias striatus* is the only chipmunk species which has been studied experimentally, Dobroruka (1972) suggests that urine marking in the Siberian chipmunk (*Eutamias sibiricus*) is related to courtship behaviours. The role of olfactory cues in the courtship and mating behaviours of this and other chipmunk species merits further study.

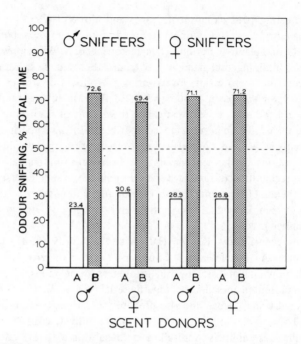

Fig. 10.6. Differences in the percentage of total time spent by male and female eastern chipmunks investigating male and female conspecific odours A and B.

Table 10.2. Number of male and female eastern chipmunks showing a preference for male versus female odours.

| | Experimental subjects | |
	♂	♀
No. showing preference for ♂	1	13
No. showing preference for ♀	14	2
Total	15	15

Conclusions

In spite of the evolutionary and taxonomic importance of the family Sciuridae, relatively little is known about the significance of biological odours in the social behaviours of these rodents. The paucity of information is particularly obvious when one considers that, even after all studies related to chemical communication are taken into account, there are data available on only ten genera and on a relatively small number of species (see Table 10.1). Scent glands are present in all sciurids which have been examined, and scent-marking behaviours are pre-valent in almost all social and gregarious species. In spite of this, even in those species which have been most extensively studied and in which complex scent glands and scent marking behaviours have been described (e.g. *Spermophilus* spp), our understanding of the significance of social odours is still only partial, at best. Clearly, additional, more incisive, and more critical studies on a broader variety of sciurid species are essential if we are adequately to understand the importance of chemical communication in this family.

Acknowledgements

I thank Thomas M. Keevin for first arousing my interest in sciurid chemical communication and for valuable assistance in locating references. Margaret Harris and Jan Murie generously provided unpublished data on *Spermophilus colum-bianus*. David A. Chisholm did the drawing for the figures. Kivett and Murie graciously provided permission for the use of their illustrations. The University of Missouri, St. Louis provided financial support.

References

Armitage, K. B. (1962). Social behaviour of a colony of the yellow-bellied marmot (*Marmota flaviventris*). *Anim. Behav.* **10**, 319-31.
— (1974). Male behaviour and territoriality in the yellow-bellied marmot. *J. Zool., Lond.* **172**, 233-65.
— (1976). Scent marking by yellow-bellied marmots. *J. Mammal.* **57**, 583-4.
Balph, D. F. and Stokes, A. W. (1963). On the ethology of a population of Unita ground squirrels. *Am. Midl. Nat.* **69**, 106-26.
Barash, D. P. (1973). The social biology of the Olympic marmot. *Anim. Behav. Monogr.* **6**, 171-245.
Barrows, E. M., Bell, W. J., and Michener, C. D. (1975). Individual odor differences and their social functions in insects. *Proc. natn. Acad. Sci. USA* **72**, 2824-8.
Benson, B. N. (1975). Dominance relationships, mating behavior and scent marking in fox squirrels (*Sciurus niger rufiventer*) Ph.D. thesis, Southern Illinois University, Carbondale.
— (1980). Dominance relationships, mating behavior, and scent marking in fox squirrels (*Sciurus niger*). *Mammalia* **44**, 143-60.
Brockie, R. (1976). Self-anointing by wild hedgehogs, *Erinaceus europaeus*, New Zealand. *Anim. Behav.* **24**, 68-71.
Brown, R. E. (1979). Mammalian social odors: a critical review. *Adv. Study Behav.* **10**, 103-62.
Burton, M. (1957). Hedgehog self-anointing. *Proc. zool. Soc. Lond.* **129**, 452-3.
Dobroruka, L. J. (1972). Scent marking and courtship in Siberian chipmunk, *Tamias sibiricus lineatus* (Siebold, 1824), with notes on the taxonomic relations
— (1958). Das Verhalten der Nagetiere. *Handb. Zool.* **8** (10), 13, 1-88.
Eibl-Eibesfeldt, I. (1951). Zur Fortpflanzungsbiologie und Jugendentwicklung des Eichhornchens. *Z. Tierpsychol.* **8**, 370-400.
Elliott, L. (1978). Social behavior and foraging ecology of the Eastern chipmunk (*Tamias striatus*) in the Adirondack Mountains. *Smithson. Contrib. Zool.* **265**, 1-107.

Ewer, R. F. (1968). *Ethology of mammals.* Logos Press, London.

Ferron, J. (1974). Etude ethologique d l'ecureuil roux d'Amerique, *Tamaiasciurus hudsonicus.* Ph.D. thesis, Universite de Montreal, Montreal, Quebec.

— (1977). Le comportement de marquage chez le spermophile a mante dorée (*Spermophilus lateralis*). *Naturaliste Can.* 104, 407–18.

Figulla, R. (1933). Untersuchungen über die Existenz und den Bau einer 'Lippen-platte' (Schumacher). *Anat. Anz.* 76, 65–79.

Halpin, Z. T. (1980). Individual odors and individual recognition: review and commentary. *Biol. Behav.* 5, 233–48.

Harris, M. A. and Murie, J. O. (1979). Responses to scent in Columbian ground squirrels. (Abstr.) *59th Ann. Meet. Am. Soc. Mammal.*

Haslett, G. W. (1973). The significance of anal scent marking in the Eastern woodchuck. (Abstr.) *Bull. ecol. Soc. Am.* 54, 43–4.

Hatt, R. T. (1926). A new dorsal gland in ground squirrels, *Callospermophilus*, with a note on its anal gland. *J. Morph. Physiol.* 42, 441–51.

Jones, T. R. and Plakke, R. K. (1981). The histology and histochemistry of the perianal scent gland of the reproductively quiescent black-tailed prairie dog (*Cynomys ludovicianus*). *J. Mammal.* 62, 362–8.

Keevin, T. M., Halpin, Z. T., and McCurdy, N. (1981). Individual and sex-specific odors in male and female eastern chipmunks (*Tamias striatus*). *Biol. Behav.* 6, 329–38.

King, J. A. (1955). Social behavior, social organization, and population dynamics in a black-tailed prairie dog town in the Black Hills of South Dakota. *Contr. Lab. vertebr. Biol. Univ. Mich.* 67, 1–123.

Kivett, V. K. (1975). Variations in integumentary gland activity and scent marking in Columbian ground squirrels (*Spermophilus c. columbianus*). Ph.D. thesis, University of Alberta, Edmonton.

— (1978). Integumentary glands of Columbian ground squirrels (*Spermophilus columbianus*): Sciuridae. *Can. J. Zool.* 56, 374–81.

— Murie, J. O., and Steiner, A. L. (1976). A comparative study of scent-gland location and related behavior in some northwestern nearctic ground squirrel species (Sciuridae): an evolutionary approach. *Can. J. Zool.* 54, 1294–306.

Koenig, L. (1957). Beobachtungen über Reiermarkierung sowie Droh-Kampf-und Abwehrverhalten des Murmeltieres (*Marmota marmota* L.). *Z. Tier-psychol.* 14, 510–21.

Layne, J. N. (1954). The biology of the red squirrel, *Tamiasciurus hudsonicus loquax* (Bangs) in central New York. *Ecol. Monogr.* 24, 227–67.

Michener, G. (1980). Kin identification, matriarchies, and the evolution of sociality in ground-dwelling sciurids. Paper presented at Symposium on Mammal Behavior, National Zoological Park. Washington, D.C. August 1980.

Münch, H. (1958). Zur Okologie und Psychologie von *Marmota m. marmota. Z. Säugetierk.* 23, 129–38.

Muul, I. (1970). Day length and food caches. In *Field studies in natural history.* Van Nostrand Reinhold, New York.

Quanstrom, W. R. (1971). Behavior of Richardson's ground squirrel, *Spermophilus richardsonii. Anim. Behav.* 19, 646–52.

Quay, W. B. (1965). Comparative survey of the sebaceous and sudoriferous glands of the oral lips and angle in rodents. *J. Mammal.* 46, 23–37.

Ralls, K. (1971). Mammalian scent marking. *Science, NY* 171, 443–9.

Rausch, R. L. and Rausch, V. R. (1971). The somatic chromosomes of some North American marmots (Sciuridae), with remarks on the relationships of *Marmota broweri* Hall and Gilmore. *Mammalia* 35, 85–101.

Schumacher, S. (1924). Eine 'Lippenplatte' beim Eichhornchen (*Sciurus vulgaris* L.). *Anat. Anz.* 58, 75–80.

Sleggs, G. F. (1926). The adult anatomy and histology of the anal glands of the Richardson ground squirrel, *Citellus richardsonii* Sabine. *Anat. Rec.* 32, 1–44.

Smith, C. C. (1968). The adaptive nature of social organization in the genus of tree squirrels *Tamiasciurus*. *Ecol. Monogr.* 38, 31–63.

Smith, J. D. and Hearn, G. W. (1979). Ultrastructure of the apocrine-sebaceous anal scent gland of the woodchuck, *Marmota monax*: evidence for apocrine and merocrine secretion by a single cell type. *Anat. Rec.* 193, 269–92.

Steiner, A. L. (1974). Body-rubbing, marking, and other scent-related behavior in some ground squirrels (Sciuridae), a descriptive study. *Can. J. Zool.* 52, 889–906.

— (1975). 'Greeting' behavior in some sciuridae, from an ontogenetic, evolutionary and socio-behavioral perspective. *Naturaliste can.* 102, 737–51.

Taylor, J. C. (1968). The use of marking points by grey squirrels. *J. Zool., Lond.* 155, 246–7.

— (1977). The frequency of Grey squirrel (*Sciurus carolinensis*) communication by the use of scent marking points. *J. Zool., Lond.* 183, 543–5.

Thomas, N. and Alexander, K. M. (1980). A pilot report on the scent marking of the Indian palm squirrel, *Funambulus palmarum*. *Proc. Symp. environ. Biol. Trivandrum*, pp. 166–72.

Vaughan, T. A. (1972). *Mammalogy*. Saunders, Philadelphia.

Watton, D. G. and Keenleyside, M. H. A. (1974). Social behaviour of the arctic ground squirrel, *Spermophilus undulatus*. *Behaviour* 50, 77–99.

Wilson, E. O. (1975). *Sociobiology: the new synthesis*. Belknap Press/Harvard University Press, Cambridge, Mass.

Yahner, R. H. (1978). The adaptive nature of the social system and behavior of the eastern chipmunk, *Tamias striatus*. *Behav. Ecol. Sociol.* 3, 397–427.

— Allen, B. L., and Peterson, W. J. (1979). Dorsal and anal glands in the eastern chipmunk, *Tamias striatus*. *Ohio J. Sci.* 79, 40–3.

11 The rodents IV: suborder Hystricomorpha

DAVID W. MACDONALD

Introduction

The suborder Hystricomorpha includes a diversity of intriguing rodents ranging from the Old World porcupines to inconspicuous forest-dwelling spiny rats and semi aquatic capybaras which, at 60 kg or more, are the largest rodents in the world. Apart from anatomical distinctions, the members of the suborder differ from other rodents in their behaviour. and especially in their reproductive biology (Kleiman 1974; Weir 1974; Kleiman, Eisenberg, and Maliniak 1979). Indeed, the South American hystericomorphs (often distinguished from the Old World forms as a separate suborder, the Caviomorpha) are said to have radiated further than any other New World mammals (Hershkovitz 1969). An aspect of that radiation of especial interest with respect to social communication is the parallel evolution between the ungulates of the African grasslands and the larger rodents, such as maras, *Dolichotis patagonum*, and capybaras, *Hydrochoerus hydrochaeris*, of the South American plains (Dubost 1968). Furthermore, Kleiman (1974) suggested that the hystricomorphs tend to a gregarious lifestyle, and recent studies, for example, of the capybara (Macdonald 1981; Schaller and Crawshaw 1981) confirm the compexity of the social behaviour of at least some species. Following Corbet and Hill (1980) the Hystricomorpha include 16 families, 56 genera, and 171 species and so it is sad to confess that there is nothing more than fragmentary information on the social odours of many of these species in the wild (Table 11.1). However, the domestic guinea pig, *Cavia porcellus*, is a member of the family Caviidae and the olfactory communication of this species has been studied extensively under laboratory conditions. Indeed, it was research into the social odours of this species that led Berüter, Beauchamp, and Muetterties (1973) to conclude that '. . . these studies establish a complexity in chemical signals that has not been previously documented for any other species'.

The location of possible scent glands has only been described for a small minority of hystricomorph species, but, to the extent that one may generalize, the typical arrangement involves paired anal sacs lying on either side of the anus and opening via ducts just within or immediately outside the anal aperture. In some species glands open above (e.g. mara (Family Caviidae) and agouti (family Dasyproctidae)), or beside (e.g. paca (family Dasyproctidae)) the anus, but in others such as capybaras, the glands are somewhat ventral to the anus, and in some cases they feed into, or lie close by, an additional highly glandular area lying between the anus and genitalia. In the male guinea pig, *Cavia aperea*, this region, the so-called perineal gland, is a capacious pocket between the anus and genitals. Porcupines of the Old World genera *Hystrix* and *Atherura* (family Hystricidae) are said to have a pair of large, tumid anal glands opening by a slit-like orifice just within the anus and equally developed in both sexes (although Mohr (1965) disagrees for *Hystrix*, for which she describes larger anal glands among males). However, among New World porcupines of the family Erethizontidae, Thullberg (in Schaffer 1940) could not find any trace of anal glands opening at the edge of the anus, but pointed to the existence of a depression between the anus and the prepuce. The early descriptions of hystricomorphs' glands were by Thullberg and by Pocock (1922), both of whom were reviewed in Schaffer (1940). They mention median glands or ducts for degus (family Octodontidae), hutias and coypus (family Capromyidae), two genera of spiny rats, *Echimys* and *Kannabateomys* (family Echimyidae), and for chinchilla rats (family Abrocomyidae). From these descriptions my understanding of the evolution of these glands is that the median or perineal condition has derived from ancestral paired anal glands. For example, Pocock's drawing on the perineal region of the plains viscacha, *Lagostomus* (family Chinchillidae), shows the apertures of two anal glands opening ventral to the anus and at the base of a triangle of glandular skin which runs from their openings to the genitalia (although Weir, pers. comm., has not seen these). Indeed Pocock (1922) conjectures that the perineally situated anal gland which he describes for the tree porcupine, *Coendou* (family Erithizontidae) might correspond to the joined and partially degenerated paired anal glands of, for example, *Hystrix*.

Aside from these scanty descriptions there is little basic information on the other glandular endowments of hystricomorphs: male cuis, *Galea musteloides*, have a chin gland, the secretions of which are rubbed onto the females' rumps and may stimulate oestrus (Weir 1971, 1973). The paca, *Cuniculus*, may have glands in the folds leading to its cheek pouches (Smythe, in Kleiman 1974); Rewell (1950) and Macdonald *et al.* (in press) describe the large nasal gland of capybaras (Plate 11.1) and both *Erethizon* and *Dinomys*, the pacarana, rub their noses on objects. The nasal secretion of the pacarana drains from the eyes (Shadle, Smelzer, and Metz 1946; Collins and Eisenberg 1972). In his description of capromyid features, Pocock (1926) does mention the absence of sweat glands in the pads of the Bahaman hutia and the coypu, but nothing is really known about the possibility of interdigital or other foot glands in members of this suborder.

As with most other mammals, the importance of generalized body odour during the sniffings associated with fleeting social encounters is unknown. Nevertheless, most species sniff and even touch each other during meetings. The communicative role of urine is almost unknown, and Bellamy and Weir's (1972) analysis of its composition for six hystricomorphs is chiefly concerned with kidney function. However, the practice of enurination, i.e. spraying urine onto a conspecific, is widespread amongst these rodents and the fossorial degu, *Spalacopus*, urinates by cocking its leg against vertical surfaces (Kleiman 1974; Reig, 1970). Finally, Shadle (1952) notes that a male porcupine, *Erethizon dorsatum*, took consider-

Table 11.1. Sources of social odours amongst the families of hystricomorph rodents. This table follows the same format as Tables 9.1 and 10.1 so that the three together comprise a complete tabulation of sources of odours in the Rodentia. Further, detailed, references regarding hystricomorphs are given in the text.

Common name and species	Odour source	Sex	Marking posture	Implied function	References
CAVIIDAE					
Guinea pig	coccygeal	M,F		sex, rut	Martan (1962)
Cavia porcellus & *C. aperea*	perineal	M,F		dominance	Beruter et al. (1974)
	urine	M,F	spraying/enurination	sex, aggression	Beauchamp (1974)
					Beauchamp et al. (1979)
					Rood (1972) Kleiman (1971)
Desert cavy					
Microcavia australis	urine	M,F	enurination	sex, aggression	Rood (1972)
	perineal	M,F	anal drag	dominance	Rood (1972)
Cuis					
Galea musteloides	chin	M	press on female	sex	Weir (1971, 1973)
	perineal	M,F	anal drag		Rood (1972)
	urine	M,F	enurination	sex, rut	Rood (1972)
Patagonian cavy					
Dolichotis patagonica	anal	M,F	anal drag	rut	Pocock (1922)
	urine	M,F	enurination	rut	Kirschofer (1960)
					Genest and Dubost (1974)
					Taber and Macdonald (in prep.)
Salt desert cavy					
Pediolagus salinicola	urine	M,F	enurination		Wilson and Kleiman (1974)

Family / Species	Source	Sex	Behaviour	Function	Reference
HYDROCHOERIDAE					
Capybara	anal	M,F	anal drag		Schaller and Crawshaw (1981)
Hydrochoerus hydrochaeris	urine	M,F		dominance	Macdonald *et al.* (1983)
	morrillo	M,F	nasal mark		Ojasti (1973)
					Rewell (1950)
DASYPROCTIDAE					
Green acouchi	anal	M	anal drag	territorial	Morris (1962)
Myoprocta pratti	urine	M,F	enurination	sex, rut	Kleiman (1971)
Agouti	anal	M,F	anal drag	territorial	Smythe (1978)
Dasyprocta punctata				anti-pred	Murie (1977)
				foraging	
CHINCHILLIDAE					
Chinchilla	urine	M,F	enurination	sex, aggression	Kleiman (1971)
Chinchilla laniger					
CAPROMYIDAE					
Bahamian huta	anal	M,F		sex	Howe (1974)
Geocapromys ingrahami	urine	M,F		sex, aggression	Gosling (1977)
Coypu	anal	M,F			
Myocastor coypu					
OCTODONTIDAE					
Degus	urine	M,F		sex, aggression	Kleiman (1975)
Octodon degus					
ERETHIZONTIDAE					
Porcupine	urine	M,F	enurination	sex	Kleiman (1971)
Erethizon dorsatum					Shadle (1952)

able interest in a thick odorous yellowish discharge from the vagina of an oestrous female. All that is known of the scent-marking of gundis (Ctenodactylidae) is that they use communal latrines (George 1974).

In the following paragraphs I will summarize the fragmentary behavioural information available on social odours of hystricomorph rodents from seven of the 16 families. For the remainder there is nothing to be said; this applies to the 12 species of Old World porcupines (Hystricidae), nine species of New World porcupine (Erethizontidae), the pacarana (Dinomyidae), 32 species of tuco-tuco (Ctenomyidae), two species of chinchilla rats (Abrocomyidae), 45 species of spiny rats (Echimyidae), two species of cane rat (Thryonomyidae), and the African rock rat (Petromyidae). Finally, the nine species of African mole-rats (Bathyergidae) are of uncertain taxonomic affinity. One species, *Heterocephalus glaber*, lives in colonies of up to 80 members, amongst whom only a dominant pair reproduce. Subordinate males have low testosterone levels and develop nipples. There is evidence that the urine of the dominant female is responsible for preventing reproduction amongst the subordinates of both sexes (Jarvis 1984, in press).

Family Caviidae

Amongst its number the family Caviidae includes the laboratory guinea pig, *Cavia porcellus*, which has been relatively well studied. The guinea pig has several perianal glands, most notably a perineal pouch, and a supracaudal gland. The perineal pouch is a median groove which, according to Pocock (1922) '. . . resolves itself in the female into a pair of pits separated by a low partition, and situated between the anus and the vulva'. In the male the pits are larger and longer, and when distended fully they appear as a single capacious pouch producing a yellowish tallow-like secretion. The coccygeal or supracaudal gland of the guinea pig lies on the midline of the rump, just above the tail vertebrae (Pinkus 1910; Sprinz 1912). It produces a thick yellowish secretion which, in the case of males, has a musky odour. The female guinea pig's supracaudal gland is much smaller and less active and, according to Martan (1962) has no detectable odour. Martan examined these glands histologically and found that those of male animals formed a highly vascularized oval structure forming an elevated button of skin on the rump which continued to the scrotal raphe. Externally the skin was covered with a sticky secretion, and within there was a complex of large, crowded sebaceous glands and sparse hair follicles. Each sebaceous gland consisted of 10-15 acini which completely surrounded a hair follicle, itself enormously enlarged with sebum and forming the duct for the gland. The cells within the acini have large vesicular nuclei, and contain lipid droplets. At the centre of a mature acinus the size of the fat droplets is greater, the nuclei become pyknotic and cell boundaries indistinct. These sebaceous glands develop from cells similar to normal skin glands, beginning to proliferate at about 23 days of age and reaching adult appearance at about 50 days. Martan discovered that their development was androgen-dependent. Indeed, Harmsen and Sittig (1973) found that a 45-day course of testosterone injections administered to female guinea pigs converted their entire perianal glandular complex to a condition indistinguishable from that of males.

Although guinea pigs appear to defecate at random locations and without communicative significance, urine is clearly important socially. Beauchamp (1973) coins the term social urination (see enurination, Eibl-Eibesfeldt 1958) to describe the behaviour of an unreceptive female guinea pig who, when investigated by a male, raises her hindquarters and squirts urine which may splash the male in the face. Male guinea pigs often spray females with urine during their courtship.

Field studies

Cavies In Rood's substantial monograph, published in 1972, he compared the behaviour of three species of Argentine cavy (subf. Caviinae). Socially, all three lived in loose, promiscuous communities wherein social ties were largely limited to liaisons between courting animals. The cuis, *Galea musteloides*, is a small animal which feeds on grasses and herbs as does the wild guinea pig, *Cavia aperea*, the probable ancestor of domestic *C. porcellus* (Weir 1974b). Wild guinea pigs occupied home-ranges of about 0.1 ha, in contrast to the larger (0.2–0.3 ha) ranges of Rood's third subject, the desert cavy, *Microcavia australis*, which fed on coarser vegetation, principally the leaves of thorn bushes (see also Rood 1970). Almost every aspect of Rood's account is punctuated by reference to behaviour that is probably of olfactory significance, emphasizing the importance of social odours to all three species. Members of each genus indulged in perineal dragging, which involved squatting and dragging the anogenital region along the ground. Both *Galea* and *Microcavia* roll on the ground and urinate at sites where they drag their perineums, and rub that region back and forth during agonistic encounters. A similar behaviour, especially amongst roused males, was common among *Cavia*, but in their case involved side-to-side motions of the lower rump and hence the coccygeal gland unique to that genus. However, the rump may not touch the ground during this 'rumba'.

The tantalizing pattern which emerges from Rood's study of these superficially very similar species is that while there is much in common between them regarding olfactory communication (they all, for example, thrust their snouts into the perineal region of conspecifics with such gusto as to lift the recipient off its feet), there are also striking differences. Not only is *Cavia* the only genus with a coccygeal gland, but male *Galea* have a chin gland. From laboratory work (Weir, 1973) one may suppose that at least part of the function of this chin gland is to apply odour to the rump of females and juveniles, during a so-called chin–rump-follow sequence where two animals move in tandem, the male behind, his chin resting on the rump of the animal head. However, *Microcavia* also exhibits this behaviour, but apparently does not have the gland, and even female *Micro-cavia* chin-rump-follow on young.

Many of the elements of the caviid ethogram have clearly been fashioned to deploy odours: both *Galea* and *Cavia* perform 'stand-threats' which involve two protagonists, normally of disputed status, lurching upwards, their dorsal hair raised, their rumps swivelled somewhat towards each other and their perineal pockets being everted to expose flashes of pink skin within. Why *Microcavia* were not seen to perform this behaviour when they also have perineal pockets is a mystery. *Cavia* and *Galea*, but not *Microcavia* engage in 'tail-up' behaviour, which is sometimes associated with enurination. Tail-up occurs when one animal is being followed, perhaps harrassed, by another. The one being followed halts, and raises its rear end towards the advancing nose of the follower, and may even spray urine in its face. Rood points out that among the members of his captive colony of *Galea*, the only animal that was never seen to display a tail-up was the most dominant male. Hence it seems probable that this display is linked with submission, and Rood speculates that enurination, which seems to deter the recipient of the spray, is a defence mechanism. The animal that is sprayed stops to clean itself while the sprayer makes good its escape.

Another behaviour involving odours is the rumba, most commonly performed by *Cavia*. This involves rhythmically oscillating the hindquarters while making a burbling noise and flashing the pink skin of the anal glands. This can lead to 'rumping' whereupon one animal swings its rear across the flank of another, presumably smearing it with odours from all or some of the perineal glands, urine, and, in the case of *Cavia* the coccygeal gland.

Conspicuous, rather time-consuming behaviours such as stand-threat, tail-up, and rumba are all rare or absent in the repertoire of *Microcavia* in comparison to *Cavia* and *Galea*. Rood proposes that this may relate to the open, arid habitat of *Microcavia*, where such conspicuous actions might attract the attention of predators.

Although faeces accumulate where cavies spend the most time, Rood's impression was that their distribution was without social significance. The communicative functions of odour amongst these species must be manifold, to judge by the diverse circumstances in which behaviour associated with their glands occur. For example, infant *Microcavia* raise their perineal regions as they suckle, males of all genera seemed to track down oestrous females by the scent of their trails and a very common form of social interaction was nose-to-nose sniffing. The communicative properties of some caviid odours have been discovered during laboratory studies of domestic *C. porcellus* (see below).

Maras   There are two species of mara or Patagonian cavy, the small species sometimes called the salt desert cavy, *Pediolagus (Dolichotis) salinicola*, and the larger (15 kg) mara, *D. patagonum* from the scrubland of southern Argentina. Wilson and Kleiman (1974) studied play in the smaller mara (and in two genera of degus (Octodontidae)) and concluded that the frisky-hopping (cf. Rood 1972) characteristic of their infants was partly prompted by odours. In the case of the *P. salinicola*, juveniles and adult males combined frisky-hopping with rolling (often in urine-saturated sand), urine-marking, anogenital dragging, and sniffing conspecifics, all of which may lead to a shared odour of all the animals involved. Kleiman (1971) describes the same behaviour in the green acouchi (Dasyproctidae).

Recently, Taber and Macdonald (in prep.) have revised Pocock's (1922) description of the glands of the anogenital region of *D. patagonum*. The paired anal glands open through rather fleshy labia on either side of, and above the anus in both sexes. In the female, they consist of a flattened cavity, no more than 4 mm in maximum internal width. In both female specimens whose glands they sectioned, Taber and Macdonald found a small lip partially creating a second, small pocket near the mouth of each gland. Opening through the walls of the pocket are the ducts of rather sparsely scattered apocrine glands and small sebacous acini. In marked contrast, the openings of the male's anal pockets are wider (c. 8 mm), tumid, and more circular. Within the rim of each opening the skin is studded with small protuberances and inside the pocket these develop into the papillae mentioned by Pocock (1922). On the dead specimen the pockets were everted and it seems probable that they are evertible in life.

Serial section revealed that the papillae and protuberances were hollow, root-like structures whose interior was divided into a multitude of dendritic passages, each filled with traces of secretion (Plate 11.2). Many of these passages exited through pores in the papillae walls. Taber and Macdonald speculate that these extraordinary structures are designed to disperse scent slowly over a wide surface area. Too little is known of the mara's behaviour to shed more light on this suggestion, although Taber and Macdonald do report the preliminary results of a field study which corroborates the findings in captivity of Mohr (1949), Kirchshoffer (1960), and Genest and Dubost (1974). In summary, wild maras appear to live in long-term monogamous pairs, which have the unusual trait of maintaining their young in communal crêches. However, pairs whose young are in the same crêche have little else in common, and probably tend almost exclusively to their own infants. Females visiting the den sniff carefully all the cubs that approach them, but only let one or two nurse. The combined observations of studies in captivity and the wild indicate two principal sources of social odour:

Anal drag   The mara sits upright, arches its back, and flattens its anogenital region against the ground (Fig. 11.1(a)). Females sometimes rock from side to side, shifting their weight from one hind leg to the other; otherwise they simply press their rumps to the ground. In contrast, males drag themselves forward with their front legs, and in so doing they scrape their stubby tail along the ground, leaving a 10–20 cm long scratch and 2–3 droppings (see also Mohr 1949). Anal drags follow careful sniffing and are generally made on patches of bare soil.

Fig. 11.1. (a) Male mara anal dragging, and (b) enurinating on to a female. (c) female mara enurinating on to approaching male.

Enurination   Both male and female maras enurinate. Males do so having approached a female (not necessarily their own mate) from the rear. An enurinating male rears up on his hind-legs and sprays a jet of urine onto the female's rump (Fig. 11.1(b)). Males enurinate (and anal drag) often after repulsing the approach of a strange male towards their female. Female enurination is usually backwardly, but may be forwardly, directed. Most commonly it is prompted by a male's approach, particularly if he then enurinates onto her rump. The female will back towards the approaching male, present her genitals and squirt urine backwards into his face (Fig. 11.1(c)). Males try to evade the stream of urine. In different circumstances females will also spray urine forwards from a stance more typical of males. For example, Taber and Macdonald report four instances where females enurinated over cubs who were pestering them to nurse. In three of those four cases the recalcitrant female enurinated forwards onto the young.

Quantitative aspects of both anal dragging and enurination are summarized in the findings of Taber and Macdonald's field study, spanning four months, from July to November and hence include the peak of births in the study area in September. Their data include over 35 000 scan samples (at two-min intervals) of the mara's behaviour including almost 15 000 on adult maras of known sex (cf. Altmann 1974). These data revealed that although scent-marking by adults with enurination or anal dragging was rather rare (only 50/14 762 of the scan observations) it was much more common around the period of female oestrus (c. 5.5 h post-partum, Dubost and Genest 1974). Observations of the seven pairs utilizing a single communal den in which all the pairs' pups were born within a two-week period, showed that of 10 893 behavioural scans up to and including the females' oestrus there were 18 bouts of scent-marking. In contrast in the following period (also six weeks in duration) there were only 17 such bouts in 25 781 scans, a difference which is significant. Indeed most anal drags and enurinations by males of this communal den were made during the two-week birth period. Overall, males anal dragged and enurinated significantly more often than females, both in terms of numbers of marks and, more importantly, numbers of bouts of marking.

Of those bouts of anal dragging by paired males which were not prompted by an approaching male, most (5/7) immediately followed anal dragging, defecating, or rolling by the female. Three of the five anal drags done by females occurred during courtship, the remaining two were by females as they left a communal den after nursing their cubs.

Maras are also known to defecate preferentially on conspicuous sites, such as bare patches of earth, but nothing is known of the distribution or function of these sites. However, dung piles are important in the communication of ungulates Gosling, see Chapter 14) with which maras are convergent (Dubost 1968). Finally Pocock (1922) describes a deep glandular pouch lying just below the glands of the male mara's penis.

Laboratory studies

In conjunction with the observations which Rood (1972) made of cuis, *Galea musteloides*, in the wild, Rood and Weir (1970) suggested that oestrus was induced by the male in this species. Weir (1971, 1973) investigated this possibility by monitoring the reproductive state of all the members of a small group of female cuis and of two solitary females, all of whom were isolated from males. To judge by periods of perforation of the vaginal membrane, only two females came into oestrus at all. In contrast they all did when introduced to a male. Weir also maintained 15 females in contact with a male who lived in their midst, but within a cage and was thus isolated from their touch. This prohibited him from smearing their rumps with his chin gland (Rood 1972). Only two of these females came into oestrus, suggesting that the male's influence required direct physical contact. Assuming that chin-gland odour is instrumental in oestrus induction, perhaps smearing it on the female is a way of maintaining a protracted and high concentration around her. The female might also eat the secretion while grooming. Rood (1972) had described physical contact in the courtship of the cuis in the wild, where the males indulge in so-called chin–rump-follows, walking behind the female and resting their chins on the female's rump, and do so with increasing frequency as the female approaches oestrus. Indeed in Weir's experiment all 15 females came into oestrus when they were allowed to touch the male. Weir describes the glandular patches present on both sexes of cuis under the chin. A comparable patch of bare skin on the throat is apparently not glandular. The necessity for the male to have physical contact with the female to induce oestrus led Weir to conclude that a primer pheromone, and not a signalling one, is involved.

The remaining laboratory studies of caviids have been of *C. porcellus*, and although many unknowns remain, these studies demonstrate several discriminations of which guinea pigs are capable on the basis of odour. As mentioned above, social odours among guinea pigs include urine and the secretions of the vagina and the perineal and coccygeal glands.

Guinea pigs urinate frequently during social and sexual encounters, and engage in enurination. Beauchamp (1973) demonstrated that, during four-minute trials, sexually experienced males spent longer sniffing within 1 cm of glass plates smeared with 0.1 ml of the urine of females (outside oestrus) than that of adult males (which was itself investigated for longer than water). Indeed, during the choice experiments, these experienced males spent longer sniffing the urine of ovariectomized females and castrated males than they did of intact adult males, and the duration of their interest in a mixture of urine from males and females fell between the results for pure urine from either sex. These male guinea pigs thus demonstrated the ability to discriminate sex on the basis of urine, and furthermore, by sniffing the urine of an adult male *C. porcellus* more than they did that of a non-receptive female *G. musteloides* they showed that this ability was also species specific. This preference for female urine emerged when males were 18–20 days of age. Possibly female guinea pigs can also discriminate sex from urine since that of males has a primer effect on their oestrous cycle (Jesel and Aron 1974; Jesel, Chateau, and Aron 1976). Although Beauchamp's (1973) experiments did not concern individual identity *per se*, he did show that males investigated the urine of male strangers for longer than they did their own urine, hence showing an ability to distinguish self from other, if not necessarily to recognize individuals. Beauchamp and Berüter's (1973) findings showed that male guinea pigs sniffed the urine of oestrous and dioestrous females for equivalent lengths of time. Subsequently Ruddy (1980) proved that both males and females could distinguish oestrous and dioestrous females on the basis of the odours of a generalized swab from the anogenital region.

Berüter *et al.* (1973) demonstrated the resilience of the ability of the male guinea pig to discriminate sex on the basis of urine when they showed that they could make the same distinction on each of a variety of fractions of the test urine. These authors thus concluded that male versus female urine discrimination was probably based on differences in patterns of urinary components. This view is now widely accepted for mammalian social odours in general, and for example, probably applies amongst other rodents to the flank glands of water voles, *Arvicola terrestris* (Stoddart, Aplin, and Wood 1975) and, amongst hystricomorphs, to the nasal and anal glands of capybaras (Macdonald *et al.*, 1984). In the instance of the guinea pig's urine, Berüter *et al.* (1973) suggested that the diagnostic differences in pattern arose from sex differences in metabolism.

Differences in the representation of various metabolites in guinea pig urine presumably explain how male guinea pigs can distinguish between the urine of others of either sex on the basis of their diet. Beauchamp (1976) demonstrated this ability by showing that the test males spent longer sniffing the urine of donors who ate one type of food (as it happens, the food with which the test males were familiar) than of those who ate a different food. Apart from highlighting the importance of diet during controlled experiments these results show at least a potential for investigating the diet (and hence behavioural and ecological correlates of diet) of animals on the basis of their urine.

The perineal glands, which lie in a pocket between the penis and the anus of male guinea pigs, were studied by Beauchamp (1974). He found that the sebum production of isolated animals was less than that of those kept in social groups, and that within each social group dominant males secreted much more sebum than did their subordinates. Furthermore, dominant males engaged more frequently in perineal dragging behaviour than did subordinates. The perineal glands of males that were castrated regressed. Champy and Kritch (1929) noted that males that were castrated prior to puberty failed to develop perineal glands. In captivity, male guinea pigs form linear dominance hierarchies (Jacobs, Beauchamp, and Hess 1971; King 1956) and engage in perineal dragging when involved in agonistic encounters with other males, when they find unfamiliar objects in their cage and, as Kunkel and Kunkel (1964) describe, during courtship. By presenting adult males with paired glass slides on which either their own perineal gland secretion or that of a strange male had been smeared, Berüter *et al.* (1974) showed that the test animals could distinguish between their own secretion and that of the stranger and could continue to do so until the mark was three days old. The findings of Berüter *et al.* (1974) suggest that the primary volatiles of male guinea pig perineal secretion are aliphatic acids, however, at least in the case of urine, Beauchamp (1973) noted that guinea pigs put their nose in the liquid or licked it, presumably indicating that non-volatile components were important. Indeed, Wysocki, Wellington, and Beauchamp (1980) found when guinea pigs investigated urine stained with rhodamine (a non-volatile fluorescent dye) the fluorescence was not recovered from the olfactory epithelium, but from the vomeronasal and septal organs. Beauchamp, Wellington, Wysocki, Brand, Kubie, and Smith (1980) have developed these studies of the communicative significance of non-volatile compounds in guinea pigs' urine (i.e. those with a value of 400 g/mol or greater), and their findings have wide implications for the study of mammalian communication. They repeated the earlier choice tests in which males were tested concerning their ability to discriminate between the urine of males and females. When the test urine was placed out of tactile reach, the ability to discriminate was greatly attenuated. These authors evaporated urine on frosted glass plates and allowed the sediment to dry for periods of 5 h to 90 days. Selecting samples of different ages they repeated the choice experiments and demonstrated that the test male subjects could distinguish the dried smears of males and females until the smears were up to 60 days of age (see also Beauchamp, Criss, and Wellington 1979; Wellington, Beauchamp, and Smith 1981). In earlier trials with non-volatile components of urine Berüter *et al.* (1973) had also shown that males could distinguish the residues of male and female samples, spending longer smelling the samples from females. In contrast, when the more volatile components alone were used, the male guinea pigs reversed their behaviour, sniffing the volatiles from the male samples for longer (Beauchamp and Berüter 1973). This series of experiments culminates elegantly with a series of choice experiments on each of three fractions of the high-molecular-weight components of urine. These were prepared on a gel filtration column which fractionates polypeptides with molecular weights under 1500 g/mol. Chromatography showed that the three peaks were similar in male and female samples. However, males could distinguish the sex of the animal from which each of two of the three fractions originated (Beauchamp *et al.* 1980).

Martan and Price (1967) investigated the effect of gonadectomy on the coccygeal gland of guinea pigs of both sexes. They monitored post-operative changes on three groups of sebaceous tissue: the glandular cells associated with the rudimentary hairs of the coccygeal gland, those associated with mammae, and those associated with body hair. Gonadectomy reduced the sebaceous development around the male's supracaudal glands, but had little effect on the smaller supracaudal glands of females. The operation also reduced sebaceous development around the female's nipples.

In his discussion of social odours of wild cavies, Rood (1972) speculated that the distinction between male and female odours of domestic guinea pigs was less obvious than the difference between the sexes of wild *Cavia aperea*. The reason for this suggestion was that wild males appeared to have difficulty in telling apart domesticated males and females. Beauchamp *et al.* (1979) took up this suggestion and showed that males selected adult female conspecifics as mates in preference to females of other species. In a series of cross-fostering experiments Beauchamp and Wellington (1981) then reared baby male *C. aperea* with non-lactating female *C. porcellus*. When adult, and in contrast to adult male *C. aperea* reared by their own species, these cross-fostered males spent more time sniffing at the urine of female *C. porcellus*, when given the choice between this and the urine of female *C. aperea*. However, these males sniffed for longer at the urine of male *C. aperea* than that of male *C. porcellus*. These results do indicate that olfactory imprinting affected mate choice; however, the combined results of Beauchamp *et al.* (1979) and Beauchamp and Wellington (1982) lead to some confusion since it appears that a hybrid male *C. porcellus* X *C. aperea* that was reared with its 'wild' father and domestic mother eventually selected F_1 female mates in preference to pure *C. porcellus* mates, like its mother. Hence, the details of olfactory imprinting remain unresolved (see also Beauchamp and Hess 1971). It is certainly clear that the early olfactory environment, and thus presumably the odours of adults, do have an influence on adult behaviour, and this seems especially to be true for males. For example, experiments in which baby guinea pigs, especially those under six days old, are reared in an atmosphere scented with a normally aversive chemical such as ethylbenzoate, or with siblings or mothers daubed in this odour, show that they can grow up to be imprinted upon it (Carter and Marr 1970; Carter 1971). In less manipulated circumstances it seems that baby guinea pigs will suck from any female which allows them to do so and will do so just as readily as from their own mother. Mother guinea pigs can recognize their own babies within 48 hours of birth (Porter, Fullerton, and Berryman 1973; Porter, Berryman, and Fullerton 1973; Fullerton, Berryman, and Porter 1974). These experiments, as with so many which involved comparisons of the times spent sniffing at, or reacting to, different stimuli may illustrate that, for example, baby guinea pigs are equally willing to suck from their mother or from a stranger, but they do not necessarily indicate that the babies cannot distinguish between the two.

It is quite clear that odours play a major role in the social behaviour of guinea pigs. Bulbectomized male guinea pigs fail to form a hierarchy within a colony, and they exhibit little inter-male aggression, are relatively inactive sexually and rarely scent mark (Beauchamp, Magnus, Shmunes, and Durham 1977; Donovan and Kopriva 1965). Some features of the role of scent have been discovered: in summary, guinea pigs can distinguish sex, reproductive state, diet, and species on the basis of the odours of all or some of urine, perineal, or coccygeal gland secretions, and in the case of the latter two, at least, there is the potential for communication of status (owing to the androgenic dependence of both glandular activity and dominance) and individual identity (owing to variations in the representation of the constituent components). However, the functioning of social odours within wild caviid societies is poorly known, and even speculation on this is sparse. Following his discussion of urine amongst captive guinea pigs, Beauchamp (1973) concludes that the odour is important in maintaining social cohesion during long (66–70 day) periods of female unreceptivity between successive post-partum oestrus periods. Kunkel and Kunkel (1964) discuss enurination among members of their colony of captive guinea pigs and conclude that '. . . *Cavia* sprays conspecifics with urine, an expression of weak or blocked aggressivity in which an escape tendency is entirely lacking'. What this means in terms of the functioning of cavy society is unknown.

Family Hydrochoeridae

The capybara, *Hydrochoerus hydrochaeris*, has two unorthodox features to its glandular armament. Most conspicuous, is the sexually dimorphic nose gland, or morrillo, which among adult males forms a bulbous, carbuncular growth aloft the snout (Plate 11.1). Second, the anal pockets of the capybara are sexually dimorphic, and those of the male are more open than the female's and studded with detachable hairs coated in a truncheon-shaped concretion of hard, crystalline, and probably secretory material.

Capybaras are large, sociable rodents, living in groups of varying sizes, but of up to 60 or more members, with an overall adult sex ratio of 1:2 ($\male:\female$). They are grazers of the inundated savannahs of northern South America and take to the water for protection, to mate, and to thermoregulate (Ojasti 1973; Macdonald 1981; Schaller and Crawshaw 1981).

Aside from some early, quaint, and largely inaccurate accounts, the anal pockets of capybaras were mentioned fleetingly by Pocock (1922) and the morrillo was described by Rewell (1950). More recently Macdonald *et al.* (in press) have described both these glands in more detail, together with behaviour associated with their use. A typical marking sequence by a male capybara involves secretions from both morrillo and anal pockets, together with urine (see Fig. 11.2). The male approaches a shrub or bush, rubs his snout up and down against it, coating the stem with the exceptionally abundant oily secretion from his morrillo. Then, moving slowly and straddling the bush, he walks over it, dragging the vegetation under his body and between his hind legs. At the same time as the leaves pass his perineal pockets the capybara urinates, rubbing his penis on the vegetation. Urine also coats his inside thighs, against which the vegetation also brushes. It seems certain that this action detaches some of the loose, secretion-coated hairs from the anal pocket. Marking may involve all or any of the morrillo, the anal pockets, and urine. It occurs most commonly without obvious social provocation, and is performed at least sometimes by most of the adult males in any group and by solitary males. Usually, but not invariably, aggression between males is followed by marking by either the victor, or by both parties, but rarely by the vanquished alone. On rare occasions, sometimes when both animals are in the water, males rub with their morrillos on the necks of females and subordinate males (Plate 11.3).

During the dry season, female capybaras mark with their poorly developed

Fig. 11.2. Marking sequences by male capybara, first with morrillo, then anal gland and urine (see text).

morrillo rather infrequently (at about one-thirtieth the frequency of males), and were never seen to use their anal pockets in the same way as males (Macdonald *et al.* 1984). However, female anal pockets secreted very actively and many social encounters involved sniffing at them (Fig. 11.3). Furthermore, during courtship, Schaller and Crawshaw (1981) found that males and females marked with equal frequency, and used both glands.

Histological examination revealed that both snout and anal gland were composed of sebaceous tissue. In the case of the male's morrillo and the female's anal pocket these were massively developed, and all the more so for larger individuals (Plate 11.5). Chemically, the secretions from both these glands were very complex. Under one set of gas chromatographic conditions, the secretions from

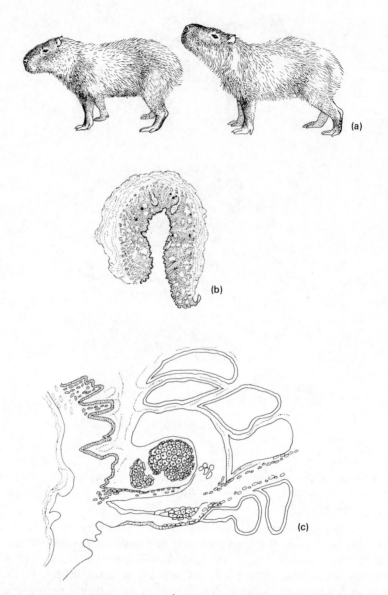

Fig. 11.3. Male capybaras often approach and sniff at the rear of females (a).
During courtship a male herds the female in front of him and towards water
(where they mate) and as he pursues he repeatedly raises his nose and sniffs. Part
of the odour he smells may come from the female's paired anal pockets, shown
(b) macroscopically and (c) microscopically.

the morrillos of adult males revealed 54 components. Each male had most, or all of these components, but in significantly different proportions. These profound differences in proportion may facilitate individual recognition, and the same applies to the secretions of female anal pockets, in which there were 30 or more components. A series of lipids including sterols (such as cholesterol) and/or terpenes, and amino acids were detected in the secretions of both glands, and the most volatile substance from the morrillo was a hydrocarbon, $C_{30}H_{50}$, whereas the bulk of the secretion was a complex mixture of esters.

The scales of the hairs of the male capybara's anal pocket are rather smooth, and the bristles are circular in cross-section (Plate 11.4). The hairs themselves are coated in layers of material which accumulate to a length of up to 8.0 mm and width of 2.5 mm. The layers are clearly visible through scanning electron microscopy, and under ×7000 magnification it is obvious that the inner ones contain more amorphous material, whereas the outer ones are a jumble of long crystals (Plate 11.4). In fact the crystals are largely a calcium salt, with lesser quantities of magnesium. Presumably successive layers derive from the hardened residue of successive bursts of secretory activity. Since the sebaceous acini of the male's anal pocket discharge into the follicles of these hairs it seems highly likely that the layers are composed of the secretions of these glands. However, since they are also doubtless doused in urine this cannot be discounted as a partial origin of the material. That these hairs are designed to be detachable seems incontestable, and, indeed, the uniform nature of the concretions makes it difficult to dismiss their production as accidental. How they function is unknown; the crystalline material could have communicative functions if licked, or it may act to release other material, either through slow diffusion, or through chemical reaction, for example with urine.

Family Dasyproctidae

Smythe (1978) has studied the agouti, *Dasyprocta punctata*, and Kleiman (1971) the green acouchi, *Myoprocta pratti*. The two species are sympatric; in captivity the green acouchi tended to be social, but in the forest it is so timid and rarely seen that its behaviour is unknown. The larger agoutis studied by Smythe lived as mated pairs within territories of about 2 ha, where they fed largely on fruit. Conspecifics, including young, and especially young females, are harassed or driven away by members of a breeding pair of agoutis (see also Roth-Kolar 1957). Within this social system Smythe concludes that odours play an important role. Both sexes have evertible anal glands with which they paste the ground and vegetation. Males mark more than females, and both sexes respond to strange objects (for example Smythe's traps) by increased marking nearby, but not on them. Similarly, the marks made by anal glands of strange agoutis stimulate higher rates of marking by a breeding pair, and this effect is especially pronounced when one male encounters the marks of another. On such occasions, the male's anal marking may be interspersed with foot thumping, and backwards scratching with the hind-feet (there is no information on foot glands). The agouti uses its anal sacs during aggressive encounters when the odour from the glands is detectable to human observers from 10 m or more. The scent similarly hangs in the air after an agouti has been surprised by a potential predator. Smythe speculates that this may serve a 'smoke-screen' function, the odour of the escaping prey confusing a predator relying heavily on scent.

Agoutis urinate in social contexts. Sexually aroused males enurinate from an upright posture upon females, and under these circumstances a female will also stand up, but urinates onto the ground. Smythe (1978) interprets enurination as a method whereby the male douses the shy female with his odour to such an extent that she becomes familiar with his scent and she will eventually cease to flee from him (cf. Kleiman 1974). Outside the context of courtship, females urinate every day or two on a particular spot outside their nest, and while the female is absent her offspring rest and roll upon this spot as if it were a surrogate mother (Kleiman 1971, 1972; *Pediolagus* behaves similarly, Wilson and Kleiman 1974). Indeed, agouti babies rub their bodies over their mothers in a way which Smythe presumes ensures a shared odour.

There are many similarities between Smythe's account of agouti behaviour and Kleiman's of acouchis. Acouchi males court their females by pursuing them, sniffing their anogenital region and, although it is not a specifically precopulatory behaviour, rearing on their hind-legs to squirt, with a fully erect penis, a shower of urine over the female, which can travel for almost a metre. The female responds to this by stopping to lick the urine from her fur, and thereby she presents a still target to the male who showers her with further urine. For two days after mating enurination disappears from the male's repertoire, but outside these two days the frequency of enurination remained similar before and after oestrus. Kleiman points out that male and female acouchis will lick enurinated urine from their fur or cage, whereas both sexes generally ignore normally voided urine. Thus she speculates that the urine voided during enurination may have different contents. Morris (1962) noted that *Myoprocta* urinated (and anal dragged) when placed in a new cage.

Smythe (1978) noted that cached food was very important to the agoutis on Barro Colorado Island where he studied them. Since the agoutis often sat down at the sites where they had just hidden food, Smythe wondered whether scent marks were being made passively which signposted the location of each cache. However, Murie's (1977) experiments indicate that the odour of the cached food itself is probably important and, anyway, as Murie points out, agoutis sit all over the place and not just adjacent to caches.

Family Capromyidae

With the exception of the coypu, *Myocastor coypus*, the dozen or so members of the Capromyidae are known as hutias and inhabit Carribbean Islands. Little is known of the hutias' biology, but in his description of the behaviour of the rare Bahaman hutia, *Capromys ingrahmi*, Pocock (1926) notes that their paired anal glands open just below the anus and Clough (1972) mentions that they sprinkle urine in trails 30-60 cm long on the sand and on sticks lying on the ground. Howe (1974) observed the same species in captivity, and found that animals of both sexes were more likely to urine mark on an experimental stick that was already tainted with hutia urine than they were on unscented control sticks. When a hutia marked a stick within the cage it was marked by another animal within 60 min on 64 per cent of occasions, and by a third animal on 13 per cent. Where one hutia marks, his companions will mark also, in general. Their marking was rarely associated with agonistic encounters, although these did occur within the colony, whose members were organized hierarchically. The sight of other hutias also stimulated marking by members of Howe's colony, and the oestrous females marked more intensively than any other individuals. Marking involved both urine and perineal dragging and Howe's conclusion was that the two were interchangeable, and so he did not discriminate between them in his data.

Howe's impression was that scent-marking functioned to communicate oestrus and to promote 'social cohesion' among hutias.

The protrudible anal glands of both sexes of the other capromyid, the amphibious copypu or nutria, were described by Pocock (1922) as forming '. . . a solid median mass, opening by four pairs of small papillate orifices into a sac . . . just below the anus . . .'. (see also Ortmann, 1960). The capacious pocket is bedded in thin sebaceous tissue (Plate 11.6). Gosling (1977) notes that the anal glands of male coypus (measuring 5-6 cm in length) are three times the size of those of females, and it seems that while males mark most frequently with these glands, a greater proportion of the scent-marking by females involves urine. Marking is often on raised objects or at well-used points of entry to the water. Faeces may also be left on elevated sites. Gosling (1977) describes the social organization of the coypus, living feral in eastern England, as a polygynous system based on a matriarchal female whose home-range overlaps with those of her daughters, all of which are in turn overlapped by a larger territorial male's.

Family Chinchillidae

The six species of this family include the colonial viscachas and chinchillas. Plains viscachas, *Lagostomus maximus*, live in large communal burrow systems (Pearson 1948). Pocock (1922) describes a glandular area of skin below, but adjacent to, the anus and anal glands of this species, but nothing is known about the use of the glands or the patch of skin.

Bignami and Beach (1968) studied chinchilla, *Chinchilla lanigera*, reproduction in captivity. They mention nothing of scent-marking but do describe a form of enurination displayed by unreceptive females towards over-assertive courting males. At the moment when a male and female stand, facing each other, the female springs up and from midair may squirt urine at the facing male. Indeed, if he persists she may then kick him.

Discussion

The continuing studies of Beauchamp and his colleagues have provided a wealth of information on the domestic guinea pig, which can clearly distinguish between urine samples which originate from animals of different sexes, ages, reproductive conditions, and diets, and from individuals belonging to different genera or species. Furthermore, other studies on the perineal and coccygeal glands of this species have indicated that both play a role in sexual behaviour. Yet, the functional implications of these feats of discrimination are largely unknown even amongst guinea pigs, and understanding of the role of social odours within hystricomorph societies is sadly wanting. Indeed, knowledge of rodent societies in general, is scant (Eisenberg 1967). Even the interpretation of the results of choice experiments can be confusing, since they hinge upon the time spent sniffing odours. Obviously, if two odours are sniffed for similar, or different, durations this does not necessarily imply that they transmit similar, or different, information, nor that they have the same or different impacts upon the sniffer. Nevertheless it is clear, as Beauchamp *et al.* (1980) write, that 'If urine samples from one class of individuals (e.g. females) are preferred to urine samples from another class (e.g. males) then we conclude that information is available in urine which differentiates the classes (i.e. females from males)'. In this usage, 'preference' is a short-hand for 'sniffed at for longer' and need not imply that the animal doing the sniffing actually likes or dislikes the odours. One could imagine that one odour might be smelt for a long time because it was genuinely pleasing, another because it was hard to decipher, another because it contained information worth memorizing. Although it does not affect the broader logical impediments of interpreting or ranking the durations of sniffing, Beauchamp *et al.* (1980) do not believe that male guinea pigs sniff for a long time at 'preferred' odours because they are hard to read since, when diluted, such odours are sniffed at for shorter periods. This leads them to conclude that it is the 'hedonic quality of the sample' which determines the duration of the investigation by the male guinea pig. Although in the case of male guinea pigs sniffing the urine of females this interpretation has appeal, there is no *prima facie* reason to suppose, in general, that the motivation to sniff for longer at certain smells is because they are pleasurable.

Outside the laboratory, detailed descriptions, let alone functional explanations, of social odours within hystricomorph societies are few. The behaviour which has caught most attention is enurination, or urine spraying upon a conspecific.

It seems to be common amongst the Hystricomorpha but, as Rood (1972), pointed out, absent amongst the other rodent sub-orders, yet present among lagomorphs. The only species of hystricomorph studied in the field and for which enurination has not yet been recorded is the capybara (amongst whom allomarking with the morrillo does occur, albeit very infrequently, Macdonald *et al.*, in press). Kleiman (1971) noted that in the cases of the green acouchi and others, urine-spraying by males was most commonly recorded during courtship. Hence, she concluded that it was motivated by sexual arousal. Recently Taber and Macdonald (in prep.) noted that mara enurinations by males on to females occurred at all times, but most often in the weeks prior to parturition. During one 1.5-h courtship bout, a male mara enurinated only once. By and large, explanations for the motivation of enurination revolve around antagonistic aspects of courtship between individuals of species where everyday social inter-actions do not involve close proximity. Although it does not seem applicable to the maras, both Kleiman (1974) and Smythe (1978) suggest that the male douses the female in his odour and so accustoms her to his increasing proximity prior to mating. Rood (1972) speculates that the urine of a dominant male on a female's flank (also deposited during 'rumping' behaviour) may deter the approach of a subdominant male. Enurination by females onto males is seen as another feature of uneasiness between the two, in this case being interpreted as a means of repulsing the too fervent attentions of the male. The male hystrico-morph does indeed pause when squirted with urine, but since he licks and sniffs at this (see Beauchamp *et al.* 1980) the conclusion that he is mechanically repelled by the experience seems unwarranted. More likely he receives informa-tion on the female's oestrus that indicates that further courtship is inappropriate. In general, Kleiman (1966) suspects that a major impact of scent-marking is upon the animal who made the mark, who is constantly reassured by the pervading presence of its own odour. Reassurance, in these terms, is also thought to underlie the female's behaviour, since this halts temporarily the unwelcome approach of the male. Alternatively, one might speculate that the males of these species actively harrass females in order to prompt them to enurinate so that the urine may be tested for hormonal state.

There are no substantial functional explanations of any hystricomorph pattern of scent-marking. One of the few proposals, namely that agoutis mark their caches, is probably incorrect. White, Fisher, and Meunier (1982) describe communal rearing of infant degus (Octodontidae) and link this to the females' inability to discriminate between the odours of their own and other young. Weir (1974*a*) believes that tuco-tuco urine proclaims ownership of a tunnel system because they use the urine to mix with mud and so to harden the tunnel walls. Rood (1972) suggests that perineal marking by male *Cavia* promotes spacing without aggression, since subordinates who happen upon the odour of a dominant's perineal drag try to avoid an encounter. Furthermore, odour plays an important role in attracting males to female desert cavies, *Microcavia*, several

days before their post-partum oestrus (Rood 1972). Beauchamp (1973) speculates that the attraction of female urine for male *Cavia* may function to maintain
social bonds during the long period between each female's oestrus. For hystricomorphs the problem is that so little is known of these bonds or the societies
within which they operate that the understanding of olfactory communication is
made doubly difficult.

Acknowledgements

I am grateful to Drs J. Rood and B. Weir for their helpful comments on an
earlier version of this chapter. Dr L. M. Gosling kindly supplied the material
from which sections of coypu anal pockets (Plate 11.6) were cut. My work on
capybaras and maras was supported, respectively, by the Royal Society and
the H. F. Guggenheim Foundation, whose sponsorship I gratefully acknowledge.

References

Altmann, J. (1974). Observational study of behavior: sampling methods.
Behaviour 49, 227–67.
Azara, F. d' (1801). *Essais sur l'histoire naturelle des quadrupedes de la province
de Paraguay*. Paris.
Azcarate, T. (1978). Sociobiologica del Chiguire, *Hydrochoerus hydrochaeris*.
Ph.D. thesis, Universidad Complutense de Madrid.
Beauchamp, G. K. (1973). Attraction of male guinea pigs to conspecific urine.
Physiol. Behav. 10, 589–94.
— (1974). The perineal scent gland and social dominance in the male guinea
pig. *Physiol. Behav.* 13, 669–73.
— (1976). Diet influences attractiveness of urine in guinea pigs. *Nature, Lond.*
263, 587–8.
— Berüter, J. (1973). Source and stability of attractive components in guinea-
pig (*Cavis porcellus*) urine. *Behav. Biol.* 9, 43–7.
— Criss, B. R., and Wellington, J. L. (1979). Chemical communication in *Cavia*:
responses of wild (*C. aperea*), domestic (*C. porcellus*) and F_1 males to urine.
Anim. Behav. 27, 1066–72.
— and Hess, E. H. (1971). The effects of cross-species rearing on the social and
sexual preferences of guinea pigs. *Z. Tierpsychol.* 28, 69–76.
— Magnus, J. G., Shmunes, N. T., and Durham, T. (1977). Effects of olfactory
bulbectomy on social behavior of male guinea pigs, *Cavia porcellus. J. comp.
physiol. Psychol.* 91, 336–46.
— and Wellington, J. L. (1981). Cross-species rearing influences urine preferences in wild guinea pigs. *Physiol. Behav.* 26, 1121–4.
— — Wysocki, C. J., Brand, J. G., Kubie, J. L., and Smith, A. B. (1980).
Chemical communication in the guinea pig: urinary components of low volatility and their access to the vomeronasal organ. In *Chemical signals: vertebrates and aquatic invertebrates* (ed. D. Müller-Schwarze and R. M. Silverstein) pp. 327–40. Plenum Press, New York.
Bellamy, D. and Weir, B. J. (1972). Urine composition of some hystricomorph
rodents confined to metabolism cages. *Comp. Biochem. Physiol.* 42A, 759–71.
Berüter, J., Beauchamp, G. K., and Muetterties, E. L. (1973). Complexity of

chemical communication in mammals: urinary components mediating sex discrimination in male guinea pigs. *Biochem. biophys. Res. Commun.* **53**, 264–71.

—— —— — (1974). Mammalian chemical communication: perineal gland secretion of the guinea pig. *Physiol. Zool.* **47**, 130–6.

Bignami, G. and Beach, F. A. (1968). Mating behaviour in the chinchilla. *Anim. Behav.* **16**, 45–53.

Carter, C. S. (1972). Effects of olfactory experience on the behaviour of the guinea pig (*Cavia porcellus*). *Anim. Behav.* **20**, 54–60.

— and Marr, J. N. (1970). Olfactory imprinting and age variables in the guinea pig *Cavia porcellus*. *Anim Behav.* **18**, 238–44.

Champy, C. and Kritch, N. (1929). Les glandes anales due cobaye et leur influencement par les glandes genitales. *Archs Anat. microsc.* **25**, 459–70.

Clough, G. C. (1972). Biology of the Bahaman hutia, *Geocapromys ingrahami*. *J. Mammal.* **53**, 807–23.

Collins, L. W. and Eisenberg, J. F. (1972). Notes on the behaviour and breeding of pacaranas, *Dinomys branicii*, in captivity. *Int. Zoo Yb.* **12**, 108–12.

Corbet, G. B. and Hill, J. E. (1980). *A world list of mammalian species.* Brit. Mus. Nat. Hist. and Comstock Publ. Assocs. (Cornell University Press).

Donovan, B. T. and Kopriva, P. C. (1965). Effect of removal or stimulation of the olfactory bulb on the estrous cycle of the guinea pig. *Endocrinology* **77**, 213–17.

Dubost, G. (1968). Les niches ecologiques des forêts tropicales Sud-Americaines et Africaines, sources de convergences remarquables entre rongeurs et artiodactyles. *Terr et Vie* **1**, 3–28.

— and Genest, H. (1974). Le comportemente social d'une colonie de maras, *Dolichotis patagonum* Z. dans le Parc de Branfere. *Z. Tierpsychol.* **35**, 225–302.

Eibl-Eibesfeldt, I. (1958). Das Verhalten der Nagetiere. *Handb. Zool.* **8** (10), 13, 1–88.

Eisenberg, J. F. (1967). A comparative study in rodent ethology with emphasis on evolution of social behaviour. *Proc. U.S. natn. Mus.* **122**, 1–51.

Fullerton, C., Berryman, J. C., and Porter, R. H. (1974). On the nature of mother–infant interactions in the guinea pig (*Cavia porcellus*). *Behaviour* **48**, 144–56.

Genest, H. and Dubost, G. (1974). Pair living in the mara (*Dolichotis patagonum*). *Mammalia* **38**, 155–62.

George, W. (1974). Notes on the ecology of gundis (*F. ctenodactylidae*). *Symp. zool. Soc. Lond.* **34**, 143–60.

Gosling, L. M. (1977). Coypu. In *The handbook of British mammals*, 2nd edn (ed. G. B. Corbett and H. N. Southern). Blackwell, Oxford.

Harmsen, R. and Sittig, N. (1973). The effect of testosterone on the development of the perianal glands of the guinea pig *Cavia porcellus* (L). *J. exp. Zool.* **186**, 269–72.

Hershkovitz, P. (1969). The recent mammals of the neotropical region: a zoogeographic and ecological review. *Q. Rev. Biol.* **44**, 1–70.

Howe, R. J. (1974). Marking behaviour of the Bahaman hutia (*Geocapromys ingrahami*). *Anim. Behav.* **22**, 645–9.

Jacobs, W. W., Beauchamp, G. K., and Hess, E. H. (1971). Male rank order in a colony of guinea pigs (*Cavia porcellus*). *Am. Zool.* **11**, 635.

Jarvis, J. U. M. (1981). Ensociality in a mammal: co-operative breeding in naked mole–rat colonies. *Science, NY* **212**, 571–3.

—— Chemical control of colony reproductive state in naked mole-rats, *Hetevocephalus glaber. Acta zool. fenn.* (In press.)

Jesel, L. and Aron, C. (1974). Action of the odor of urine on estrous cycle duration in the guinea pig. *C. r. Séanc. Soc. Biol. Filial,* **168**, 819–23.

—— Chateau, D., and Aron, C. (1976). Changes in estrous cycle duration in female rats or guinea pigs exposed to the odor originating from castrated animals of the same species. *C. r. Séanc. Soc. Biol. Filial.* **170**, 982–6.

King, J. A. (1956). Social relations of the domestic guinea pigs living under semi-natural conditions. *Ecology* **32**, 221–8.

Kirchshoffer, R. (1960). Über das Harnspritzen (enurination) der grossen Mara (*Dolichotis patagonum*). *Z. Säugetierk.* **25**, 112–27.

Kleiman, D. G. (1966). Scent marking in the Canidae. *Symp. zool. Soc. Lond.* **18**, 167–77.

—— (1971). The courtship and copulatory behavior of the green acouchi, *Myoprocta pratti. Z. Tierpsychol.* **29**, 259–78.

—— (1972). Maternal behavior of the green acouchi (*Myoprocta pratti* Pocock), a South American Caviomorph rodent. *Behaviour* **43**, 48–84.

—— (1974). Patterns of behavior in hystricomorph rodents. *Symp. zool. Soc. London* **34**, 171–209.

—— (1975). The effects of exposure to conspecific urine on urine-marking in male and female degus (*Octodon degus*). *Behav. Biol.* **14**, 519–26.

—— Eisenberg, J. F., and Maliniak, E. (1979). Reproductive parameters and productivity of caviomorph rodents. In *Vertebrate ecology in the north neotropics* (ed. J. F. Eisenberg) pp. 173–82. Smithsonian Institution Press, Washington, D.C.

Kunkel, P. and Kunkel, I. (1964). Beiträge zur Ethologie des Hausmeerschweinchens, *Cavia operea f. porcellus* (L). *Z. Tierpsychol.* **21**, 602–41.

Macdonald, D. W. (1981). Dwindling resources and the social behaviour of capybaras (*Hydrochoerus hydrochaeris*) (Mammalia). *J. Zool., Lond.* **194**, 371–91.

—— Kranz, K., and Aplin, R. T. (1984). Behavioural, anatomical and chemical aspects of scent marking amongst capybaras, *Hydrochoerus hydrochaeris*, (Rodentia: caviomorpha). *J. Zool. London.* **202**, 341–60.

Martan, J. (1962). Effect of castration and androgen replacement on the supracaudal gland of the male guinea pig. *J. Morph.* **110**, 285–98.

—— and Price, D. (1967). Comparative responsiveness of supracaudal and other sebaceous glands in male and female guinea pigs to hormones. *J. Morph.* **121**, 209–21.

Mills, M. G. L., Gorman, M. L., and Mills, M. E. J. (1980). The scent marking behaviour of the brown hyaena, *Hyaena brunnea. S. Afr. J. Zool.* **15**, 240–8.

Mohr, E. (1949). Einiges vom grossen and vom kleinen Mara (*Dolichotis patagonum*) Zimm. und Salinicola Burm. *Zool. Gart. Lpz* **16**, 111–33.

Mohr, E. (1965). Altweltliche Stachelschweine. *Neue. Brehm. Buch.* No. 350.

Morris, D. (1962). The behavior of the green acouchi (*Myoprocta pratti*) with special reference to scatter hoarding. *Proc. zool. Soc. Lond,* **139**, 701–32.

Murie, J. O. (1977). Cues used for cache finding by agoutis *Dasyprocta punctata. J. Mammal.* **58**, 95–6.

Ojasti, J. (1973). *Estudio biologico del chiguire o capibara.* Fondo Nac. Invest. Agropec., Caracas.

Ortmann, R. (1960). Die Analregion des Säugetiere. *Handb. Zool.* **3**(7), 1–68.

Pearson, P. O. (1948). Life history of mountain viscachas in Peru. *J. Mammal.* **29**, 345–74.

Pereira, J. N., McEvan Jenkinson, D., and Finley, J. N. (1980). The structure of the skin of the capybara. *Acta cient. Venezolana* 31, 361–4.

Pinkus, F. (1910). Steissdrüse des Meerschweinchens. *Dermatologia* 17, 584.

Pocock, R. I. (1922). On the external characters of some Hystricomorph rodents. *Proc. zool. Soc., Lond.* 365–427.

— (1926). The external characters of the Jamaican hutia (*Capromys brownii*). *Proc. zool. Soc., Lond.* 413–18.

Porter, R. H., Berryman, J. C., and Fullerton, C. (1973). Exploration and attachment behavior in infant guinea pigs. *Behaviour* 45, 315–22.

— Fullerton, C., and Berryman, J. C. (1973). Guinea pig maternal–young attachment behaviour. *Z. Tierpsychol.* 32, 489–95.

Reig, O. (1970). Ecological notes on the flossorial octodont rodent, *Spalaiopus cyanus* (Molina). *J. Mammal.* 51, 592–600.

Rewell, R. E. (1950). Hypertrophy of sebaceous glands on the snout as a secondary male sexual character in the Capybara, *Hydrochoerus hydrochaeris*. *Proc. zool. Soc., Lond.* 119, 817–19.

Rood, J. P. (1970). Ecology and social behavior of the desert cavy (*Microcavia australis*). *Am. midl. Nat.* 83, 415–54.

— (1972). Ecological and behavioural comparisons of three genera of Argentine cavies. *Anim. Behav. Monogr.* 5, 1–83.

— and Weir, B. J. (1970). Reproduction in female wild guinea pigs. *J. Reprod. Fert.* 23, 393–409.

Roth-Kolar, H. (1957). Beitrage zu einem Aktionssystem des Aguti (*Dasyprocta aguti aguti* L.). *Z. Tierpsychol.* 14, 362–75.

Ruddy, L. L. (1980). Discrimination among colony mates' anogenital odors by guinea pigs (*Cavia porcellus*). *J. comp. physiol. Psychol.* 94, 767–74.

Schaffer, J. (1940.). *Die Hautdrüsenorgane der Säugetiere*. Urban & Schwarzenberg. Berlin.

Schaller, G. B. and Crawshaw, P. G. (1981). Social organisation in a capybara population. *Säugetierk. Mitt.* 29, 3–16.

Shadle, A. R. (1952). Sexual maturity and first recorded copulation of a 16-month male porcupine, *Erethizon dorsatum dorsatum*. *J. Mammal.* 33, 239–41.

— Smelzer, M., and Metz, M. (1946). The sex reactions of porcupines (*Erethizon dorsatum*) before and after copulation. *J. Mammal.* 27, 116–21.

Smythe, N. (1978). The natural history of the Central American agouti (*Dasyprocta punctata*). *Smithsonian Contribs. Zool.* 257, 1–52.

Sprinz, O. (1912). Über die Glandula candalis bei, *Cavia cobaya*. *Dermat. Wochenschr.* 55, 1371–80.

Stoddart, D. M., Aplin, R. T., and Wood, M. J. (1975). Evidence for social difference in the flank organ secretion of *Arvicola terrestris* (Rodentia: Microtinae). *J. Zool., Lond.* 177, 529–40.

Taber, A. B. and Macdonald, D. W. (1984). Scent dispersing papillae and associated behaviour of the mara, *Dolichotis patagonum* (Rodentia: caviomorpha). *J. Zool., Lond.* 203, 298–301.

Weir, B. J. (1971). The evolution of oestrus in the cuis, *Galea musteloides*. *J. Reprod. Fert.* 26, 405–8.

— (1973). The role of the male in the evocation of oestrus in the cuis, *Galea musteloides* (Rodentia: Hystricomorpha). *J. Reprod. Fert.* Suppl. 19, 421–32.

— (1974a). The tuco-tuco and plains viscacha. *Symp. Zool. Soc. Lond.*, 34, 113–30.

— (1974*b*). Notes on the origin of the domestic guinea pig, *Symp. Zool. Soc. Lond.* **34**, 437–46.

— (1974*c*). Reproductive characteristics of hystricomorph rodents. *Symp. Zool. Soc., Lond.* **34**, 265–301.

Wellington, J. L., Beauchamp, G. K., and Smith, A. B. III (1981). Stability of chemical communicants of gender in guinea pig urine. *Behav. neural Biol.* **32**, 364–75.

White, P., Fisher, R., and Meunier, G. (1982). The lack of recognition of lactating females by infant *Octodon degus. Physiol. Behav.* **28**, 623–5.

Wilson, S. C. and Kleiman, D. G. (1974). Eliciting play: a comparative study (*Octodon, Octodontomys, Pediolagus, Phoca, Choeropsis, Ailuropoda*). *Am. Zool.* **14**, 341–70.

Wysocki, G. J., Wellington, J. L., and Beauchamp, G. K. (1980). Access of urinary nonvolatiles to the mammalian vomeronasal organ. *Science, NY* **207**, 781–3.

Author index

Page numbers in **bold type** refer to Volume 1, those in roman type to Volume 2.

Index of common names

Page numbers in **bold type** refer to Volume 1, those in roman type to Volume 2.

Index of Linnean names

Page numbers in bold type refer to Volume 1, those in roman type refer to Volume 2.

Index of odour sources

Page numbers in **bold type** refer to Volume 1, those in roman type to Volume 2.